DC Electrical Circuits Analysis

Mehdi Rahmani-Andebili

DC Electrical Circuits Analysis

Practice Problems, Methods, and Solutions

Second Edition

Springer

Mehdi Rahmani-Andebili
Department of Electrical and Computer
Engineering
University of Alabama
Tuscaloosa, AL, USA

ISBN 978-3-032-00252-5 ISBN 978-3-032-00253-2 (eBook)
https://doi.org/10.1007/978-3-032-00253-2

Ievgenii Meyer/Shutterstock.com

This Springer imprint is published by the registered company Springer Nature Switzerland AG
The registered company address is: Gewerbestrasse 11, 6330 Cham, Switzerland

Preface of Book

Electrical Circuit Analysis is one of the most fundamental subjects of electrical engineering major that is taught for sophomore students at universities and colleges worldwide in two or three courses in successive semesters under the names of "DC Electrical Circuit Analysis", "AC Electrical Circuit Analysis", and "Advanced Electrical Circuit Analysis" or "Electrical Circuit Analysis I, II, and III". This textbook covers all subjects of DC electrical circuits analysis including basic concepts and laws of electrical circuits analysis; methods of electrical circuits analysis; theorems of electrical circuits analysis; energy storage elements and circuit analysis under steady-state condition; analysis of first-order circuits; and analysis of second-order circuits.

Problem Types: In each chapter of the textbook, different types of problems and exercises have been presented that are categorized as follows.

- ***Problems with detailed solution***: They have been designed to teach students the subjects in detail, step-by-step, and with different methods.
- ***Partially solved exercises***: They have been designed to encourage students to practice problems while guiding them through the problem-solving procedure and hinting the required formulas.
- ***Exercises with final answer***: They have been designed to encourage students to practice more by themselves while hinting them by the final answer as well as to help instructors give tests or quizzes.

Clasification: The problems have been categorized in different levels based on their difficulty levels (easy, normal, and hard) and calculation amounts (small, normal, and large). This classification helps students study the textbook in the most efficient way.

Notes Box: A "Notes Box" has been added after the solution of each problem that includes all the formulas and important notes that are needed to solve the problem.

Traps Box: A "Traps Box" has been added that discloses the traps that have been designed in the choices to challenge students' minds and encourage them to deeply think about the problem solutions as well as discourage them from careless selection of choices. The "Traps Box", in fact, predicts the possible confusions and tricks that might be encountered by students in real tests.

Comprehensiveness and Self-Study Features: In this textbook, the problems have been fully solved, in detail, step-by-step, and with various methods and techniques. Hence, the textbook can be used as a self-study and practicing study guide by students and a supplementary teaching source by instructors. Moreover, since the problems include detailed solutions, the textbook is appropriate for under-prepared students and beneficial for knowledgeable students because it includes advanced problems and exercises.

Educational Support: Students and instructors are welcome to reach out to the author at mehdi.rahmani.andebili@gmail.com to inquire about the educational support related to the textbook, available exclusively from the author. They can also visit the author's weblog at https://RahmaniAndebiliBooks.blogspot.com for more and updated information.

In the following, the description of each chapter of the textbook is briefly presented.

Chapters 1 and 2 cover the basic concepts and laws of electrical circuits analysis. The subjects include resistance, Ohm's law, equivalent resistance of series, parallel, and series-parallel connections, wye-delta transformation and heuristic methods for equivalent resistance calculation, voltage and current division rules, ideal independent voltage source, ideal independent current source, ideal dependent voltage and current sources, Kirchhoff's current law (KCL), Kirchhoff's voltage law (KVL), and power.

Chapters 3 and 4 teach the nodal analysis, mesh analysis, and heuristic circuit analysis.

Chapters 5 and 6 study the theorems of electrical circuits analysis that include input resistance, Thevenin theorem, Norton theorem, source transformation theorem, superposition theorem, and maximum power transfer theorem.

Chapters 7 and 8 review the energy storage elements and circuit analysis under steady-state condition. The subjects include equivalent capacitance and inductance of series, parallel, and series-parallel connections as well as circuit analysis under steady-state condition.

Chapters 9 and 10 are related to the analysis of first-order circuits. The subjects include time constant, source-free response of an RC circuit, step response of an RC circuit, complete response of an RC circuit, source-free response of an RL circuit, step response of an RL circuit, complete response of an RL circuit, heuristic methods for first-order circuits, first-order circuits including ramp and singularity functions, differential equation of first-order circuits, and energy stored in first-order circuits.

Chapters 11 and 12 are concerned with the analysis of second-order circuits. The subjects include differential equation of second-order circuits, response types of second-order circuits, general response of second-order circuits, analysis of second-order circuits during switching operation, second-order circuits including singularity functions, and energy stored in second-order circuits.

The subjects covered in the book "*AC Electrical Circuits Analysis: Practice Problems, Methods, and Solutions, 2nd Edition*" are as follows.

- Basic Concepts and Laws of Electrical Circuits Analysis in Sinusoidal Steady-State
- Sinusoidal Steady-State Analysis
- Application of Electrical Circuit Theorems in Sinusoidal Steady-State
- Average Power, Reactive Power, Apparent Power, Complex Power, and Power Factor
- Resonance, Frequency Response, Bandwidth, and Quality Factor
- Sinusoidal Steady-State Analysis of Circuits Including Magnetically Coupled Inductors
- Sinusoidal Steady-State Analysis of Circuits Including Transformers

Moreover, the subjects covered in the book "*Advanced Electrical Circuits Analysis: Practice Problems, Methods, and Solutions, 2nd Edition*" are as follows.

- State Equations of Electrical Circuits
- Circuit Analysis in Laplace Domain
- Network Function and Its Applications and Theorems in Laplace Domain
- Natural Frequencies, Zeros and Poles, and Damping Status of Electrical Circuits
- Linear Time-Invariant Networks
- Two-Port Networks and their Series, Parallel, and Cascade Connections

Other Books Published by the Author

AC Electrical Circuits Analysis: Practice Problems, Methods, and Solutions, 2nd Ed.

Advanced Electrical Circuits Analysis: Practice Problems, Methods, and Solutions, 2nd Ed.

Analysis and Design of Feedback Control Systems with MATLAB Applications: Practice Problems, Methods, and Solutions, 2nd Ed.

Analysis of AC Electric Machines: Practice Problems, Methods, and Solutions, 2nd Ed.

Analysis of DC Electric Machines, Electromechanical Systems, and Magnetic Circuits: Practice Problems, Methods, and Solutions, 2nd Ed.

Analysis of Power Systems: Practice Problems, Methods, and Solutions, 2nd Ed.

Applications of Artificial Intelligence in Planning and Operation of Smart Grid

Applications of Fuzzy Logic in Planning and Operation of Smart Grids

Calculus I: Practice Problems, Methods, and Solutions, 2nd Ed.

Calculus II: Practice Problems, Methods, and Solutions

Calculus III: Practice Problems, Methods, and Solutions

DC Electrical Circuits Analysis: Practice Problems, Methods, and Solutions, 2nd Ed.

Design, Control, and Operation of Microgrids in Smart Grids

Digital Logic Principles: A Comprehensive Tutorial and Lessons

Electromagnetics: Practice Problems, Methods, and Solutions

Electronics: A Comprehensive Tutorial and Lessons

General Physics I: Practice Problems, Methods, and Solutions

General Physics II: Practice Problems, Methods, and Solutions

Mathematics of Engineering and Science: Practice Problems, Methods, and Solutions

MATLAB Lessons, Examples, and Exercises: A Tutorial for Beginners and Experts

Operation of Smart Homes

Ordinary Differential Equations: Practice Problems, Methods, and Solutions, 2nd Ed.

Planning and Operation of Electric Vehicles in Smart Grid

Planning and Operation of Plug-in Electric Vehicles: Technical, Geographical, and Social Aspects

Power System Analysis: Comprehensive Lessons

Precalculus: Practice Problems, Methods, and Solutions, 2nd Ed.

Tuscaloosa, AL, USA Mehdi Rahmani-Andebili

Contents

About the Author

Mehdi Rahmani-Andebili is an Assistant Professor with the Department of Electrical and Computer Engineering at the University of Alabama. He received his first M.Sc. and Ph.D. degrees in Electrical Engineering from Tarbiat Modares University and Clemson University in 2011 and 2016, respectively, and his second Ph.D. (2nd M.Sc.) degree in Physics and Astronomy from the University of Alabama in Huntsville in 2019. Moreover, he was a Postdoctoral Fellow at the Sharif University of Technology during 2016–2017. As a professor, he has taught a wide selection of courses and labs including Power System Analysis, Electric Machines, Feedback Control Systems Analysis and Design, Signals and Systems, Electromagnetics, Electronics, Digital Logic, Industrial Electronics, Renewable Distributed Generation and Storage, AC Electrical Circuits Analysis, DC Electrical Circuits Analysis, Electrical Circuits and Devices, Fundamental of Electrical Engineering, Essentials of Electrical Engineering Technology, and Algebra and Calculus-Based Physics. Dr. Rahmani-Andebili has published over 300 single-author or first-author scholarly works including peer-reviewed journal papers, conference papers, textbooks, books, and book chapters in the fields of Electrical Engineering, Power Engineering, Physics, Mathematics, and Computer Programming. He is an IEEE Senior Member and a permanent reviewer of many reputable journals. His research areas include Smart Grid, Applications of Artificial Intelligence in Planning and Operation of Power Systems, Integration of Renewables and Energy Storages into Power Systems, Energy Scheduling and Demand-Side Management, Electric Vehicles and Distributed Generation, and Advanced Optimization Techniques in Power System Studies.

Basic Concepts and Laws of Electrical Circuits Analysis: Part A

1

Abstract

In this chapter, the problems concerned with the basic concepts and laws of electrical circuits analysis are studied. The subjects include resistance, Ohm's law, equivalent resistance of series, parallel, and series-parallel connections, wye-delta transformation and heuristic methods for equivalent resistance calculation, voltage and current division rules, ideal independent voltage source, ideal independent current source, ideal dependent voltage and current sources, Kirchhoff's current law (KCL), Kirchhoff's voltage law (KVL), and power consumption and generation. Herein, different types of problems and exercises are presented that are categorized as follows.

- ***Problems with detailed solution***: They have been designed to teach students the subjects in detail. Moreover, they have been categorized in different levels based on their difficulty levels (easy, normal, and hard) and calculation amounts (small, normal, and large).
- ***Partially solved exercises***: They have been designed to encourage students to practice problems while guiding them through the problem-solving procedure and hinting the required formulas.
- ***Exercises with final answer***: They have been designed to encourage students to practice more by themselves while hinting them by the final answer as well as to help instructors to give tests or quizzes.

1.1 Resistance

Problem

1.1. The diameter of the circular cross-sectional area, length, and resistance of a wire are 1 mm, 2 m, and 0.1 Ω, respectively. Calculate the special resistance (resistivity) of the wire [1–9].

Difficulty level ● Easy ○ Normal ○ Hard
Calculation amount ● Small ○ Normal ○ Large

1) 3.92×10^{-8} Ωm
2) 1.56×10^{-7} Ωm
3) 3.92×10^{-7} Ωm
4) 1.56×10^{-8} Ωm

F. X. Malcata, *DC Electrical Circuits Analysis*, https://doi.org/10.1007/978-3-032-00253-2_1

Partially Solved Exercise

1.2. The radius of the circular cross-sectional area, length, and special resistance (resistivity) of a wire are 2 mm, 3 m, and 1.5 Ωm, respectively. Calculate the resistance of the wire.

Solution

As we know, the resistance of a wire can be calculated as follows.

$$R = \rho \frac{l}{A}$$

Herein, ρ, l, and A, are the special resistance (resistivity), length, and cross-sectional area of the wire, respectively. Thus, for this problem, we can write:

$$R = \rho \frac{l}{\pi r^2}$$

$$\Rightarrow R = (\qquad) \times \frac{(\quad)}{\pi(\qquad)^2}$$

$$\Rightarrow R = 358 \times 10^3 \ \Omega$$

$$\Rightarrow R = 358 \ \text{k}\Omega$$

Problem

1.3. Consider two wires. The length of the first one is twice the length of the second wire. Moreover, the diameter of circular cross-sectional area of the first wire is half of the one of the second wire. In addition, the special resistance (resistivity) of the first wire is one-third of the one of the second wire. Calculate the ratio of resistance of the first wire to the second one.

Difficulty level ● Easy ○ Normal ○ Hard

Calculation amount ● Small ○ Normal ○ Large

1) $\frac{1}{6}$
2) 24
3) $\frac{4}{3}$
4) $\frac{8}{3}$

Exercise

1.4. Calculate the ratio of resistance of the second wire to the first one if $l_2 = 2l_1$, $r_2 = 3r_1$, and $\rho_2 = 4.5\rho_1$.

Final Answer

$$\frac{R_2}{R_1} = 1$$

Note: Read the "educational support" part of the preface of the textbook.

1.2 Ohm's Law

Problem

1.5. Calculate the resistance of an electrical device that draws 0.1 A current at 110 V.

Difficulty level ● Easy ○ Normal ○ Hard

Calculation amount ● Small ○ Normal ○ Large

1) 110 Ω
2) 1100 Ω
3) 11 Ω
4) 0.001 Ω

Exercise

1.6. On the catalogue of a vacuum cleaner, 4 A and 120 V are written. What is the equivalent resistance of the vacuum cleaner?

Final Answer

$R = 30\ \Omega$

Note: *Read the "educational support" part of the preface of the textbook.*

Problem

1.7. Calculate the current of a 1 kΩ resistor connected to a 10 V power supply.

Difficulty level ● Easy ○ Normal ○ Hard

Calculation amount ● Small ○ Normal ○ Large

1) 100 A
2) 10000 A
3) 0.01 A
4) 1 A

Exercise

1.8. A 12 Ω, 120 V washing machine is available. What current does it draw from the power outlet?

Final Answer

$I = 10\ \text{A}$

Note: *Read the "educational support" part of the preface of the textbook.*

Problem

1.9. What voltage will appear across a 15 Ω resistance if it carries 100 mA current?

Difficulty level ● Easy ○ Normal ○ Hard

Calculation amount ● Small ○ Normal ○ Large

1) 150 V
2) 0.006 V
3) 0.15 V
4) 1.5 V

Exercise

1.10. Calculate the nominal voltage of an electrical device that its rated resistance and current are 5 kΩ and 10 mA, respectively.

Final Answer

$V = 50$ V

Note: *Read the "educational support" part of the preface of the textbook.*

1.3 Equivalent Resistance of Series, Parallel, and Series-Parallel Connections

Problem

1.11. In the circuit shown in Fig. 1.1, the resistors are connected in series. Calculate the equivalent resistance of the circuit.

Difficulty level ● Easy ○ Normal ○ Hard

Calculation amount ● Small ○ Normal ○ Large

1) 1 Ω
2) 11 Ω
3) 10 Ω
4) 4 Ω

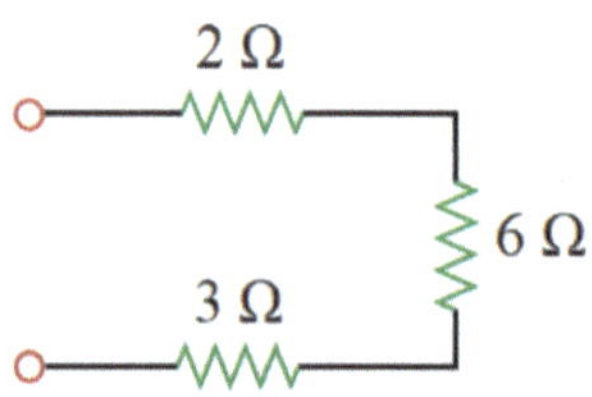

Fig. 1.1 The circuit of problem 1.11

Problem

1.12. In the circuit shown in Fig. 1.2, the resistors are connected in parallel. Calculate the equivalent resistance of the circuit.

Difficulty level ● Easy ○ Normal ○ Hard

Calculation amount ● Small ○ Normal ○ Large

1) 1 Ω
2) 11 Ω
3) 10 Ω
4) 4 Ω

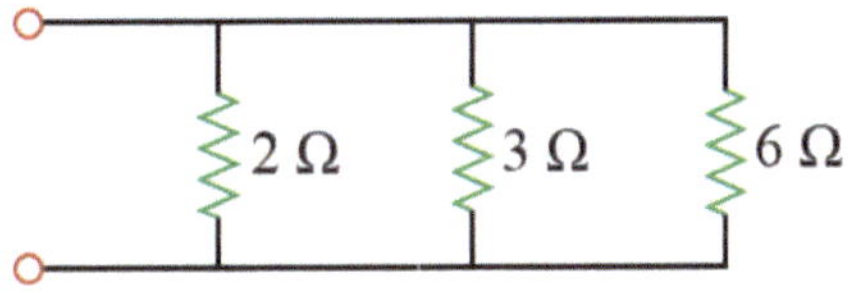

Fig. 1.2 The circuit of problem 1.12

Partially Solved Exercise

1.13. In the circuit of Fig. 1.3, calculate the equivalent resistance seen from the terminal.

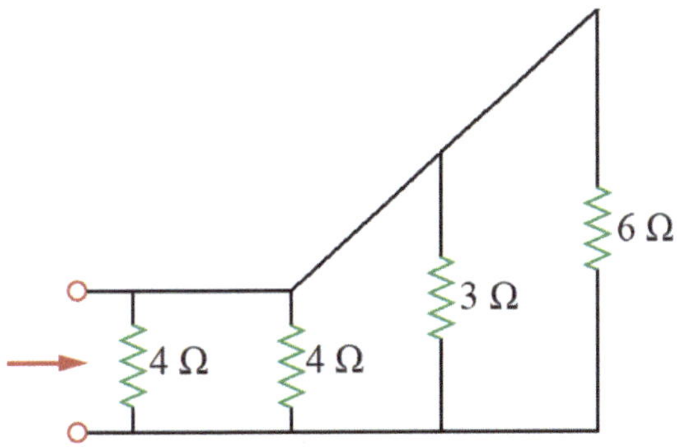

Fig. 1.3 The circuit of partially solved exercise 1.13

Solution

As is shown in Fig. 1.4, the circuit has been simplified. Now, it is noticed that all the resistors are in parallel. Hence:

$$R_{eq} = (4||4)||(3||6)$$

$$\Rightarrow R_{eq} = \left(\frac{(\quad)}{2}\right)||\left(\frac{(\quad)\times(\quad)}{(\quad)+(\quad)}\right)$$

$$\Rightarrow R_{eq} = (\quad)||(\quad)$$

$$\Rightarrow R_{eq} = \frac{(\quad)}{2}$$

$$\Rightarrow R_{eq} = 1\ \Omega$$

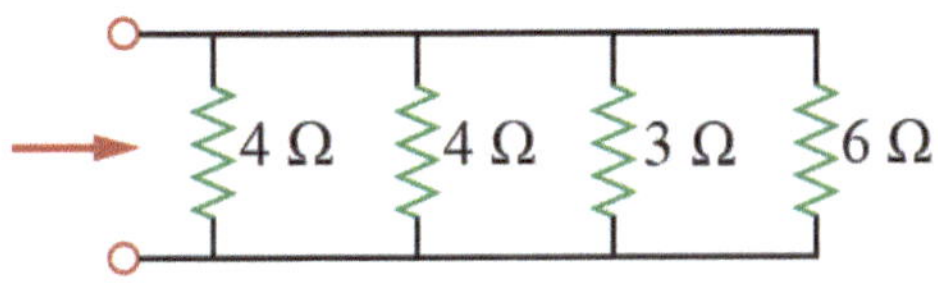

Fig. 1.4 The circuit of partially solved exercise 1.13

Problem

1.14. In the circuit shown in Fig. 1.5, the resistors are connected in series–parallel from. Calculate the equivalent resistance of the circuit.

Difficulty level ● Easy ○ Normal ○ Hard

Calculation amount ● Small ○ Normal ○ Large

1) 4 Ω
2) 1 Ω
3) 0.4 Ω
4) 2.5 Ω

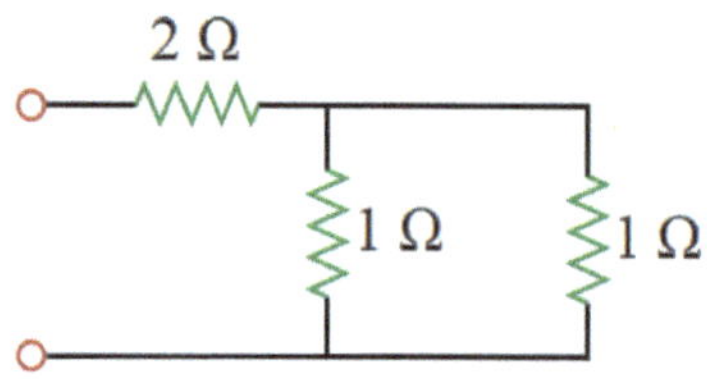

Fig. 1.5 The circuit of problem 1.14

Partially Solved Exercise

1.15. Calculate the equivalent resistance of the circuit shown in Fig. 1.6.

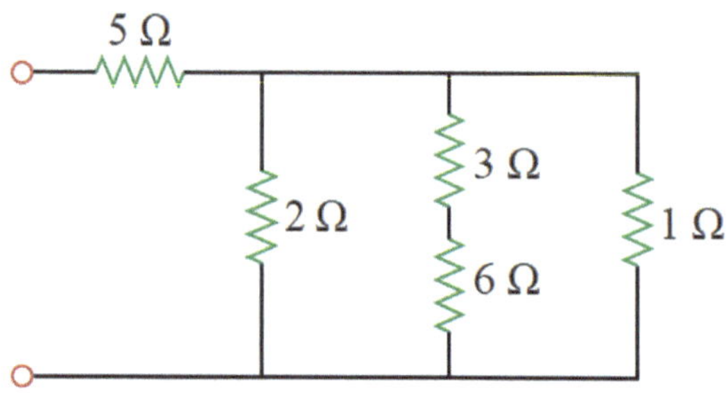

Fig. 1.6 The circuit of partially solved exercise 1.15

Solution

The equivalent resistance of the circuit can be calculated as follows. First, the series connection of the middle branch should be simplified as follows.

$$R_{3,6} = R_3 + R_6 = (\quad) + (\quad) = 9\ \Omega$$

Then, the three parallel branches can be combined as follows.

$$\frac{1}{R_{2,9,1}} = \frac{1}{R_2} + \frac{1}{R_9} + \frac{1}{R_1}$$

$$\Rightarrow \frac{1}{R_{2,9,1}} = \frac{1}{(\quad)} + \frac{1}{(\quad)} + \frac{1}{(\quad)}$$

$$\Rightarrow \frac{1}{R_{2,9,1}} = \frac{(\quad) + (\quad) + (\quad)}{18}$$

$$\Rightarrow R_{2,9,1} = \frac{18}{29}\ \Omega$$

Finally, there are only two resistors connected in series. Thus:

$$R_{eq} = (\quad) + (\quad) = \frac{145 + 18}{29}$$

$$\Rightarrow R_{eq} = \frac{163}{29}\ \Omega$$

Exercise

1.16. Calculate the equivalent resistance of the circuit shown in Fig. 1.7.

Final Answer

$R_{eq} = 14.4\ \Omega$

Note: Read the "educational support" part of the preface of the textbook.

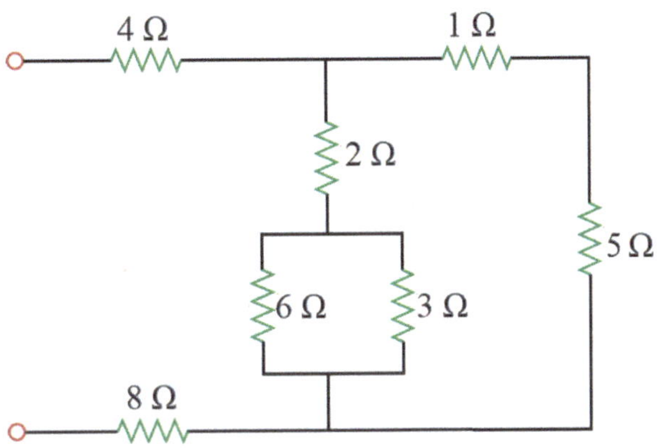

Fig. 1.7 The circuit of exercise 1.16

Problem

1.17. In the circuit of Fig. 1.8, the resistance of each resistor is 1 Ω. Calculate the equivalent resistance seen from the terminal.

Difficulty level ○ Easy ● Normal ○ Hard

Calculation amount ● Small ○ Normal ○ Large

1) 1 Ω
2) 2 Ω
3) 3 Ω
4) 4 Ω

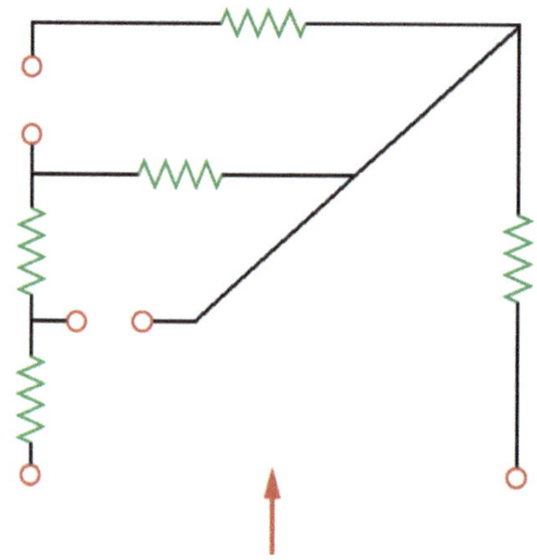

Fig. 1.8 The circuit of problem 1.17

Exercise

1.18. Calculate the equivalent resistance seen from the terminal of the circuit shown in Fig. 1.9. The circuit includes an infinite number of resistors connected in series and the resistance of each of them is one Ohm.

Final Answer

$R_{eq} = \infty$

Note: *Read the "educational support" part of the preface of the textbook.*

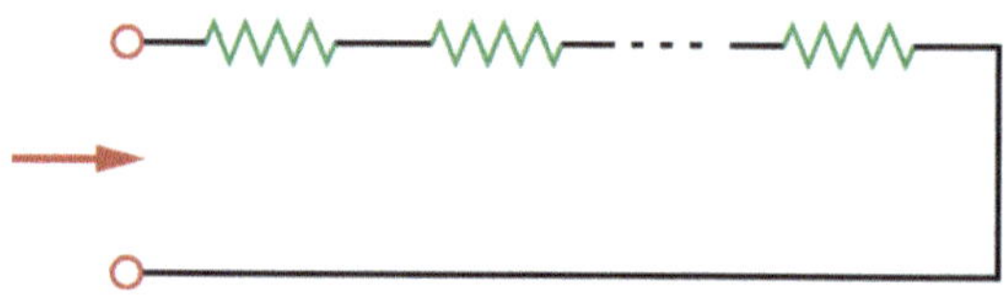

Fig. 1.9 The circuit of exercise 1.18

Exercise

1.19. Calculate the equivalent resistance seen from the terminal of the circuit shown in Fig. 1.10. The circuit includes an infinite number of resistors connected in parallel and the resistance of each of them is 1000 Ω.

Final Answer

$R_{eq} = 0\ \Omega$

Note: *Read the "educational support" part of the preface of the textbook.*

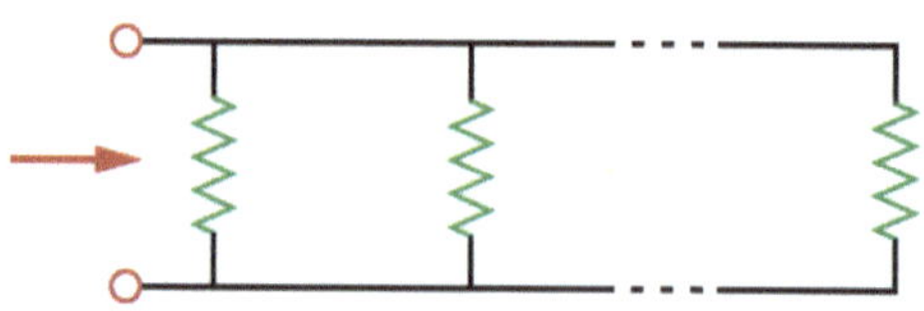

Fig. 1.10 The circuit of exercise 1.19

Problem

1.20. In the circuit of Fig. 1.11, calculate the equivalent resistance seen from the terminal.

Difficulty level ○ Easy ● Normal ○ Hard

Calculation amount ● Small ○ Normal ○ Large

1) 1 Ω
2) 2 Ω
3) 3 Ω
4) 4 Ω

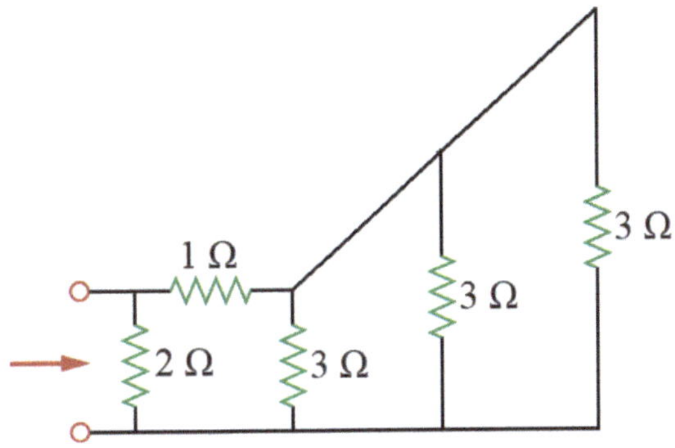

Fig. 1.11 The circuit of problem 1.20

Exercise

1.21. Calculate the equivalent resistance seen from the terminal of the circuit shown in Fig. 1.12. Herein, the resistance of each element is one Ohm.

Final Answer

$R_{eq} = 0.75\ \Omega$

Note: Read the "educational support" part of the preface of the textbook.

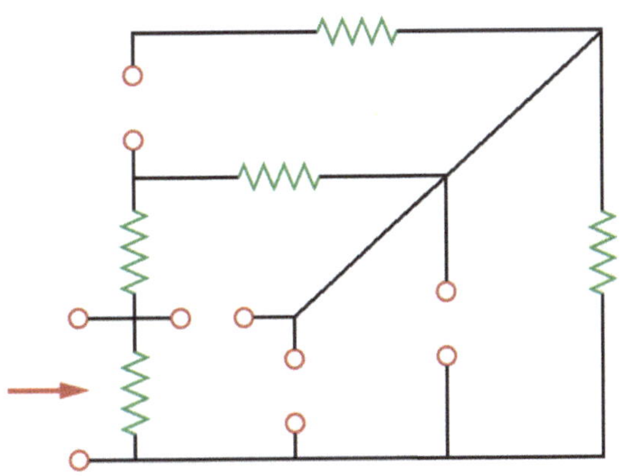

Fig. 1.12 The circuit of exercise 1.21

1.4 Wye-Delta Transformation for Equivalent Resistance Calculation

1.22. In the circuit of Fig. 1.13, calculate the equivalent resistance seen from the terminal.

Difficulty level ○ Easy ● Normal ○ Hard

Calculation amount ○ Small ● Normal ○ Large

1) 8 Ω
2) 28 Ω
3) 20 Ω
4) 12 Ω

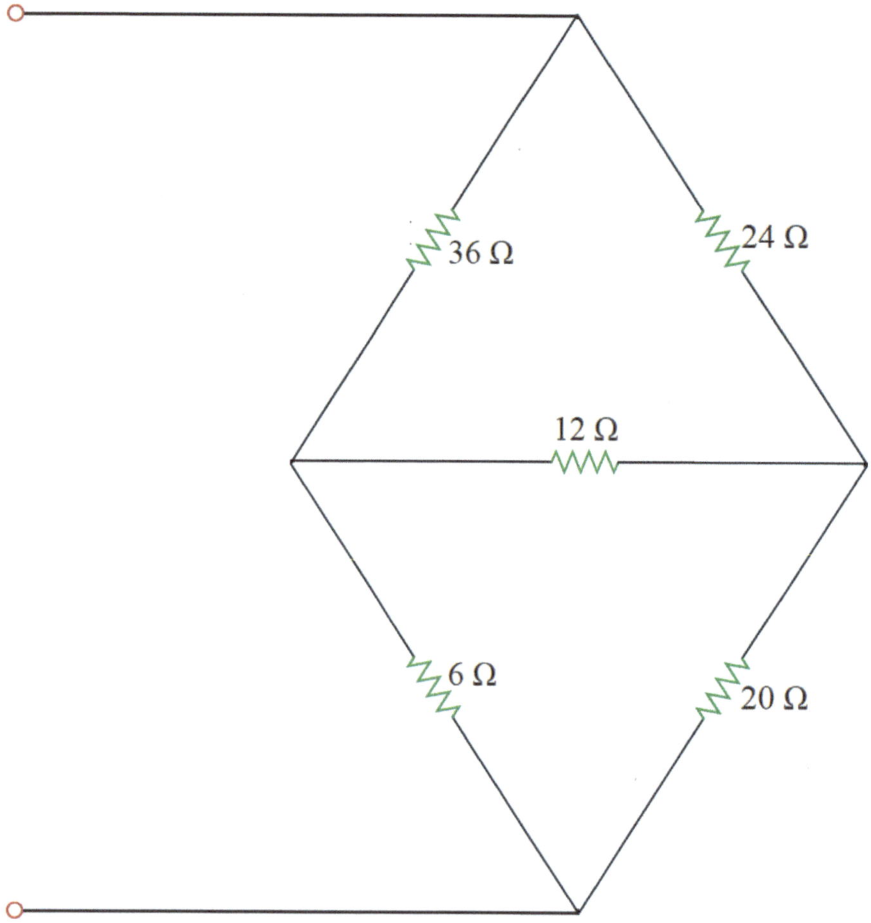

Fig. 1.13 The circuit of problem 1.22

Partially Solved Exercise

1.23. In the circuit shown in Fig. 1.14, we have $R_A = 10\ \Omega$, $R_B = 20\ \Omega$, and $R_C = 40\ \Omega$. Calculate the value of R_{AB}, R_{BC}, and R_{AC}.

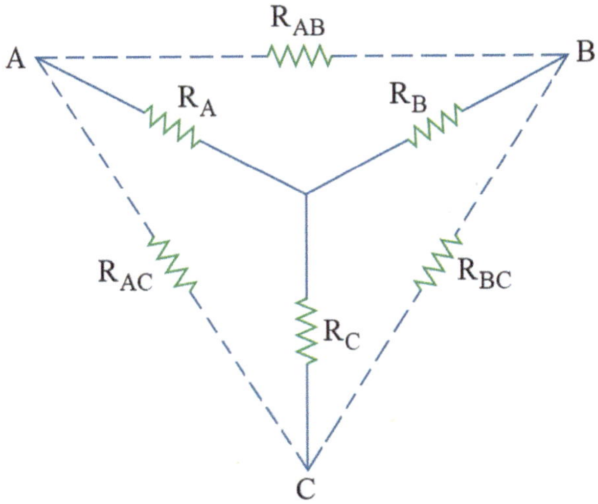

Fig. 1.14 The circuit of partially solved exercise 1.23

Solution

As we know, the relations below are valid when a wye-connection (called Y-connection, T-connection, or star-connection) is converted to a delta-connection (called Δ-connection or π-connection).

$$R_{AB} = \frac{R_A R_B + R_B R_C + R_A R_C}{R_C}$$

$$R_{BC} = \frac{R_A R_B + R_B R_C + R_A R_C}{R_A}$$

$$R_{AC} = \frac{R_A R_B + R_B R_C + R_A R_C}{R_B}$$

Therefore, for this problem, we can write as follows.

$$R_{AB} = \frac{(\quad)(\quad) + (\quad)(\quad) + (\quad)(\quad)}{(\quad)}$$

$$\Rightarrow R_{AB} = 140\ \Omega$$

$$R_{BC} = \frac{(\quad)(\quad) + (\quad)(\quad) + (\quad)(\quad)}{(\quad)}$$

$$\Rightarrow R_{BC} = 70\ \Omega$$

$$R_{AC} = \frac{(\quad)(\quad) + (\quad)(\quad) + (\quad)(\quad)}{(\quad)}$$

$$\Rightarrow R_{AC} = 35\ \Omega$$

Partially Solved Exercise

1.24. In the circuit shown in Fig. 1.15, we have $R_{AB} = 25\ \Omega$, $R_{BC} = 15\ \Omega$, and $R_{AC} = 10\ \Omega$. Calculate the value of R_A, R_B, and R_C.

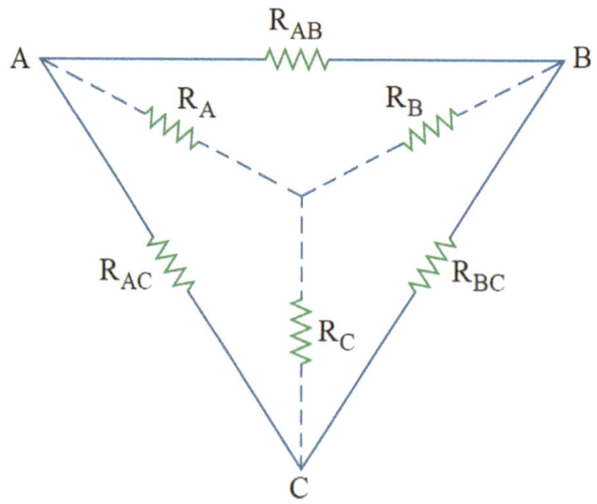

Fig. 1.15 The circuit of partially solved exercise 1.24

Solution

As we know, the following relations are valid when a delta-connection (Δ-connection or π-connection) is converted to a wye-connection (Y-connection, T-connection, or star-connection).

$$R_A = \frac{R_{AC}R_{AB}}{R_{AB} + R_{BC} + R_{AC}}$$

$$R_B = \frac{R_{AB}R_{BC}}{R_{AB} + R_{BC} + R_{AC}}$$

$$R_C = \frac{R_{AC}R_{BC}}{R_{AB} + R_{BC} + R_{AC}}$$

Therefore, for this problem, we can write as follows.

$$R_A = \frac{(\quad)(\quad)}{(\quad)+(\quad)+(\quad)}$$

$$\Rightarrow R_A = 5\ \Omega$$

$$R_B = \frac{(\quad)(\quad)}{(\quad)+(\quad)+(\quad)}$$

$$\Rightarrow R_B = 7.5\ \Omega$$

$$R_C = \frac{(\quad)(\quad)}{(\quad)+(\quad)+(\quad)}$$

$$\Rightarrow R_C = 3\ \Omega$$

Exercise

1.25. In the circuit of Fig. 1.16, calculate the equivalent resistance seen from the terminal.

Final Answer

$R_{eq} = 27\ \Omega$

Note: Read the "educational support" part of the preface of the textbook.

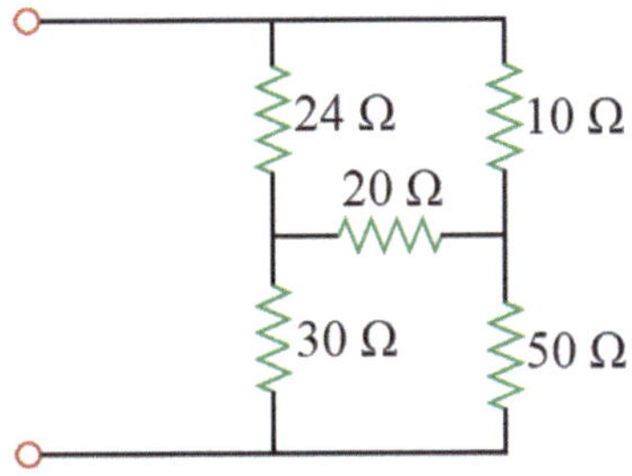

Fig. 1.16 The circuit of exercise 1.25

1.5 Heuristic Methods for Equivalent Resistance Calculation

Problem

1.26. In the circuit shown in Fig. 1.17, the resistor R is an Ohmic resistor and can take any value. In other words:

$$0 \leq R < \infty$$

Calculate the range of input resistance (R_{in}) seen from terminal a-b.

Difficulty level ○ Easy ○ Normal ● Hard

Calculation amount ○ Small ● Normal ○ Large

1) $0 < R_{in} < \infty$
2) $0.75 < R_{in} < \infty$
3) $0 < R_{in} < 0.83$
4) $0.78 < R_{in} < 0.83$

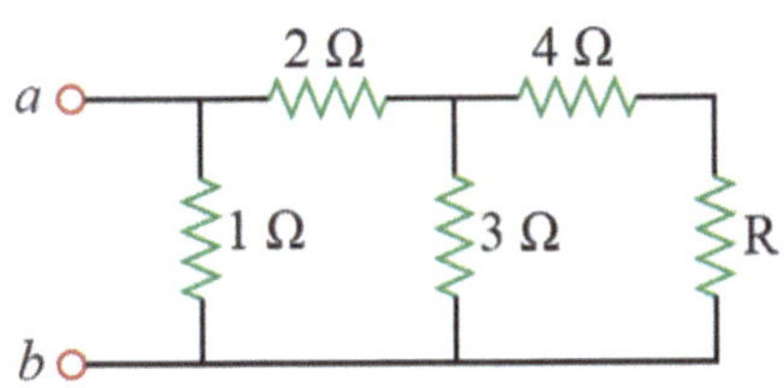

Fig. 1.17 The circuit of problem 1.26

Exercise

1.27. Calculate the range of resistance seen from the input terminal of the circuit illustrated in Fig. 1.18.

Final Answer

$0.66 < R_{in} < 1$

Note: Read the "educational support" part of the preface of the textbook.

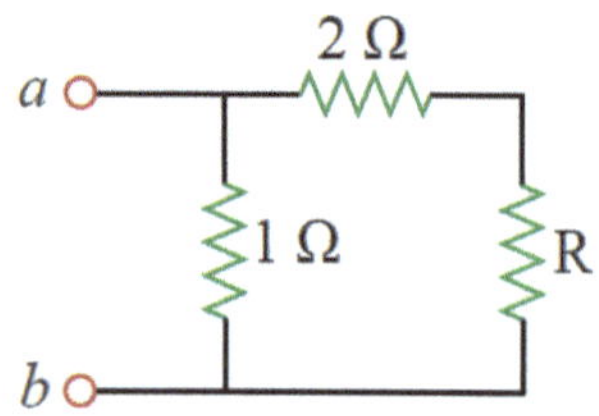

Fig. 1.18 The circuit of exercise 1.27

Problem

1.28. In the circuit illustrated in Fig. 1.19, all the resistors are 1 Ω. Calculate the equivalent resistance between the nodes "A" and "B".

Difficulty level ○ Easy ○ Normal ● Hard
Calculation amount ○ Small ● Normal ○ Large

1) $\frac{1}{4}\ \Omega$
2) $\frac{1}{3}\ \Omega$
3) $\frac{1}{2}\ \Omega$
4) $\frac{3}{4}\ \Omega$

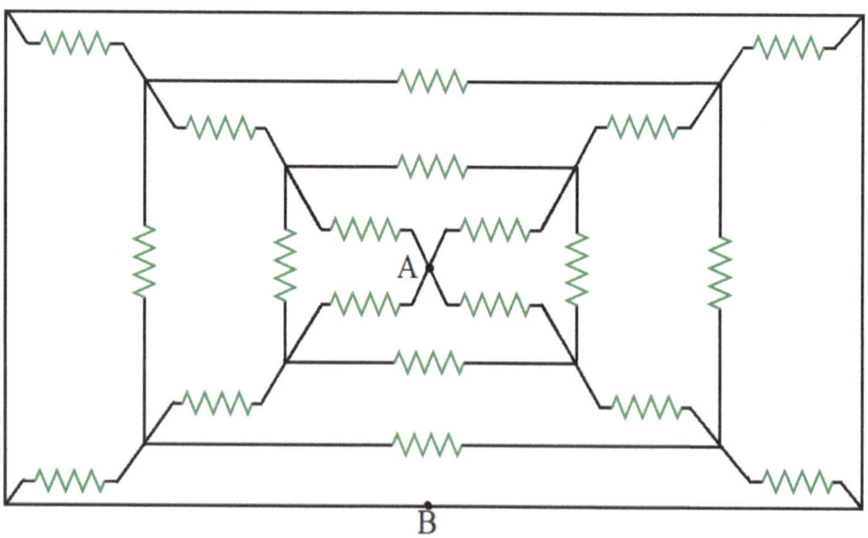

Fig. 1.19 The circuit of problem 1.28

Exercise

1.29. Calculate the equivalent resistance between the nodes "A" and "B" in the circuit of Fig. 1.20. Herein, all the resistors are 2 Ω.

Final Answer

$R_{AB} = 0$

Note: *Read the "educational support" part of the preface of the textbook.*

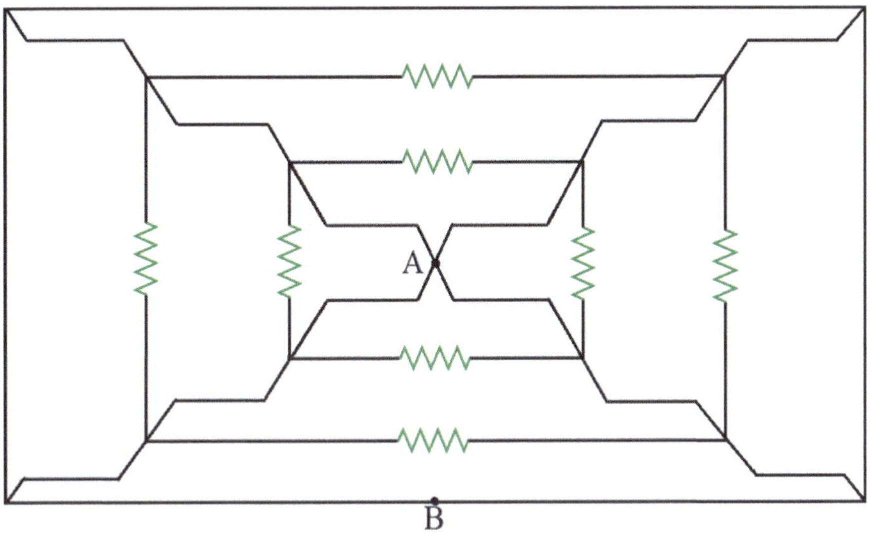

Fig. 1.20 The circuit of exercise 1.29

Problem

1.30. In the circuit of Fig. 1.21, suppose that the resistance of each resistor is R. Calculate the input resistance of the circuit.

Difficulty level ○ Easy ○ Normal ● Hard

Calculation amount ○ Small ● Normal ○ Large

1) R
2) 0.618 R
3) 0.5 R
4) 0.382 R

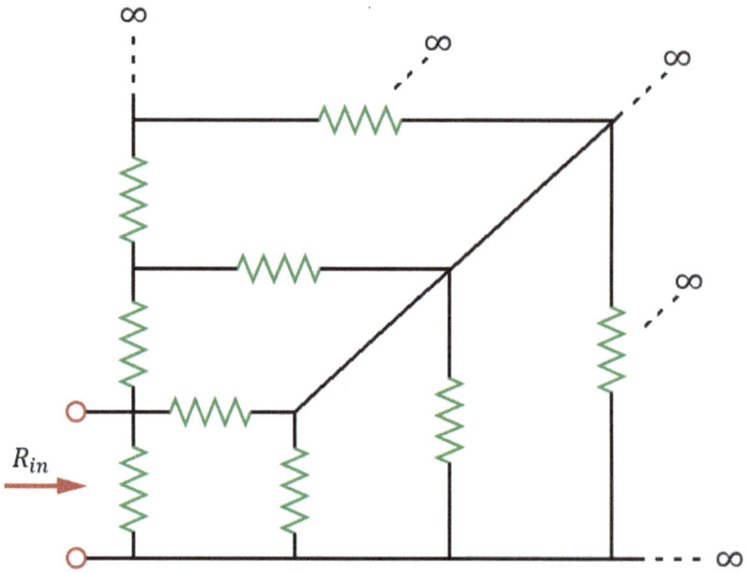

Fig. 1.21 The circuit of problem 1.30

Exercise

1.31. Calculate the equivalent resistance seen from the terminal in the circuit of Fig. 1.22. Herein, each resistor is 1 Ω.

Final Answer

$R_{in} \approx 1.62\ \Omega$

Note: *Read the "educational support" part of the preface of the textbook.*

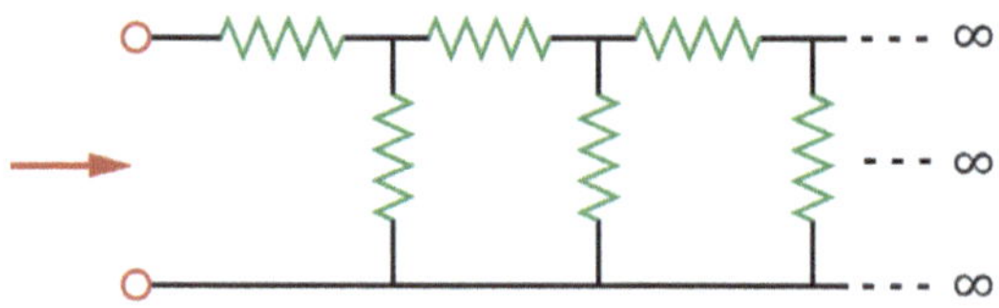

Fig. 1.22 The circuit of exercise 1.31

Problem

1.32. Calculate the equivalent resistance seen from terminal A-B in the circuit of Fig. 1.23. Assume that each resistor has 1 Ω resistance.

Difficulty level ○ Easy ○ Normal ● Hard

Calculation amount ○ Small ● Normal ○ Large

1) $\frac{1}{3}\ \Omega$
2) $\frac{2}{3}\ \Omega$
3) $1\ \Omega$
4) $\frac{1}{2}\ \Omega$

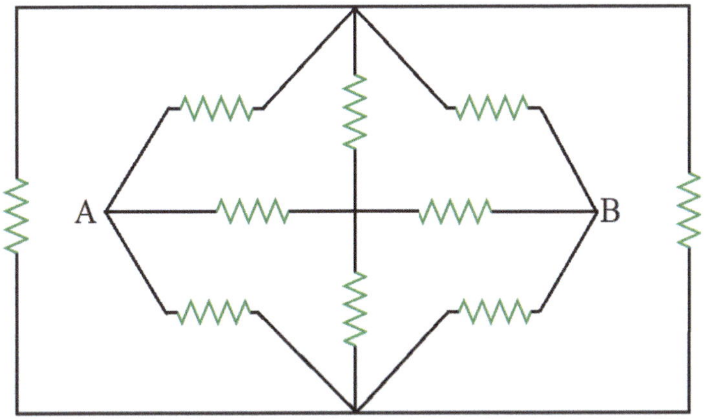

Fig. 1.23 The circuit of problem 1.32

Problem

1.33. In the circuit of Fig. 1.24, calculate the equivalent resistance between the nodes "A" and "O".

Difficulty level ○ Easy ○ Normal ● Hard

Calculation amount ○ Small ○ Normal ● Large

1) 9 Ω
2) 1.8 Ω
3) 2.15 Ω
4) 7.5 Ω

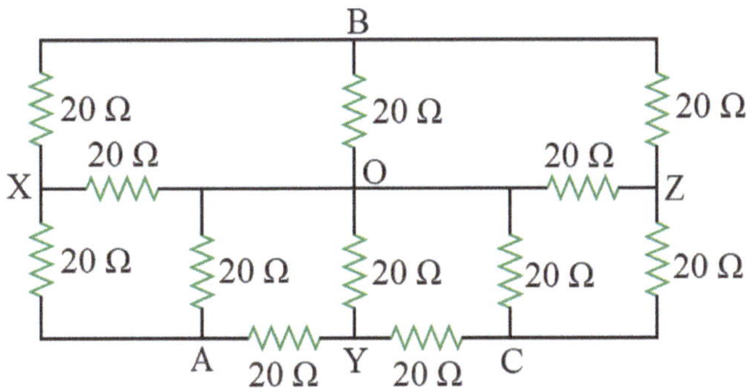

Fig. 1.24 The circuit of problem 1.33

1.6 Voltage and Current Division Rules

Problem

1.34. Which one of the following choices is wrong about the series circuit shown in Fig. 1.25.

Difficulty level ● Easy ○ Normal ○ Hard

Calculation amount ● Small ○ Normal ○ Large

1) $I_1 = \frac{V_1}{R_1}, I_2 = \frac{V_2}{R_2}, I = \frac{V}{R_{eq}}$
2) $I_1 = I_2 = I$
3) $V_1 = \frac{R_1}{R_1 + R_2} V, V_2 = \frac{R_2}{R_1 + R_2} V$
4) $V_1 = V_2 = V$

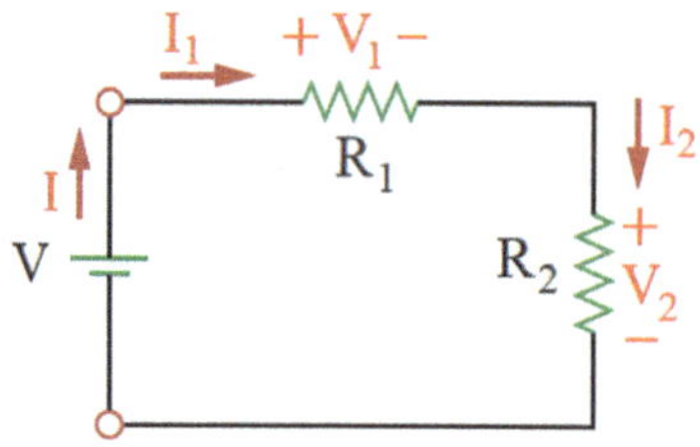

Fig. 1.25 The circuit of problem 1.34

Problem

1.35. Which one of the following choices is wrong about a parallel circuit shown in Fig. 1.26.

Difficulty level ● Easy ○ Normal ○ Hard
Calculation amount ● Small ○ Normal ○ Large

1) $V_1 = I_1 R_1,\ V_2 = I_2 R_2, V = I R_{eq}$
2) $V_1 = V_2 = V$
3) $I_1 = \frac{R_2}{R_1 + R_2} I,\ I_2 = \frac{R_1}{R_1 + R_2} I$
4) $I_1 = I_2 = I$

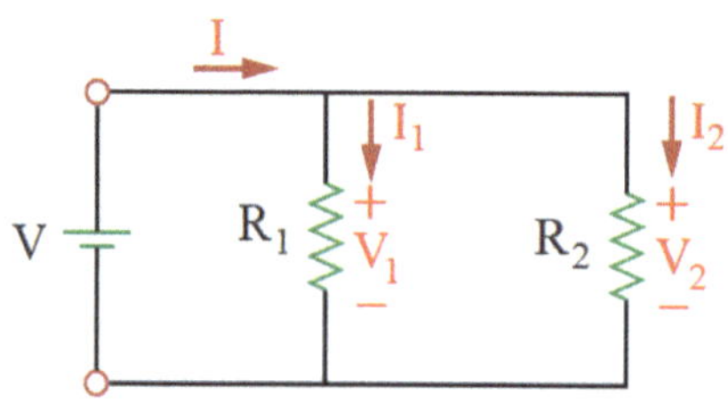

Fig. 1.26 The circuit of problem 1.35

Problem

1.36. In the circuit shown in Fig. 1.27, calculate the current of each resistor.

Difficulty level ● Easy ○ Normal ○ Hard
Calculation amount ● Small ○ Normal ○ Large

1) $I_1 = 10$ A, $I_2 = 20$ A
2) $I_1 = 20$ A, $I_2 = 10$ A
3) $I_1 = 20$ A, $I_2 = 20$ A
4) $I_1 = 10$ A, $I_2 = 10$ A

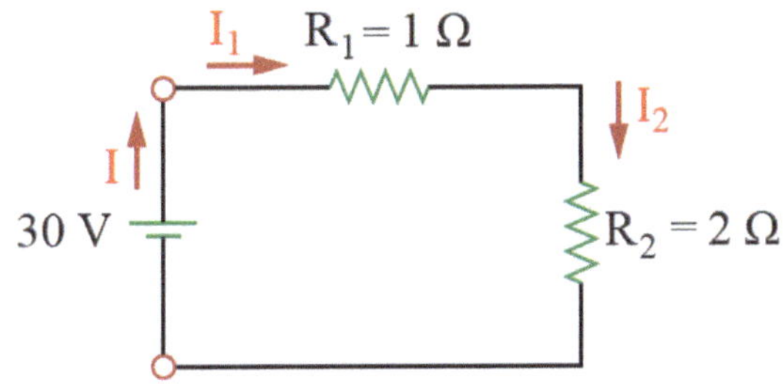

Fig. 1.27 The circuit of problem 1.36

Problem

1.37. In the circuit shown in Fig. 1.28, calculate the voltage of each resistor.

Difficulty level ● Easy ○ Normal ○ Hard

Calculation amount ● Small ○ Normal ○ Large

1) $V_1 = 10\ \text{V},\ V_2 = 10\ \text{V}$
2) $V_1 = 20\ \text{V},\ V_2 = 20\ \text{V}$
3) $V_1 = 10\ \text{V},\ V_2 = 20\ \text{V}$
4) $V_1 = 20\ \text{V},\ V_2 = 20\ \text{V}$

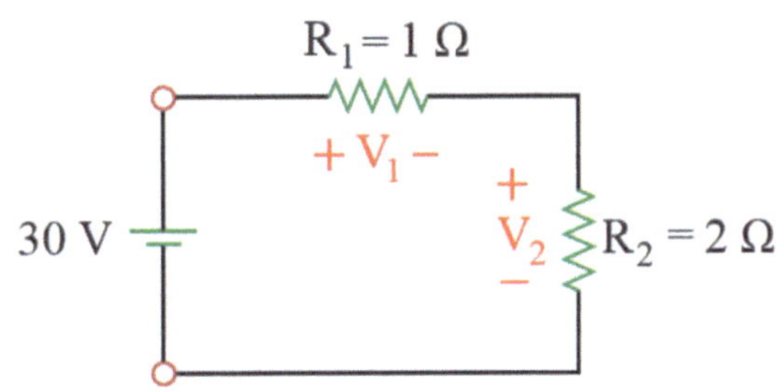

Fig. 1.28 The circuit of problem 1.37

Exercise

1.38. Calculate the voltage of the third resistor in the circuit shown in Fig. 1.29.

Final Answer

$V_3 = 30\ \text{V}$

Note: *Read the "educational support" part of the preface of the textbook.*

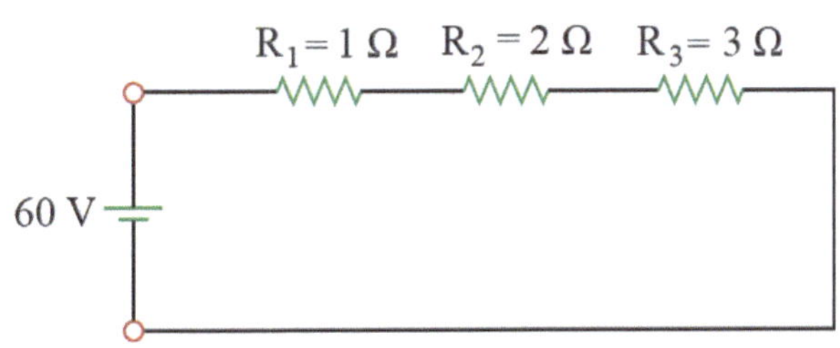

Fig. 1.29 The circuit of exercise 1.38

Problem

1.39. In the circuit shown in Fig. 1.30, calculate the voltage of each resistor.

Difficulty level ● Easy ○ Normal ○ Hard

Calculation amount ● Small ○ Normal ○ Large

1) $V_1 = 30$ V, $V_2 = 30$ V
2) $V_1 = 10$ V, $V_2 = 20$ V
3) $V_1 = 20$ V, $V_2 = 10$ V
4) $V_1 = 30$ V, $V_2 = 15$ V

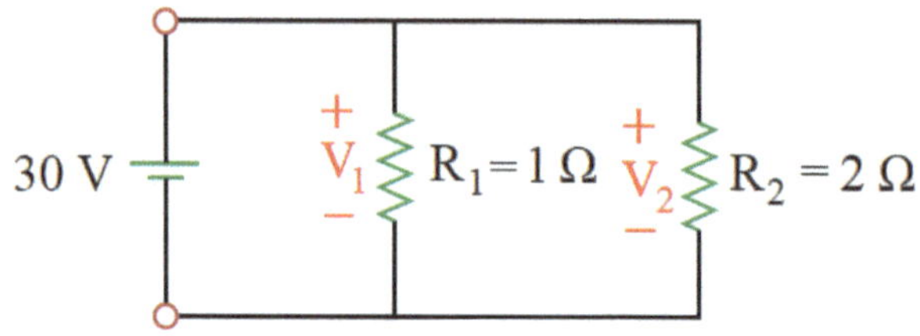

Fig. 1.30 The circuit of problem 1.39

Problem

1.40. In the circuit shown in Fig. 1.31, calculate the current of each resistor.

Difficulty level ● Easy ○ Normal ○ Hard

Calculation amount ● Small ○ Normal ○ Large

1) $I_1 = 30$ A, $I_2 = 30$ A
2) $I_1 = 30$ A, $I_2 = 15$ A
3) $I_1 = 10$ A, $I_2 = 20$ A
4) $I_1 = 20$ A, $I_2 = 10$ A

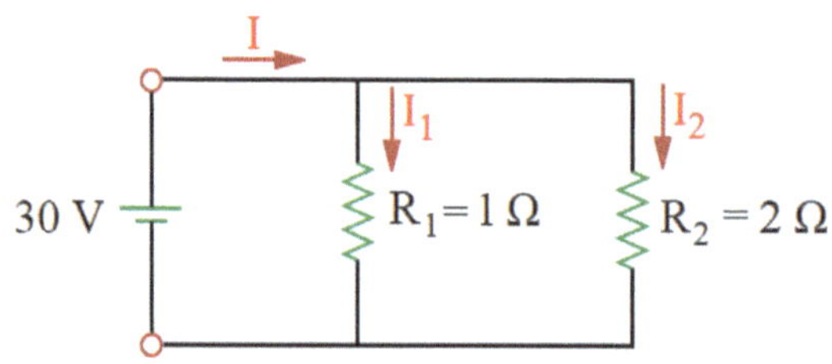

Fig. 1.31 The circuit of problem 1.40

Exercise

1.41. Calculate the current of the second resistor in the circuit shown in Fig. 1.32.

Final Answer

$I_2 = 5$ A

Note: *Read the "educational support" part of the preface of the textbook.*

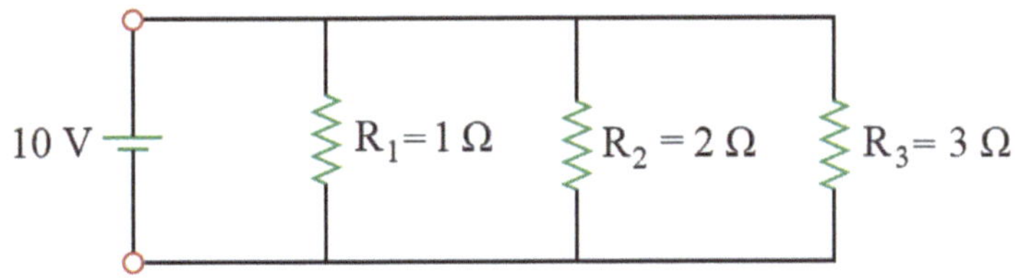

Fig. 1.32 The circuit of exercise 1.41

1.7 Ideal Independent Voltage Source

Problem

1.42. In the circuit shown in Fig. 1.33, simplify the circuit and calculate its equivalent ideal independent voltage source.

Difficulty level ● Easy ○ Normal ○ Hard

Calculation amount ● Small ○ Normal ○ Large

1) 15 V
2) 1 V
3) 8 V
4) 19 V

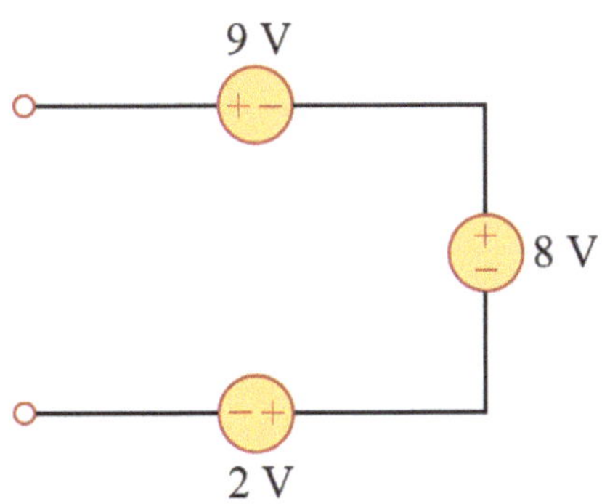

Fig. 1.33 The circuit of problem 1.42

Exercise

1.43. In the circuit shown in Fig. 1.34, simplify the circuit and calculate its equivalent ideal independent voltage source.

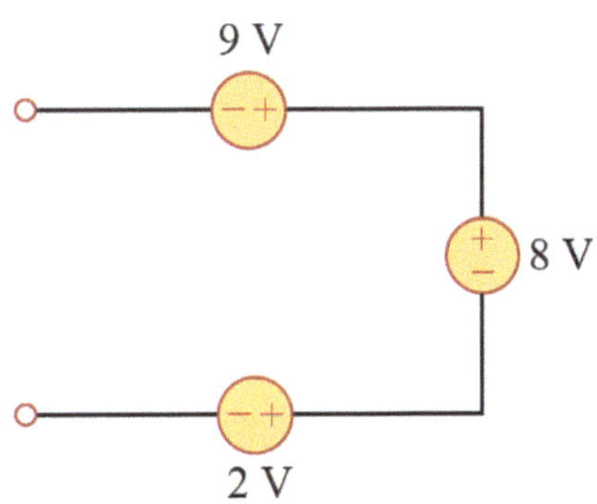

Fig. 1.34 The circuit of exercise 1.43

Final Answer

The answer is shown in Fig. 1.35.

Note: Read the "educational support" part of the preface of the textbook.

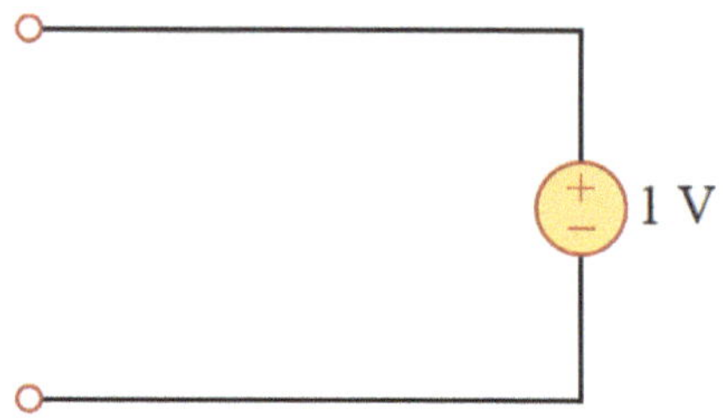

Fig. 1.35 The circuit of exercise 1.43

Problem

1.44. In the circuit shown in Fig. 1.36, simplify the circuit and calculate the voltage of its equivalent ideal independent voltage source if $V_{s1} = V_{s2}$.

Difficulty level	● Easy	○ Normal	○ Hard
Calculation amount	● Small	○ Normal	○ Large

1) V_{s1}
2) $V_{s1} + V_{s2}$
3) 0
4) This parallel connection is theoretically wrong.

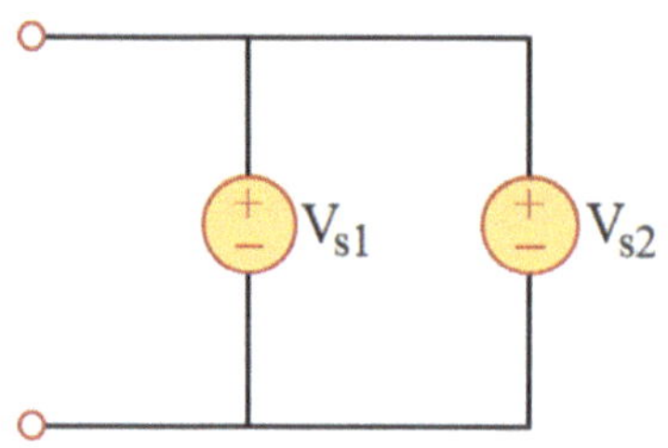

Fig. 1.36 The circuit of problem 1.44

Problem

1.45. In the circuit shown in Fig. 1.37, simplify the circuit and calculate the voltage of its equivalent ideal independent voltage source if $V_{s1} \neq V_{s2}$.

Difficulty level	● Easy	○ Normal	○ Hard
Calculation amount	● Small	○ Normal	○ Large

1) V_{s1}
2) $V_{s1} + V_{s2}$
3) V_{s2}
4) This parallel connection is theoretically wrong.

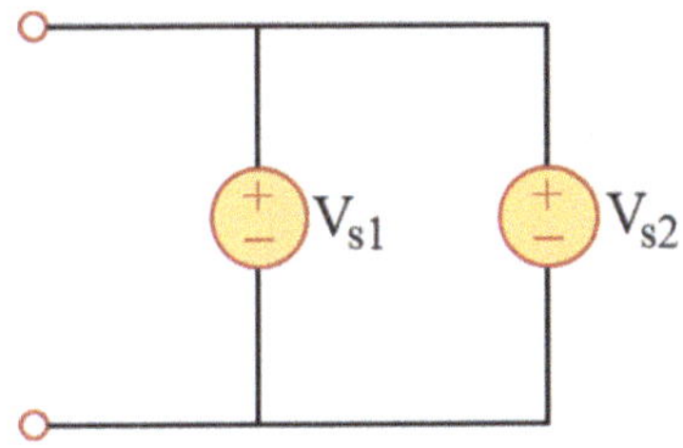

Fig. 1.37 The circuit of problem 1.45

1.8 Ideal Independent Current Source

Problem

1.46. In the circuit shown in Fig. 1.38, simplify the circuit and calculate the current of its equivalent ideal independent current source.

Difficulty level ● Easy ○ Normal ○ Hard
Calculation amount ● Small ○ Normal ○ Large

1) 5 A
2) 2 A
3) 1 A
4) 6 A

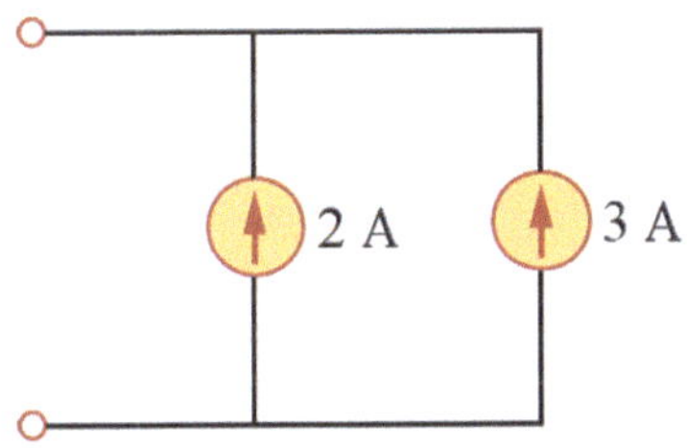

Fig. 1.38 The circuit of problem 1.46

Exercise

1.47. In the circuit shown in Fig. 1.39, simplify the circuit and calculate the current of its equivalent ideal independent current source.

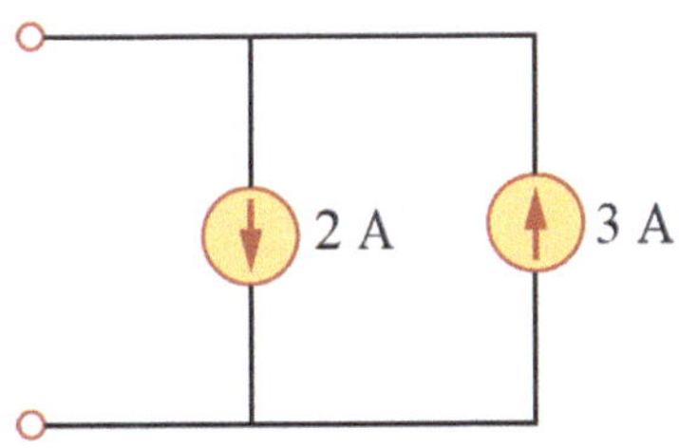

Fig. 1.39 The circuit of exercise 1.47

Final Answer

The answer is shown in Fig. 1.40.

Note: *Read the "educational support" part of the preface of the textbook.*

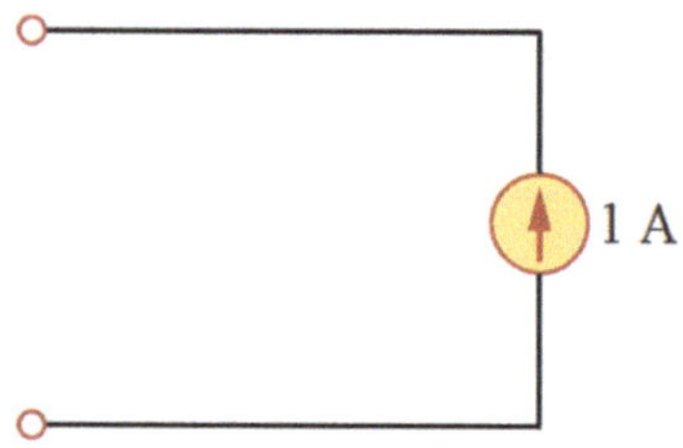

Fig. 1.40 The circuit of exercise 1.47

Problem

1.48. In the circuit shown in Fig. 1.41, simplify the circuit and calculate the current of its equivalent ideal independent current source if $I_{s1} = I_{s2}$.

Difficulty level ● Easy ○ Normal ○ Hard
Calculation amount ● Small ○ Normal ○ Large

1) I_{s1}
2) $I_{s1} + I_{s2}$
3) 0
4) This series connection is theoretically wrong.

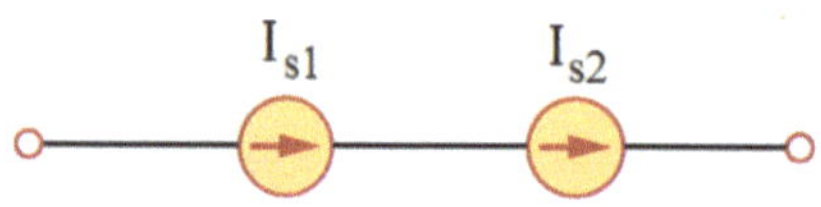

Fig. 1.41 The circuit of problem 1.48

Problem

1.49. In the circuit shown in Fig. 1.42, simplify the circuit and calculate the current of its equivalent ideal independent current source if $I_{s1} \neq I_{s2}$.

Difficulty level ● Easy ○ Normal ○ Hard
Calculation amount ● Small ○ Normal ○ Large

1) I_{s1}
2) $I_{s1} + I_{s2}$
3) I_{s2}
4) This series connection is theoretically wrong.

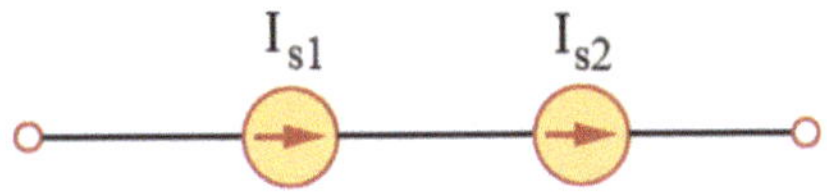

Fig. 1.42 The circuit of problem 1.49

1.9 Ideal Dependent Voltage and Current Sources

Problem

1.50. In the circuit shown in Fig. 1.43, determine the type of the ideal independent source.

Difficulty level ● Easy ○ Normal ○ Hard
Calculation amount ● Small ○ Normal ○ Large

1) Voltage-controlled voltage source (VCVS)
2) Current-controlled voltage source (CCVS)
3) Voltage-controlled current source (VCCS)
4) Current-controlled current source (CCCS)

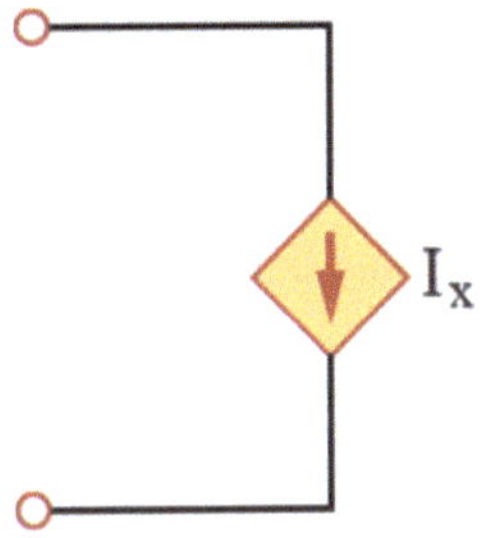

Fig. 1.43 The circuit of problem 1.50

Problem

1.51. In the circuit shown in Fig. 1.44, determine the type of the ideal independent source.

Difficulty level ● Easy ○ Normal ○ Hard
Calculation amount ● Small ○ Normal ○ Large

1) Voltage-controlled voltage source (VCVS)
2) Current-controlled voltage source (CCVS)
3) Voltage-controlled current source (VCCS)
4) Current-controlled current source (CCCS)

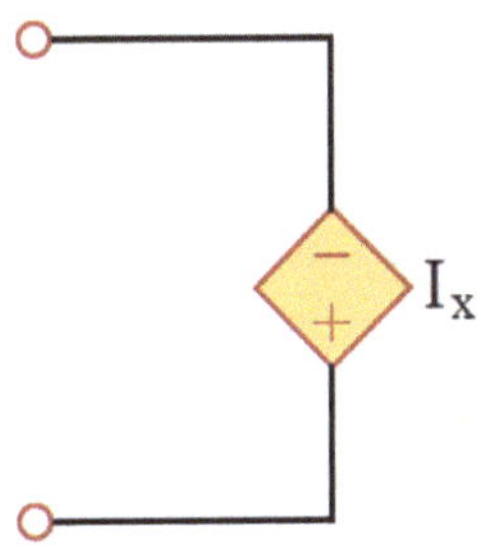

Fig. 1.44 The circuit of problem 1.51

Problem

1.52. In the circuit shown in Fig. 1.45, determine the type of the ideal independent source.

Difficulty level ● Easy ○ Normal ○ Hard

Calculation amount ● Small ○ Normal ○ Large

1) Voltage-controlled voltage source (VCVS)
2) Current-controlled voltage source (CCVS)
3) Voltage-controlled current source (VCCS)
4) Current-controlled current source (CCCS)

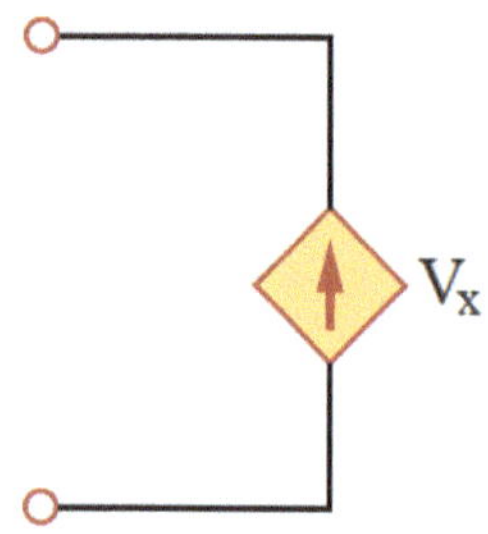

Fig. 1.45 The circuit of problem 1.52

Problem

1.53. In the circuit shown in Fig. 1.46, determine the type of the ideal independent source.

Difficulty level ● Easy ○ Normal ○ Hard

Calculation amount ● Small ○ Normal ○ Large

1) Voltage-controlled voltage source (VCVS)
2) Current-controlled voltage source (CCVS)
3) Voltage-controlled current source (VCCS)
4) Current-controlled current source (CCCS)

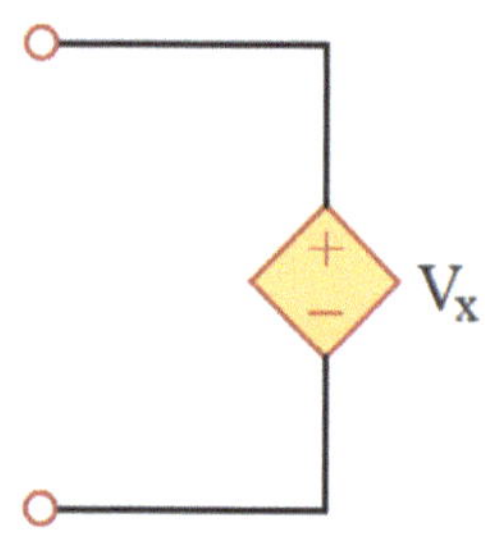

Fig. 1.46 The circuit of problem 1.53

1.10 Kirchhoff's Current Law (KCL)

Problem

1.54. In the circuit of Fig. 1.47, calculate the current shown by I.

Difficulty level ● Easy ○ Normal ○ Hard

Calculation amount ● Small ○ Normal ○ Large

1) 15 A
2) 1 A
3) – 1 A
4) – 15 A

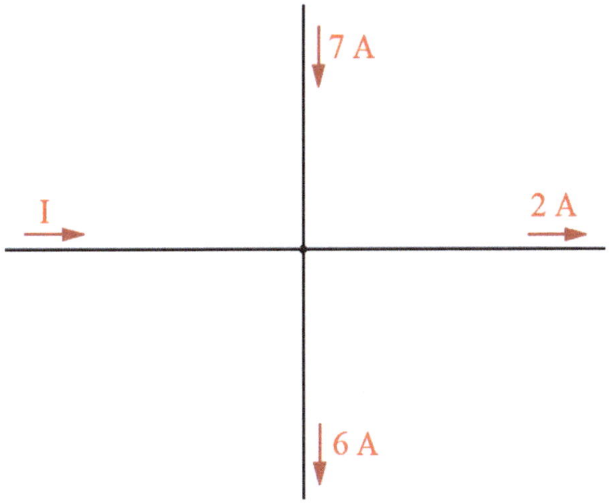

Fig. 1.47 The circuit of problem 1.54

Exercise

1.55. In the circuit shown in Fig. 1.48, calculate the current I.

Final Answer

$I = 0$

Note: *Read the "educational support" part of the preface of the textbook.*

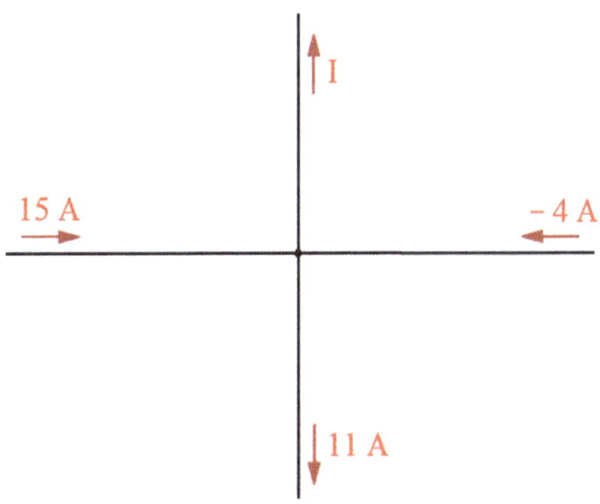

Fig. 1.48 The circuit of exercise 1.55

Partially Solved Exercise

1.56. In the circuit shown in Fig. 1.49, calculate the currents I_1 and I_2.

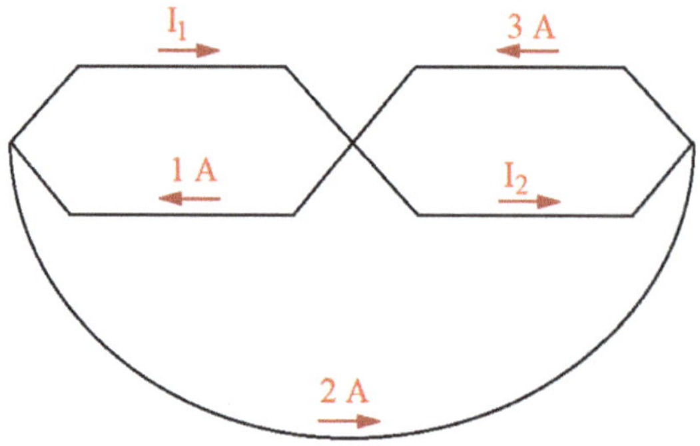

Fig. 1.49 The circuit of partially solved exercise 1.56

Solution

By applying KCL for the left-side node, we have:

$$I_1 + (\quad) = (\quad)$$

$$\Rightarrow I_1 = -1\text{ A}$$

Moreover, by applying KCL for the right-side node, we have:

$$I_2 + (\quad) = (\quad)$$

$$\Rightarrow I_2 = 1\text{ A}$$

Problem

1.57. In the circuit shown in Fig. 1.50, calculate the output voltage V_o.

1) $\frac{5}{3}$ V
2) $-\frac{5}{3}$ V
3) $\frac{10}{3}$ V
4) $-\frac{10}{3}$ V

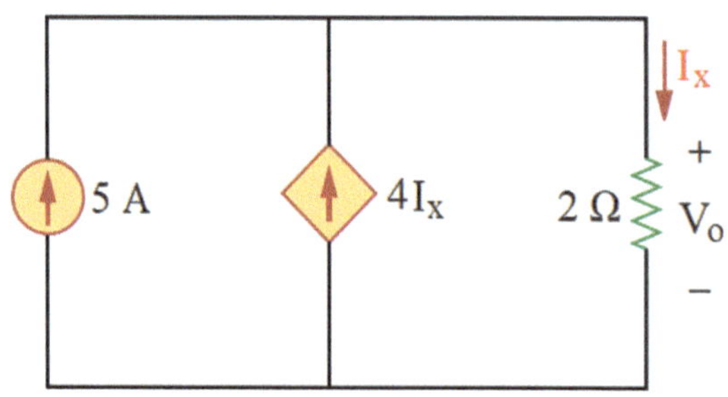

Fig. 1.50 The circuit of problem 1.57

Exercise

1.58. In the circuit shown in Fig. 1.51, calculate the output voltage V_o.

Final Answer

$V_o = 4.5$ V

Note: Read the "educational support" part of the preface of the textbook.

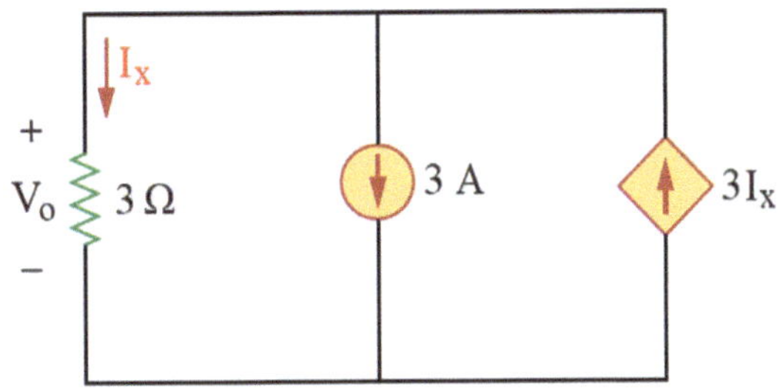

Fig. 1.51 The circuit of exercise 1.58

1.11 Kirchhoff's Voltage Law (KVL)

Problem

1.59. In the circuit illustrated in Fig. 1.52, calculate the voltage V_1 by using KVL.

Difficulty level ● Easy ○ Normal ○ Hard

Calculation amount ● Small ○ Normal ○ Large

1) 8 V
2) 4 V
3) 12 V
4) 6 V

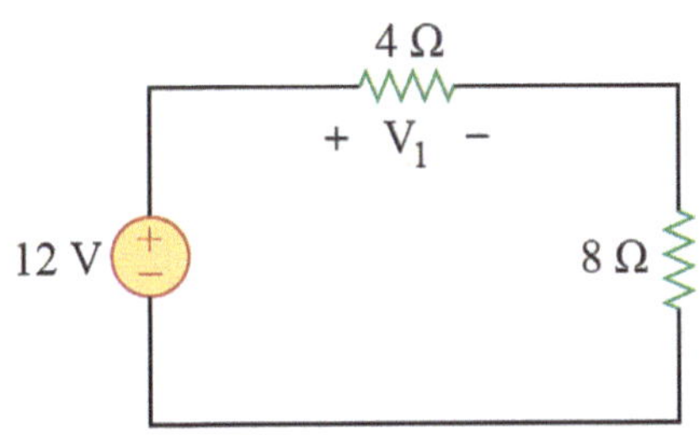

Fig. 1.52 The circuit of problem 1.59

Partially Solved Exercise

1.60. In the circuit illustrated in Fig. 1.53, calculate the voltage V_2 by using KVL.

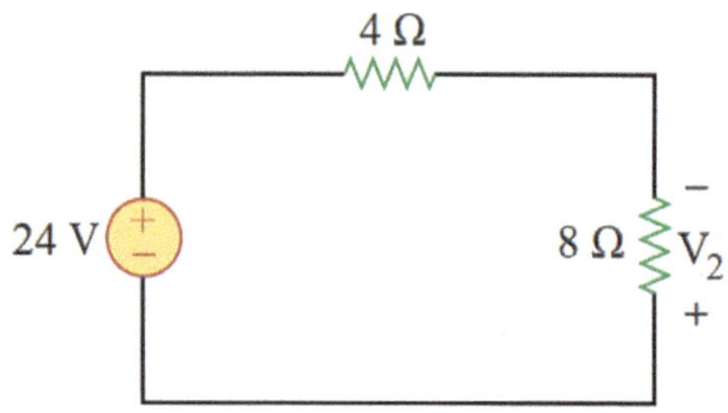

Fig. 1.53 The circuit of partially solved exercise 1.60

Solution

Since the voltages of the circuit components are unknown, a current needs to be assigned to the circuit as is shown in Fig. 1.54. Now, by applying KVL, we have:

$$(\quad)+(\quad)+(\quad)=0$$

$$\Rightarrow (\quad)+(\quad)=0$$

$$\Rightarrow I=2\ \mathrm{A}$$

In addition, by applying Ohm's law for the 8-Ω resistor, we have:

$$V_2=(\quad)$$

Therefore:

$$V_2=-16\ \mathrm{V}$$

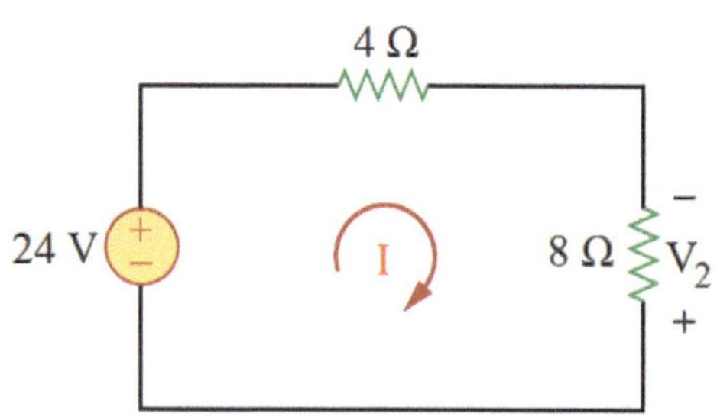

Fig. 1.54 The circuit of partially solved exercise 1.60

Exercise

1.61. In the circuit illustrated in Fig. 1.55, calculate the voltages V_1 and V_2 by using KVL.

Final Answer

$V_1 = -3\ \text{V},\ V_2 = 1\ \text{V}$

Note: Read the "educational support" part of the preface of the textbook.

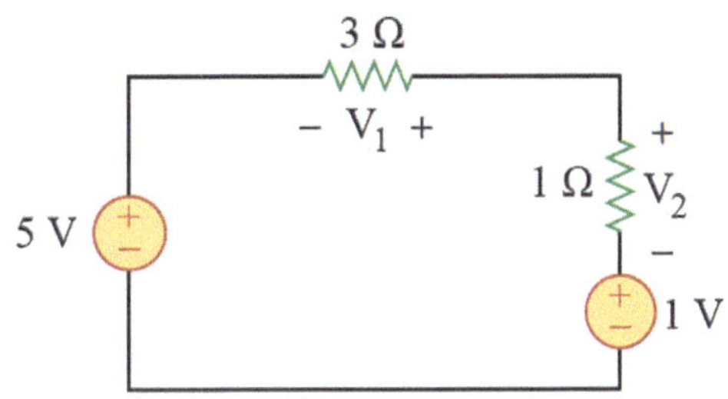

Fig. 1.55 The circuit of exercise 1.61

Problem

1.62. In the circuit illustrated in Fig. 1.56, calculate the voltages V_1 and V_2 by using KVL.

Difficulty level ○ Easy ● Normal ○ Hard

Calculation amount ○ Small ● Normal ○ Large

1) $V_1 = \frac{24}{11}\ \text{V},\ V_2 = \frac{8}{11}\ \text{V}$
2) $V_1 = \frac{12}{11}\ \text{V},\ V_2 = \frac{4}{11}\ \text{V}$
3) $V_1 = 12\ \text{V},\ V_2 = 4\ \text{V}$
4) $V_1 = \frac{3}{11}\ \text{V},\ V_2 = \frac{1}{11}\ \text{V}$

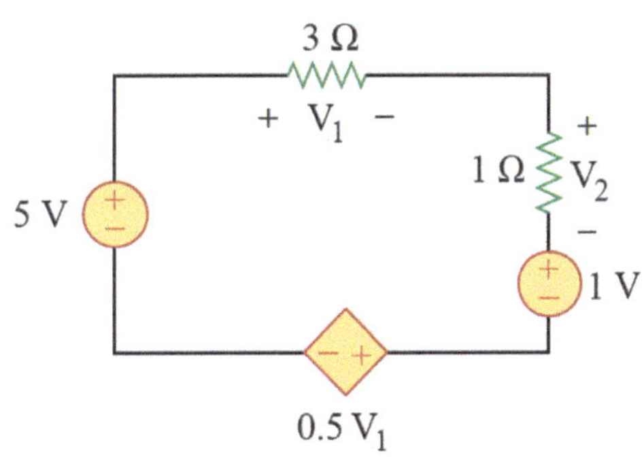

Fig. 1.56 The circuit of problem 1.62

Exercise

1.63. In the circuit illustrated in Fig. 1.57, calculate the voltages V_1 and V_2 by using KVL.

Final Answer

$V_1 = \frac{8}{3}\ \text{V},\ V_2 = \frac{8}{9}\ \text{V}$

Note: Read the "educational support" part of the preface of the textbook.

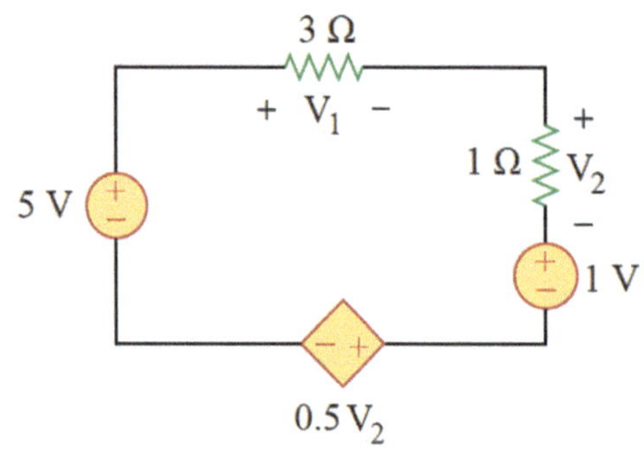

Fig. 1.57 The circuit of exercise 1.63

1.12 Power Consumption and Generation

Problem

1.64. In the circuit shown in Fig. 1.58, calculate the power wasted by each resistor.

Difficulty level ● Easy ○ Normal ○ Hard

Calculation amount ● Small ○ Normal ○ Large

1) $P_1 = 100$ W, $P_2 = 200$ W
2) $P_1 = 10$ W, $P_2 = 20$ W
3) $P_1 = 20$ W, $P_2 = 10$ W
4) $P_1 = 200$ W, $P_2 = 100$ W

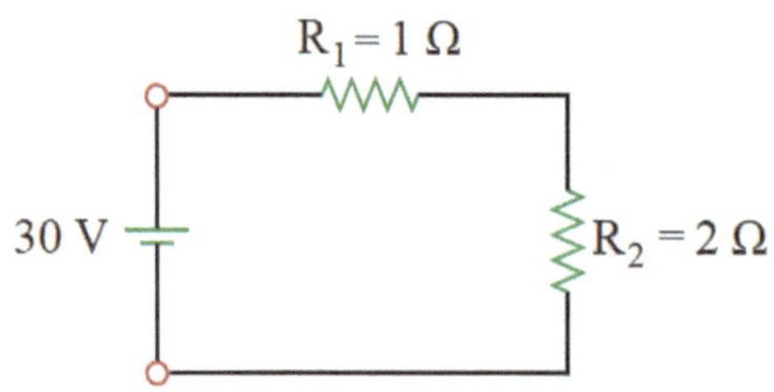

Fig. 1.58 The circuit of problem 1.64

Problem

1.65. In the circuit shown in Fig. 1.59, calculate the consumption power of each resistor.

Difficulty level ● Easy ○ Normal ○ Hard

Calculation amount ● Small ○ Normal ○ Large

1) $P_1 = 30\ \text{W},\ P_2 = 15\ \text{W}$
2) $P_1 = 15\ \text{W},\ P_2 = 30\ \text{W}$
3) $P_1 = 450\ \text{W},\ P_2 = 900\ \text{W}$
4) $P_1 = 900\ \text{W},\ P_2 = 450\ \text{W}$

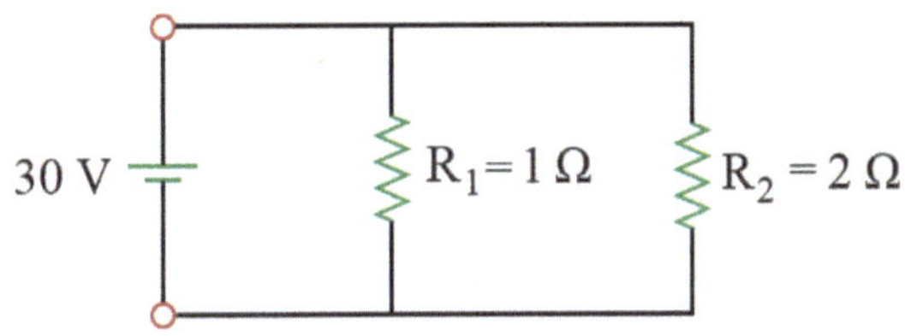

Fig. 1.59 The circuit of problem 1.65

Partially Solved Exercise

1.66. Calculate the power wasted by each resistor in the circuit shown in Fig. 1.60.

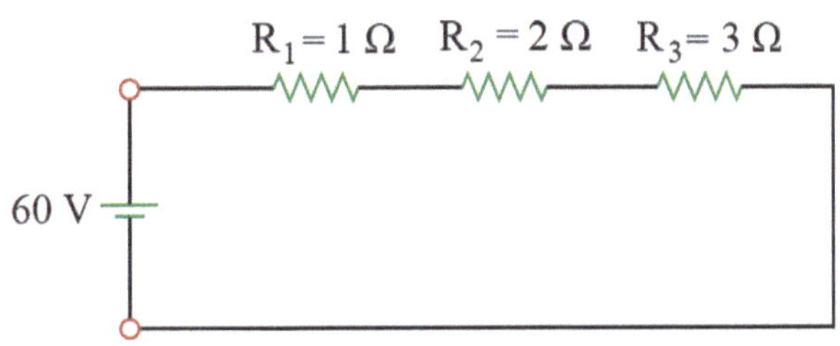

Fig. 1.60 The circuit of partially solved exercise 1.66

Solution

The resistors are connected in series; therefore, their currents can be calculated as follows.

$$I_1 = I_2 = I_3 = \frac{(\quad)}{(\quad) + (\quad) + (\quad)}$$

$$\Rightarrow I_1 = I_2 = I_3 = (\quad)\ \text{A}$$

The power wasted or consumed by a resistor can be calculated by using one of the following relations.

$$P = RI^2$$

$$P = VI$$

$$P = \frac{V^2}{R}$$

For this problem, the first relation is more appropriate. Hence:

$$P_1 = R_1 I_1^2 = (\quad) \times (\quad)^2 \Rightarrow P_1 = 100 \text{ W}$$

$$P_2 = R_2 I_2^2 = (\quad) \times (\quad)^2 \Rightarrow P_2 = 200 \text{ W}$$

$$P_3 = R_3 I_3^2 = (\quad) \times (\quad)^2 \Rightarrow P_3 = 300 \text{ W}$$

Partially Solved Exercise

1.67. Calculate the power wasted by each resistor in the circuit shown in Fig. 1.61.

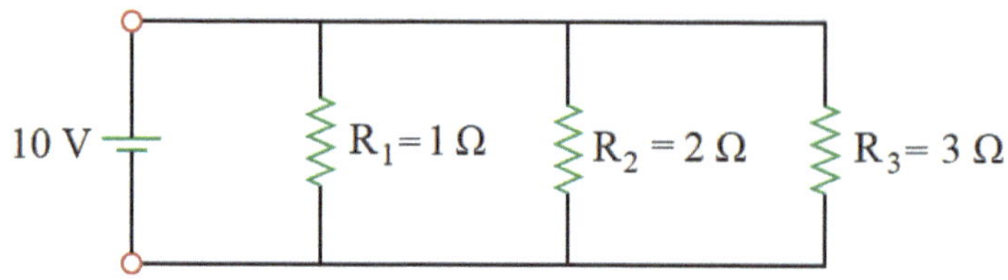

Fig. 1.61 The circuit of partially solved exercise 1.67

Solution

The circuit is a parallel circuit; hence, the voltage of each resistor is equal to the voltage of the source. In other words:

$$V_1 = V_2 = V_3 = (\quad) \text{ V}$$

The power wasted or consumed by a resistor can be calculated by using one of the following relations.

$$P = RI^2$$

$$P = VI$$

$$P = \frac{V^2}{R}$$

For this problem, the third relation is more appropriate. Hence:

$$P_1 = \frac{V_1^2}{R_1} = \frac{(\quad)^2}{(\quad)} \Rightarrow P_1 = 100 \text{ W}$$

$$P_2 = \frac{V_2^2}{R_2} = \frac{(\quad)^2}{(\quad)} \Rightarrow P_2 = 50 \text{ W}$$

$$P_3 = \frac{V_3^2}{R_3} = \frac{(\quad)^2}{(\quad)} \Rightarrow P_3 = 33.33 \text{ W}$$

Problem

1.68. In the circuit of Fig. 1.62, which one of the sources is a power consumer?

Difficulty level ○ Easy ● Normal ○ Hard

Calculation amount ● Small ○ Normal ○ Large

1) The left-side current source
2) The left-side voltage source
3) The right-side current source
4) The right-side voltage source

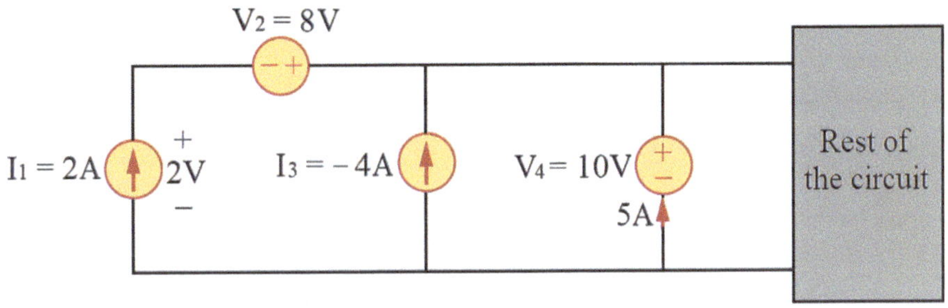

Fig. 1.62 The circuit of problem 1.68

Problem

1.69. The power of some of the components is given in the circuit of Fig. 1.63. Calculate the current I shown in the circuit.

Difficulty level ○ Easy ○ Normal ● Hard

Calculation amount ● Small ○ Normal ○ Large

1) 1 A
2) 2 A
3) 0.5 A
4) 0.25 A

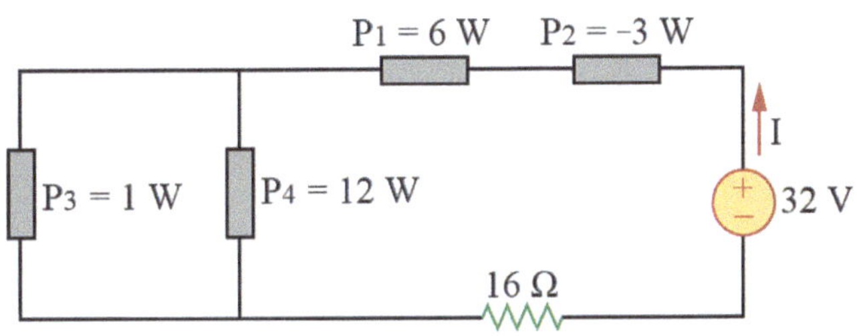

Fig. 1.63 The circuit of problem 1.69

Problem

1.70. Calculate the amount and type (generation or consumption) of power of the $5 - V$ voltage source in the circuit of Fig. 1.64.

Difficulty level ○ Easy ● Normal ○ Hard

Calculation amount ● Small ○ Normal ○ Large

1) 15 W, generation.
2) 15 W, consumption.
3) 20 W, consumption.
4) 20 W, generation.

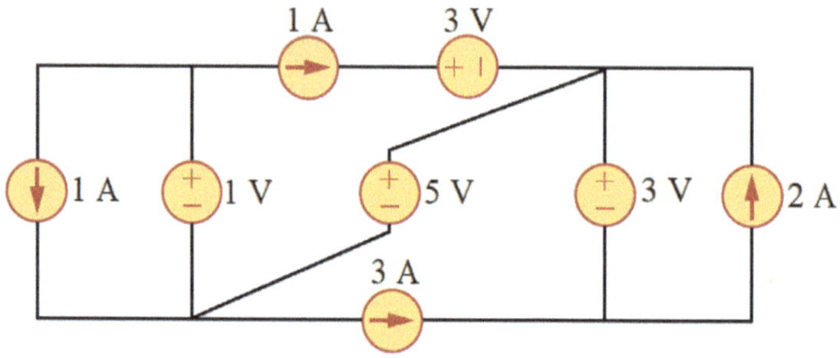

Fig. 1.64 The circuit of problem 1.70

References

1. Rahmani-Andebili, M., DC Electrical Circuit Analysis - Practice Problems, Methods, and Solutions (1st Ed.), Springer Nature, 2020.
2. Rahmani-Andebili, M., General Physics II - Practice Problems, Methods, and Solutions, Sringer Nature, 2025.
3. Rahmani-Andebili, M., General Physics I - Practice Problems, Methods, and Solutions, Springer Nature, 2025.
4. Rahmani-Andebili, M., Mathematics of Engineering and Science - Comprehensive Lessons, Springer Nature, 2024.
5. Rahmani-Andebili, M., Differential Equations – Practice Problems, Methods, and Solutions, Springer Nature, 2022.
6. Rahmani-Andebili, M., Calculus III – Practice Problems, Methods, and Solutions, Springer Nature, 2023.
7. Rahmani-Andebili, M., Calculus II – Practice Problems, Methods, and Solutions, Springer Nature, 2023.
8. Rahmani-Andebili, M., Calculus I (2nd Ed.) – Practice Problems, Methods, and Solutions, Springer Nature, 2023.
9. Rahmani-Andebili, M., Precalculus (2nd Ed.) – Practice Problems, Methods, and Solutions, Springer Nature, 2024.

Basic Concepts and Laws of Electrical Circuits Analysis: Part B

2

Abstract

In this chapter, the problems of the first chapter are fully solved, in detail, step-by-step, and with different methods. Herein, a "Notes Box" has been added after the solution of each problem that includes all the formulas and important notes that are needed to solve the problem. Moreover, a "Traps Box" has been added that discloses the traps that have been designed in the choices to challenge students' minds and encourage them to deeply think about the problem solutions as well as discourage them from careless selection of choices. The "Traps Box", in fact, predicts the possible confusions and tricks that might be encountered by students in real tests.

2.1 Resistance

2.1. Based on the information given in the problem, we have [1–9]:

$$2r = 1 \text{ mm}$$

$$l = 2 \text{ m}$$

$$R = 0.1 \ \Omega$$

The resistance of a wire can be calculated as follows.

$$R = \rho \frac{l}{A}$$

Herein, ρ, l, and A, are the special resistance (resistivity), length, and cross-sectional area of the wire, respectively.

Thus, for this problem, we can write:

$$R = \rho \frac{l}{\pi r^2}$$

$$\Rightarrow 0.1 = \rho \frac{2}{\pi (0.5 \times 10^{-3})^2}$$

F. X. Malcata, *DC Electrical Circuits Analysis*, https://doi.org/10.1007/978-3-032-00253-2_2

$$\Rightarrow \rho = \frac{0.1 \times 0.25\pi \times 10^{-6}}{2}$$

$$\Rightarrow \rho = 3.92 \times 10^{-8}\ \Omega\text{m}$$

Choice (1) is the answer.

Traps

- You would choose Choice (2) if you had mistakenly considered the diameter as the radius.

2.3. Based on the information given in the problem, we have:

$$l_1 = 2l_2$$

$$d_1 = 0.5d_2 \Rightarrow r_1 = 0.5r_2$$

$$\rho_1 = \frac{1}{3}\rho_2$$

As we know, the resistance of a wire can be calculated as follows.

$$R = \rho\frac{l}{A}$$

Herein, ρ, l, and A, are the special resistance (resistivity), length, and cross-sectional area of the wire, respectively. Thus, for this problem, we can write:

$$R = \rho\frac{l}{\pi r^2}$$

$$\Rightarrow \frac{R_1}{R_2} = \frac{\rho_1}{\rho_2} \times \frac{l_1}{l_2} \times \left(\frac{r_2}{r_1}\right)^2$$

$$\Rightarrow \frac{R_1}{R_2} = \frac{1}{3} \times 2 \times \left(\frac{1}{0.5}\right)^2$$

$$\Rightarrow \frac{R_1}{R_2} = \frac{8}{3}$$

Choice (4) is the answer.

Traps

- You would choose Choice (1) if you had mistakenly considered a direct proportionality between resistance and radius.
- You would choose Choice (2) if you had mistakenly considered a reverse proportionality between resistance and resistivity.

2.2 Ohm's Law

2.5. Based on the information given in the problem, we have:

$$I = 0.1 \text{ A}$$

$$V = 110 \text{ V}$$

Ohm's law presents a relation between the voltage (V), current (A), and resistance (Ω) of an Ohmic resistor as follows.

$$R = \frac{V}{I}$$

Hence, for this problem, we can write as follows.

$$R = \frac{110}{0.1}$$

$$\Rightarrow R = 1100 \ \Omega$$

Choice (2) is the answer.

Traps

- You would choose Choice (3) or (4) if you had forgotten Ohm's law.

2.7. Based on the information given in the problem, we have:

$$R = 1000 \ \Omega$$

$$V = 10 \text{ V}$$

Ohm's law presents a relation between the voltage (V), current (A), and resistance (Ω) of an Ohmic resistor as follows.

$$R = \frac{V}{I}$$

Hence, for this problem, we can write as follows.

$$1000 = \frac{10}{I}$$

$$\Rightarrow I = \frac{10}{1000}$$

$$\Rightarrow I = 0.01 \text{ A}$$

Choice (3) is the answer.

Traps

- You would choose Choice (1) or (2) if you had forgotten Ohm's law.

2.9. Based on the information given in the problem, we have:

$$R = 15\ \Omega$$

$$I = 0.1\ \text{A}$$

Ohm's law presents a relation between the voltage (V), current (A), and resistance (Ω) of an Ohmic resistor as follows.

$$R = \frac{V}{I}$$

Hence, for this problem, we can write as follows.

$$15 = \frac{V}{0.1}$$

$$\Rightarrow V = 15 \times 0.1$$

$$\Rightarrow V = 1.5\ \text{V}$$

Choice (4) is the answer.

Traps

- You would choose Choice (1) or (2) if you had forgotten Ohm's law.

2.3 Equivalent Resistance of Series, Parallel, and Series-Parallel Connections

2.11. The total resistance of the circuit can be calculated as follows.

$$R_{eq} = 2 + 6 + 3$$

$$\Rightarrow R_{eq} = 11\ \Omega$$

Choice (2) is the answer (Fig. 2.1).

Traps

- You would choose Choice (1) if you had mistakenly used the formula concerned with parallel resistors.

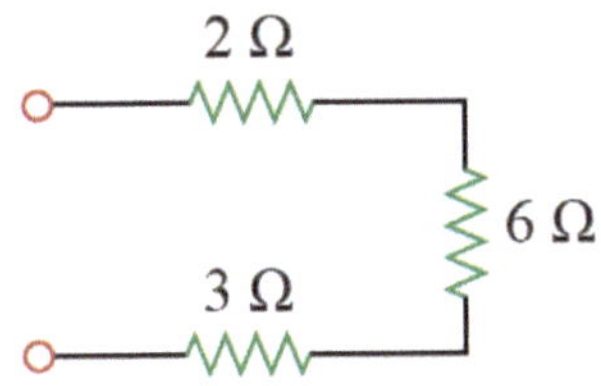

Fig. 2.1 The circuit of problem 2.11

Notes

The equivalent resistance of n series resistors can be calculated as follows.

$$R_{eq} = R_1 + \ldots + R_n$$

If $R_1 = \ldots = R_n = R$, we have:

$$R_{eq} = nR$$

2.12. The total resistance of the circuit can be calculated as follows.

$$\frac{1}{R_{eq}} = \frac{1}{2} + \frac{1}{3} + \frac{1}{6}$$

$$\Rightarrow \frac{1}{R_{eq}} = \frac{3+2+1}{6} = \frac{6}{6}$$

$$\Rightarrow R_{eq} = 1\ \Omega$$

Choice (1) is the answer (Fig. 2.2).

Traps

- You would choose Choice (2) if you had mistakenly used the formula concerned with series resistors.

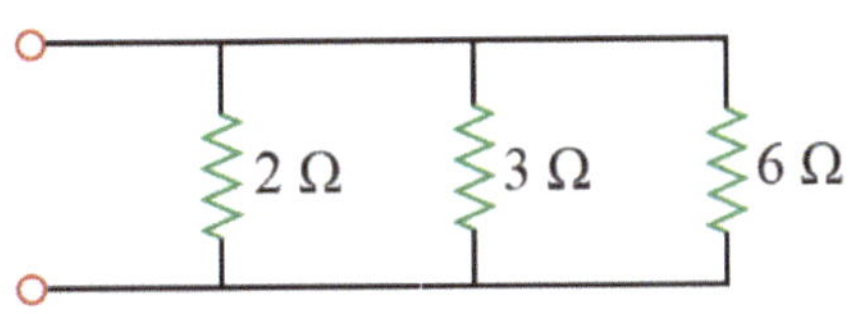

Fig. 2.2 The circuit of problem 2.12

Notes

The equivalent resistance of n parallel resistors can be calculated as follows.

$$\frac{1}{R_{eq}} = \frac{1}{R_1} + \ldots + \frac{1}{R_n}$$

If $R_1 = \ldots = R_n = R$, we have:

$$R_{eq} = \frac{R}{n}$$

If $n = 2$, we have:

$$\frac{1}{R_{eq}} = \frac{1}{R_1} + \frac{1}{R_2} \Rightarrow R_{eq} = \frac{R_1 R_2}{R_1 + R_2}$$

2.14. In the circuit, the resistors are connected in series–parallel form. Therefore:

$$R_{eq} = 2 + 1||1$$

$$\Rightarrow R_{eq} = 2 + \frac{1 \times 1}{1 + 1}$$

$$\Rightarrow R_{eq} = 2.5\ \Omega$$

Choice (4) is the answer (Fig. 2.3).

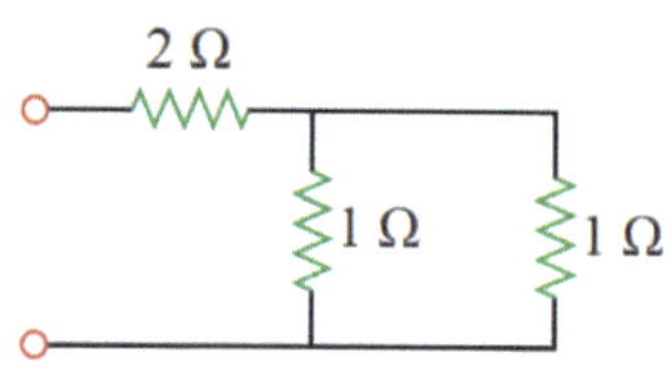

Fig. 2.3 The circuit of problem 2.14

Notes

The equivalent resistance of n series resistors can be calculated as follows.

$$R_{eq} = R_1 + \ldots + R_n$$

If $R_1 = \ldots = R_n = R$, we have:

$$R_{eq} = nR$$

The equivalent resistance of n parallel resistors can be calculated as follows.

$$\frac{1}{R_{eq}} = \frac{1}{R_1} + \ldots + \frac{1}{R_n}$$

If $R_1 = \ldots = R_n = R$, we have:

$$R_{eq} = \frac{R}{n}$$

If $n = 2$, we have:

$$\frac{1}{R_{eq}} = \frac{1}{R_1} + \frac{1}{R_2} \Rightarrow R_{eq} = \frac{R_1 R_2}{R_1 + R_2}$$

2.17. Based on the information given in the problem, we have:

$$R = 1\ \Omega$$

As is shown in Fig. 2.4, by removing the free-end branches, the circuit becomes simpler. Now, the circuit includes four series resistors. Therefore:

$$R_{eq} = 4R = 4 \times 1$$

$$\Rightarrow R_{eq} = 4\ \Omega$$

Choice (4) is the answer.

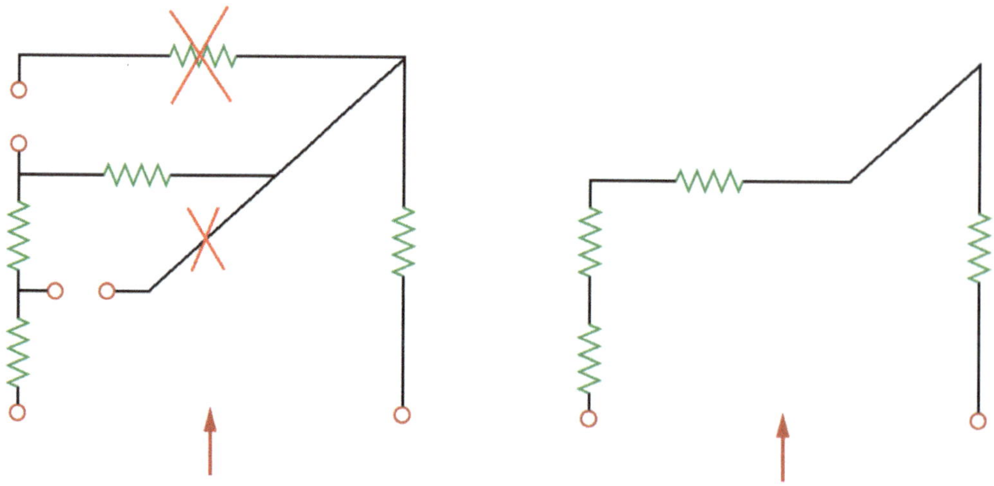

Fig. 2.4 The circuit of problem 2.17

Notes

The equivalent resistance of n series resistors can be calculated as follows.

$$R_{eq} = R_1 + \ldots + R_n$$

If $R_1 = \ldots = R_n = R$, we have:

$$R_{eq} = nR$$

2.20. As can be seen in Fig. 2.5, the 3-Ω resistors are connected in parallel. Hence:

$$R_{eq} = 2||(1 + 3||3||3)$$

$$\Rightarrow R_{eq} = 2||\left(1 + \frac{3}{3}\right)$$

$$\Rightarrow R_{eq} = 2||2$$

$$\Rightarrow R_{eq} = \frac{2}{2}$$

$$\Rightarrow R_{eq} = 1\ \Omega$$

Choice (1) is the answer.

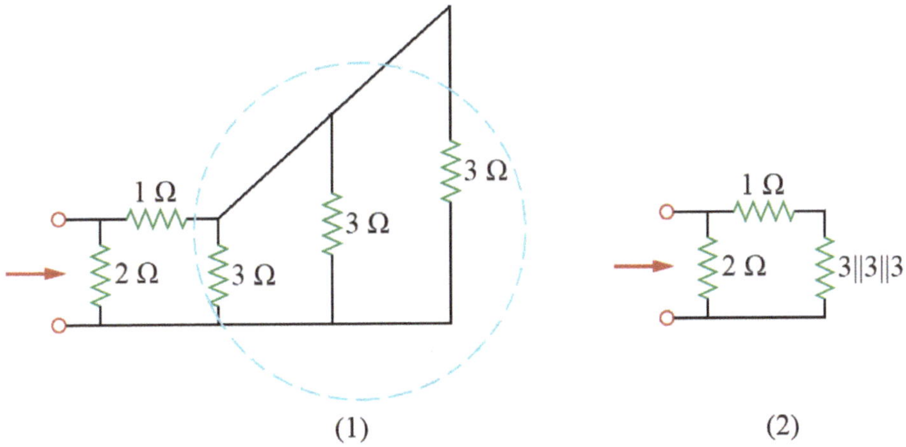

Fig. 2.5 The circuit of problem 2.20

Notes

The equivalent resistance of n series resistors can be calculated as follows.

$$R_{eq} = R_1 + \ldots + R_n$$

If $R_1 = \ldots = R_n = R$, we have:

$$R_{eq} = nR$$

The equivalent resistance of n parallel resistors can be calculated as follows.

$$\frac{1}{R_{eq}} = \frac{1}{R_1} + \ldots + \frac{1}{R_n}$$

If $R_1 = \ldots = R_n = R$, we have:

$$R_{eq} = \frac{R}{n}$$

2.4 Wye-Delta Transformation for Equivalent Resistance Calculation

2.22. As can be noticed, the resistors are neither series nor parallel; therefore, the series-parallel rule is not applicable for this problem.

The circuit includes a delta-connection of resistors that can be easily solved if the type of connection is converted to the wye-connection as follows.

$$R_A = \frac{24 \times 36}{24 + 36 + 12} = 12\ \Omega$$

$$R_B = \frac{24 \times 12}{24 + 36 + 12} = 4\ \Omega$$

$$R_C = \frac{36 \times 12}{24 + 36 + 12} = 6\ \Omega$$

Now, the primary circuit is simplified as is shown in Fig. 2.6. Hence, the equivalent resistance seen from the terminal can be calculated as follows.

$$R_{eq} = 12 + (6+6)||(4+20)$$

$$\Rightarrow R_{eq} = 12 + 12||24$$

$$\Rightarrow R_{eq} = 12 + \frac{12 \times 24}{12 + 24}$$

$$\Rightarrow R_{eq} = 12 + 8$$

$$\Rightarrow R_{eq} = 20\ \Omega$$

Choice (3) is the answer.

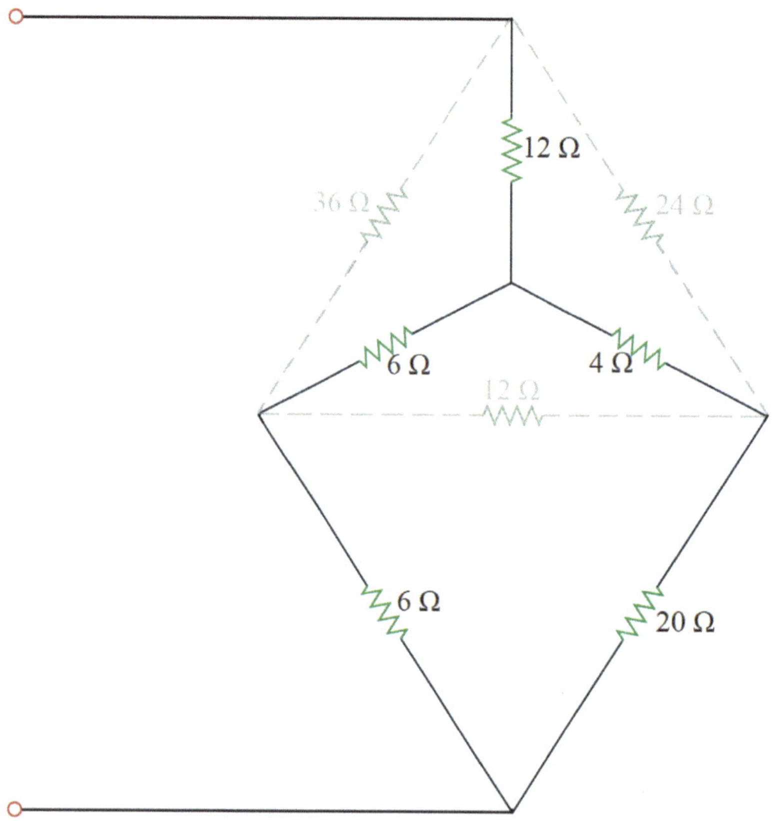

Fig. 2.6 The circuit of problem 2.22

Notes

The relations below are used when a wye-connection (called Y-connection, T-connection, or star-connection) is converted to a delta-connection (called Δ-connection or π-connection) shown in Fig. 2.7.

$$R_{AB} = \frac{R_A R_B + R_B R_C + R_A R_C}{R_C}$$

$$R_{BC} = \frac{R_A R_B + R_B R_C + R_A R_C}{R_A}$$

$$R_{AC} = \frac{R_A R_B + R_B R_C + R_A R_C}{R_B}$$

If $R_A = R_B = R_C = R$, then we have:

$$R_{AB} = R_{BC} = R_{AC} = 3R$$

Or:

$$R_\Delta = 3R_Y$$

Moreover, the following relations are valid when a delta-connection (Δ-connection or π-connection) is converted to a wye-connection
(Y-connection, T-connection, or star-connection) illustrated in Fig. 1.

$$R_A = \frac{R_{AC} R_{AB}}{R_{AB} + R_{BC} + R_{AC}}$$

$$R_B = \frac{R_{AB} R_{BC}}{R_{AB} + R_{BC} + R_{AC}}$$

$$R_C = \frac{R_{AC} R_{BC}}{R_{AB} + R_{BC} + R_{AC}}$$

If $R_{AB} = R_{BC} = R_{AC} = R$, then we have:

$$R_A = R_B = R_C = \frac{1}{3} R$$

Or:

$$R_Y = \frac{1}{3} R_\Delta$$

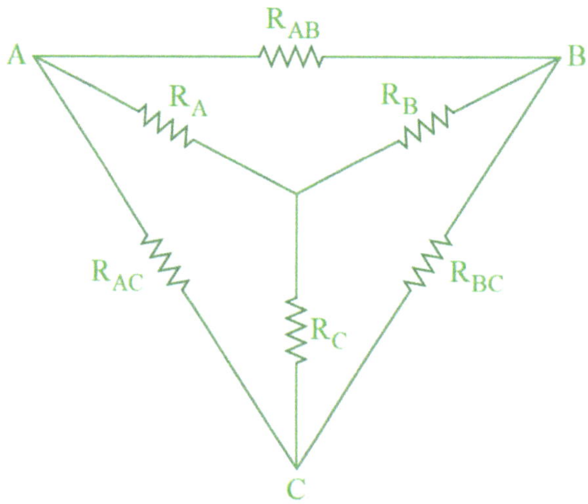

Fig. 2.7 The circuit of problem 2.22

2.5 Heuristic Methods for Equivalent Resistance Calculation

2.26. First, we need to calculate the input resistance of the circuit by using series-parallel rule as follows.

$$R_{in} = 1||(2 + (3||(4 + R)))$$

$$\Rightarrow R_{in} = 1||\left(2 + \frac{3R + 12}{R + 7}\right)$$

$$\Rightarrow R_{in} = 1||\frac{5R + 26}{R + 7}$$

$$\Rightarrow R_{in} = \frac{\frac{5R + 26}{R + 7}}{\frac{6R + 33}{R + 7}}$$

$$\Rightarrow R_{in} = \frac{5R + 26}{6R + 33}$$

The latest term is a rational function that always has an ascending or descending trend. Therefore, to find its range, we need to calculate its value for the lowest and highest values of its parameter (R).

On the other hand, we know that the range of resistance of an Ohmic resistor is $[0, \infty)$.
Therefore:

$$\lim_{R \to 0} R_{in} = \lim_{R \to 0} \frac{5R + 26}{6R + 33} = \frac{26}{33} = 0.78\ \Omega$$

$$\lim_{R \to \infty} R_{in} = \lim_{R \to \infty} \frac{5R + 26}{6R + 33} = \frac{5}{6} = 0.83\ \Omega$$

Thus:

$$0.78 < R_{in} < 0.83$$

Choice (4) is the answer (Fig. 2.8).

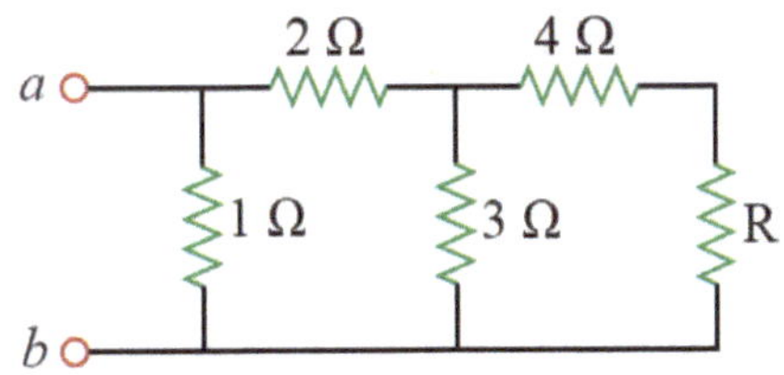

Fig. 2.8 The circuit of problem 2.26

2.28. Solving this problem will be easy if we consider the symmetric structure of the circuit. As is shown in Fig. 2.9, the nodes "C_1", "C_2", "C_3", and "C_4" have equal electric potential because the equivalent resistance between the nodes "A" and "B" is of interest. Therefore, these nodes can be short-circuited. Hence, the resistors, indicated by "f", are removed and the resistors, indicated by "e", become parallel.

Similarly, the nodes "D_1", "D_2", "D_3", and "D_4" have equal electric potential, and thus the resistors indicated by "h" are eliminated and the resistors, indicated by "g", as well as the resistors, indicated by "i", become in parallel separately. Therefore, the equivalent resistance seen from terminal A-B can be calculated as follows.

$$R_{AB} = \frac{1}{4} + \frac{1}{4} + \frac{1}{4}$$

$$\Rightarrow R_{AB} = \frac{3}{4}\ \Omega$$

Choice (4) is the answer.

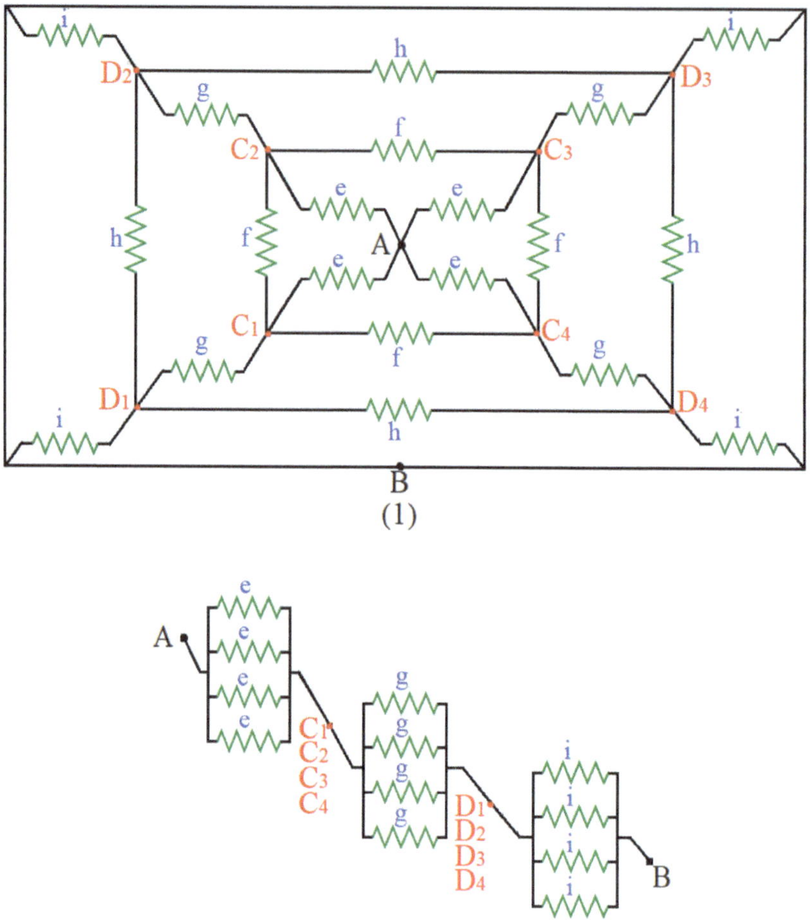

Fig. 2.9 The circuit of problem 2.28

2.30. As can be noticed, there are some symmetries in the circuit. Blue-color boundary in Fig. 2.10.1 is a parallel circuit with infinite number of components. Therefore, its equivalent resistance is zero, as can be seen in the following.

$$R_{Blue} = \underbrace{R||R||\ldots||R||R}_{n\,times} = \frac{R}{n} = \frac{R}{\infty} = 0 \qquad (1)$$

Thus, this part of circuit can be replaced by a short circuit branch, as is illustrated in Fig. 2.10.2.

Fig. 2.10.3 is the simplified version of Fig. 2.10.2. This type of circuit is called infinite ladder-circuit.

As can be seen in Fig. 2.10.3, R_{in} is just a parallel connection of R and R_{Red}.

$$R_{in} = R||R_{Red}$$

$$\Rightarrow R_{in} = \frac{RR_{Red}}{R + R_{Red}} \qquad (2)$$

Moreover, the relation below is held between R_{Red} and R_{Green}.

$$R_{Red} = R||(R + R_{Green})$$

$$\Rightarrow R_{Red} = \frac{R \times (R + R_{Green})}{R + (R + R_{Green})}$$

$$\Rightarrow R_{Red} = \frac{R^2 + RR_{Green}}{2R + R_{Green}} \tag{3}$$

However, since the ladder-circuit has an infinite size, it is concluded that:

$$R_{Green} \approx R_{Red} \tag{4}$$

Solving (3) and (4):

$$\Rightarrow R_{Red} = \frac{R^2 + RR_{Red}}{2R + R_{Red}}$$

$$\Rightarrow 2RR_{Red} + R_{Red}^2 = R^2 + RR_{Red}$$

$$\Rightarrow R_{Red}^2 + RR_{Red} - R^2 = 0$$

$$\Rightarrow R_{Red} = \frac{-R \pm \sqrt{R^2 + 4R^2}}{2}$$

$$\Rightarrow R_{Red} = R\left(\frac{-1 \pm \sqrt{5}}{2}\right)$$

$$\Rightarrow R_{Red} = -1.618\ \mathrm{R}, 0.618\ \mathrm{R}$$

Herein, only 0.618 R is acceptable. Thus:

$$R_{Red} = 0.618\ \mathrm{R} \tag{5}$$

Solving (2) and (5):

$$R_{in} = \frac{0.618\ \mathrm{R}^2}{1.618\ \mathrm{R}}$$

$$\Rightarrow R_{in} = 0.382\ \mathrm{R}$$

Choice (4) is the answer.

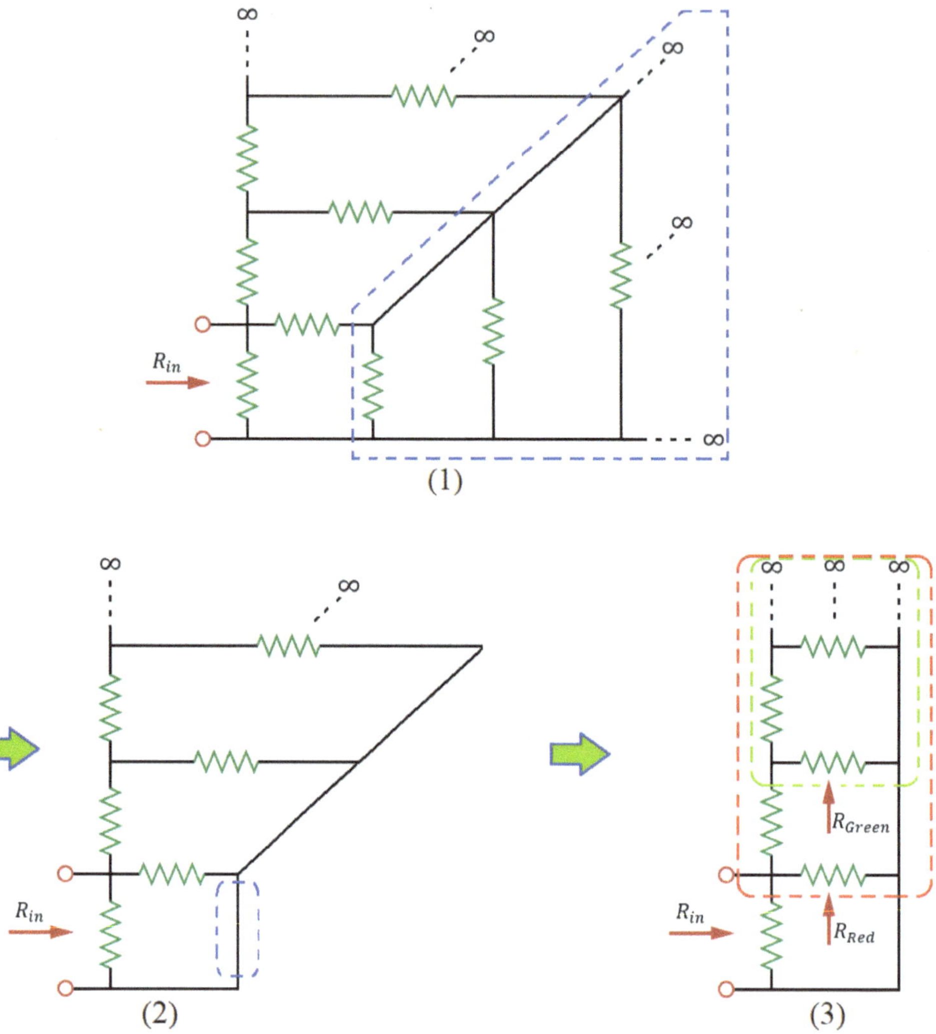

Fig. 2.10 The circuit of problem 2.30

Notes

In this problem, the relation below has been used.

$$ax^2 + bx + c = 0 \Rightarrow x = \frac{-b \pm \sqrt{b^2 - 4ac}}{2a}$$

2.32. The circuit is symmetrically seen from terminal A-B and the AB axis is its symmetry axis. Therefore, the circuit can be folded through AB axis, as is shown in Fig. 2.11.2. As a result, the vertical resistors on the left and right sides are eliminated since their two ends are connected to each other. Additionally, the resistance of three resistors is halved, due to the parallelism.

Now, as can be seen in Fig. 2.11.2, the three 0.5 Ω resistors have wye-connection. By changing their connection to delta-connection, the circuit is converted to Fig. 2.11.3 with the following resistors.

$$R_{AC} = 3 \times \frac{1}{2} = \frac{3}{2} \ \Omega$$

$$R_{AB} = 3 \times \frac{1}{2} = \frac{3}{2} \ \Omega$$

$$R_{BC} = 3 \times \frac{1}{2} = \frac{3}{2} \ \Omega$$

Now, the circuit of Fig. 2.11.3 can be simplified like the one shown in Fig. 2.11.4 where each parallel branch is simplified as follows.

$$1 || \frac{3}{2} = \frac{1 \times \frac{3}{2}}{1 + \frac{3}{2}} = \frac{3}{5} \ \Omega$$

Finally, we have:

$$R_{AB} = \left(\frac{3}{5} + \frac{3}{5}\right) || \left(\frac{3}{2}\right)$$

$$\Rightarrow R_{AB} = \frac{6}{5} || \frac{3}{2}$$

$$\Rightarrow R_{AB} = \frac{\frac{6}{5} \times \frac{3}{2}}{\frac{6}{5} + \frac{3}{2}} = \frac{\frac{18}{10}}{\frac{12 + 15}{10}}$$

$$\Rightarrow R_{AB} = \frac{2}{3} \ \Omega$$

Choice (2) is the answer.

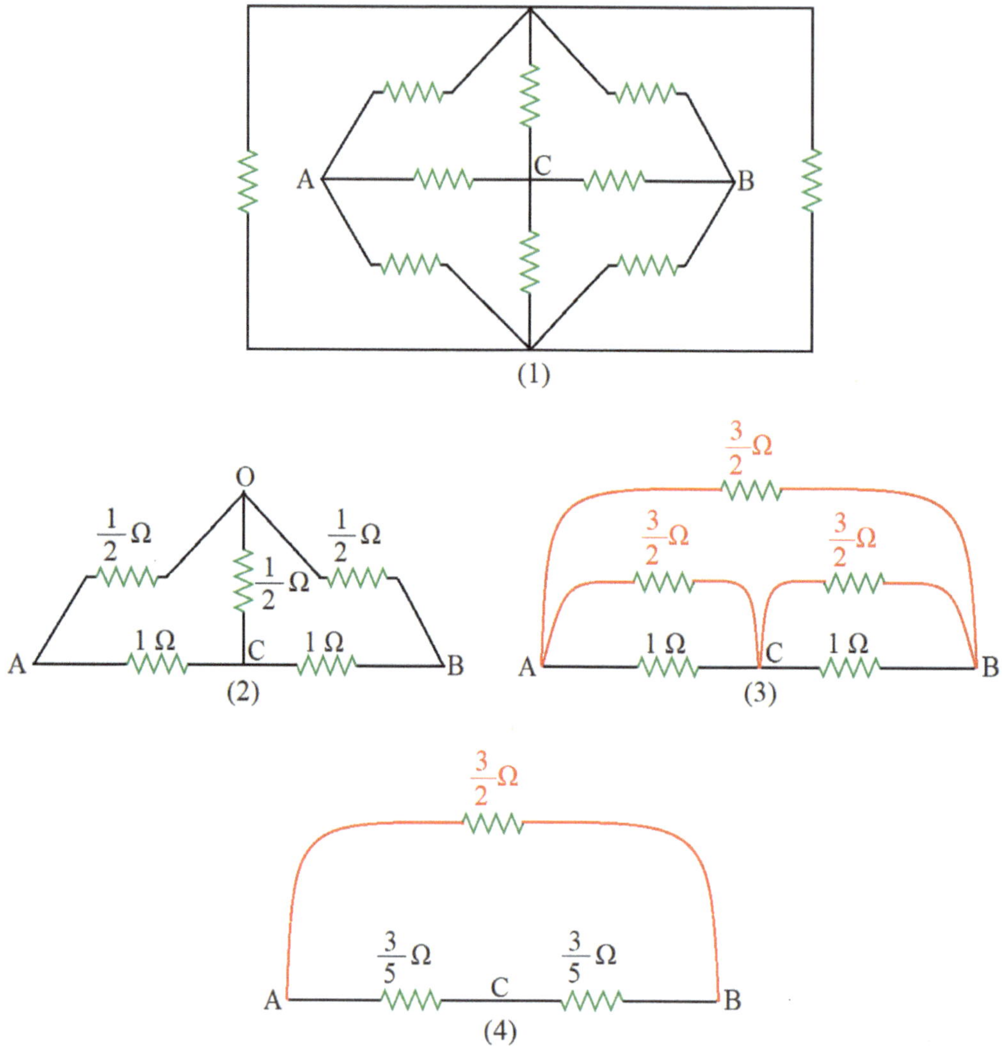

Fig. 2.11 The circuit of problem 2.32

Notes

The relations below are valid when a wye-connection (called Y-connection, T-connection, or star-connection) is converted to a delta-connection (called Δ-connection or π-connection) (Fig. 2.12).

$$R_{AB} = \frac{R_A R_B + R_B R_C + R_A R_C}{R_C}$$

$$R_{BC} = \frac{R_A R_B + R_B R_C + R_A R_C}{R_A}$$

$$R_{AC} = \frac{R_A R_B + R_B R_C + R_A R_C}{R_B}$$

If $R_A = R_B = R_C = R$, then we have:

$$R_{AB} = R_{BC} = R_{AC} = 3R$$

Or:

$$R_\Delta = 3R_Y$$

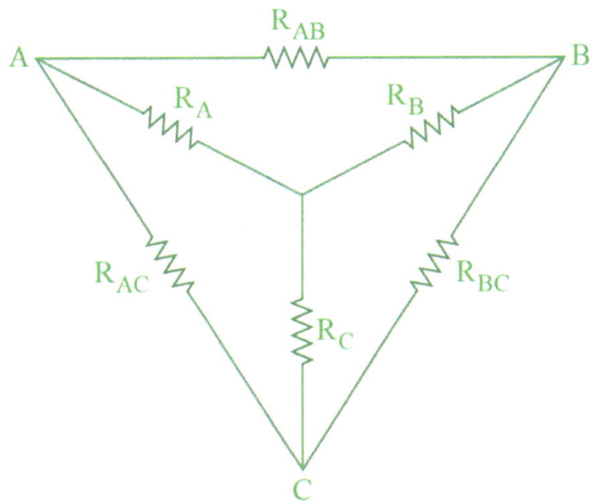

Fig. 2.12 The circuit of problem 2.32

2.33. The circuit seems to be complicated, but it is not because there is a symmetry in the structure of the circuit. Since the equivalent resistance between the nodes "A" and "O" is requested, the A-O-Z axis is the symmetry axis of the circuit. Therefore, the symmetric nodes "B" and "C" as well as the other the symmetric nodes, that is, "X" and "Y" have equal voltages if a power source is connected between the nodes "A" and "O". Therefore, the circuit can be folded through the A-O-Z axis, as is shown in Fig. 2.13.2. Herein, all the branches, except the branches between A-O and O-Z, include two resistors in parallel. Thus, the resistance of each of them is 10 Ω now.

Next, the loop B-C-O-Z can be simplified, as is shown in Fig. 2.13.3 and in the following.

$$R_{BCOZ} = (10 + 20)||(10) = \frac{30 \times 10}{30 + 10}$$

$$\Rightarrow R_{BCOZ} = 7.5\ \Omega \tag{1}$$

After that, the circuit can be simplified further by using series–parallel rules as follows.

Fig. 2.13.3 and Fig. 2.13.4:

$$10 + 7.5 = 17.5\ \Omega \tag{2}$$

Fig. 2.13.4 and Fig. 2.13.5:

$$17.5 \parallel 10 = \frac{17.5 \times 10}{17.5 \times 10} = \frac{70}{11}\ \Omega \tag{3}$$

Fig. 2.13.5 and Fig. 2.13.6:

$$\frac{70}{11} + 10 = \frac{180}{11}\ \Omega \tag{4}$$

Fig. 2.13.6 and Fig. 2.13.7:

$$R_{AO} = \frac{180}{11}||20$$

$$\Rightarrow R_{AO} = \frac{\frac{180}{11} \times 20}{\frac{180}{11} + 20}$$

$$\Rightarrow R_{AO} = 9\ \Omega$$

Choice (1) is the answer.

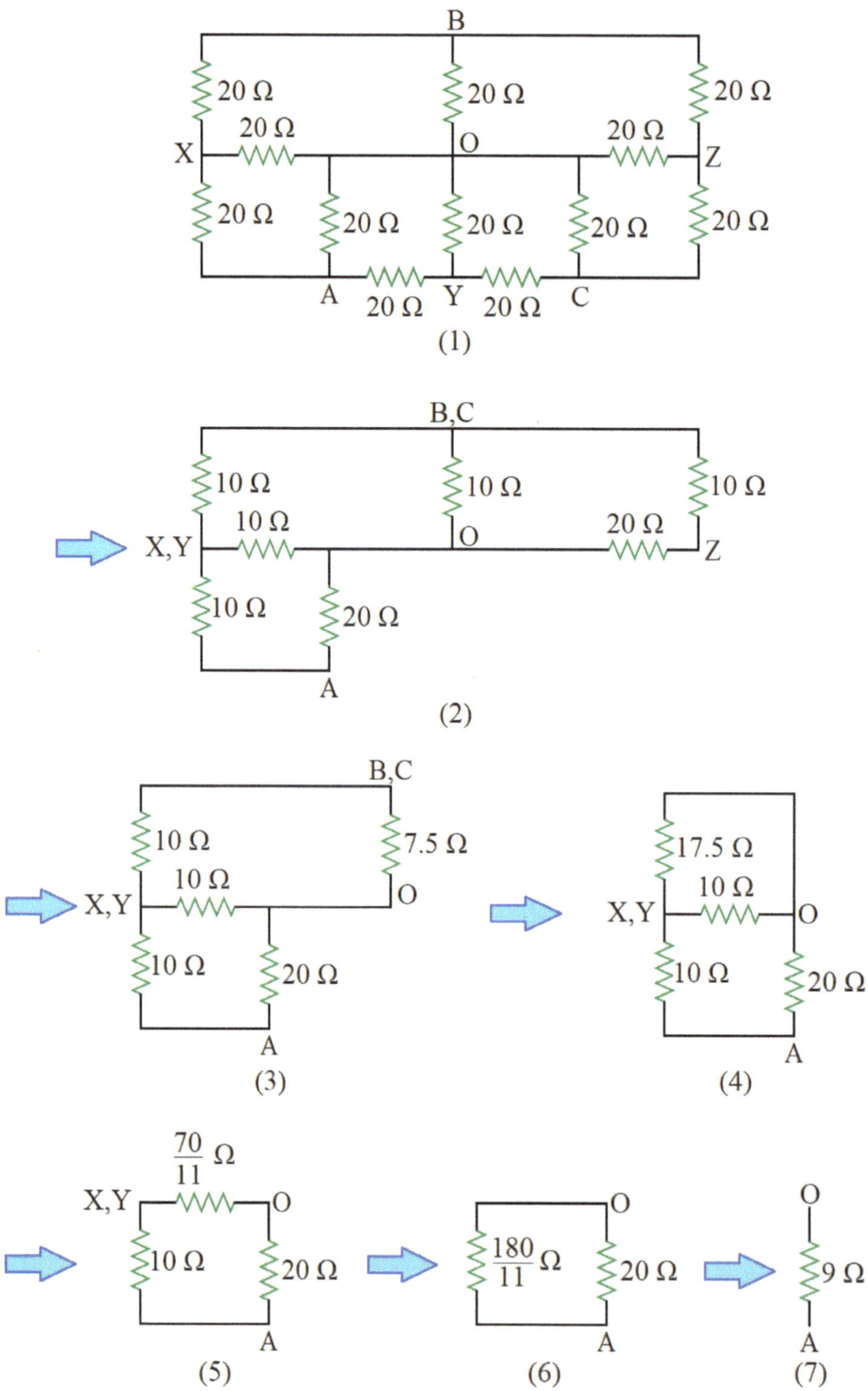

Fig. 2.13 The circuit of problem 2.33

2.6 Voltage and Current Division Rules

2.34. Choice (1) is correct because it shows the correct relationship between the current, voltage, and resistance of each resistor as well as for the equivalent resistor.

Choice (2) is correct because in a series circuit, the current of each resistor is equal, and they are equal to the total current of the circuit.

Choice (3) is correct because it shows the voltage division rule. In a series circuit, the voltage of each resistor is proportional to its size.

Choice (4) is wrong based on the rule mentioned in Choice (3).

Choice (4) is the answer (Fig. 2.14).

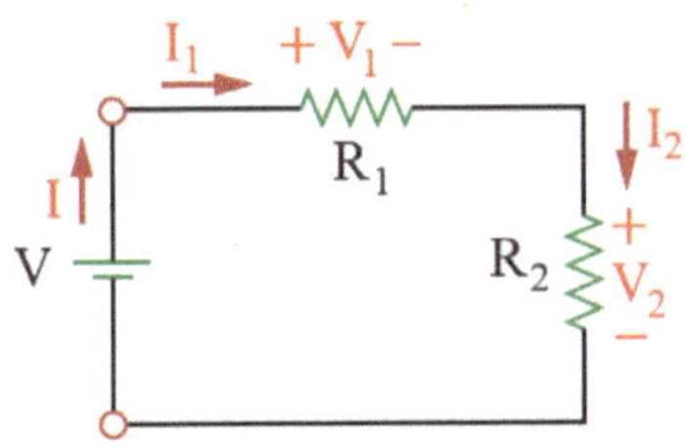

Fig. 2.14 The circuit of problem 2.34

2.35. Choice (1) is correct because it shows the correct relationship between the current, voltage, and resistance of each resistor as well as for the equivalent resistor.

Choice (2) is correct because in a parallel circuit, the voltage of each resistor is equal, and they are equal to the voltage of the source.

Choice (3) is correct because it shows the current division rule. In a parallel circuit, the current of each resistor is inversely proportional to its size.

Choice (4) is wrong based on the rule mentioned in Choice (3).

Choice (4) is the answer (Fig. 2.15).

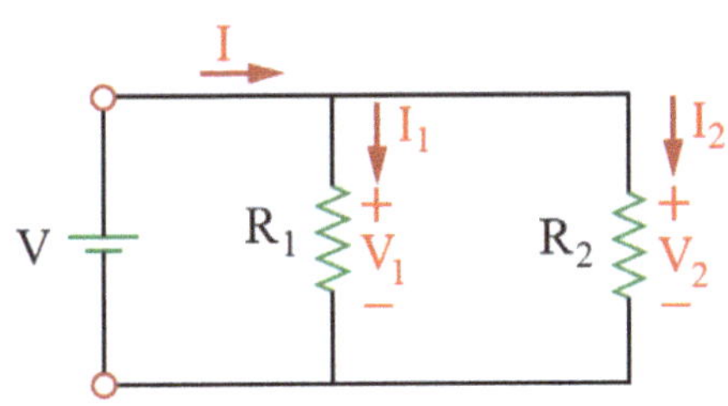

Fig. 2.15 The circuit of problem 2.35

2.36. In the circuit illustrated in Fig. 2.16, the resistors are connected in series. Therefore, their currents are equal and can be calculated as follows.

$$I_1 = I_2 = I = \frac{30}{1+2}$$

$$\Rightarrow I_1 = I_2 = 10 \text{ A}$$

Choice (4) is the answer.

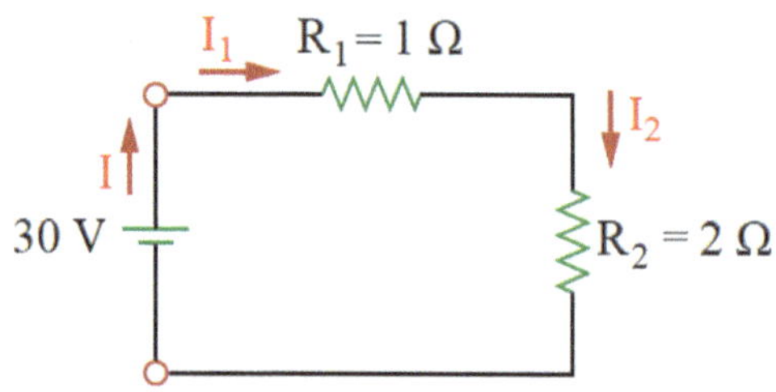

Fig. 2.16 The circuit of problem 2.36

2.37. The circuit is a series circuit; therefore, the voltage of each resistor can be calculated by using voltage division rule as follows.

$$V_1 = \frac{R_1}{R_1 + R_2} V$$

$$V_2 = \frac{R_2}{R_1 + R_2} V$$

Hence:

$$V_1 = \frac{1}{1+2} \times 30 \Rightarrow V_1 = 10 \text{ V}$$

$$V_2 = \frac{2}{1+2} \times 30 \Rightarrow V_2 = 20 \text{ V}$$

Choice (3) is the answer (Fig. 2.17).

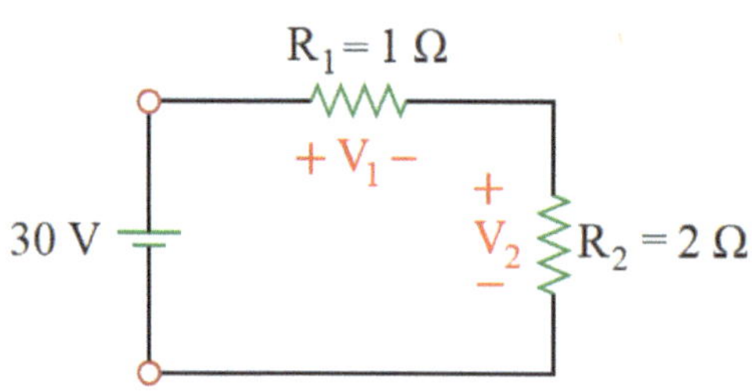

Fig. 2.17 The circuit of problem 2.37

2.39. The circuit is a parallel circuit; therefore, the voltage of each resistor is equal to the voltage of the source. Therefore:

$$V_1 = V_2 = 30 \text{ V}$$

Choice (1) is the answer (Fig. 2.18).

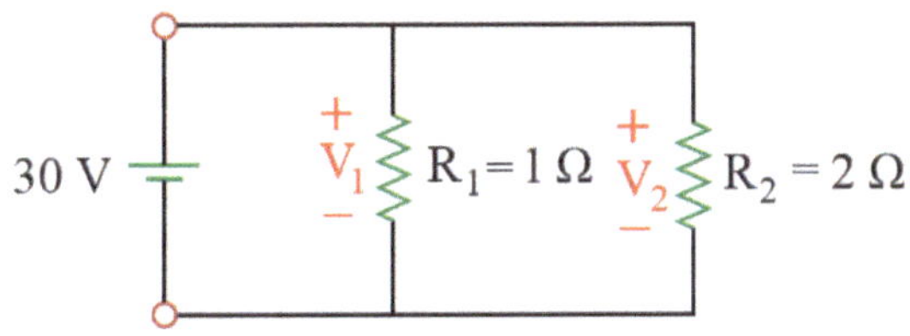

Fig. 2.18 The circuit of problem 2.39

2.40. The circuit is a parallel circuit; therefore, the current of resistors can be calculated by using Ohm's law as follows.

$$I_1 = \frac{V_1}{R_1}$$

$$I_2 = \frac{V_2}{R_2}$$

Therefore:

$$I_1 = \frac{30}{1} \Rightarrow I_1 = 30 \text{ A}$$

$$I_1 = \frac{30}{2} \Rightarrow I_1 = 15 \text{ A}$$

Choice (2) is the answer (Fig. 2.19).

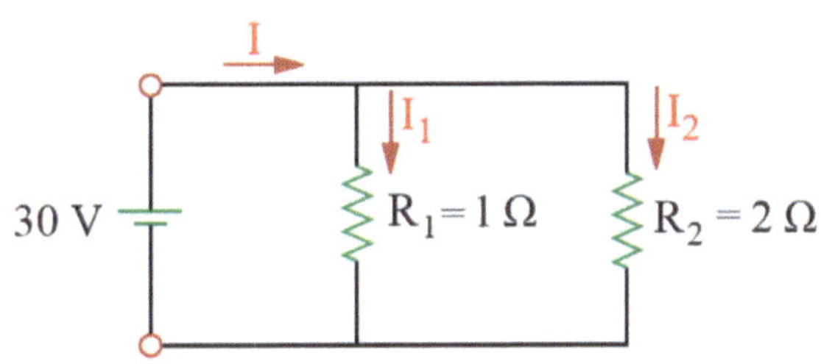

Fig. 2.19 The circuit of problem 2.40

2.7 Ideal Independent Voltage Source

2.42. As is shown in Fig. 2.20, by moving all the ideal independent voltage sources to the vertical branch, the circuit is simplified. Now, the voltage of equivalent ideal independent voltage source of the circuit can be calculated by the algebraic sum of the voltages of the ideal independent voltage sources as follows.

$$V = 9 + 8 + 2$$

$$\Rightarrow V = 19 \text{ V}$$

Choice (4) is the answer.

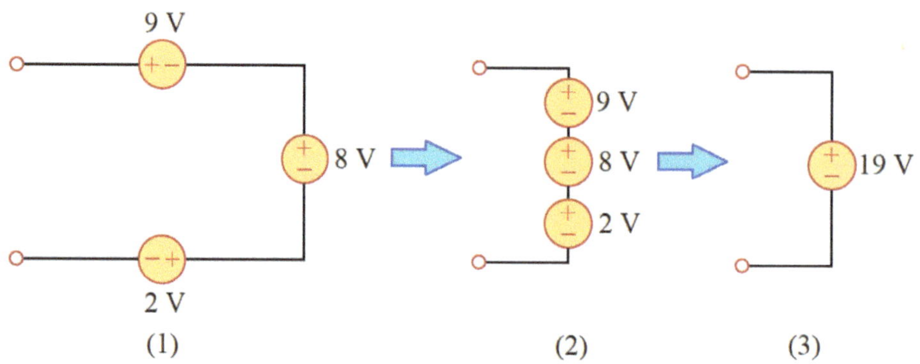

Fig. 2.20 The circuit of problem 2.42

2.44. Since $V_{s1} = V_{s2}$, the voltage of equivalent ideal independent voltage source of the circuit is V_{s1} or V_{s2}. Choice (1) is the answer (Fig. 2.21).

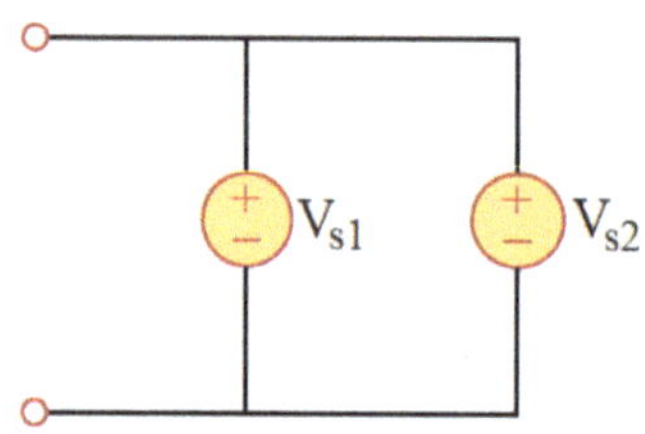

Fig. 2.21 The circuit of problem 2.44

2.45. Since $V_{s1} \neq V_{s2}$, their parallel connection is theoretically wrong because Kirchhoff's voltage law (KVL) is not satisfied. Choice (4) is the answer (Fig. 2.22).

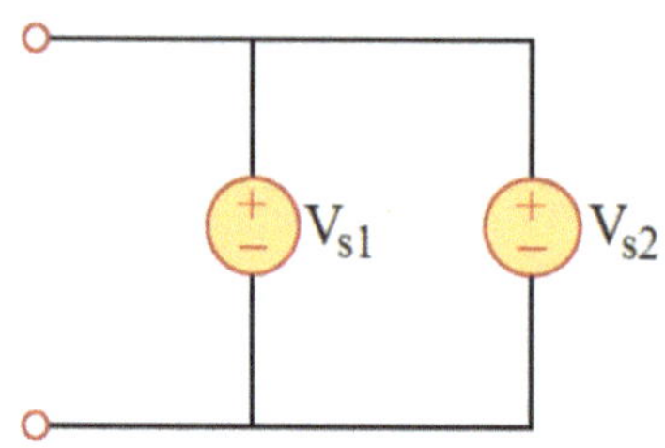

Fig. 2.22 The circuit of problem 2.45

2.8 Ideal Independent Current Source

2.46. The current of equivalent ideal independent current source of the circuit can be calculated by the algebraic sum of the currents of the ideal independent current sources as follows.

$$I = 2 + 3$$

$$\Rightarrow I = 5\ \text{A}$$

Choice (1) is the answer (Fig. 2.23).

Fig. 2.23 The circuit of problem 2.46

2.48. Since $I_{s1} = I_{s2}$, the current of equivalent ideal independent current source of the circuit is I_{s1} or I_{s2}. Choice (1) is the answer (Fig. 2.24).

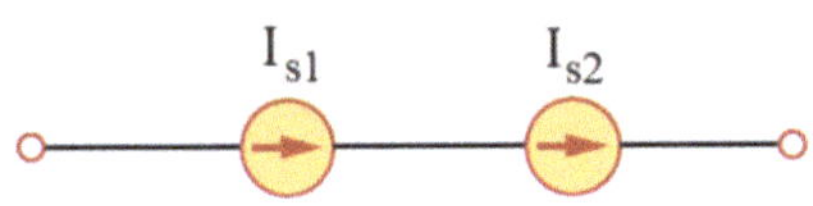

Fig. 2.24 The circuit of problem 2.48

2.49. Since $I_{s1} \neq I_{s2}$, their series connection is theoretically wrong because Kirchhoff's current law (KCL) is not satisfied. Choice (4) is the answer (Fig. 2.25).

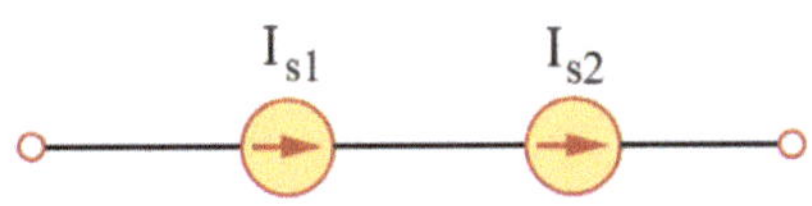

Fig. 2.25 The circuit of problem 2.49

2.9 Ideal Dependent Voltage and Current Sources

2.50. As can be noticed, the circuit component is an ideal independent current source which is controlled by the current of another component. Therefore, it is a current-controlled current source (CCCS). Choice (4) is the answer (Fig. 2.26).

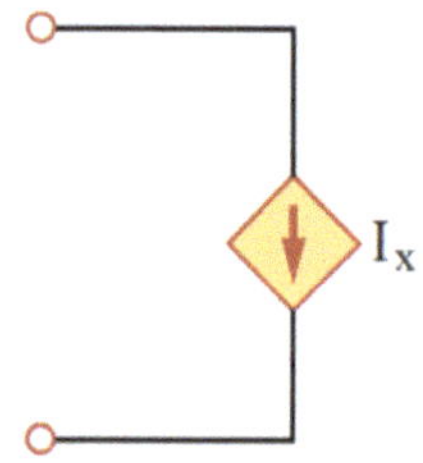

Fig. 2.26 The circuit of problem 2.50

2.51. As can be noticed, the circuit component is an ideal independent voltage source which is controlled by the current of another component. Therefore, it is a current-controlled voltage source (CCVS). Choice (2) is the answer (Fig. 2.27).

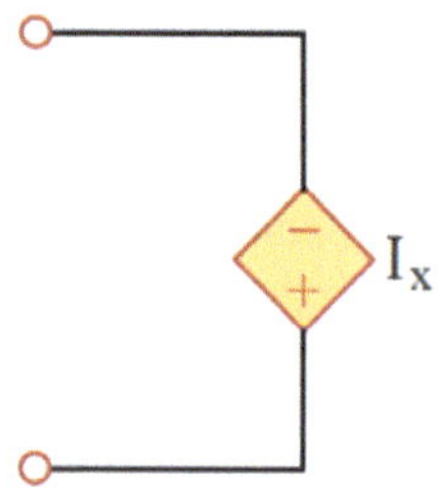

Fig. 2.27 The circuit of problem 2.51

2.52. As can be noticed, the circuit component is an ideal independent current source which is controlled by the voltage of another component. Therefore, it is a voltage-controlled current source (VCCS). Choice (3) is the answer (Fig. 2.28).

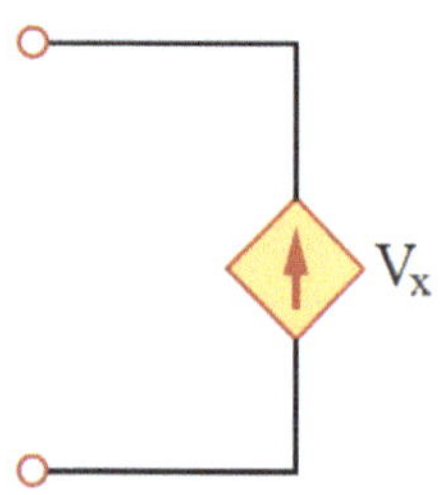

Fig. 2.28 The circuit of problem 2.52

2.53. As can be noticed, the circuit component is an ideal independent voltage source which is controlled by the voltage of another component. Therefore, it is a voltage-controlled voltage source (VCVS). Choice (1) is the answer (Fig. 2.29).

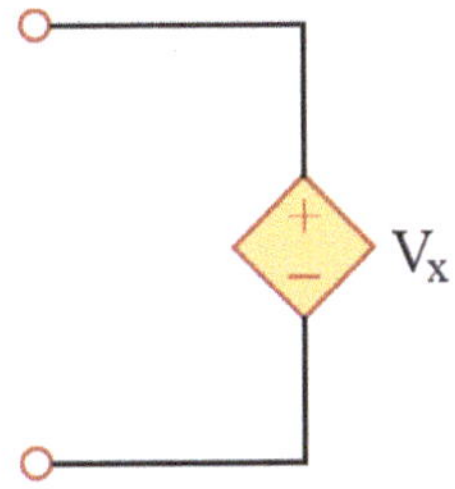

Fig. 2.29 The circuit of problem 2.53

2.10 Kirchhoff's Current Law (KCL)

2.54. By applying KCL for the node, we have:

$$I + 7 = 2 + 6$$

$$\Rightarrow I = 1 \text{ A}$$

Choice (2) is the answer (Fig. 2.30).

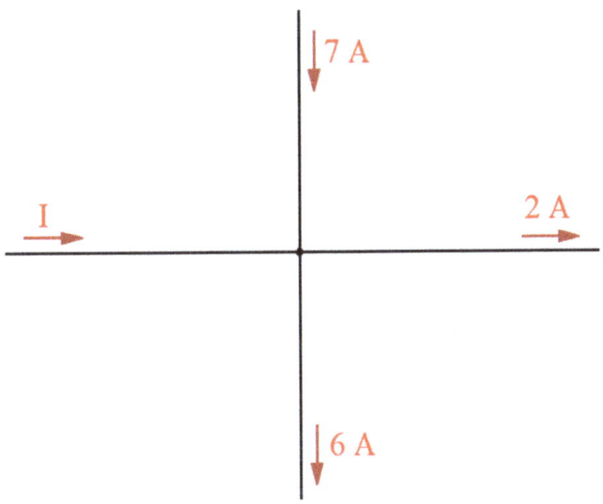

Fig. 2.30 The circuit of problem 2.54

2.57. By applying KCL for the top node, we can write:

$$5 + 4I_x = I_x$$

$$\Rightarrow I_x = -\frac{5}{3} \text{ A}$$

In addition, by applying Ohm's law for the resistor, we have:

$$V_o = 2I_x$$

Therefore:

$$V_o = -\frac{10}{3}\ \mathrm{V}$$

Choice (4) is the answer (Fig. 2.31).

Traps

- You would choose Choice (3) if you had not correctly considered the voltage and current polarities of the resistor.

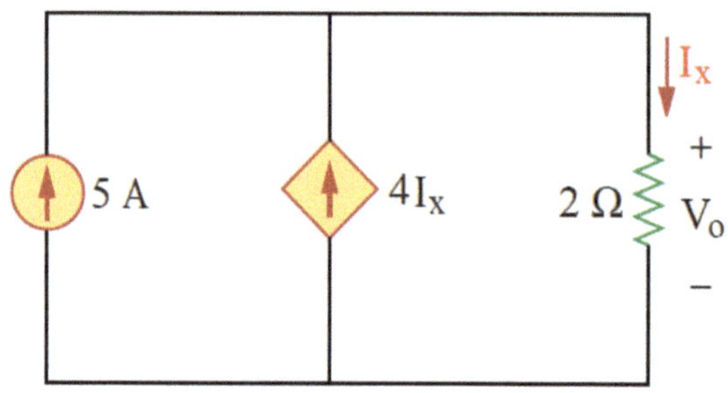

Fig. 2.31 The circuit of problem 2.57

2.11 Kirchhoff's Voltage Law (KVL)

2.59. Since the voltages of the circuit components are unknown, a current needs to be assigned to the circuit as is shown in Fig. 2.32. Now, by applying KVL, we can write as follows.

$$-12 + 4I + 8I = 0$$

$$\Rightarrow -12 + 12I = 0$$

$$\Rightarrow I = 1\ \mathrm{A}$$

In addition, by applying Ohm's law for the 4-Ω resistor, we have:

$$V_1 = 4I$$

Therefore:

$$V_1 = 4\ \mathrm{V}$$

Choice (2) is the answer.

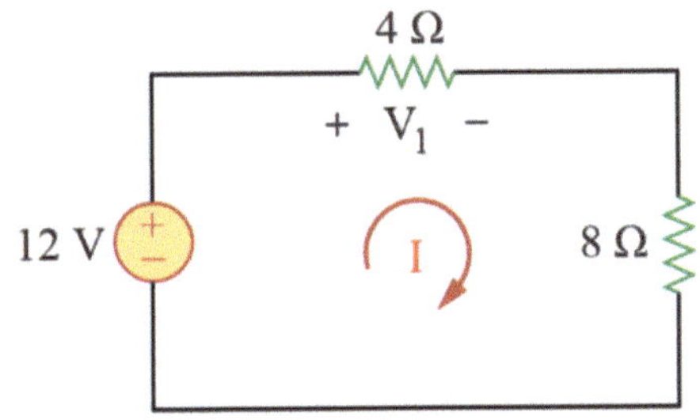

Fig. 2.32 The circuit of problem 2.59

2.62. First, a current is assigned to the circuit as is shown in Fig. 2.33. Now, by applying KVL, we have:

$$-5 + 3I + I + 1 + 0.5V_1 = 0$$

$$\Rightarrow -4 + 4I + 0.5V_1 = 0 \tag{1}$$

In addition, by applying Ohm's law for the 3-Ω resistor, we have:

$$V_1 = 3I \tag{2}$$

Solving (1) and (2):

$$-4 + 4I + 0.5(3I) = 0$$

$$\Rightarrow -4 + 4I + 1.5I = 0$$

$$\Rightarrow -4 + 5.5I = 0$$

$$\Rightarrow I = \frac{8}{11}\text{ A} \tag{3}$$

Solving (2) and (3):

$$V_1 = \frac{24}{11}\text{ V}$$

Applying Ohm's law for the 1-Ω resistor:

$$V_2 = I \tag{4}$$

Solving (3) and (4):

$$V_2 = \frac{8}{11}\text{ V}$$

Choice (1) is the answer.

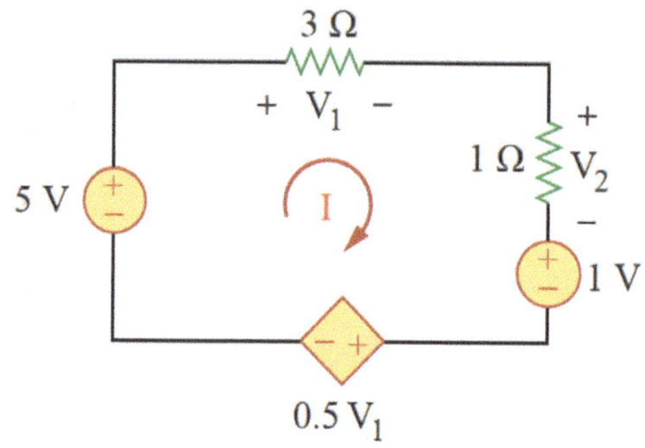

Fig. 2.33 The circuit of problem 2.62

2.12 Power Consumption and Generation

2.64. In the circuit illustrated in Fig. 2.34, the resistors are connected in series. Therefore, their currents can be calculated as follows.

$$I_1 = I_2 = \frac{30}{1+2}$$

$$\Rightarrow I_1 = I_2 = 10 \text{ A}$$

The power wasted or consumed by a resistor can be calculated by using one of the following relations.

$$P = RI^2$$

$$P = VI$$

$$P = \frac{V^2}{R}$$

For this problem, the first relation is more appropriate. Hence:

$$P_1 = R_1 I_1^2 = 1 \times 10^2 \Rightarrow P_1 = 100 \text{ W}$$

$$P_2 = R_2 I_2^2 = 2 \times 10^2 \Rightarrow P_2 = 200 \text{ W}$$

Choice (1) is the answer.

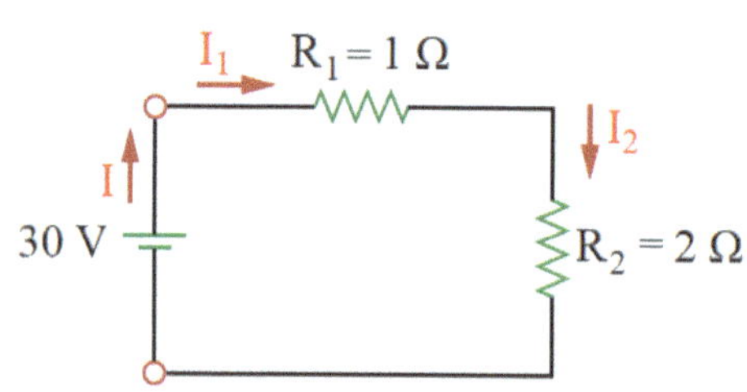

Fig. 2.34 The circuit of problem 2.64

2.65. The circuit is a parallel circuit; therefore, the voltage of each resistor is equal to the voltage of the source. In other words:

$$V_1 = V_2 = 30 \text{ V}$$

The power wasted or consumed by a resistor can be calculated by using one of the following relations.

$$P = RI^2$$

$$P = VI$$

$$P = \frac{V^2}{R}$$

For this problem, the third relation is more appropriate. Hence:

$$P_1 = \frac{V_1^2}{R_1} = \frac{30^2}{1} \Rightarrow P_1 = 900 \text{ W}$$

$$P_2 = \frac{V_2^2}{R_2} = \frac{30^2}{2} \Rightarrow P_2 = 450 \text{ W}$$

Choice (4) is the answer (Fig. 2.35).

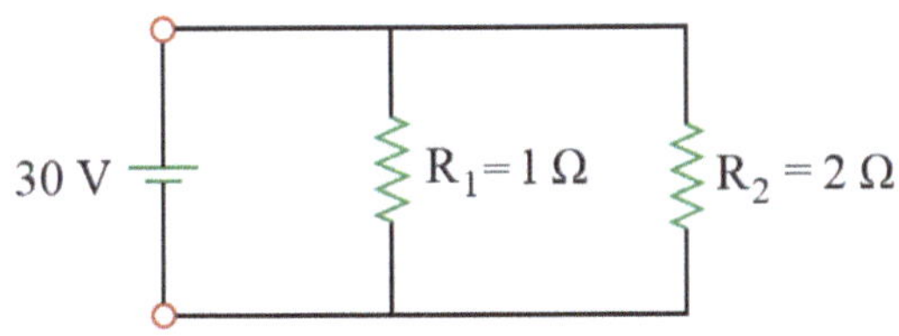

Fig. 2.35 The circuit of problem 2.65

2.68. A circuit component absorbs or consumes power if the value of multiplication of its voltage and current (*VI*) is positive while an associated reference direction for its voltage and current (current enters the positive terminal of component) is considered. Now, let us check the value of *VI* for each component of the circuit.

$$P_1 = V_1I_1 = -2 \times 2 \Rightarrow P_1 < 0$$

$$P_2 = V_2I_2 = -8 \times 2 \Rightarrow P_2 < 0$$

$$P_3 = V_3I_3 = 10 \times 4 \Rightarrow P_3 > 0$$

$$P_4 = V_4I_4 = -10 \times 5 \Rightarrow P_4 < 0$$

As can be seen, the third power source is a power absorber. Choice (3) is the answer (Fig. 2.36).

Traps

- You would choose Choice (1), (2), or (4) if you had not correctly used the associated reference direction for the voltage and current of the complements.

Notes

When an associated reference direction for the voltage and current of a circuit component is considered, that is, when the current enters the positive terminal of the component:
(a) the component absorbs or consumes power if the value of multiplication of its voltage and current (VI) is positive, and
(b) the component delivers or generates power if the value of multiplication of its voltage and current (VI) is negative.

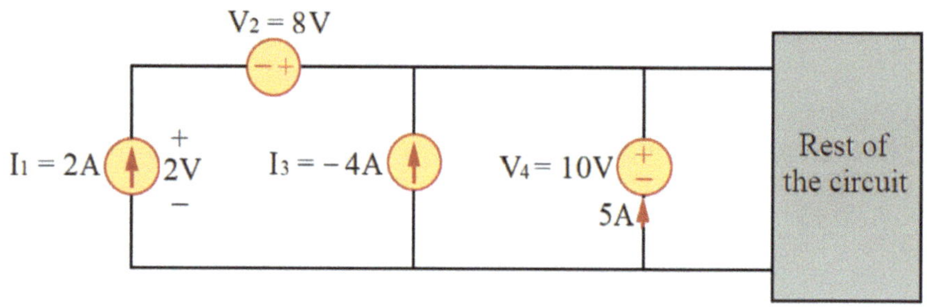

Fig. 2.36 The circuit of problem 2.68

2.69. Based on the energy conservation law, the algebraic sum of power of circuit components in any circuit at any time is zero if a reference direction is considered for the components. In other words:

$$\sum_{i=1}^{N} P_i = 0$$

Therefore:

$$P_1 + P_2 + P_3 + P_4 + RI^2 - 32I = 0$$

$$\Rightarrow 6 + (-3) + 1 + 12 + 16I^2 - 32I = 0$$

$$\Rightarrow 16I^2 - 32I + 16 = 0$$

$$\Rightarrow I^2 - 2I + 1 = 0$$

$$\Rightarrow I = 1 \text{ A}$$

Choice (1) is the answer (Fig. 2.37).

Notes

The power consumed by a resistor can be calculated by using one of the following relations.

$$P = RI^2$$

$$P = VI$$

$$P = \frac{V^2}{R}$$

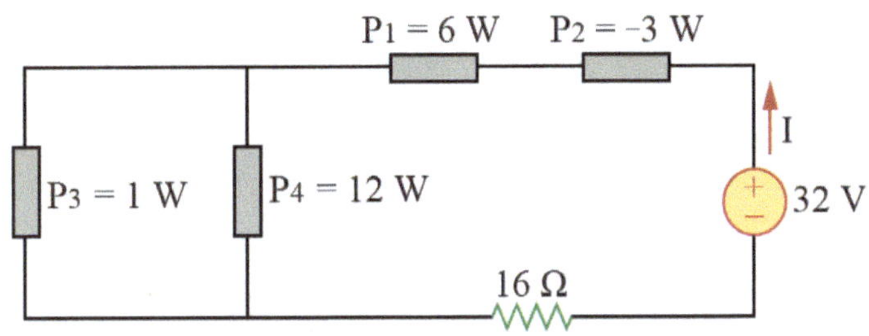

Fig. 2.37 The circuit of problem 2.69

2.70. To calculate the power of a circuit component, the voltage and current of the component must be known. The current of the $5 - V$ voltage source can be calculated by applying KCL for the region shown in Fig. 2.38, as follows.

$$1 + 3 = I_{5-Vsource}$$

$$\Rightarrow I_{5-Vsource} = 4 \text{ A}$$

Thus, the power of the $5 - V$ voltage source is as follows:

$$P = (V_{5-Vsource})(I_{5-Vsource}) = 5 \times 4$$

$$\Rightarrow P = 20 \text{ W}$$

Herein, since $P > 0$, the $5 - V$ voltage source is a power consumer. Choice (3) is the answer.

Traps

- You would choose Choice (4) if you had not correctly used the associated reference direction for the voltage and current of the complement.

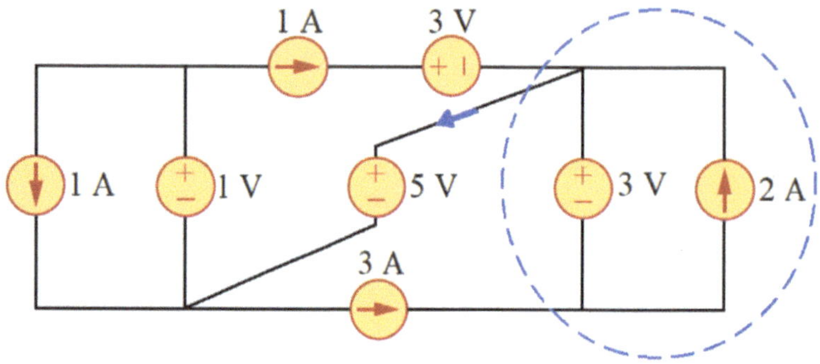

Fig. 2.38 The circuit of problem 2.70

Notes

When an associated reference direction for the voltage and current of a circuit component is considered, that is, when the current enters the positive terminal of the component:
(a) the component absorbs or consumes power if the value of multiplication of its voltage and current (VI) is positive, and
(b) the component delivers or generates power if the value of multiplication of its voltage and current (VI) is negative.

References

1. Rahmani-Andebili, M., DC Electrical Circuit Analysis - Practice Problems, Methods, and Solutions (1st Ed.), Springer Nature, 2020.
2. Rahmani-Andebili, M., General Physics II - Practice Problems, Methods, and Solutions, Sringer Nature, 2025.
3. Rahmani-Andebili, M., General Physics I - Practice Problems, Methods, and Solutions, Springer Nature, 2025.
4. Rahmani-Andebili, M., Mathematics of Engineering and Science - Comprehensive Lessons, Springer Nature, 2024.
5. Rahmani-Andebili, M., Differential Equations – Practice Problems, Methods, and Solutions, Springer Nature, 2022.
6. Rahmani-Andebili, M., Calculus III – Practice Problems, Methods, and Solutions, Springer Nature, 2023.
7. Rahmani-Andebili, M., Calculus II – Practice Problems, Methods, and Solutions, Springer Nature, 2023.
8. Rahmani-Andebili, M., Calculus I (2nd Ed.) – Practice Problems, Methods, and Solutions, Springer Nature, 2023.
9. Rahmani-Andebili, M., Precalculus (2nd Ed.) – Practice Problems, Methods, and Solutions, Springer Nature, 2024.

3 Methods of Electrical Circuits Analysis: Part A

Abstract

In this chapter, the basic and advanced problems of nodal analysis, mesh analysis, and heuristic analysis are studied. Herein, different types of problems and exercises are presented that are categorized as follows.

- ***Problems with detailed solution***: They have been designed to teach students the subjects in detail. Moreover, they have been categorized in different levels based on their difficulty levels (easy, normal, and hard) and calculation amounts (small, normal, and large).
- ***Partially solved exercises***: They have been designed to encourage students to practice problems while guiding them through the problem-solving procedure and hinting the required formulas.
- ***Exercises with final answer***: They have been designed to encourage students to practice more by themselves while hinting them by the final answer as well as to help instructors to give tests or quizzes.

3.1 Nodal Analysis

Problem

3.1. In the circuit shown in Fig. 3.1, calculate the value of current I_a [1–9].

Difficulty level ○ Easy ● Normal ○ Hard

Calculation amount ● Small ○ Normal ○ Large

1) 0.5 A
2) 1 A
3) 2 A
4) 1.5 A

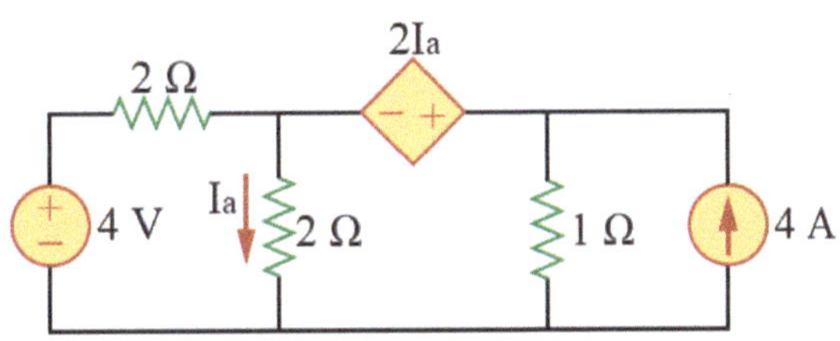

Fig. 3.1 The circuit of problem 3.1

V. T. Davis, *DC Electrical Circuits Analysis*, https://doi.org/10.1007/978-3-032-00253-2_3

Partially Solved Exercise

3.2. In Problem 3.1, calculate the power consumed by the 1-Ω resistor.

Solution

From the solution of Problem 3.1, we have the following information.

$$e_1 = (\quad) \text{ V}$$

$$e_2 = (\quad) \times e_1$$

As we know, the power wasted or consumed by a resistor can be calculated by using one of the following relations.

$$P = RI^2$$

$$P = VI$$

$$P = \frac{V^2}{R}$$

For this problem, the third relation is more appropriate. Hence:

$$P = \frac{(e_2)^2}{R}$$

$$\Rightarrow P = \frac{(\quad)^2}{(\quad)}$$

$$\Rightarrow P = 16 \text{ W}$$

Exercise

3.3. In the circuit illustrated in Fig. 3.2, calculate the value of node voltages e_1 and e_2.

Final Answer

$e_1 = 26.66$ V, $e_2 = 40$ V

Note: *Read the "educational support" part of the preface of the textbook.*

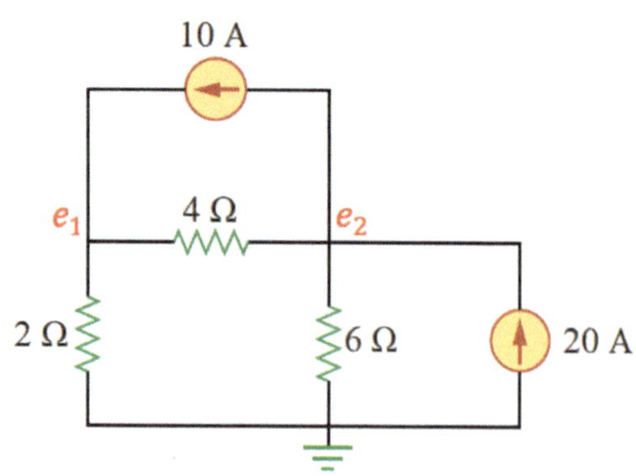

Fig. 3.2 The circuit of exercise 3.3

Problem

3.4. In the circuit of Fig. 3.3, calculate the voltages V_x and V_y.

Difficulty level ○ Easy ● Normal ○ Hard
Calculation amount ● Small ○ Normal ○ Large

1) $V_x = 8$ V, $V_y = 2$ V
2) $V_x = -8$ V, $V_y = 2$ V
3) $V_x = 8$ V, $V_y = -2$ V
4) $V_x = -8$ V, $V_y = -2$ V

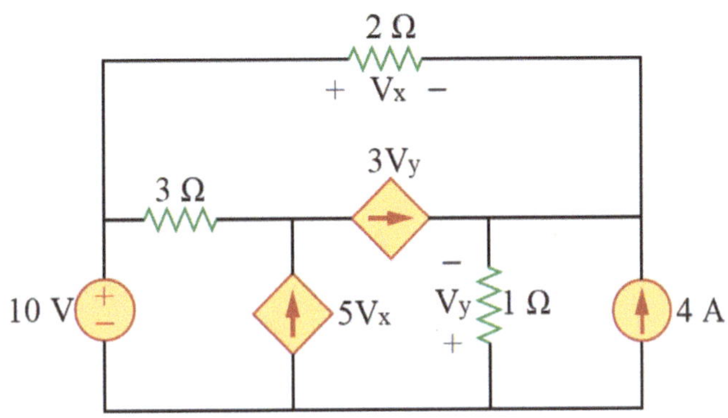

Fig. 3.3 The circuit of problem 3.4

Exercise

3.5. In the circuit illustrated in Fig. 3.4, calculate the value of node voltages e_1 and e_2.

Final Answer

$e_1 = 1$ V, $e_2 = 7$ V

Note: *Read the "educational support" part of the preface of the textbook.*

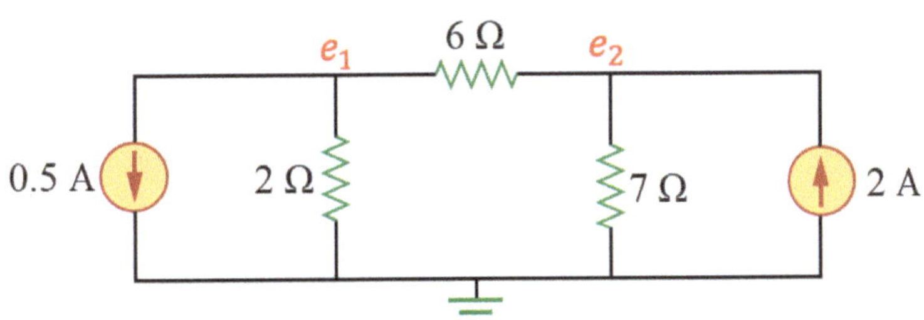

Fig. 3.4 The circuit of exercise 3.5

Partially Solved Exercise

3.6. Calculate the power of the independent current source in the circuit of Fig. 3.5.

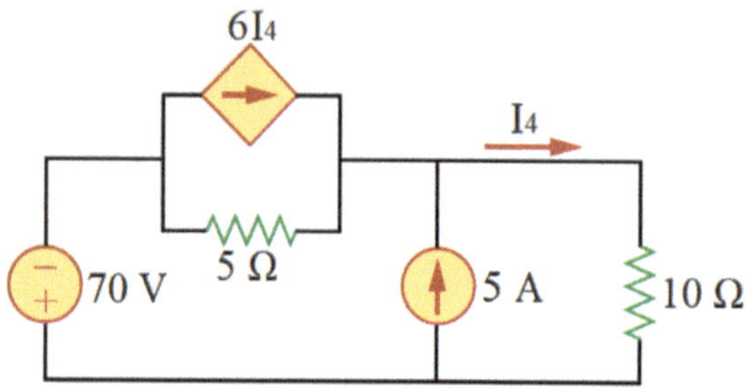

Fig. 3.5 The circuit of partially solved exercise 3.6

Solution

To calculate the power of the independent current source, its voltage must be known. In this regard, the conventional polarity for the voltage and current of the component (associate reference direction) is used, that is, the current must enter the positive terminal of the component.

Step 1 (determining the type of circuit analysis)

As can be noticed from Fig. 3.6, there is only one independent node with unknown voltage (e_1) and three independent meshes with unknown currents. Thus, nodal analysis is the best method to solve this problem.

Step 2 (formulating the circuit analysis equations)

KCL in node 1:

$$-6I_4 + \frac{e_1 - (\quad)}{(\quad)} - 5 + \frac{e_1}{(\quad)} = 0$$

$$\Rightarrow (\quad)I_4 + (\quad)e_1 + (\quad) = 0 \quad (1)$$

Using Ohm's law to determine the relation between the voltage and current of the 10-Ω resistor:

$$I_4 = \frac{(\quad)}{(\quad)} \quad (2)$$

Step 3 (solving the system of equations)

Solving (1) and (2):

$$(\quad)e_1 + (\quad)e_1 + (\quad) = 0$$

$$\Rightarrow (\quad)e_1 = (\quad)$$

$$\Rightarrow e_1 = (\quad)\text{ V}$$

Step 4 (calculating the value of required parameter)

As we know, the power of a circuit component can be calculated as follows.

$$P = VI$$

Hence:

$$P = -(e_1)(\quad)$$

$$\Rightarrow P = -(\quad)(\quad)$$

$$\Rightarrow P = -150 \text{ W}$$

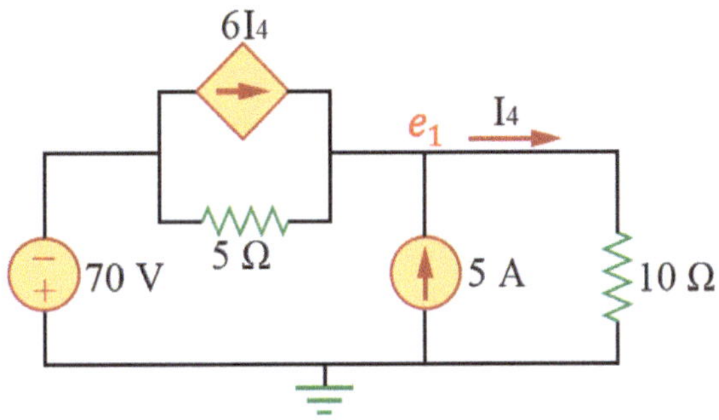

Fig. 3.6 The circuit of partially solved exercise 3.6

Problem

3.7. In the circuit of Fig. 3.7, calculate the power of the independent current source.

Difficulty level ○ Easy ● Normal ○ Hard

Calculation amount ○ Small ● Normal ○ Large

1) It generates 32 W.
2) It consumes 32 W.
3) It generates 67 W.
4) It consumes 64 W.

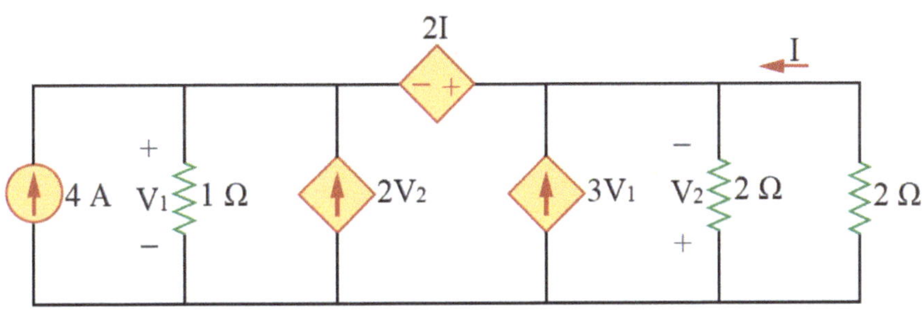

Fig. 3.7 The circuit of problem 3.7

Exercise

3.8. In the circuit illustrated in Fig. 3.8, calculate the value of node voltages e_1 and e_2.

Final Answer

$e_1 = -1.2$ V, $e_2 = 16.8$ V

Note: Read the "educational support" part of the preface of the textbook.

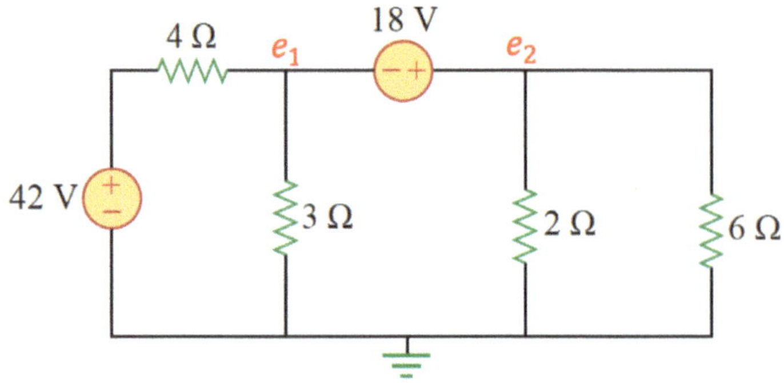

Fig. 3.8 The circuit of exercise 3.8

Problem

3.9. Calculate the node voltages in the circuit of Fig. 3.9.

Difficulty level ○ Easy ● Normal ○ Hard

Calculation amount ○ Small ● Normal ○ Large

1) $e_1 = \frac{4}{5}$ V, $e_2 = \frac{8}{5}$ V
2) $e_1 = 4$ V, $e_2 = 8$ V
3) $e_1 = 2$ V, $e_2 = 4$ V
4) $e_1 = \frac{2}{5}$ V, $e_2 = \frac{4}{5}$ V

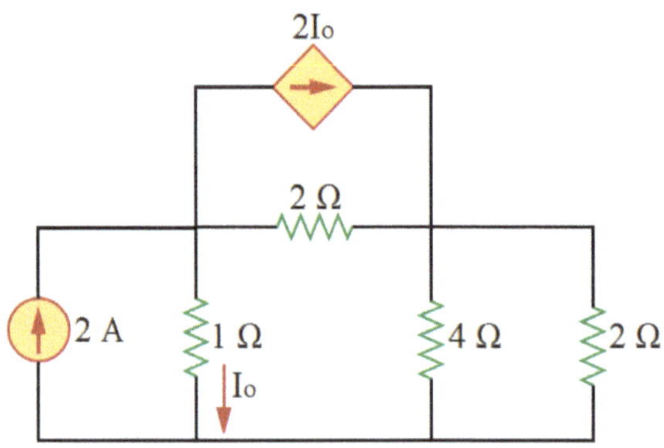

Fig. 3.9 The circuit of problem 3.9

Partially Solved Exercise

3.10. In Problem 3.9, calculate the power of the dependent current source.

Solution

To calculate the power of a component, we need to know its voltage and current considering the conventional polarity for the voltage and current of the component (associate reference direction). Therefore, the current must enter the positive terminal of the component.

As we know, the power of a circuit component can be calculated as follows.

$$P = VI$$

On the other hand, we know the value of I_0, e_1, and e_2 from the solution of Problem 3.9. Hence:

$$P = (e_1 - e_2)(2I_0)$$

$$\Rightarrow P = ((\quad) - (\quad))(\quad)$$

$$\Rightarrow P = -1.28 \text{ W}$$

Problem

3.11. Apply nodal analysis to calculate the value of current I in the circuit of Fig. 3.10.

Difficulty level ○ Easy ● Normal ○ Hard

Calculation amount ○ Small ● Normal ○ Large

1) $\frac{1}{3}$ mA
2) $\frac{2}{3}$ mA
3) $\frac{1}{2}$ mA
4) $\frac{4}{3}$ mA

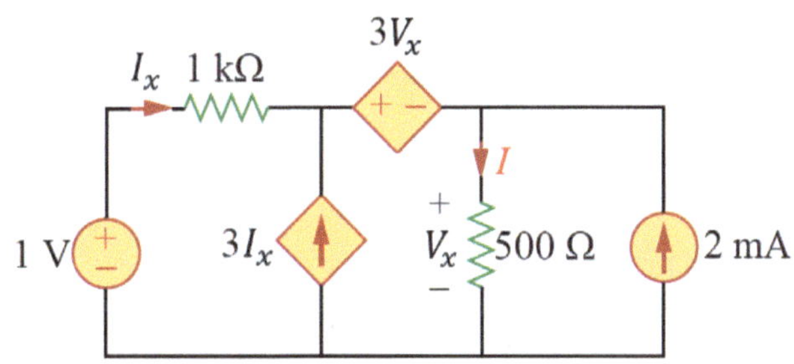

Fig. 3.10 The circuit of problem 3.11

Problem

3.12. In the circuit of Fig. 3.11, the quantities indicated for the resistors present their conductance in mho (siemens). Determine the condition on α to adjust the circuit as an amplifier. In other words, the value of $\frac{V_o}{V_s}$ must be greater than one.

Difficulty level ○ Easy ○ Normal ● Hard

Calculation amount ○ Small ● Normal ○ Large

1) $\alpha > 10$
2) $\alpha > 12$
3) $\alpha > 14$
4) Because of the presence of a dependent source in the circuit, the circuit will always behave as an amplifier.

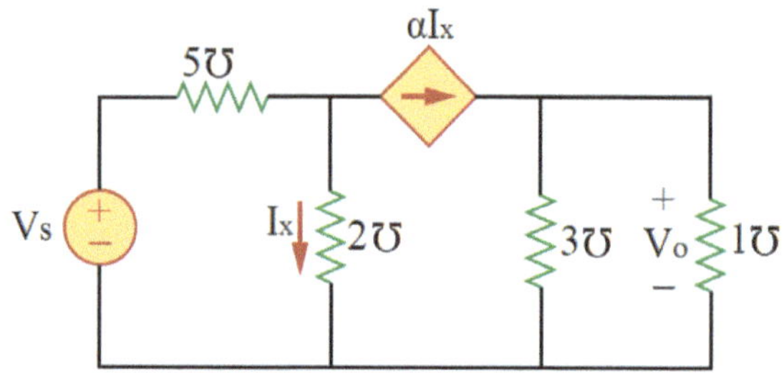

Fig. 3.11 The circuit of problem 3.12

Problem

3.13. In the circuit of Fig. 3.12, calculate the value of current I_{sc}.

Difficulty level ○ Easy ○ Normal ● Hard

Calculation amount ○ Small ● Normal ○ Large

1) 1 A
2) 2 A
3) 4 A
4) 8 A

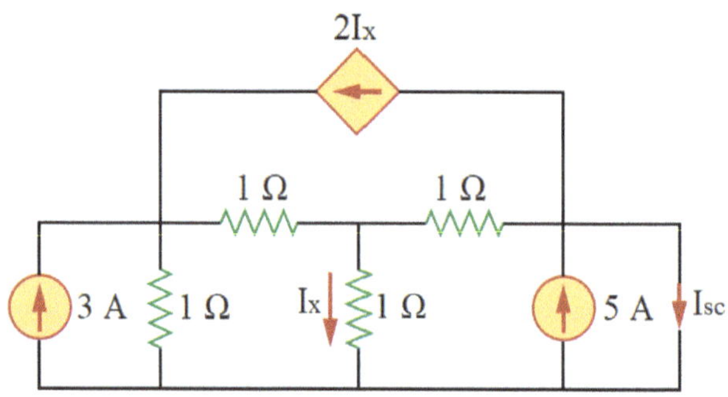

Fig. 3.12 The circuit of problem 3.13

Problem

3.14. Calculate the value of current I in the circuit shown in Fig. 3.13.

Difficulty level ○ Easy ○ Normal ● Hard

Calculation amount ○ Small ● Normal ○ Large

1) 5 A
2) 6 A
3) −5 A
4) −6 A

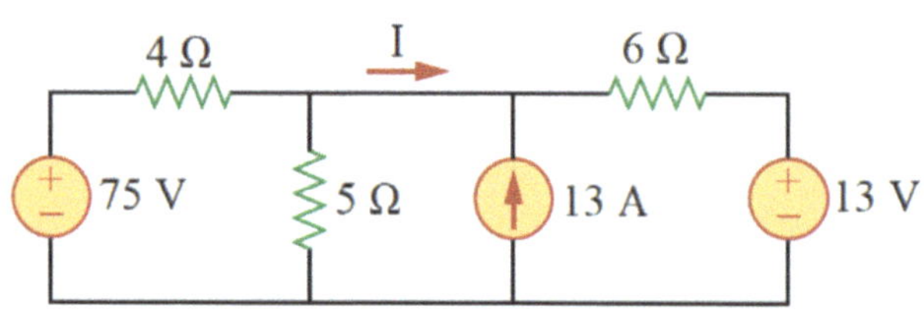

Fig. 3.13 The circuit of problem 3.14

3.2 Mesh Analysis

Problem

3.15. Calculate the value of current i_1 in the circuit of Fig. 3.14.

Difficulty level ○ Easy ● Normal ○ Hard

Calculation amount ○ Small ● Normal ○ Large

1) $\frac{40}{21}$ A
2) $\frac{40}{7}$ A
3) $\frac{80}{7}$ A
4) $\frac{80}{21}$ A

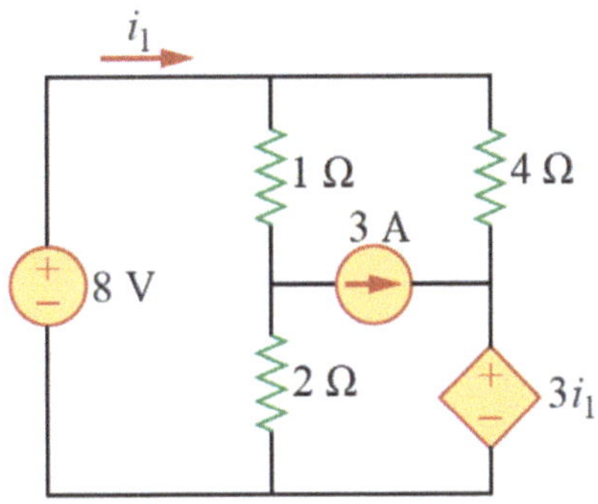

Fig. 3.14 The circuit of problem 3.15

Exercise

3.16. In Problem 3.15, calculate the power of the dependent voltage source.

Final Answer

$P \approx 24.49$ W

Note: Read the "educational support" part of the preface of the textbook.

Exercise

3.17. Calculate the value of mesh currents (clockwise) in the circuit of Fig. 3.15.

Final Answer

$I_1 = 1$ A, $I_2 = 0$ A

Note: Read the "educational support" part of the preface of the textbook.

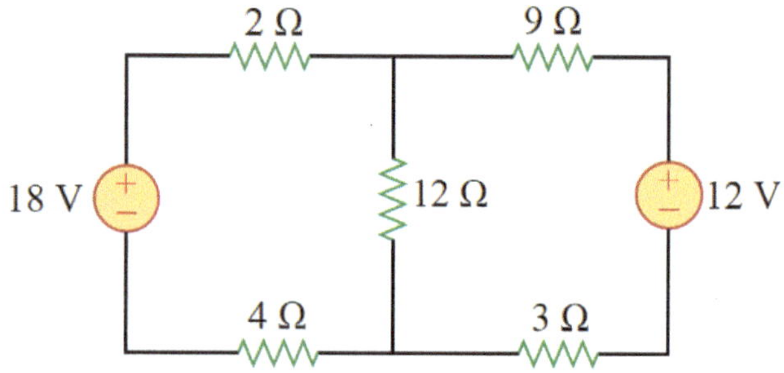

Fig. 3.15 The circuit of exercise 3.17

Partially Solved Exercise

3.18. In the circuit of Fig. 3.16, write the mesh analysis equations in the form of a matrix based on the given mesh currents.

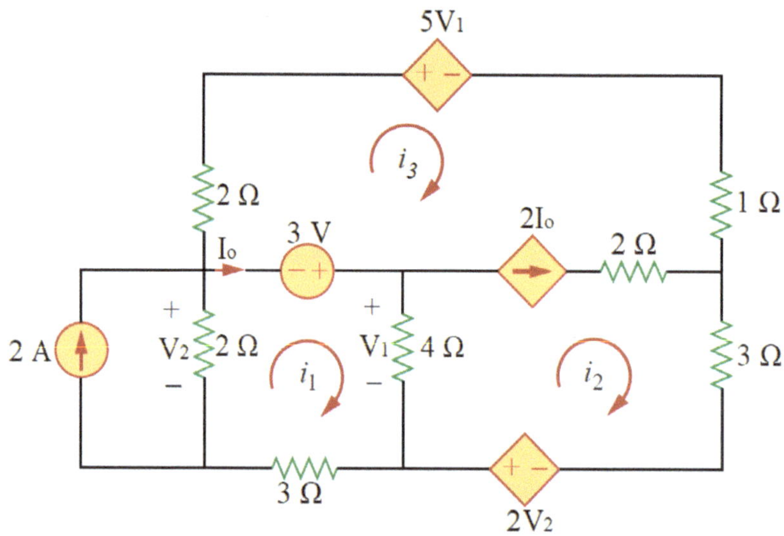

Fig. 3.16 The circuit of partially solved exercise 3.18

Solution

Step 1 (determining the type of circuit analysis)

We need to apply KVL in the indicated meshes and organize the resultant equations based on the mesh currents.

Step 2 (formulating the circuit analysis equations)

Since there is a current source between meshes 2 and 3, we need to apply KVL for their supermesh.

KVL in mesh 1:

$$(\quad)(i_1 - (\quad)) - 3 + (\quad)(i_1 - i_2) + 3i_1 = 0$$

$$\Rightarrow (\quad)i_1 - (\quad)i_2 = (\quad) \quad (1)$$

KVL in the supermesh including meshes 2 and 3:

$$(\quad)i_2 - 2V_2 + (\quad)(i_2 - i_1) + 3 + (\quad)i_3 + 5V_1 + (\quad)i_3 = 0$$

$$\Rightarrow (\quad)i_1 + (\quad)i_2 + (\quad)i_3 + (\quad)V_2 + (\quad)V_1 + (\quad) = 0 \quad (2)$$

The relation between the current of current source and the involved mesh currents (auxiliary equation) is as follows.

$$2I_o = (\quad) - (\quad) \quad (3)$$

Defining V_1 based on the mesh currents by using Ohm's law for the 4-Ω resistor:

$$V_1 = (\quad)((\quad) - (\quad)) \quad (4)$$

Defining V_2 based on the mesh currents by using Ohm's law for the 2-Ω resistor:

$$V_2 = (\quad)((\quad) - (\quad)) \quad (5)$$

Defining I_o based on the mesh currents:

$$I_o = (\quad) - (\quad) \quad (6)$$

Apply (4) and (5) in (2):

$$(\quad)i_1 + (\quad)i_2 + (\quad)i_3 - (\quad)(2(2 - i_1)) + (\quad)(4(i_1 - i_2)) + 3 = 0$$

$$\Rightarrow 20i_1 - 13i_2 + 3i_3 = 5 \quad (7)$$

Apply (6) in (3):

$$(\quad)((\quad) - (\quad)) = (\quad) - (\quad)$$

$$\Rightarrow 2i_1 - i_2 - i_3 = 0 \quad (8)$$

Equations (1), (7), (8) can be written in the form of matrices, as is seen in the following (Fig. 3.17).

$$\begin{bmatrix} 9 & -4 & 0 \\ 20 & -13 & 3 \\ 2 & -1 & -1 \end{bmatrix} \begin{bmatrix} i_1 \\ i_2 \\ i_3 \end{bmatrix} = \begin{bmatrix} 7 \\ 5 \\ 0 \end{bmatrix}$$

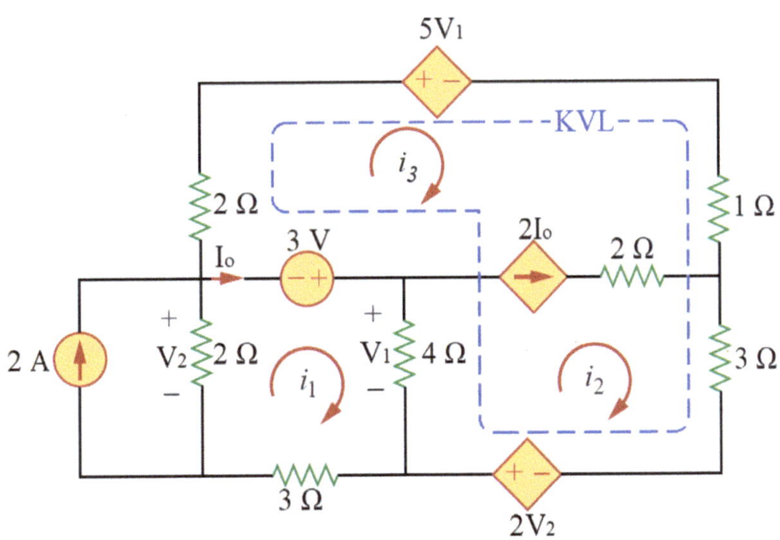

Fig. 3.17 The circuit of partially solved exercise 3.18

Problem

3.19. Calculate the value of V_x and V_y in the circuit of Fig. 3.18.

Difficulty level ○ Easy ● Normal ○ Hard

Calculation amount ○ Small ○ Normal ● Large

1) $V_x = 20$ V, $V_y = -7.5$ V
2) $V_x = 20$ V, $V_y = 7.5$ V
3) $V_x = 7.5$ V, $V_y = -15$ V
4) $V_x = 7.5$ V, $V_y = 15$ V

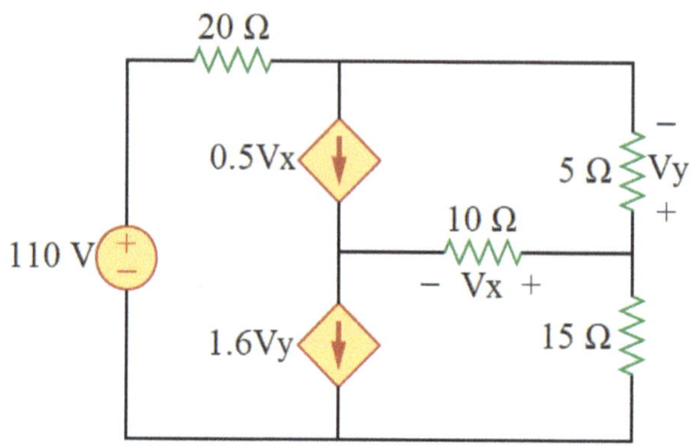

Fig. 3.18 The circuit of problem 3.19

Problem

3.20. Apply mesh analysis to calculate the value of current I in the circuit of Fig. 3.19.

Difficulty level ○ Easy ● Normal ○ Hard

Calculation amount ○ Small ● Normal ○ Large

1) 2 mA
2) $\frac{2}{3}$ mA
3) $-\frac{4}{3}$ mA
4) $\frac{4}{3}$ mA

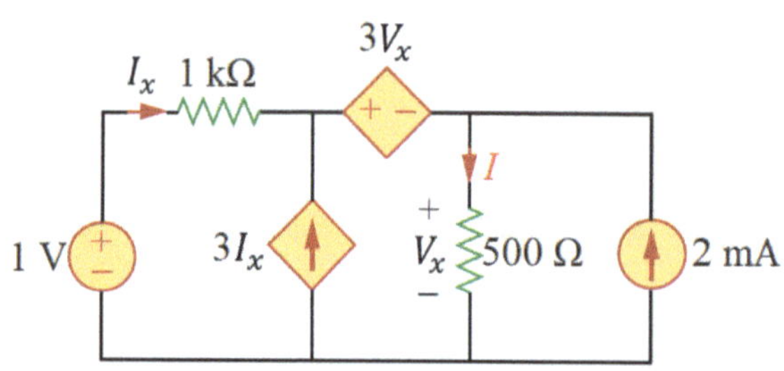

Fig. 3.19 The circuit of problem 3.20

Problem

3.21. In the circuit of Fig. 3.20, calculate the power lost in the vertical 4 – Ω resistor.

Difficulty level ○ Easy ● Normal ○ Hard

Calculation amount ○ Small ● Normal ○ Large

1) 1.44 W
2) 36 W
3) 23.04 W
4) 12.96 W

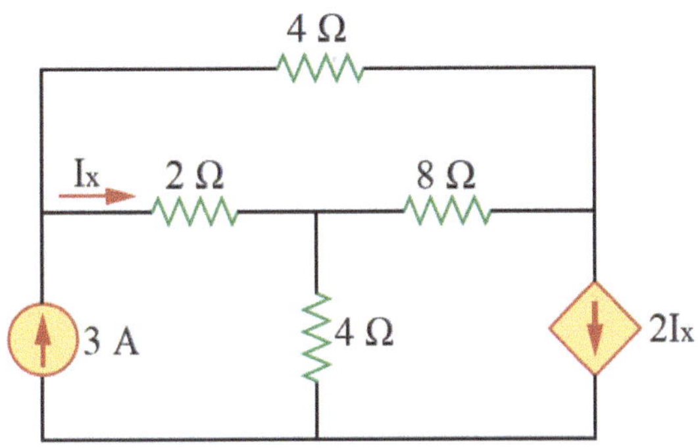

Fig. 3.20 The circuit of problem 3.21

Exercise

3.22. In Problem 3.21, calculate the power consumed by the horizontal 4 – Ω resistor.

Final Answer

$P = 12.96$ W

Note: Read the "educational support" part of the preface of the textbook.

Partially Solved Exercise

3.23. In the circuit of Fig. 3.21, calculate the power of the independent voltage source.

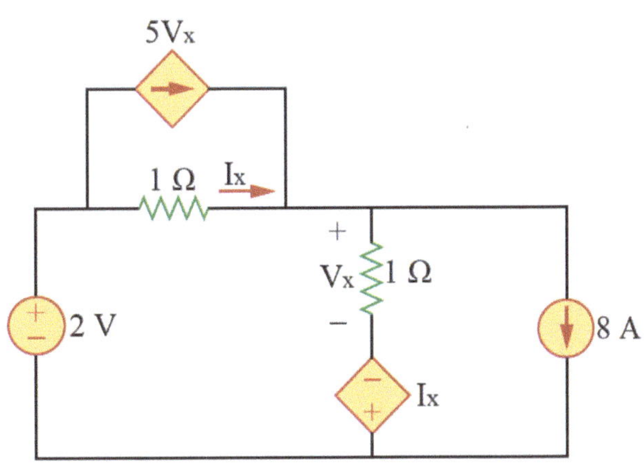

Fig. 3.21 The circuit of partially solved exercise 3.23

Solution

To calculate the power of a component, we need to multiply its voltage and current while considering the voltage and current polarities. Herein, the conventional polarity for the voltage and current of the component (associate reference direction) is used, that is, the current must enter the positive terminal of the component.

Step 1 (determining the type of circuit analysis)

Mesh analysis is the best method to solve this problem.

Step 2 (formulating the circuit analysis equations)

KVL in mesh 1 (see Fig. 3.22):

$$(\quad) + I_x(\quad) + (\quad) + (\quad) = 0$$

$$\Rightarrow -2 + V_x = 0$$

$$\Rightarrow V_x = 2 \quad (1)$$

Using Ohm's law for the 1-Ω resistor in the vertical branch to define V_x based on the mesh current:

$$V_x = (\quad)((\quad) - (\quad))$$

$$\Rightarrow V_x = i_1 - 8 \quad (2)$$

Step 3 (solving the system of equations)

Solving (1) and (2):

$$(\quad) = (\quad) - (\quad)$$

$$\Rightarrow i_1 = 10 \quad (3)$$

Step 4 (calculating the value of required parameter(s))

The power of the independent voltage source can be calculated as follows.

$$P = VI$$

$$\Rightarrow P = -(\quad)(\quad) \quad (4)$$

The minus sign is applied in the power formula because the current is leaving the positive terminal of the voltage of the component.

Solving (3) and (4):

$$P = -(\quad)(\quad)$$

$$\Rightarrow P = -20 \text{ W}$$

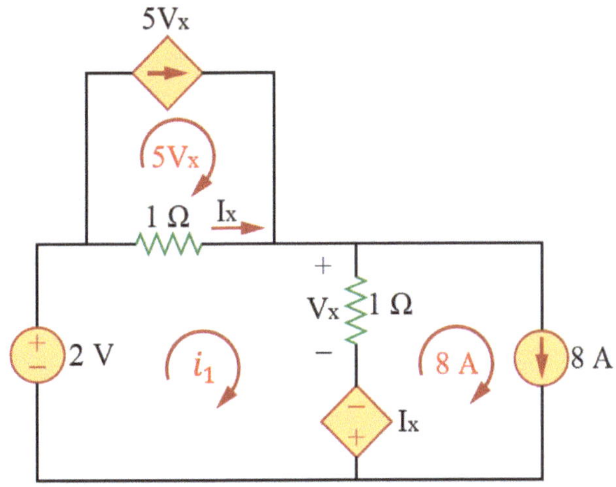

Fig. 3.22 The circuit of partially solved exercise 3.23

3.3 Circuit Analysis by Using Heuristic Methods

Problem

3.24. In the circuit of Fig. 3.23, calculate the voltage across the 5-Ω resistor.

Difficulty level ● Easy ○ Normal ○ Hard
Calculation amount ● Small ○ Normal ○ Large

1) 10 V
2) 15 V
3) 20 V
4) −30 V

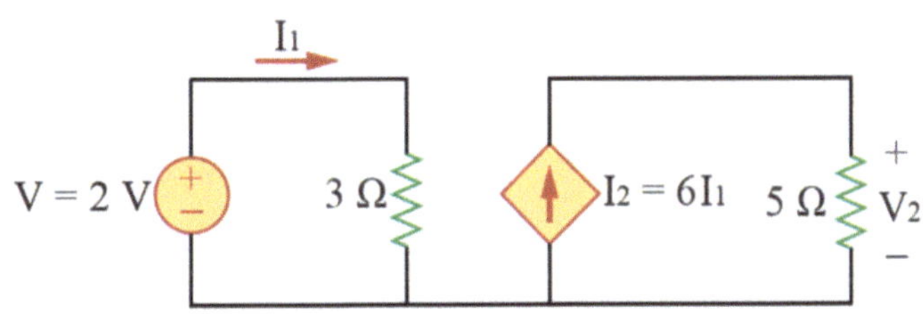

Fig. 3.23 The circuit of problem 3.24

Problem

3.25. Calculate the value of current *I* in the circuit of Fig. 3.24.

Difficulty level ● Easy ○ Normal ○ Hard
Calculation amount ● Small ○ Normal ○ Large

1) 1.5 A
2) 2 A
3) 2.5 A
4) −0.5 A

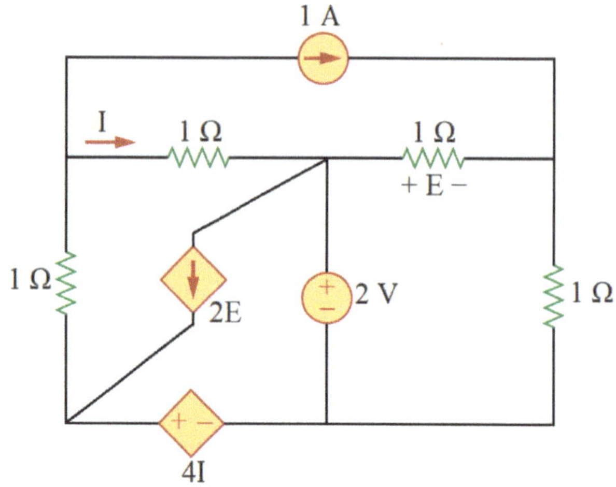

Fig. 3.24 The circuit of problem 3.25

Problem

3.26. Calculate the value of parameter V in the circuit of Fig. 3.25 so that we have $V_x = 0$.

Difficulty level ○ Easy ● Normal ○ Hard

Calculation amount ● Small ○ Normal ○ Large

1) 3 V
2) 6 V
3) 12 V
4) 18 V

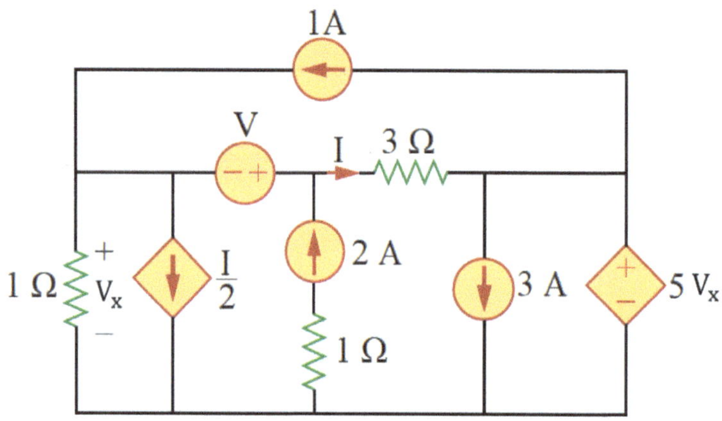

Fig. 3.25 The circuit of problem 3.26

Problem

3.27. Calculate the value of output voltage V_o in the circuit of Fig. 3.26.

Difficulty level ○ Easy ● Normal ○ Hard

Calculation amount ○ Small ● Normal ○ Large

1) $\frac{27}{5}$ V
2) 5 V
3) $\frac{24}{5}$ V
4) 6 V

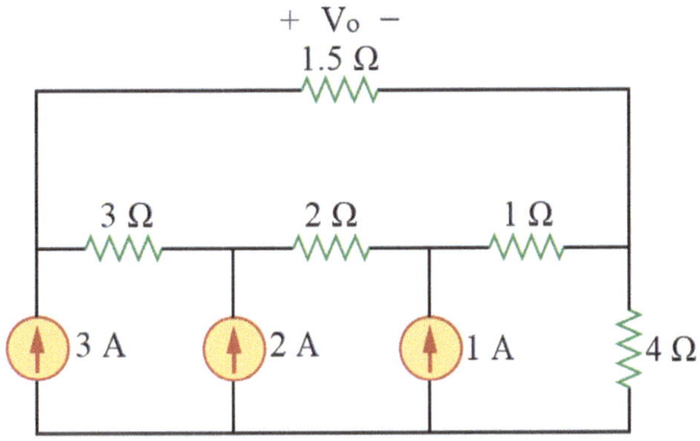

Fig. 3.26 The circuit of problem 3.27

Problem

3.28. Calculate the value of voltage V_{ab} in the circuit of Fig. 3.27.

Difficulty level ○ Easy ● Normal ○ Hard

Calculation amount ○ Small ● Normal ○ Large

1) 1.25 V
2) 1.75 V
3) 2.5 V
4) 3.25 V

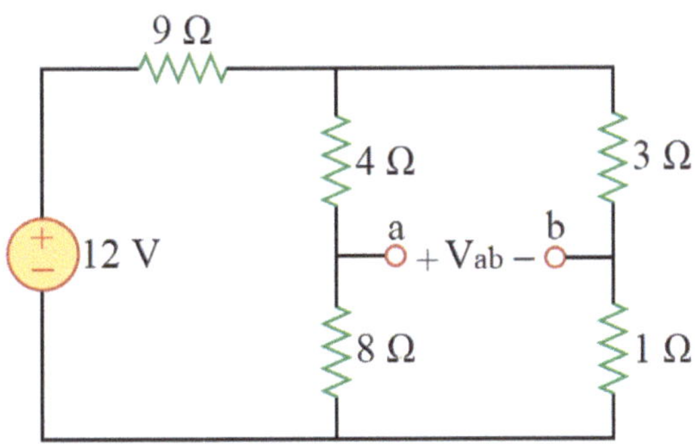

Fig. 3.27 The circuit of problem 3.28

References

1. Rahmani-Andebili, M., DC Electrical Circuit Analysis - Practice Problems, Methods, and Solutions (1st Ed.), Springer Nature, 2020.
2. Rahmani-Andebili, M., General Physics II - Practice Problems, Methods, and Solutions, Sringer Nature, 2025.
3. Rahmani-Andebili, M., General Physics I - Practice Problems, Methods, and Solutions, Springer Nature, 2025.
4. Rahmani-Andebili, M., Mathematics of Engineering and Science - Comprehensive Lessons, Springer Nature, 2024.
5. Rahmani-Andebili, M., Differential Equations – Practice Problems, Methods, and Solutions, Springer Nature, 2022.
6. Rahmani-Andebili, M., Calculus III – Practice Problems, Methods, and Solutions, Springer Nature, 2023.
7. Rahmani-Andebili, M., Calculus II – Practice Problems, Methods, and Solutions, Springer Nature, 2023.
8. Rahmani-Andebili, M., Calculus I (2nd Ed.) – Practice Problems, Methods, and Solutions, Springer Nature, 2023.
9. Rahmani-Andebili, M., Precalculus (2nd Ed.) – Practice Problems, Methods, and Solutions, Springer Nature, 2024.

4 Methods of Electrical Circuits Analysis: Part B

Abstract

In this chapter, the problems of the third chapter are fully solved, in detail, step-by-step, and with different methods. Herein, a "Notes Box" has been added after the solution of each problem that includes all the formulas and important notes that are needed to solve the problem. Moreover, a "Traps Box" has been added that discloses the traps that have been designed in the choices to challenge students' minds and encourage them to deeply think about the problem solutions as well as discourage them from careless selection of choices. The "Traps Box", in fact, predicts the possible confusions and tricks that might be encountered by students in real tests.

4.1 Nodal Analysis

4.1 *Step 1 (determining the type of circuit analysis)*

As can be noticed from Fig. 4.1, there are two independent nodes with unknown voltages (e_1 and e_2) and two independent meshes with unknown currents. Hence, there is no preference between nodal analysis and mesh analysis. However, herein, nodal analysis is used to solve this problem.

Step 2 (formulating the circuit analysis equations)

As can be seen in the circuit, the dependent voltage source is placed between two nodes, thus we must use KCL for the supernode of these two nodes [1–9].

KCL in the supernode:

$$\frac{e_1 - 4}{2} + \frac{e_1}{2} + \frac{e_2}{1} - 4 = 0$$

$$\Rightarrow e_1 + e_2 = 6 \quad (1)$$

The relation between the voltage of voltage source and its node voltages (auxiliary equation) is as follows.

$$e_2 - e_1 = 2I_a \quad (2)$$

Defining I_a based on the node voltages:

$$I_a = \frac{e_1}{2} \quad (3)$$

V. T. Davis, *DC Electrical Circuits Analysis*, https://doi.org/10.1007/978-3-032-00253-2_4

Step 3 (solving the system of equations)

Solving (2) and (3):

$$e_2 - e_1 = 2\left(\frac{e_1}{2}\right)$$

$$\Rightarrow e_2 = 2e_1 \tag{4}$$

Solving (1) and (4):

$$e_1 + 2e_1 = 6$$

$$\Rightarrow e_1 = 2 \text{ V} \tag{5}$$

Step 4 (calculating the value of required parameter)

Solving (3) and (5):

$$I_a = \frac{2}{2}$$

$$\Rightarrow I_a = 1 \text{ A}$$

Choice (2) is the answer.

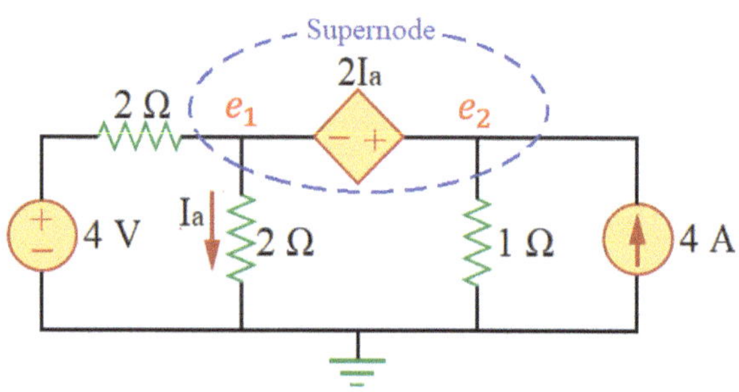

Fig. 4.1 The circuit of problem 4.1

4.4 *Step 1 (determining the type of circuit analysis)*

As can be noticed from Fig. 4.2, there are only two independent nodes with unknown voltages (e_1 and e_2) and four independent meshes with unknown currents. Thus, nodal analysis is the best method to solve this problem.

Step 2 (formulating the circuit analysis equations)

KCL in the second node:

$$-3V_y + \frac{e_2}{1} - 4 + \frac{e_2 - 10}{2} = 0$$

$$\Rightarrow -3V_y + \frac{3}{2}e_2 - 9 = 0 \tag{1}$$

Defining V_x and V_y based on the voltage of independent nodes (e_1 and e_2):

$$V_y = -e_2 \tag{2}$$

$$V_x = 10 - e_2 \tag{3}$$

Step 3 (solving the system of equations)

Solving (1) and (2):

$$3e_2 + \frac{3}{2}e_2 - 9 = 0$$

$$\Rightarrow \frac{9}{2}e_2 - 9 = 0$$

$$\Rightarrow e_2 = 2 \tag{4}$$

Step 4 (calculating the value of required parameter)

Solving (2) and (4):

$$V_y = -2 \text{ V}$$

Solving (3) and (4):

$$V_x = 10 - 2$$

$$\Rightarrow V_x = 8 \text{ V}$$

Choice (3) is the answer.

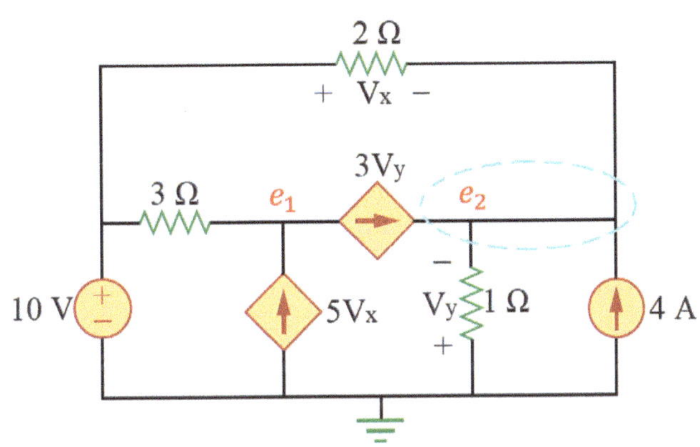

Fig. 4.2 The circuit of problem 4.4

4.7 To calculate the power of the independent current source, its voltage must be known. In this regard, the conventional polarity for the voltage and current of the component (associate reference direction) can be used, that is, the current must enter the positive terminal of the component.

Step 1 (determining the type of circuit analysis)

As can be noticed from Fig. 4.3, there are only two independent nodes with unknown voltages (V_1 and V_2) and four independent meshes with unknown currents. Thus, nodal analysis is the best method to solve this problem.

Step 2 (formulating the circuit analysis equations)

Using Ohm's law to define I based on the node voltages:

$$I = \frac{V_2}{2} \tag{1}$$

Defining the voltage of the dependent voltage source based on the node voltages:

$$2I = -V_2 - V_1$$

$$\overset{(1)}{\Rightarrow} 2\left(\frac{V_2}{2}\right) = -V_2 - V_1$$

$$\Rightarrow V_2 = -\frac{1}{2}V_1 \tag{2}$$

KCL in the supernode:

$$-4 + \frac{V_1}{1} - 2V_2 - 3V_1 + \frac{-V_2}{2} + \frac{-V_2}{2} = 0$$

$$\Rightarrow -4 - 2V_1 - 3V_2 = 0 \tag{3}$$

Step 3 (solving the system of equations)

Solving (2) and (3):

$$-4 - 2V_1 - 3\left(-\frac{1}{2}V_1\right) = 0$$

$$\Rightarrow -4 - \frac{1}{2}V_1 = 0$$

$$\Rightarrow V_1 = -8 \tag{4}$$

Step 4 (calculating the value of required parameter)

As we know, the power of a circuit component can be calculated as follows.

$$P = VI$$

Hence:

$$P = -V_1 \times 4$$

$$\overset{(4)}{\Rightarrow} P = -(-8) \times 4$$

$$\Rightarrow P = 32 \text{ W}$$

Since the value of power is a positive quantity, the independent current source is a power consumer. Choice (2) is the answer.

Traps

- You would choose Choice (1) if you had not correctly used the associated reference direction for the voltage and current of the complement.

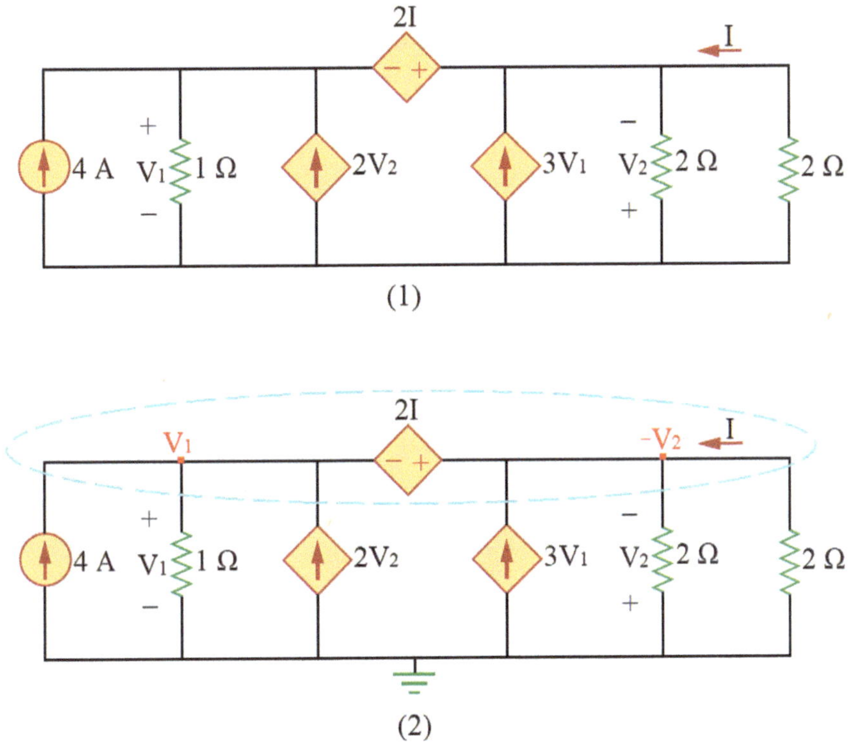

Fig. 4.3 The circuit of problem 4.7

> **Notes**
>
> When an associated reference direction for the voltage and current of a circuit component is considered, that is, when the current enters the positive terminal of the component:
>
> (a) The component absorbs or consumes power if the value of multiplication of its voltage and current (VI) is positive, and
>
> (b) The component delivers or generates power if the value of multiplication of its voltage and current (VI) is negative.

4.9 ***Step 1 (determining the type of circuit analysis)***

As can be noticed from Fig. 4.4, there are only two independent nodes with unknown voltage (e_1 and e_2) and four independent meshes with unknown currents. Thus, nodal analysis is the best method to solve this problem.

Step 2 (formulating the circuit analysis equations)

Defining I_0 based on the node voltages:

$$I_0 = \frac{e_1}{1}$$

$$\Rightarrow I_0 = e_1 \tag{1}$$

KCL in node 1:

$$-2 + \frac{e_1}{1} + \frac{e_1 - e_2}{2} + 2I_0 = 0$$

$$\overset{(1)}{\Rightarrow} -2 + \frac{e_1}{1} + \frac{e_1 - e_2}{2} + 2e_1 = 0$$

$$\Rightarrow -2 + \frac{7}{2}e_1 - \frac{1}{2}e_2 = 2 \qquad (2)$$

KCL in node 2:

$$\frac{e_2}{2} + \frac{e_2}{4} + \frac{e_2 - e_1}{2} - 2I_0 = 0$$

$$\overset{(1)}{\Rightarrow} \frac{e_2}{2} + \frac{e_2}{4} + \frac{e_2 - e_1}{2} - 2e_1 = 0$$

$$\Rightarrow -\frac{5}{2}e_1 + \frac{5}{4}e_2 = 0$$

$$\Rightarrow e_2 = 2e_1 \qquad (3)$$

Step 3 (solving the system of equations)

Solving (2) and (3):

$$-2 + \frac{7}{2}e_1 - \frac{1}{2}(2e_1) = 0$$

$$\Rightarrow -2 + \frac{5}{2}e_1 = 0$$

$$\Rightarrow e_1 = \frac{4}{5}\ \text{V} \qquad (4)$$

Solving (3) and (4):

$$e_2 = \frac{8}{5}\ \text{V} \qquad (5)$$

Choice (1) is the answer.

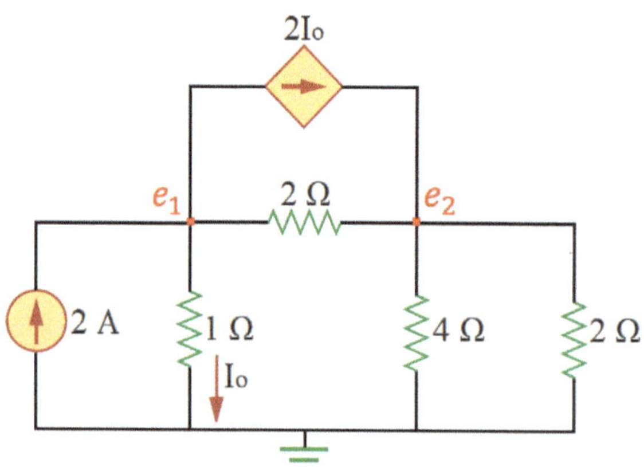

Fig. 4.4 The circuit of problem 4.9

4.11 In this problem, it is convenient to consider the units of kΩ, *mA*, and *V* for the resistances, the currents, and the voltages, respectively.

Step 1 (determining the type of circuit analysis)

Nodal analysis has been recommended in the problem.

Step 2 (formulating the circuit analysis equations)

The unknown voltages of the nodes are indicated by e_1 and V_x, as can be seen in Fig. 4.5. To solve the problem by using nodal analysis, we need to apply KCL in each independent node. Since there is a voltage source between two nodes, we must apply KCL for the supernode including these two nodes.

KCL in the supernode:

$$\frac{e_1 - 1}{1} - 3I_x + \frac{V_x}{0.5} - 2 = 0 \tag{1}$$

Also, we need to identify the relation between the voltage of voltage source and these node voltages (e_1 and e_2).

$$e_1 - V_x = 3V_x$$

$$\Rightarrow e_1 = 4V_x \tag{2}$$

Using Ohm's laws for the 1-kΩ resistor, to define I_x based on the node voltages:

$$I_x = \frac{1 - e_1}{1}$$

$$\Rightarrow I_x = 1 - e_1 \tag{3}$$

Step 3 (solving the system of equations)

Put (3) in (1) and simplify it:

$$e_1 - 1 - 3(1 - e_1) + \frac{V_x}{0.5} - 2 = 0$$

$$\Rightarrow e_1 - 1 - 3 + 3e_1 + 2V_x - 2 = 0$$

$$\Rightarrow 4e_1 + 2V_x = 6 \tag{4}$$

Solving (2) and (4):

$$4(4V_x) + 2\ V_x = 6$$

$$\Rightarrow V_x = \frac{1}{3}\ \text{V}$$

Step 4 (calculating the value of required parameter)

Using Ohm's laws for the 500-Ω resistor:

$$I = \frac{V_x}{0.5}$$

$$\Rightarrow I = \frac{\frac{1}{3}V}{0.5k\,\Omega}$$

$$\Rightarrow I = \frac{2}{3}\text{ mA}$$

Choice (2) is the answer.

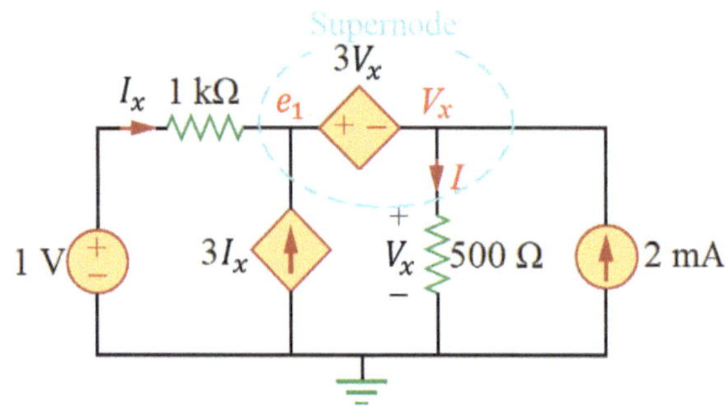

Fig. 4.5 The circuit of problem 4.11

4.12 *Step 1 (determining the type of circuit analysis)*

As can be noticed from Fig. 4.6, there are only two independent nodes with unknown voltages (e_1 and V_o) and three independent meshes with unknown currents. Thus, nodal analysis is the best method to solve this problem.

Step 2 (formulating the circuit analysis equations)

KCL in the node with the node voltage e_1:

$$5(e_1 - V_s) + 2e_1 + \alpha I_x = 0$$

$$\Rightarrow 7e_1 - 5V_s + \alpha I_x = 0 \tag{1}$$

KCL in the node with the node voltage V_o:

$$-\alpha I_x + 3V_o + V_o = 0$$

$$\Rightarrow I_x = \frac{4}{\alpha}V_o \tag{2}$$

On the other hand, we can define I_x based on the node voltage by using Ohm's law as follows.

$$I_x = 2e_1 \tag{3}$$

Step 3 (solving the system of equations)

Solving (2) and (3):

$$2e_1 = \frac{4}{\alpha}V_o$$

$$\Rightarrow e_1 = \frac{2}{\alpha} V_o \tag{4}$$

Solving (1), (2), and (4):

$$7\left(\frac{2}{\alpha} V_o\right) - 5V_s + \alpha\left(\frac{4}{\alpha} V_o\right) = 0$$

$$\Rightarrow \frac{14 + 4\alpha}{\alpha} V_o = 5V_s$$

$$\Rightarrow V_o = \frac{5\alpha}{14 + 4\alpha} V_s$$

Step 4 (calculating the value of required parameter)

The value of $\frac{V_o}{V_s}$ must be greater than one to adjust the circuit as an amplifier. Thus:

$$\frac{V_o}{V_s} = \frac{5\alpha}{14 + 4\alpha}$$

$$\Rightarrow \frac{5\alpha}{14 + 4\alpha} > 1$$

$$\Rightarrow 5\alpha > 14 + 4\alpha$$

$$\Rightarrow \alpha > 14$$

Choice (3) is the answer.

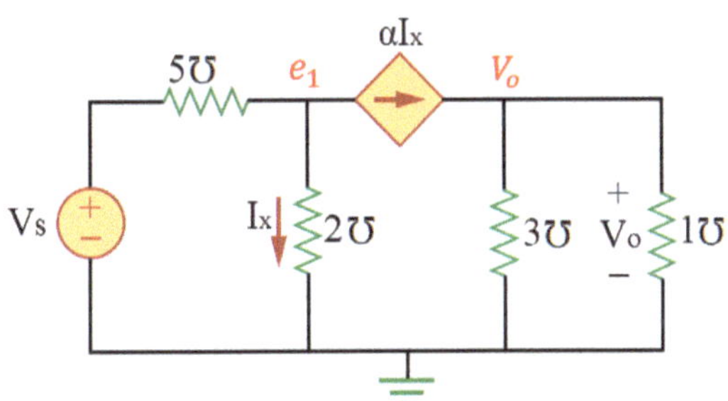

Fig. 4.6 The circuit of problem 4.12

4.13 *Step 1 (determining the type of circuit analysis)*

As can be noticed from Fig. 4.7, there are only two independent nodes with unknown voltages (e_1 and e_2) and three independent meshes with unknown currents. Thus, nodal analysis is the best method to solve this problem.

Step 2 (formulating the circuit analysis equations)

The voltage of the third node is zero ($e_3 = 0$ V) since it is grounded. Moreover, it should be noted that a short circuit branch never eliminates a current source.

KCL in node 1:

$$-3+\frac{e_1}{1}+\frac{e_1-e_2}{1}-2I_x=0$$

$$\Rightarrow 2e_1-e_2-2I_x=3 \tag{1}$$

Defining I_x based on the node voltages:

$$I_x=\frac{e_2-0}{1}$$

$$\Rightarrow I_x=e_2 \tag{2}$$

Solving (1) and (2):

$$2e_1-3e_2=3 \tag{3}$$

KCL in node 2:

$$\frac{e_2-e_1}{1}+\frac{e_2}{1}+\frac{e_2-0}{1}=0$$

$$\Rightarrow -e_1+3e_2=0$$

$$\Rightarrow e_1=3e_2 \tag{4}$$

Step 3 (solving the system of equations)

Solving (3) and (4):

$$2(3e_2)-3e_2=3$$

$$\Rightarrow e_2=1\text{ V} \tag{5}$$

Step 4 (calculating the value of required parameter)

KCL in node 3.

$$2I_x+\frac{0-e_2}{1}-5+I_{sc}=0$$

$$\Rightarrow 2I_x-e_2-5+I_{sc}=0$$

$$\overset{(2)}{\Rightarrow} I_{sc}=5-e_2$$

$$\overset{(5)}{\Rightarrow} I_{sc}=5-1$$

$$\Rightarrow I_{sc}=4\text{ A}$$

Choice (3) is the answer.

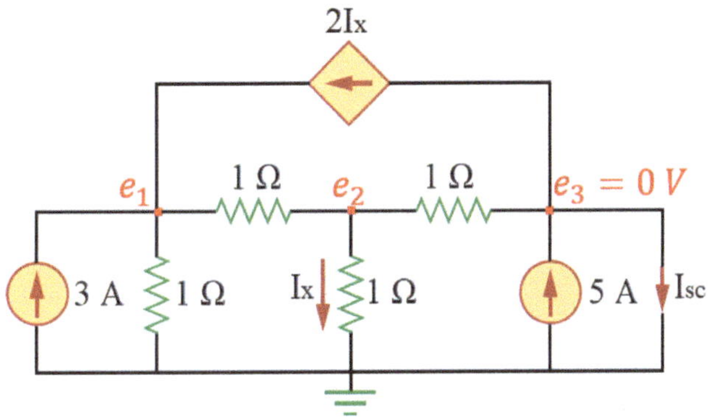

Fig. 4.7 The circuit of problem 4.13

4.14 ***Step 1 (determining the type of circuit analysis)***

As can be noticed from Fig. 4.8, there is only one independent node with unknown voltage (e) and three independent meshes with unknown currents. Thus, nodal analysis is the best method to solve this problem.

Step 2 (formulating the circuit analysis equations)

KCL for the supernode:

$$\frac{e-75}{4}+\frac{e}{5}-13+\frac{e-13}{6}=0$$

$$\frac{15+12+10}{60}e+\frac{-450-312-52}{24}=0$$

$$\Rightarrow \frac{37}{60}e=\frac{814}{24}$$

$$\Rightarrow e=55\text{ V} \tag{1}$$

Step 3 (solving the system of equations)

The value of unknown parameters of the system of equations has already been calculated.

Step 4 (calculating the value of required parameter)

KCL in node 1:

$$\frac{e-75}{4}+\frac{e}{5}+I=0 \tag{2}$$

Solving (1) and (2):

$$\frac{55-75}{4}+\frac{55}{5}+I=0$$

$$\Rightarrow I = 5 - 11$$

$$\Rightarrow I = -6\ \text{A}$$

Choice (4) is the answer.

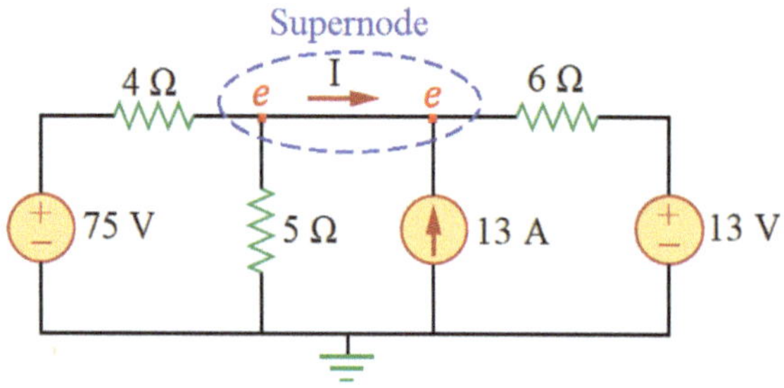

Fig. 4.8 The circuit of problem 4.14

4.2 Mesh Analysis

4.15 *Step 1 (determining the type of circuit analysis)*

Mesh analysis is the best method to solve this problem.

Step 2 (formulating the circuit analysis equations)

Since there is a current source between meshes 2 and 3, we must consider a supermesh instead of these meshes.

KVL in mesh 1:

$$-8 + 1(i_1 - i_3) + 2(i_1 - i_2) = 0$$

$$\Rightarrow 3i_1 - 2i_2 - i_3 = 8 \quad (1)$$

KVL in the supermesh including meshes 2 and 3:

$$1(i_3 - i_1) + 4i_3 + 3i_1 + 2(i_2 - i_1) = 0$$

$$\Rightarrow 2i_2 + 5i_3 = 0 \quad (2)$$

The relation between the current of current source and the involved mesh currents (auxiliary equation) for the supermesh is as follows.

$$i_2 = i_3 + 3 \quad (3)$$

Step 3 (solving the system of equations)

Solving (2) and (3):

$$2(i_3 + 3) + 5i_3 = 0$$

$$\Rightarrow i_3 = -\frac{6}{7}\text{ A} \quad (4)$$

Solving (3) and (4):

$$i_2 = \frac{15}{7}\text{ A} \quad (5)$$

Solving (1), (4), and (5):

$$3i_1 - 2\left(\frac{15}{7}\right) - \left(-\frac{6}{7}\right) = 8$$

$$\Rightarrow i_1 = \frac{80}{21}\text{ A}$$

Choice (4) is the answer (Fig. 4.9).

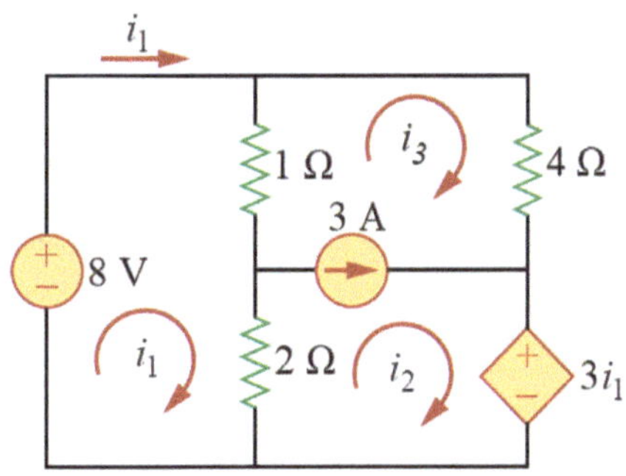

Fig. 4.9 The circuit of problem 4.15

4.19 *Step 1 (determining the type of circuit analysis)*

Mesh analysis is the best method to solve this problem.

Step 2 (formulating the circuit analysis equations)

As can be seen in the circuit of Fig. 4.10, there is one current source between meshes 1 and 2 as well as between meshes 1 and 3. Therefore, we must apply a supermesh for meshes 1, 2, and 3.

Using Ohm's law for the 5-Ω resistor to define V_y based on the mesh current:

$$V_y = -5i_2 \quad (1)$$

Using Ohm's law for the 10-Ω resistor to define V_x based on the mesh currents:

$$V_x = 10(i_2 - i_3) \quad (2)$$

KVL in the supermesh:

$$-110 + 20i_1 + 5i_2 + 15i_3 = 0 \tag{3}$$

Defining the current of the bottom dependent current source based on the mesh currents:

$$1.6V_y = i_1 - i_3 \tag{4}$$

Defining the current of the top dependent current source based on the mesh currents:

$$0.5V_x = i_1 - i_2 \tag{5}$$

Step 3 (solving the system of equations)

Solving (1) and (4):

$$1.6(-5i_2) = i_1 - i_3$$

$$\Rightarrow i_1 + 8i_2 - i_3 = 0 \tag{6}$$

Solving (2) and (5):

$$0.5(10(i_2 - i_3)) = i_1 - i_2$$

$$\Rightarrow i_1 - 6i_2 + 5i_3 = 0 \tag{7}$$

Solving (6) and (7):

$$-(8i_2 - i_3) - 6i_2 + 5i_3 = 0$$

$$\Rightarrow -14i_2 + 6i_3 = 0$$

$$\Rightarrow i_2 = \frac{3}{7}i_3 \tag{8}$$

Solving (3), (6), and (8):

$$-110 + 20\left(-\left(8\left(\frac{3}{7}i_3\right) - i_3\right)\right) + 5\left(\frac{3}{7}i_3\right) + 15i_3 = 0$$

$$\Rightarrow -110 - \frac{340}{7}i_3 + \frac{15}{7}i_3 + 15i_3 = 0$$

$$\Rightarrow -110 - \frac{220}{7}i_3 = 0$$

$$\Rightarrow i_3 = -\frac{7}{2} \tag{9}$$

Solving (8) and (9):

$$i_2 = \frac{3}{7}\left(-\frac{7}{2}\right) = -\frac{3}{2} \tag{10}$$

Step 4 (calculating the value of required parameter)

Solving (1) and (10):

$$V_y = -5\left(-\frac{3}{2}\right)$$

$$\Rightarrow V_y = 7.5 \text{ V}$$

Solving (2), (9), and (10):

$$V_x = 10\left(-\frac{3}{2} - \left(-\frac{7}{2}\right)\right)$$

$$\Rightarrow V_x = 20 \text{ V}$$

Choice (2) is the answer.

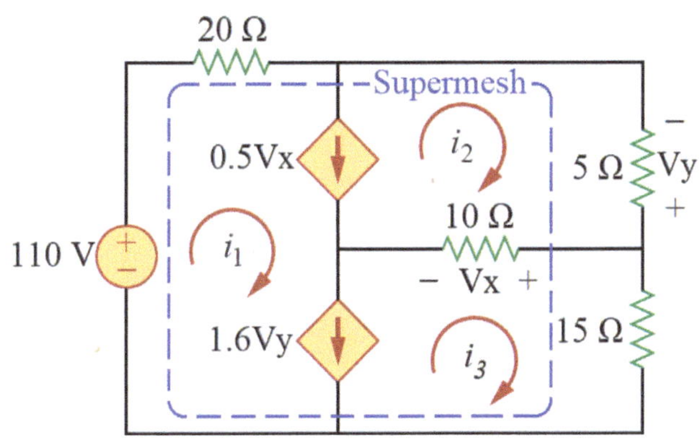

Fig. 4.10 The circuit of problem 4.19

4.20 In this problem, it is convenient to consider the units of $k\ \Omega$, mA, and V for the resistances, currents, and voltages, respectively.

Step 1 (determining the type of circuit analysis)

Mesh analysis has been recommended in the problem.

Step 2 (formulating the circuit analysis equations)

To solve the problem by using mesh analysis, we need to apply KVL in each independent mesh. Since there is a current source between the two meshes, we must apply KVL for the supermesh including these two meshes.

As can be seen in Fig. 4.11, the current of the right-side mesh is known.

$$i_3 = 2 \text{ mA} \quad (1)$$

KVL for the supermesh:

$$-1 + I_x \times 1 + 3V_x + V_x = 0$$

$$\Rightarrow -1 + I_x + 4V_x = 0 \tag{2}$$

Moreover, we need to determine the relation between the current of the current source and these mesh currents (i_1 and i_2).

$$i_2 - I_x = 3I_x$$

$$\Rightarrow I_x = \frac{i_2}{4} \tag{3}$$

Using Ohm's laws for the 500-Ω resistor, to define V_x based on the mesh currents:

$$V_x = 0.5(i_2 + i_3)$$

$$\overset{(1)}{\Rightarrow} V_x = 0.5(i_2 + 2) \tag{4}$$

Put (4) in (2) and simplify it:

$$-1 + I_x + 4(0.5(i_2 + 2)) = 0$$

$$\Rightarrow I_x + 2i_2 + 3 = 0 \tag{5}$$

Step 3 (solving the system of equations)

Solving (3) and (5):

$$\frac{i_2}{4} + 2i_2 + 3 = 0$$

$$\Rightarrow \frac{9}{4} i_2 = -3$$

$$\Rightarrow i_2 = -\frac{4}{3} \text{ mA} \tag{6}$$

Step 4 (calculating the value of required parameter)

As can be noticed from the circuit:

$$I = i_2 + i_3$$

$$\overset{(1)(6)}{\Rightarrow} I = -\frac{4}{3} + 2$$

$$\Rightarrow I = \frac{2}{3} \text{ mA}$$

Choice (2) is the answer.

Traps

- You would choose Choice (1) or (3) if you had not considered the fact the requested current is equal to sum of the second and third currents.

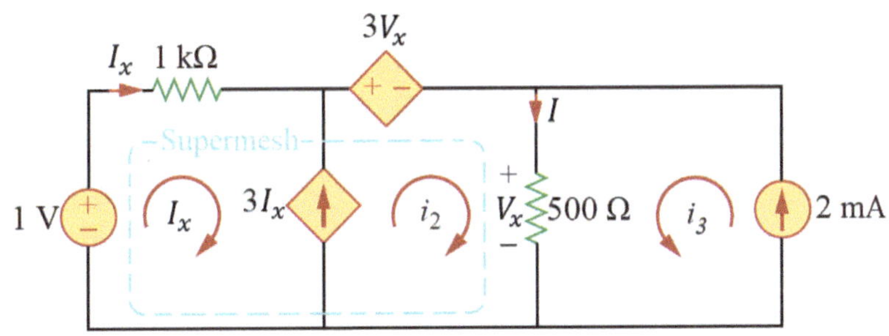

Fig. 4.11 The circuit of problem 4.20

4.21 We need to calculate the voltage and current of the resistor and then multiply them to calculate its lost power considering the voltage and current polarities.

Step 1 (determining the type of circuit analysis)

Herein, mesh analysis is applied to solve the problem.

Step 2 (formulating the circuit analysis equations)

From the circuit shown in Fig. 4.12, it is noticed that:

$$I_1 = 3 \text{ A} \quad (1)$$

$$I_2 = 2I_x \quad (2)$$

$$I_x = 3 - I_3 \quad (3)$$

Solving (2) and (3):

$$I_3 = 3 - \frac{1}{2}I_2 \quad (4)$$

KVL in the third mesh:

$$4I_3 + 8(I_3 - I_2) + 2(I_3 - I_1) = 0$$

$$\Rightarrow 14I_3 - 8I_2 - 2I_1 = 0 \quad (5)$$

Step 3 (solving the system of equations)

Solving (1), (4), and (5):

$$14\left(3 - \frac{1}{2}I_2\right) - 8I_2 - 2 \times 3 = 0$$

$$\Rightarrow 15I_2 = 36$$

$$\Rightarrow I_2 = \frac{12}{5}\ A = 2.4\ A \tag{6}$$

Step 4 (calculating the value of required parameter)

The power consumed by the vertical $4-\Omega$ resistor can be calculated as follows.

$$P = RI^2$$

$$\Rightarrow P = 4(I_1 - I_2)^2$$

$$\Rightarrow P = 4(3 - 2.4)^2$$

$$\Rightarrow P = 1.44\ W$$

Choice (1) is the answer.

Traps

- You would choose Choice (2) or (3) if you had not considered the fact the current of the resistor is equal to the subtraction of the first and second currents.

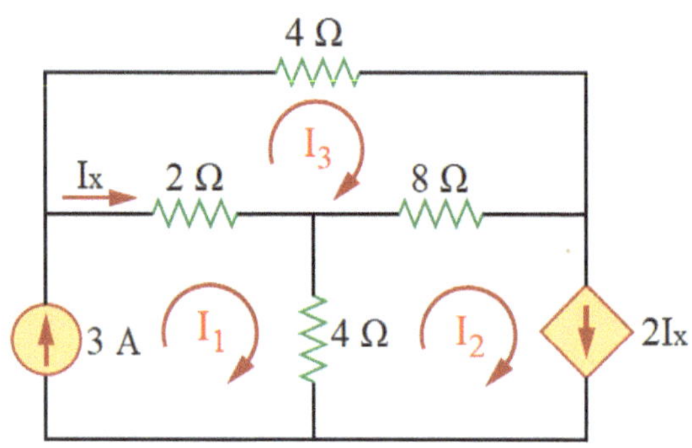

Fig. 4.12 The circuit of problem 4.21

Notes

The power consumed by a resistor can be calculated by using one of the following relations.

$$P = RI^2$$

$$P = VI$$

$$P = \frac{V^2}{R}$$

4.3 Circuit Analysis by Using Heuristic Methods

4.24 Using Ohm's law in the left-side loop:

$$I_1 = \frac{2}{3}\ \mathrm{A} \tag{1}$$

Using Ohm's law in the right-side loop:

$$V_2 = 5I_2$$

$$\Rightarrow V_2 = 5(6I_1) = 30I_1 \tag{2}$$

Solving (1) and (2):

$$V_2 = 30 \times \frac{2}{3}$$

$$\Rightarrow V_2 = 20\ \mathrm{V}$$

Choice (3) is the answer (Fig. 4.13).

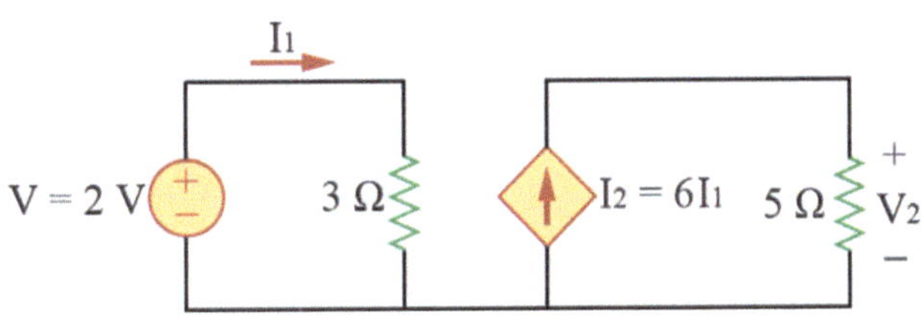

Fig. 4.13 The circuit of problem 4.24

4.25 Applying conventional methods like mesh and nodal analyses will need lots of calculations. Thus, this problem is solved by using a heuristic approach as follows.

By looking at the circuit, it can be noticed that we need to apply one KVL and one KCL, as is shown in Fig. 4.14.

$$\mathrm{KVL}: -V + I + 2 - 4I = 0 \tag{1}$$

$$\mathrm{KCL}: 1 + I + \frac{V}{1} = 0$$

$$\Rightarrow V = -I - 1 \tag{2}$$

Solving (1) and (2):

$$-(-I - 1) + I + 2 - 4I = 0$$

$$\Rightarrow 3 = 2I$$

$$\Rightarrow I = 1.5 \text{ A}$$

Choice (1) is the answer.

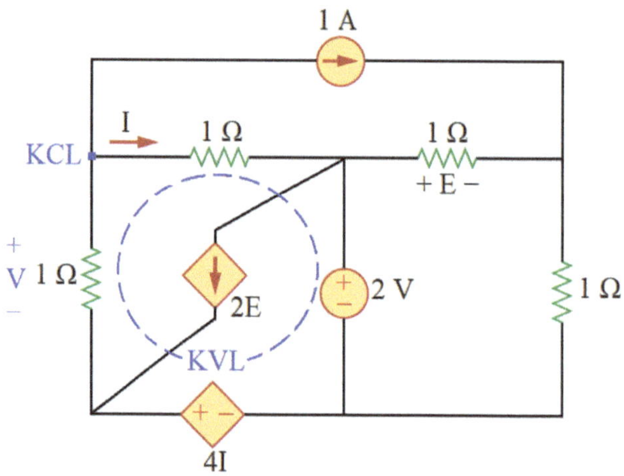

Fig. 4.14 The circuit of problem 4.25

4.26 Applying conventional methods like mesh and nodal analyses will need lots of calculations. Thus, this problem is solved by using a heuristic approach as follows.

Based on the given information $V_x = 0$. Hence, the voltage of the resistor is zero, and no current flows through that branch.

By applying KCL in the indicated supernode of the circuit of Fig. 4.15, we have:

$$-1 + 0 + \frac{I}{2} - 2 + I = 0$$

$$\Rightarrow 1.5I = 3$$

$$\Rightarrow I = 2 \text{ A} \quad (1)$$

Moreover, we need to write a KVL in the indicated loop.

$$-V + 3I - 0 = 0$$

$$\Rightarrow V = 3I \quad (2)$$

Solving (1) and (2):

$$\Rightarrow V = 3 \times 2$$

$$\Rightarrow V = 6 \text{ V}$$

Choice (2) is the answer.

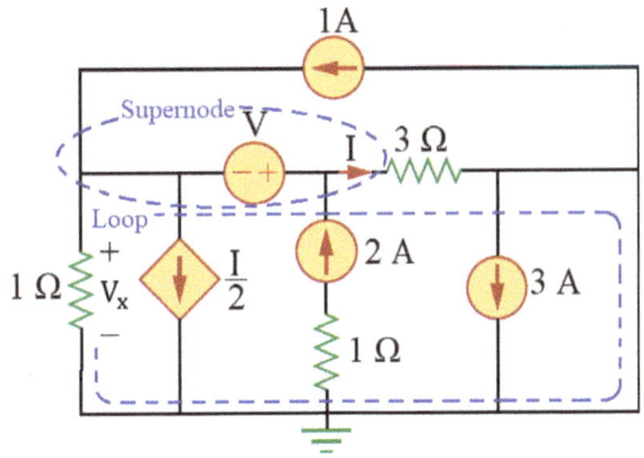

Fig. 4.15 The circuit of problem 4.26

4.27 Applying conventional methods like mesh and nodal analyses will result in a four-equation-four-unknown problem that needs lots of calculations. Hence, this problem is solved by using a heuristic approach as follows.

As can be seen, by using KCL, the current of resistors in the horizontal branches are determined, and then a KVL is applied in the top loop.

KVL in the top loop:

$$3(3-I)+2(5-I)+(6-I)-1.5I=0$$

$$\Rightarrow -7.5I+25=0$$

$$\Rightarrow I=\frac{10}{3}\text{ A} \quad (1)$$

Using Ohm' law for the 1.5-Ω resistor:

$$V_o=1.5I \quad (2)$$

Solving (1) and (2):

$$V_o=1.5\times\frac{10}{3}$$

$$\Rightarrow V_o=5\text{ V}$$

Choice (2) is the answer (Fig. 4.16).

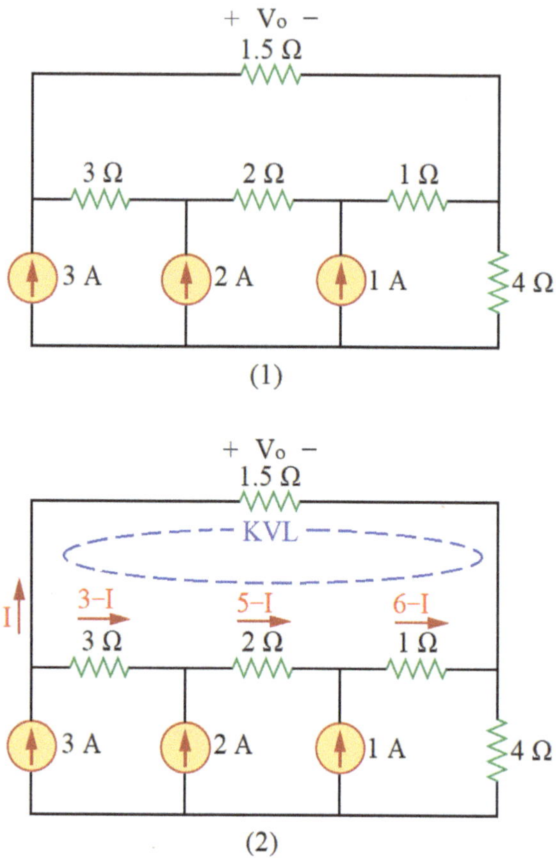

Fig. 4.16 The circuit of problem 4.27

4.28 Applying conventional methods like mesh and nodal analyses will need lots of calculations. Therefore, this problem is solved by using a heuristic approach as follows.

Herein, we can apply current division rule to find the node voltages "a" and "b", and then calculate V_{ab}.

$$I = \frac{12}{9 + (4+8)||(3+1)}$$

$$\Rightarrow I = \frac{12}{9 + \frac{12 \times 4}{12 + 4}} = \frac{12}{9+3}$$

$$\Rightarrow I = 1 \text{ A}$$

Applying current division rule:

$$I_{4,8} = \frac{(R_3 + R_1)}{(R_3 + R_1) + (R_4 + R_8)} \times I$$

$$\Rightarrow I_{4,8} = \frac{(3+1)}{(3+1) + (4+8)} \times 1$$

$$\Rightarrow I_{4,8} = 0.25 \text{ A}$$

KCL on the top node:

$$I_{3,1} = 1 - I_{4,8}$$

$$\Rightarrow I_{3,1} = 1 - 0.25$$

$$\Rightarrow I_{3,1} = 0.75 \text{ A}$$

Therefore:

$$V_{ab} = V_a - V_b$$

$$\Rightarrow V_{ab} = 8 \times I_{4,8} - 1 \times I_{3,1}$$

$$\Rightarrow V_{ab} = 8 \times 0.25 - 1 \times 0.75$$

$$\Rightarrow V_{ab} = 2 - 0.75$$

$$\Rightarrow V_{ab} = 1.25 \text{ V}$$

Choice (1) is the answer (Fig. 4.17).

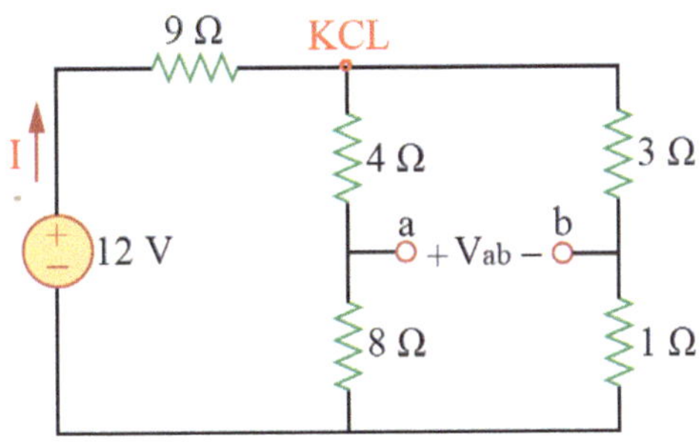

Fig. 4.17 The circuit of problem 4.28

References

1. Rahmani-Andebili, M., DC Electrical Circuit Analysis - Practice Problems, Methods, and Solutions (1st Ed.), Springer Nature, 2020.
2. Rahmani-Andebili, M., General Physics II - Practice Problems, Methods, and Solutions, Sringer Nature, 2025.
3. Rahmani-Andebili, M., General Physics I - Practice Problems, Methods, and Solutions, Springer Nature, 2025.
4. Rahmani-Andebili, M., Mathematics of Engineering and Science - Comprehensive Lessons, Springer Nature, 2024.
5. Rahmani-Andebili, M., Differential Equations – Practice Problems, Methods, and Solutions, Springer Nature, 2022.
6. Rahmani-Andebili, M., Calculus III – Practice Problems, Methods, and Solutions, Springer Nature, 2023.
7. Rahmani-Andebili, M., Calculus II – Practice Problems, Methods, and Solutions, Springer Nature, 2023.
8. Rahmani-Andebili, M., Calculus I (2nd Ed.) – Practice Problems, Methods, and Solutions, Springer Nature, 2023.
9. Rahmani-Andebili, M., Precalculus (2nd Ed.) – Practice Problems, Methods, and Solutions, Springer Nature, 2024.

Theorems of Electrical Circuits Analysis: Part A 5

Abstract

In this chapter, the basic and advanced problems related to the theorems of electrical circuits analysis are studied. The subjects include input resistance, Thevenin theorem, Norton theorem, source transformation theorem, superposition theorem, and maximum power transfer theorem. Herein, different types of problems and exercises are presented that are categorized as follows.

- ***Problems with detailed solution***: They have been designed to teach students the subjects in detail. Moreover, they have been categorized in different levels based on their difficulty levels (easy, normal, and hard) and calculation amounts (small, normal, and large).
- ***Partially solved exercises***: They have been designed to encourage students to practice problems while guiding them through the problem-solving procedure and hinting the required formulas.
- ***Exercises with final answer***: They have been designed to encourage students to practice more by themselves while hinting them by the final answer as well as to help instructors to give tests or quizzes.

5.1 Input Resistance

Problem

5.1. Calculate the input resistance seen from terminal a-b in the circuit of Fig. 5.1 [1–9].

Difficulty level ○ Easy ● Normal ○ Hard
Calculation amount ● Small ○ Normal ○ Large

1) 1 Ω
2) 1.33 Ω
3) 1.5 Ω
4) 2 Ω

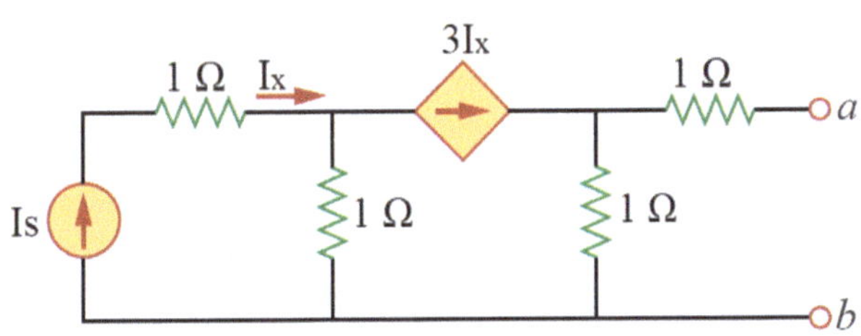

Fig. 5.1 The circuit of problem 5.1

V. T. Davis, *DC Electrical Circuits Analysis*, https://doi.org/10.1007/978-3-032-00253-2_5

Problem

5.2. In the circuit of Fig. 5.2, calculate the input resistance seen from terminal a-b.

Difficulty level ○ Easy ● Normal ○ Hard
Calculation amount ○ Small ○ Normal ● Large

1) $\frac{5}{2}\ \Omega$
2) $\frac{2}{5}\ \Omega$
3) $\frac{2k-5}{8k-26}$
4) $\frac{8k-26}{2k-5}$

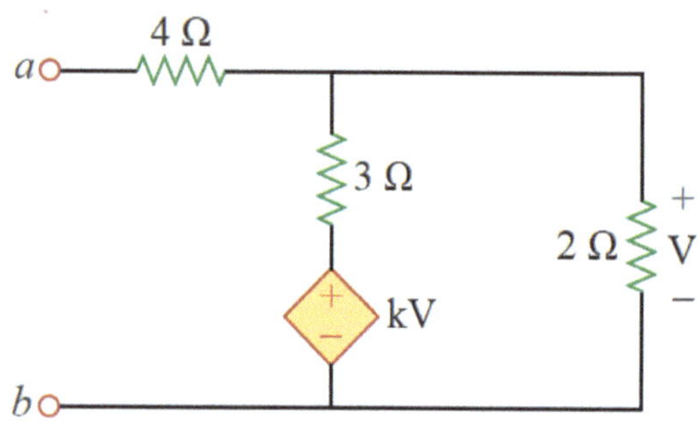

Fig. 5.2 The circuit of problem 5.2

Exercise

5.3. In Problem 5.2, calculate the value of k so that the input resistance seen from the terminal is zero.

Final Answer

$k = \frac{13}{4}$

Note: *Read the "educational support" part of the preface of the textbook.*

Partially Solved Exercise

5.4. Calculate the input resistance of the circuit of Fig. 5.3 seen from terminal a-b.

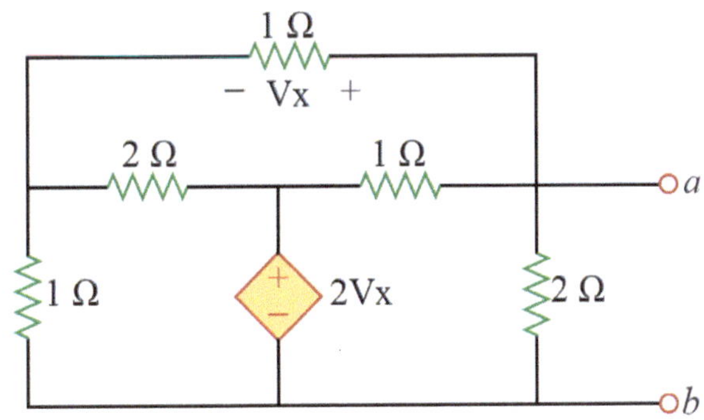

Fig. 5.3 The circuit of partially solved exercise 5.4

Solution

Step 1 (turning off all the independent sources)

If the circuit includes any independent voltage and current sources, they need to be turned off. In this regard, each independent voltage source must be replaced by a short-circuit branch and each independent current source is replaced by an open-circuit branch.

Herein, the circuit does not include any independent source.

Step 2 (determining the type of problem)

Type 1: If the circuit does not include any dependent source (dependent voltage or current source), the series–parallel rule is applied to calculate the input resistance of the circuit.

Type 2: If the circuit includes at least a dependent source (dependent voltage or current source), a test source (independent voltage or current source) with the voltage and current of V_t and I_t needs to be connected to the input terminal of the circuit. This problem is a type 2 problem.

Step 3 (determining the type of circuit analysis)

Nodal analysis is the best method to apply in this problem since it will result in the least calculation.

Step 4 (formulating the circuit analysis equations).

KCL in node 1:

$$\frac{e_1}{(\quad)} + \frac{e_1 - (\quad)}{(\quad)} + \frac{e_1 - V_t}{(\quad)} = 0$$

$$\Rightarrow (\quad)e_1 - (\quad)V_x - (\quad)V_t = 0 \quad (1)$$

KCL in node 3:

$$-I_t + \frac{V_t}{(\quad)} + \frac{V_t - 2V_x}{(\quad)} + \frac{V_t - e_1}{(\quad)} = 0$$

$$\Rightarrow -I_t - e_1 - (\quad)V_x + (\quad)V_t = 0 \quad (2)$$

Defining V_x based on the node voltages:

$$V_x = (\quad) - (\quad) \quad (3)$$

Step 5 (solving the system of equations to calculate the value of $R_{in} = \frac{V_t}{I_t}$***).***

Solving (1) and (3):

$$(\quad)e_1 - (V_t - e_1) - (\quad)V_t = 0$$

$$\Rightarrow e_1 = \frac{4}{7}V_t \quad (4)$$

Solving (2) and (3):

$$-I_t - (\quad)e_1 - (\quad)(V_t - e_1) + (\quad)V_t = 0$$

$$\Rightarrow -I_t + e_1 + 0.5V_t = 0 \quad (5)$$

Solving (4) and (5): Fig. 5.4

$$-(\quad)I_t + (\quad)V_t + (\quad)V_t = 0$$

$$\Rightarrow -(\quad)I_t + (\quad)V_t = 0$$

$$\Rightarrow \frac{V_t}{I_t} = (\quad)$$

$$\Rightarrow R_{in} = \frac{14}{15}\,\Omega$$

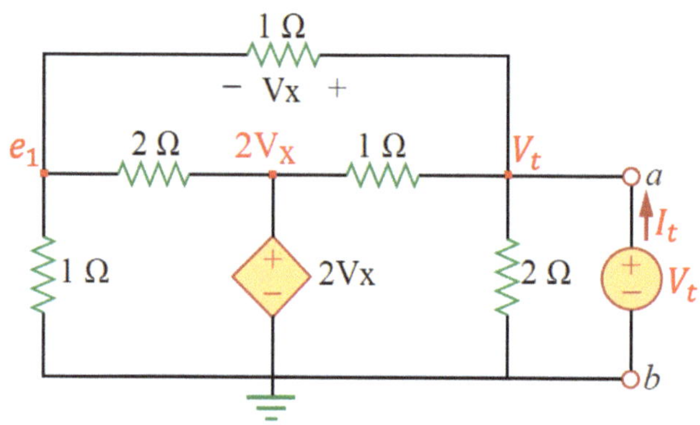

Fig. 5.4 The circuit of partially solved exercise 5.4

Problem

5.5. Calculate the input resistance of the circuit of Fig. 5.5 seen from the terminal a-b.

Difficulty level ○ Easy ● Normal ○ Hard
Calculation amount ○ Small ○ Normal ● Large

1) $\dfrac{2(\alpha-3)}{3\alpha-7}$

2) $\dfrac{3\alpha-7}{2(\alpha-3)}$

3) $\dfrac{\alpha-3}{3\alpha-7}$

4) $\dfrac{3\alpha-7}{\alpha-3}$

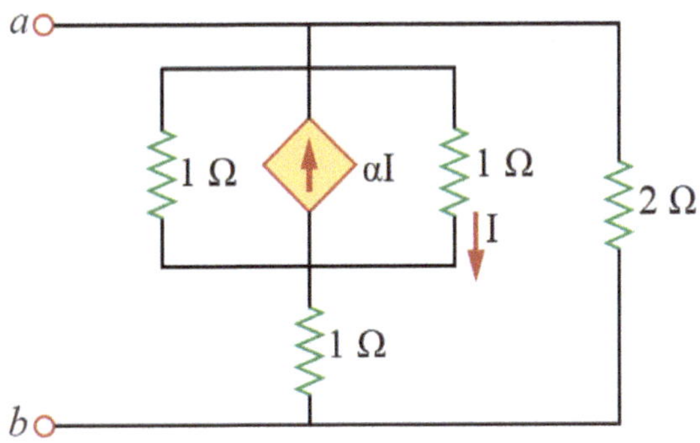

Fig. 5.5 The circuit of problem 5.5

Exercise

5.6. In problem 5.5, calculate the range of value of parameter α so that the input resistance becomes a negative quantity.

Final Answer

$\frac{7}{3} < \alpha < 3$

Note: *Read the "educational support" part of the preface of the textbook.*

Partially Solved Exercise

5.7. Calculate the resistance seen from terminal A-B in the circuit of Fig. 5.6.

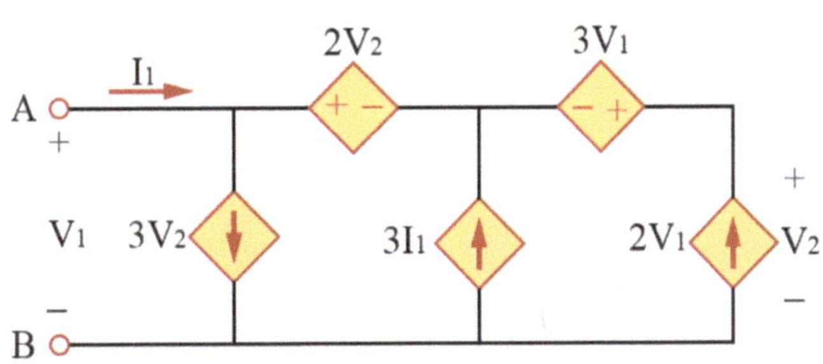

Fig. 5.6 The circuit of partially solved exercise 5.7

Solution

2.7. ***Step 1 (turning off all the independent sources)***

If the circuit includes any independent voltage and current sources, they need to be turned off. In this regard, each independent voltage source must be replaced by a short-circuit branch and each independent current source is replaced by an open-circuit branch.
Herein, the circuit does not include any independent source.

Step 2 (determining the type of problem)

Type 1: If the circuit does not include any dependent source (dependent voltage or current source), the series–parallel rule is applied to calculate the input resistance of the circuit.

Type 2: If the circuit includes at least a dependent source (dependent voltage or current source), a test source (independent voltage or current source) with the voltage and current of V_t and I_t needs to be connected to the input terminal of the circuit. This problem is a type 2 problem.

Step 3 (determining the type of circuit analysis)

Heuristic approach is the best method to apply in this problem since it will result in the least calculation.

Step 4 (formulating the circuit analysis equations)

KVL in the loop of the circuit of Fig. 5.7:

$$(\quad)+(\quad)+(\quad)+(\quad)=0$$

$$\Rightarrow V_2=\frac{4}{3}V_1 \quad (1)$$

KCL in the supernode:

$$(\quad)+(\quad)+(\quad)+(\quad)=0$$

$$\Rightarrow -4I_1+3V_2-2V_1=0 \quad (2)$$

Step 5 (solving the system of equations to calculate the value of $R_{in}=\frac{V_t}{I_t}$***).***

Solving (1) and (2):

$$(\quad)I_1+3\left(\frac{4}{3}V_1\right)+(\quad)V_1=0$$

$$\Rightarrow (\quad)I_1+(\quad)V_1=0$$

$$\Rightarrow \frac{V_1}{I_1}=(\quad)$$

$$\Rightarrow R_{AB}=2\ \Omega$$

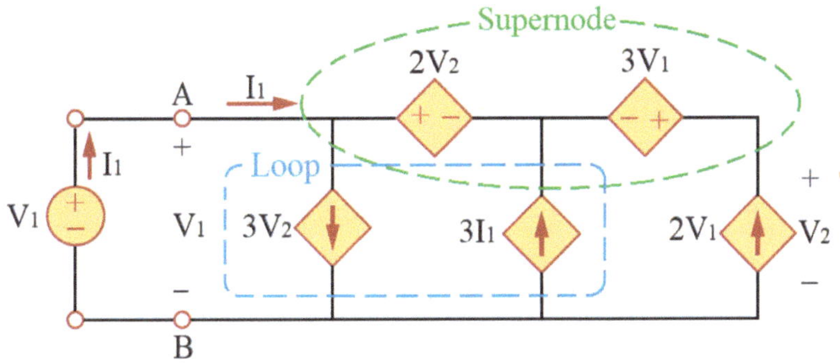

Fig. 5.7 The circuits of partially solved exercise 5.7

Problem

5.8. Calculate the input resistance seen from terminal A-B of the circuit of Fig. 5.8.

Difficulty level ○ Easy ● Normal ○ Hard
Calculation amount ● Small ○ Normal ○ Large

1) 0
2) 1 Ω
3) ∞
4) $\frac{1}{2}$ Ω

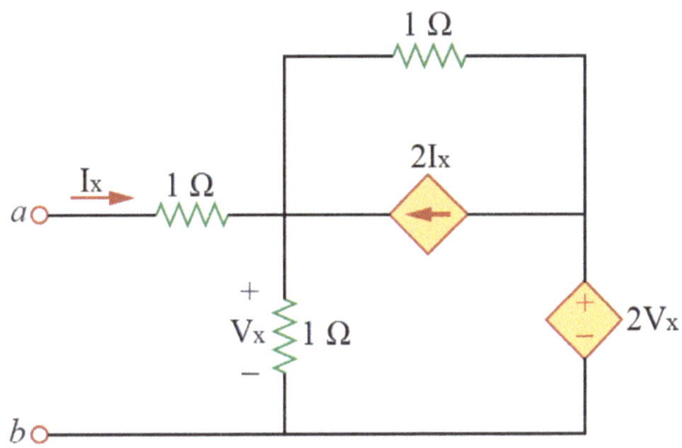

Fig. 5.8 The circuit of problem 5.8

Problem

5.9. In the circuit of Fig. 5.9, calculate the input resistance seen from terminal a-b.

Difficulty level ○ Easy ● Normal ○ Hard
Calculation amount ○ Small ● Normal ○ Large

1) R
2) $R(1 + 2\alpha\mu)$
3) $R\left(1 + \frac{\alpha\mu}{2}\right)$
4) $\frac{R}{3}$

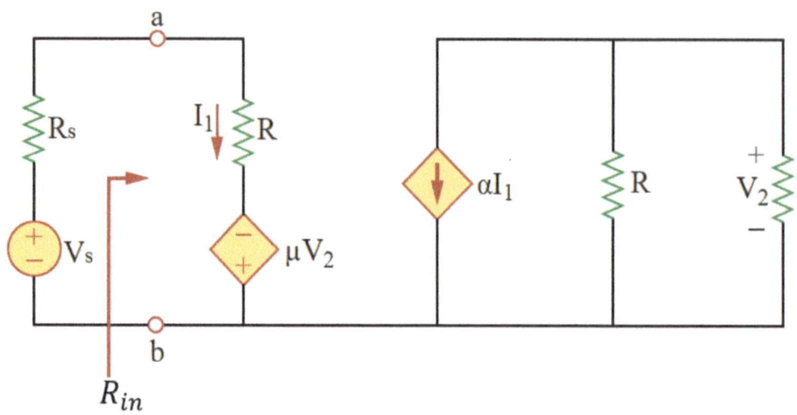

Fig. 5.9 The circuit of problem 5.9

Problem

5.10. In the circuit of Fig. 5.10, calculate the equivalent resistance R_{eq} illustrated.

Difficulty level ○ Easy ● Normal ○ Hard
Calculation amount ○ Small ● Normal ○ Large

1) $\frac{1}{3}$ R
2) 2 R
3) 3 R
4) $\frac{1}{2}$ R

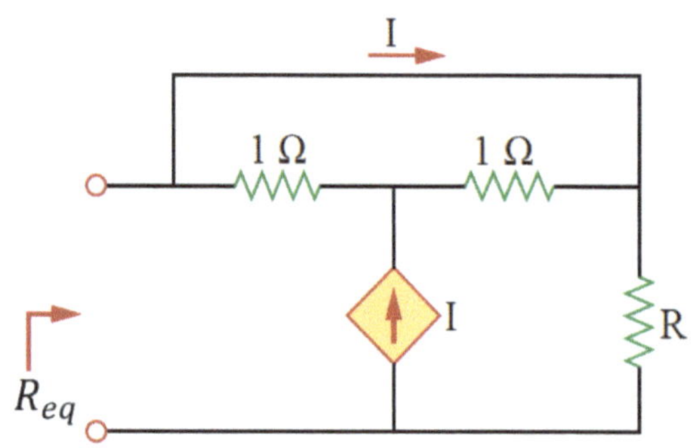

Fig. 5.10 The circuit of problem 5.10

Partially Solved Exercise

5.11. In the circuit of Fig. 5.11, calculate the input resistance R_{in} seen from terminal a-b.

1) $\frac{1}{2}$ Ω
2) $\frac{2}{3}$ Ω
3) $\frac{3}{2}$ Ω
4) $\frac{5}{2}$ Ω

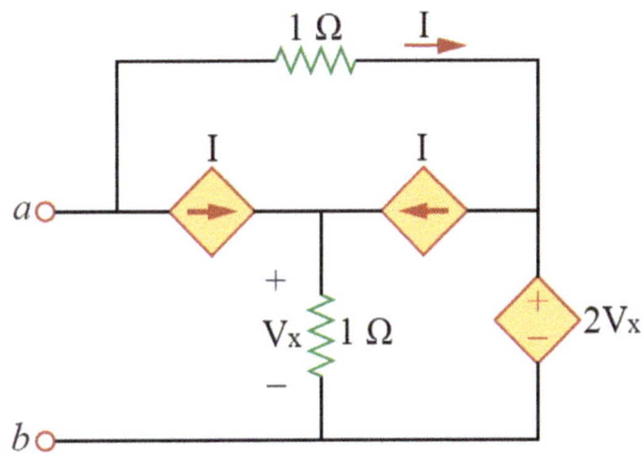

Fig. 5.11 The circuit of partially solved exercise 5.11

Solution

Step 1 (turning off all the independent sources)

If the circuit includes any independent voltage and current sources, they need to be turned off. In this regard, each independent voltage source must be replaced by a short-circuit branch and each independent current source is replaced by an open-circuit branch.

Herein, the circuit does not include any independent source.

Step 2 (determining the type of problem)

Type 1: If the circuit does not include any dependent source (dependent voltage or current source), the series–parallel rule is applied to calculate the input resistance of the circuit.

Type 2: If the circuit includes at least a dependent source (dependent voltage or current source), a test source (independent voltage or current source) with the voltage and current of V_t and I_t needs to be connected to the input terminal of the circuit. This problem is a type 2 problem.

Step 3 (determining the type of circuit analysis)

Nodal analysis is the best method to apply in this problem since it will result in the least calculation.

Step 4 (formulating the circuit analysis equations)

Defining I based on the node voltages:

$$I = \frac{(\quad) - (\quad)}{(\quad)}$$

$$\Rightarrow I = V_t - 2V_x \quad (1)$$

KCL in the left-side node:

$$(\quad) + (\quad) + (\quad) = 0$$

$$\Rightarrow I = 0.5I_t \quad (2)$$

KCL in the middle node:

$$(\quad) + (\quad) + \frac{V_x}{(\quad)} = 0$$

$$\Rightarrow V_x = 2I \quad (3)$$

Step 5 (solving the system of equations to calculate the value of $R_{in} = \frac{V_t}{I_t}$).

Solving (1) and (2):

$$0.5I_t = V_t - 2V_x \quad (4)$$

Solving (2) and (3):

$$V_x = I_t \quad (5)$$

Solving (4) and (5): Fig. 5.12

$$(\quad)I_t = (\quad)V_t + (\quad)I_t$$

$$\Rightarrow (\quad)I_t = (\quad)V_t$$

$$\Rightarrow \frac{V_t}{I_t} = (\quad)$$

$$\Rightarrow R_{in} = 2.5\ \Omega$$

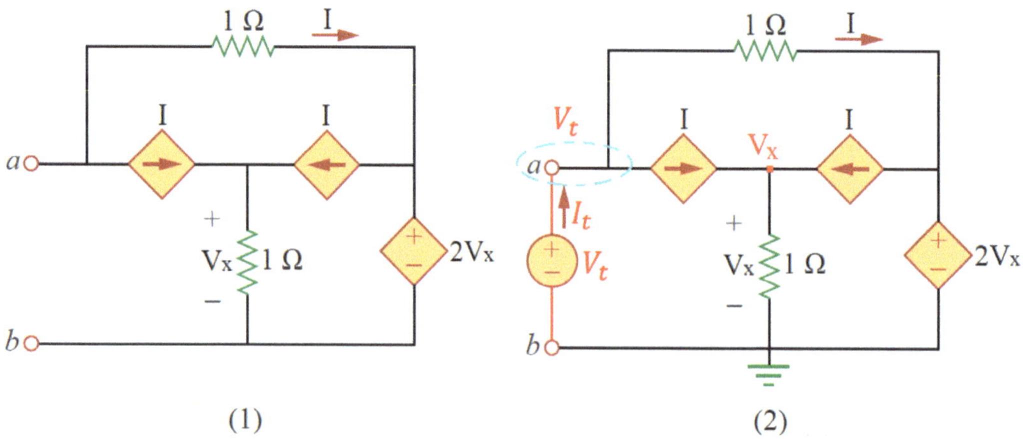

Fig. 5.12 The circuits partially solved exercise 5.11

Problem

5.12. Calculate the input resistance seen from terminal A-B of the circuit of Fig. 5.13.

Difficulty level ○ Easy ● Normal ○ Hard
Calculation amount ○ Small ● Normal ○ Large

1) 3 R + 3.2
2) 2 R − 1.5
3) 1.4 R + 3
4) 2.4 R + 2

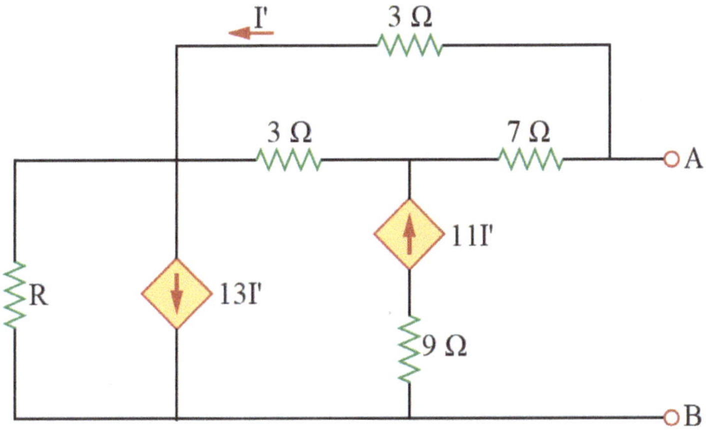

Fig. 5.13 The circuit of problem 5.12

Problem

5.13. Calculate the equivalent resistance seen from the terminal in the circuit of Fig. 5.14.

Difficulty level ○ Easy ● Normal ○ Hard
Calculation amount ○ Small ○ Normal ● Large

1) $\frac{2+\beta}{2-\beta}R_L$
2) $\frac{\beta+2}{\beta-2}R_L$
3) $\frac{2-\beta}{2+\beta}R_L$
4) $\frac{\beta-2}{\beta+2}R_L$

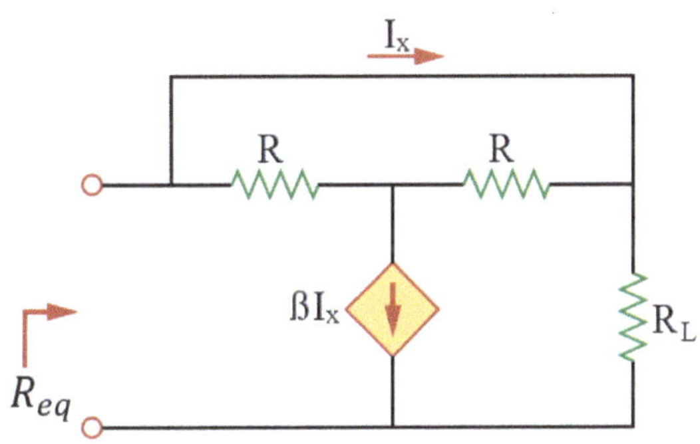

Fig. 5.14 The circuit of problem 5.13

Problem

5.14. Calculate the input resistance seen from terminal a-b in the circuit of Fig. 5.15.

Difficulty level ○ Easy ○ Normal ● Hard
Calculation amount ○ Small ● Normal ○ Large

1) $\frac{5}{11}\ \Omega$
2) $1\ \Omega$
3) $3\ \Omega$
4) $\frac{39}{11}\ \Omega$

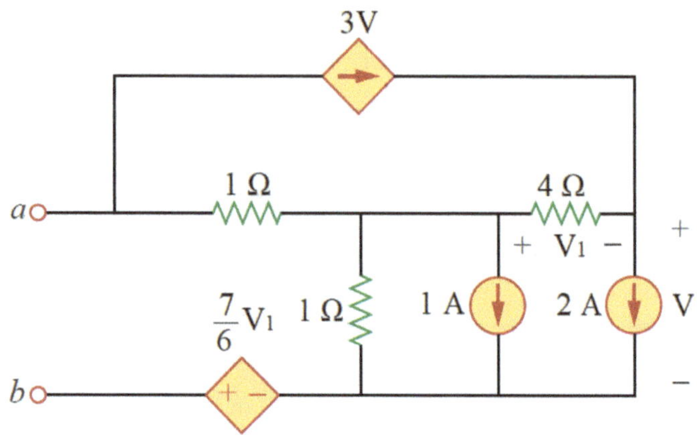

Fig. 5.15 The circuit of problem 5.14

Problem

5.15. In the circuit of Fig. 5.16, calculate the value of input resistance seen from terminal a-b.

Difficulty level ○ Easy ● Normal ○ Hard
Calculation amount ○ Small ● Normal ○ Large

1) $\frac{1-m}{1-2m}$
2) $\frac{4-2m}{4-m}$
3) $\frac{4-m}{11-2m}$
4) $\frac{11-2m}{4-m}$

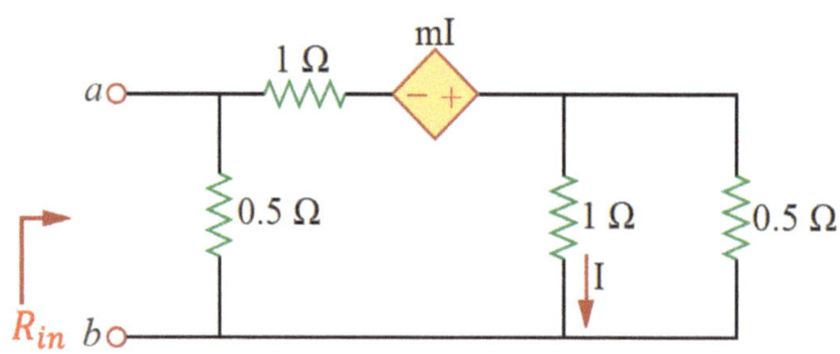

Fig. 5.16 The circuit of problem 5.15

Exercise

5.16. In Problem 5.15, for what range of "m", the value of input resistance seen from the terminal is negative?

Final Answer

$4 < m < 5.5$

Note: Read the "educational support" part of the preface of the textbook.

Problem

5.17. In the circuit of Fig. 5.17, calculate the input resistance seen from terminal a-b.

Difficulty level ○ Easy ○ Normal ● Hard
Calculation amount ○ Small ● Normal ○ Large

1) $\dfrac{R}{\alpha R + 1}$
2) $\dfrac{1}{\alpha - 1}$
3) $\dfrac{R}{(\alpha - 1)R + 1}$
4) $\dfrac{R}{\alpha R - 1}$

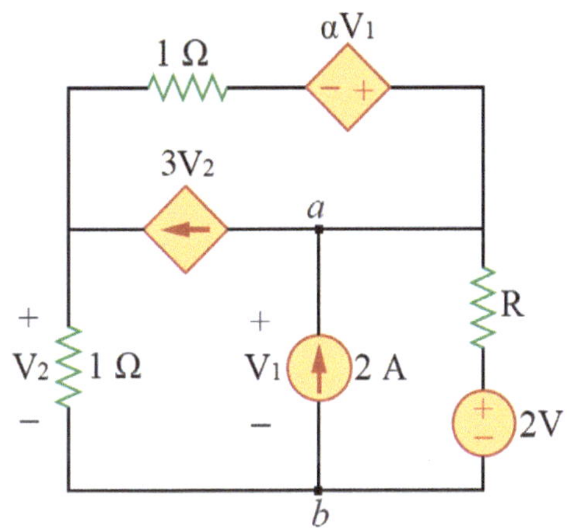

Fig. 5.17 The circuit of problem 5.17

Exercise

5.18. In Problem 5.17, for what value of α, the input resistance seen from the terminal will be equal to R? In other words, $R_{in} = R$.

Final Answer

$\alpha = 1$

Note: Read the "educational support" part of the preface of the textbook.

Problem

5.19. Calculate the input resistance in the circuit of Fig. 5.18.

Difficulty level	○ Easy	○ Normal	● Hard
Calculation amount	○ Small	○ Normal	● Large

1) $\dfrac{3-2\alpha}{3+\alpha}$

2) $\dfrac{2R(3-2\alpha)}{3+\alpha}$

3) $\dfrac{3+\alpha}{R(3-2\alpha)}$

4) $\dfrac{R(3-2\alpha)}{3+\alpha}$

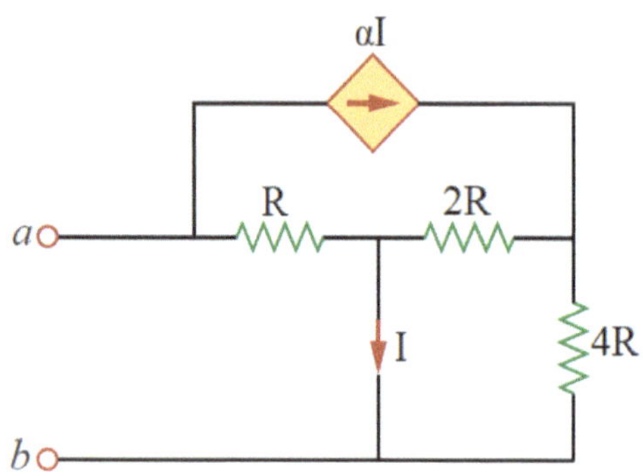

Fig. 5.18 The circuit of problem 5.19

Exercise

5.20. In Problem 5.19, calculate the value of input resistance if α goes to infinity ($\alpha \to \infty$). In other words, calculate the value of the following term.

$$\lim_{\alpha\to\infty} R_{in}$$

Final Answer

$\lim_{\alpha\to\infty} R_{in} = -2\text{ R}$

Note: *Read the "educational support" part of the preface of the textbook.*

5.2 Thevenin Theorem

Problem

5.21. Calculate the Thevenin voltage and Thevenin resistance of the circuit of Fig. 5.19 seen from terminal a-b.

Difficulty level	○ Easy	● Normal	○ Hard
Calculation amount	● Small	○ Normal	○ Large

1) $V_{Th} = 15$ V, $R_{Th} = 2\ \Omega$
2) $V_{Th} = 12$ V, $R_{Th} = 4\ \Omega$
3) $V_{Th} = 12$ V, $R_{Th} = 5\ \Omega$
4) $V_{Th} = 15$ V, $R_{Th} = 5\ \Omega$

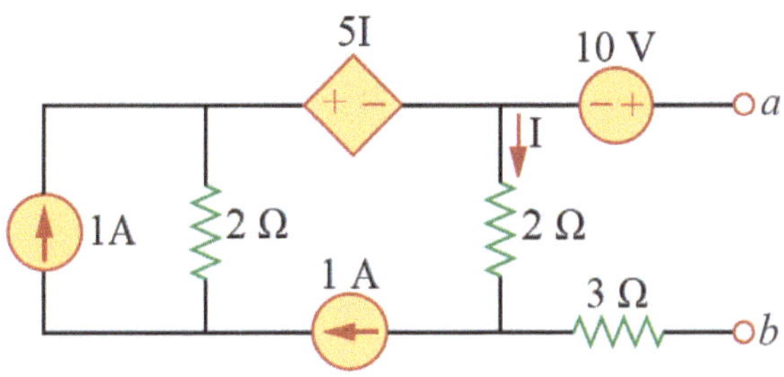

Fig. 5.19 The circuit of problem 5.21

Problem

5.22. Find a relation between V_t and I_t in the circuit of Fig. 5.20.

Difficulty level ○ Easy ● Normal ○ Hard
Calculation amount ○ Small ● Normal ○ Large

1) $V_t = 3I_t + 20$
2) $V_t = 5I_t + 30$
3) $V_t = 7I_t + 10$
4) $V_t = 7I_t + 20$

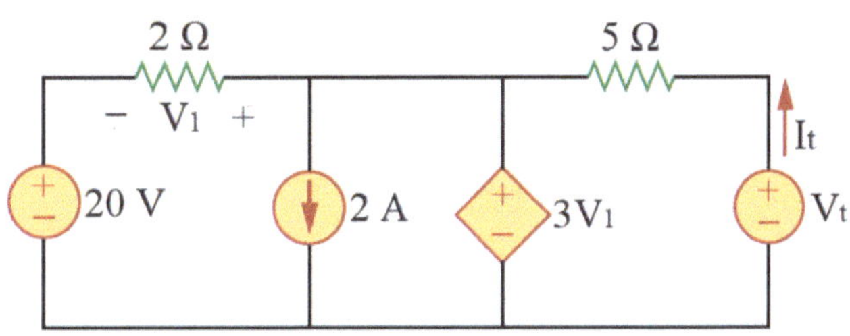

Fig. 5.20 The circuit of problem 5.22

Partially Solved Exercise

5.23. Calculate the parameters of Thevenin equivalent circuit seen from terminal a-b in the circuit of Fig. 5.21.

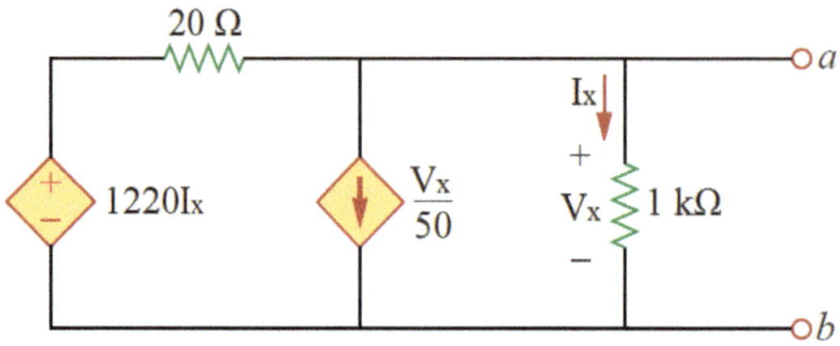

Fig. 5.21 The circuit of partially solved exercise 5.23

Solution

Step 1 (determining the method of solution)

Method 1: First, a test source (independent voltage or current source) with the voltage and current of V_t and I_t is connected to the input terminal of the circuit. Then, the circuit is analyzed by using a circuit analysis method to find a relation between the voltage and current of the test source in the following form.

$$V_t = \alpha I_t + \beta$$

Then, it is concluded that:

$$R_{Th} = \alpha$$

$$V_{Th} = \beta$$

As can be seen, in this method, the value of V_{Th} and R_{Th} are simultaneously calculated.

Method 2: In this method, the value of V_{Th} and R_{Th} are separately calculated. To calculate the value of V_{Th}, the terminal is assumed to be an open circuit branch. Then, the circuit is analyzed by using a circuit analysis method to find the value of V_{oc}. Next, we have:

$$V_{Th} = V_{oc}$$

$$R_{Th} = R_{in} = \frac{V_t}{I_t}$$

Moreover, to calculate R_{Th}, which is the same as input resistance (R_{in}), first, all the independent sources (independent voltage and current sources) are turned off (each independent voltage source is replaced by short circuit branch and each independent current source is replaced by an open circuit branch). Then, if the circuit does not include any dependent source (dependent voltage or current source), the series–parallel rule is applied to calculate the input resistance of the circuit. However, if the circuit includes at least a dependent source, a test source (independent voltage or current source) with the voltage and current of V_t and I_t is connected to the input terminal of the circuit. Next, the circuit is analyzed by using a circuit analysis method to find the value of $\frac{V_t}{I_t}$. After that, we have:

In this problem, the second method is applied because it is more straightforward.

Step 2 (solving the problem)

Calculating V_{Th}: Since there is no independent source in the circuit, it is concluded that:

$$V_{oc} = 0\,\text{V}$$

$$\Rightarrow V_{Th} = 0\,\text{V}$$

Calculating R_{Th}: KCL in the top node:

$$(\quad)+(\quad)+\frac{(\quad)}{(\quad)}+\frac{(\quad)-(\quad)}{(\quad)}=0 \quad (1)$$

We need to define I_x based on the favorable variables (V_t, I_t to find $R_{Th}=\frac{V_t}{I_t}$). Thus, by using Ohm's law, we have:

$$I_x=\frac{(\quad)}{(\quad)} \quad (2)$$

Moreover, from the circuit, it is noticed that:

$$V_x=V_t \quad (3)$$

Now, the problem can be solved as follows.

Solving (1)-(3): Fig. 5.22

$$-I_t+\frac{V_t}{1000}+\frac{V_t}{50}+\frac{V_t-1220\frac{V_t}{1000}}{20}=0$$

$$\Rightarrow -I_t+V_t\left(\frac{1}{1000}+\frac{1}{50}+\frac{1-1220\frac{1}{1000}}{20}\right)=0$$

$$\Rightarrow -I_t+V_t(\quad)=0$$

$$\Rightarrow V_t(\quad)=I_t$$

$$\Rightarrow \frac{V_t}{I_t}=(\quad)$$

$$\Rightarrow R_{in}=(\quad)$$

$$\Rightarrow R_{Th}=100\ \Omega$$

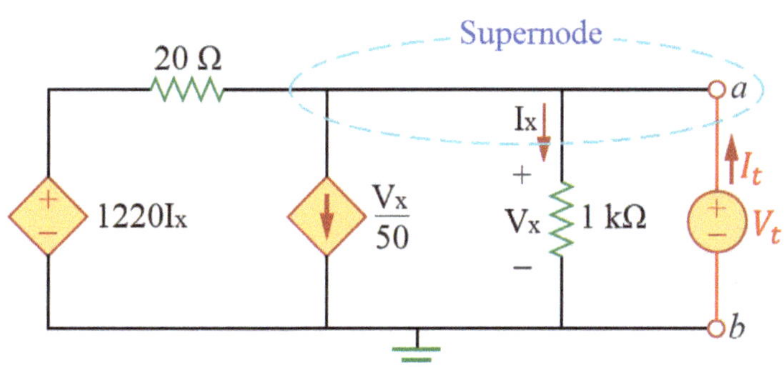

Fig. 5.22 The circuit of partially solved exercise 5.23

Problem

5.24. Calculate the Thevenin voltage seen from the terminal of the circuit of Fig. 5.23.

Difficulty level ○ Easy • Normal ○ Hard
Calculation amount ○ Small • Normal ○ Large

1) 3 V
2) -3 V
3) $\frac{1}{3}$ V
4) $-\frac{1}{3}$ V

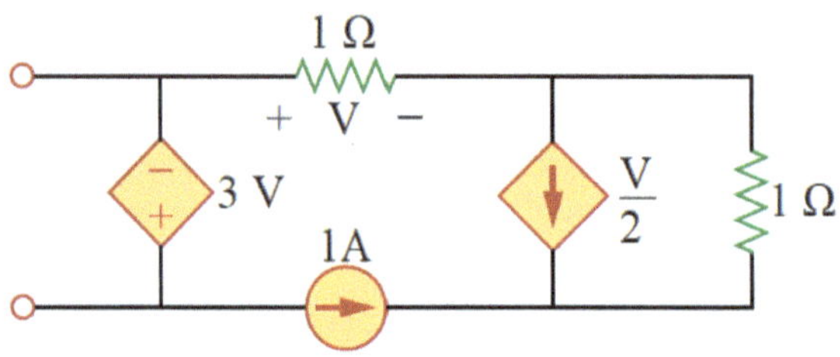

Fig. 5.23 The circuit of problem 5.24

Partially Solved Exercise

5.25. Calculate the parameters of the Thevenin equivalent circuit seen from terminal a-b in the circuit of Fig. 5.24.

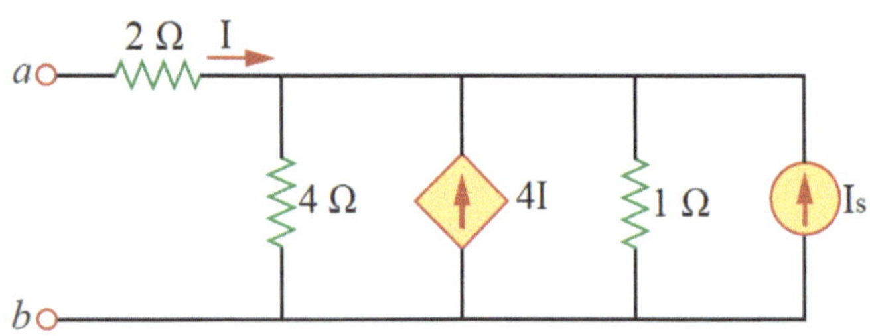

Fig. 5.24 The circuit of partially solved exercise 5.25

Solution

Step 1 (determining the method of solution)

Method 1: First, a test source (independent voltage or current source) with the voltage and current of V_t and I_t is connected to the input terminal of the circuit. Then, the circuit is analyzed by using a circuit analysis method to find a relation between the voltage and current of the test source in the following form.

$$V_t = \alpha I_t + \beta$$

Then, it is concluded that:

$$R_{Th} = \alpha$$

$$V_{Th} = \beta$$

As can be seen, in this method, the value of V_{Th} and R_{Th} are simultaneously calculated.

Method 2: In this method, the value of V_{Th} and R_{Th} are separately calculated. To calculate the value of V_{Th}, the terminal is assumed to be an open circuit branch. Then, the circuit is analyzed by using a circuit analysis method to find the value of V_{oc}. Next, we have:

$$V_{Th} = V_{oc}$$

Moreover, to calculate R_{Th}, which is the same as input resistance (R_{in}), first, all the independent sources (independent voltage and current sources) are turned off (each independent voltage source is replaced by short circuit branch and each independent current source is replaced by an open circuit branch). Then, if the circuit does not include any dependent source (dependent voltage or current source), the series–parallel rule is applied to calculate the input resistance of the circuit. However, if the circuit includes at least a dependent source, a test source (independent voltage or current source) with the voltage and current of V_t and I_t is connected to the input terminal of the circuit. Next, the circuit is analyzed by using a circuit analysis method to find the value of $\frac{V_t}{I_t}$. After that, we have:

$$R_{Th} = R_{in} = \frac{V_t}{I_t}$$

In this problem, the second method is applied because it is more straightforward.

Step 2 (solving the problem)

First, we should combine the parallel 4-Ω and 1-Ω resistors to simplify the analysis of the circuit, as is shown.

Calculating V_{Th}: KCL in the indicated node in Fig. 5.25.2:

$$(\qquad) + \frac{(\qquad)}{(\qquad)} + (\qquad) + (\qquad) = 0$$

$$\Rightarrow \frac{5}{4}e - 5I - I_s = 0 \quad (1)$$

The terminal is an open circuit branch. Hence:

$$I = 0 \quad (2)$$

Solving (1) and (2):

$$(\qquad)e + (\qquad)I_s = 0$$

$$\Rightarrow e = \frac{4}{5}I_s \quad (3)$$

KVL in the left-side mesh:

$$(\qquad) + (\qquad) + (\qquad) = 0 \quad (4)$$

Solving (2)-(4):

$$V_{oc} = (\quad) I_s$$

$$\Rightarrow V_{Th} = \frac{4}{5} I_s$$

Calculating R_{Th}: As can be noticed from the circuit of Fig. 5.25.3, we have $I = I_t$.

KCL in the indicated node:

$$(\quad) + \frac{(\quad)}{(\quad)} + (\quad) = 0$$

$$\Rightarrow e = 4I_t \quad (5)$$

KCL in node "a":

$$(\quad) + \frac{(\quad) - (\quad)}{(\quad)} = 0 \quad (6)$$

Solving (5) and (6):

$$(\quad) + \frac{(\quad) - (\quad)}{(\quad)} = 0$$

$$\Rightarrow (\quad) I_t + (\quad) V_t = 0$$

$$\Rightarrow \frac{V_t}{I_t} = 6$$

$$\Rightarrow R_{Th} = 6\ \Omega$$

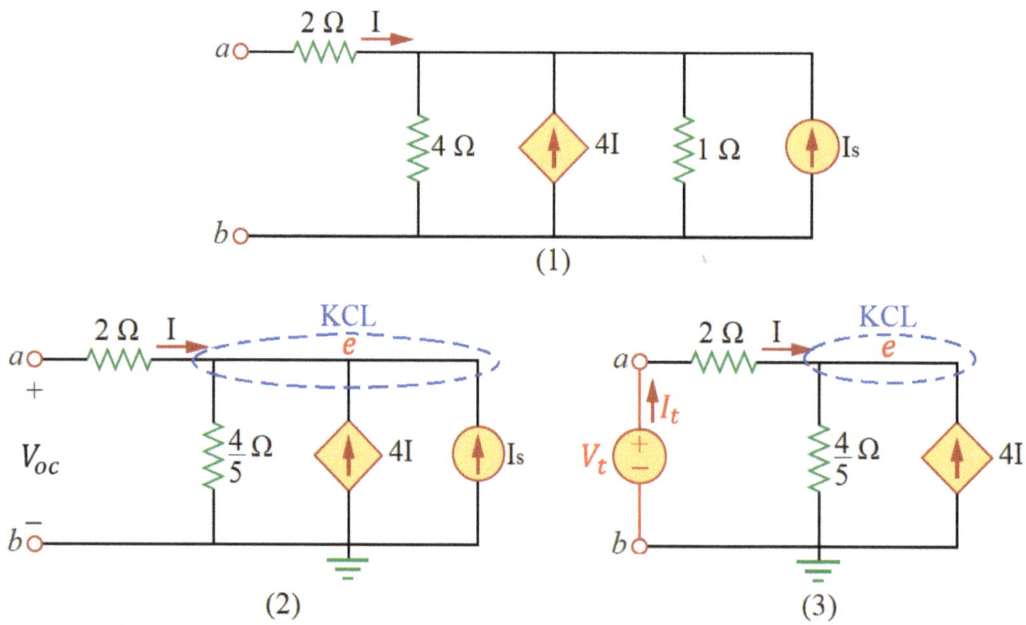

Fig. 5.25 The circuits of partially solved exercise 5.25

Problem

5.26. Calculate the Thevenin voltage and resistance seen from the terminal in the circuit of Fig. 5.26.

Difficulty level ○ Easy ● Normal ○ Hard
Calculation amount ○ Small ● Normal ○ Large

1) $V_{Th} = 2\ \text{V},\ R_{Th} = 5\ \Omega$
2) $V_{Th} = 5\ \text{V},\ R_{Th} = 2\ \Omega$
3) $V_{Th} = 25\ \text{V},\ R_{Th} = 20\ \Omega$
4) $V_{Th} = 25\ \text{V},\ R_{Th} = 25\ \Omega$

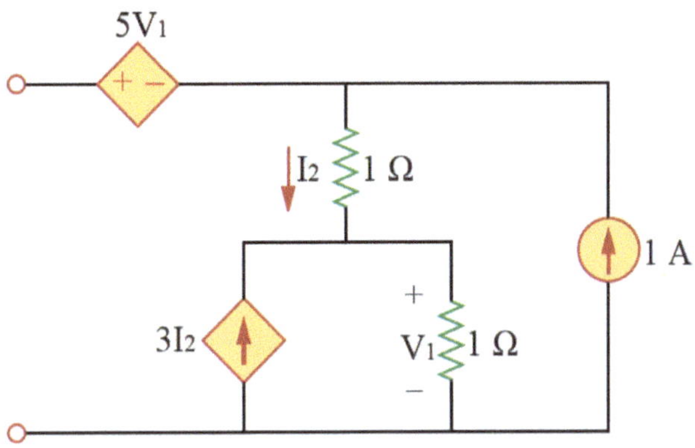

Fig. 5.26 The circuit of problem 5.26

Partially Solved Exercise

5.27. Calculate the Thevenin voltage and Thevenin resistance seen from terminal a-b in the circuit of Fig. 5.27.

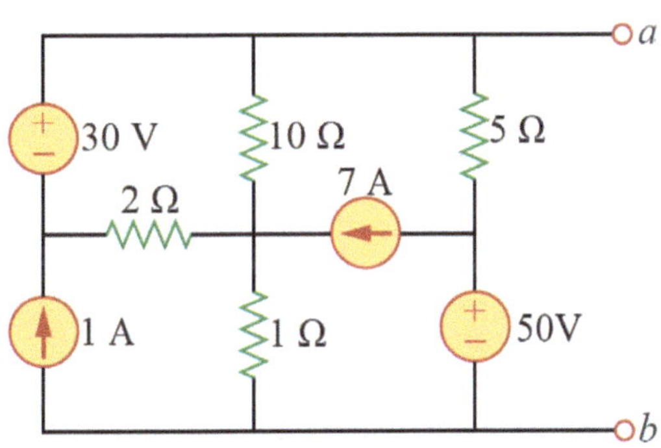

Fig. 5.27 The circuit of partially solved exercise 5.27

Solution

Step 1 (determining the method of solution)

Method 1: First, a test source (independent voltage or current source) with the voltage and current of V_t and I_t is connected to the input terminal of the circuit. Then, the circuit is analyzed by using a circuit analysis method to find a relation between the voltage and current of the test source in the following form.

$$V_t = \alpha I_t + \beta$$

Then, it is concluded that:

$$R_{Th} = \alpha$$

$$V_{Th} = \beta$$

As can be seen, in this method, the value of V_{Th} and R_{Th} are simultaneously calculated.

Method 2: In this method, the value of V_{Th} and R_{Th} are separately calculated. To calculate the value of V_{Th}, the terminal is assumed to be an open circuit branch. Then, the circuit is analyzed by using a circuit analysis method to find the value of V_{oc}. Next, we have:

$$V_{Th} = V_{oc}$$

Moreover, to calculate R_{Th}, which is the same as input resistance (R_{in}), first, all the independent sources (independent voltage and current sources) are turned off (each independent voltage source is replaced by short circuit branch and each independent current source is replaced by an open circuit branch). Then, if the circuit does not include any dependent source (dependent voltage or current source), the series–parallel rule is applied to calculate the input resistance of the circuit. However, if the circuit includes at least a dependent source, a test source (independent voltage or current source) with the voltage and current of V_t and I_t is connected to the input terminal of the circuit. Next, the circuit is analyzed by using a circuit analysis method to find the value of $\frac{V_t}{I_t}$. After that, we have:

$$R_{Th} = R_{in} = \frac{V_t}{I_t}$$

In this problem, the second method is applied because it is more straightforward.

Step 2 (solving the problem)

Calculating R_{Th}: Since the circuit does not include any dependent source, the Thevenin resistance of the circuit can be simply calculated by using series–parallel rule while all the independent sources are turned off. As is illustrated in Fig. 5.28.2, the independent voltage sources are replaced by the short circuit branches and the independent current sources are replaced by the open circuit branches.

$$R_{in} = (\quad)||((\quad) + (\quad)||(\quad))$$

$$\Rightarrow R_{in} = (\quad)||\left((\quad) + \frac{(\quad)\times(\quad)}{(\quad)+(\quad)}\right)$$

$$\Rightarrow R_{in} = (\quad)||((\quad) + (\quad))$$

$$\Rightarrow R_{in} = (\quad)||(\quad)$$

$$\Rightarrow R_{in} = \frac{(\quad)\times(\quad)}{(\quad)+(\quad)}$$

$$\Rightarrow R_{in} = \frac{\frac{(\quad)}{(\quad)}}{\frac{(\quad)}{(\quad)}}\ \Omega$$

$$\Rightarrow R_{Th} = \frac{40}{23}\ \Omega$$

Calculating V_{Th}: Finding a relation between the voltage of the 30-V independent voltage source and the node voltages:

$$(\quad) - (\quad) = (\quad)$$

$$\Rightarrow e_1 = V_{oc} - 30 \quad (1)$$

KCL in the supernode:

$$(\quad) + \frac{(\quad) - (\quad)}{(\quad)} + \frac{(\quad) - (\quad)}{(\quad)} + \frac{(\quad) - (\quad)}{(\quad)} = 0$$

$$\Rightarrow \frac{1}{2}e_1 - \frac{6}{10}e_2 + \frac{3}{10}V_{oc} - 11 = 0 \quad (2)$$

KCL in node 2:

$$\frac{(\quad) - (\quad)}{(\quad)} + \frac{(\quad)}{(\quad)} + (\quad) + \frac{(\quad) - (\quad)}{(\quad)} = 0$$

$$\Rightarrow -\frac{1}{2}e_1 + \frac{16}{10}e_2 - \frac{1}{10}V_{oc} - 7 = 0 \quad (3)$$

Now, the equations can be solved as follows:

Solving (1) and (2):

$$\frac{1}{2}(V_{oc} - 30) - \frac{6}{10}e_2 + \frac{3}{10}V_{oc} - 11 = 0$$

$$\Rightarrow -\frac{6}{10}e_2 + \frac{8}{10}V_{oc} - 26 = 0 \quad (4)$$

Solving (1) and (3):

$$-\frac{1}{2}(V_{oc} - 30) + \frac{16}{10}e_2 - \frac{1}{10}V_{oc} - 7 = 0$$

$$\Rightarrow \frac{16}{10}e_2 - \frac{6}{10}V_{oc} + 8 = 0 \quad (5)$$

Solving (4) and (5):

$$-\frac{6}{10}\left(\frac{10}{16}\right)\left(\frac{6}{10}V_{oc} - 8\right) + \frac{8}{10}V_{oc} - 26 = 0$$

$$\Rightarrow -\frac{3}{8} \times \frac{6}{10}V_{oc} + \frac{3}{8} \times 8 + \frac{8}{10}V_{oc} - 26 = 0$$

$$\Rightarrow (\quad)V_{oc} + (\quad) + (\quad)V_{oc} + (\quad) = 0$$

$$\Rightarrow (\quad)V_{oc} + (\quad) = 0$$

$$\Rightarrow V_{oc} = (\quad)$$

$$\Rightarrow V_{Th} = 40 \text{ V}$$

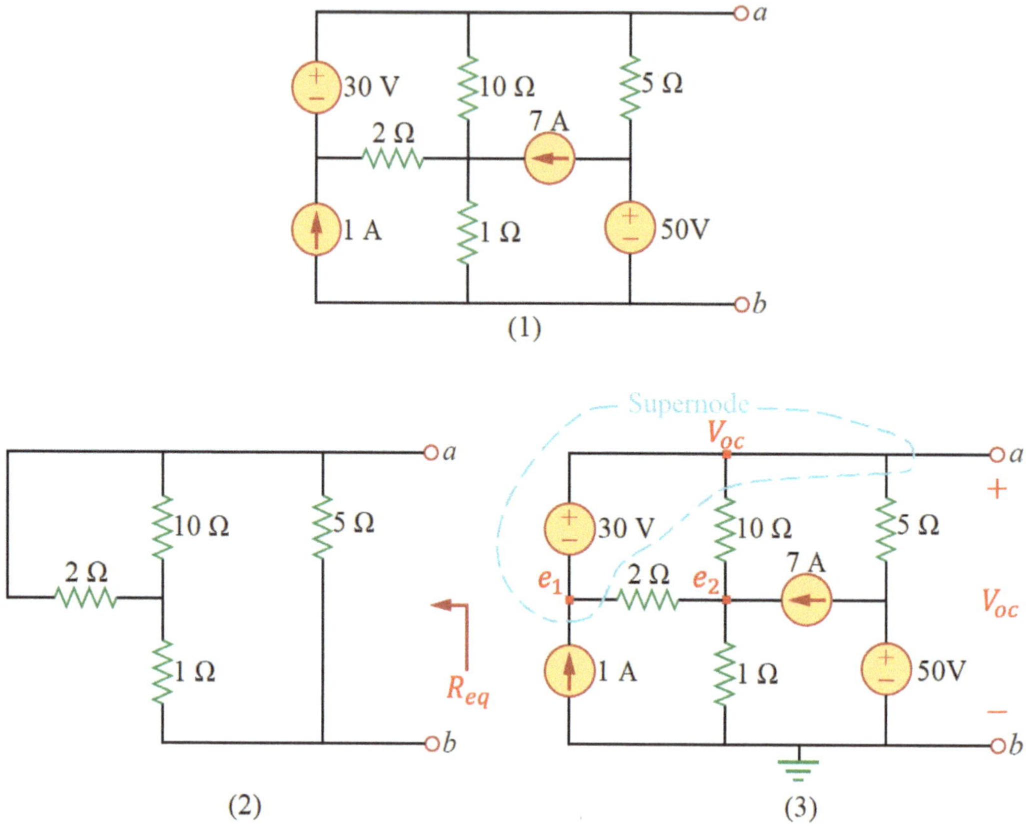

Fig. 5.28 The circuits of partially solved exercise 5.27

Problem

5.28. Calculate the Thevenin voltage and resistance seen from terminal a-b in the circuit of Fig. 5.29.

Difficulty level ○ Easy ● Normal ○ Hard
Calculation amount ○ Small ● Normal ○ Large

1) $V_{Th} = 8\ \text{V},\ R_{Th} = 6\ \Omega$
2) $V_{Th} = 12\ \text{V},\ R_{Th} = 6\ \Omega$
3) $V_{Th} = 16\ \text{V},\ R_{Th} = 2\ \Omega$
4) $V_{Th} = 24\ \text{V},\ R_{Th} = 4\ \Omega$

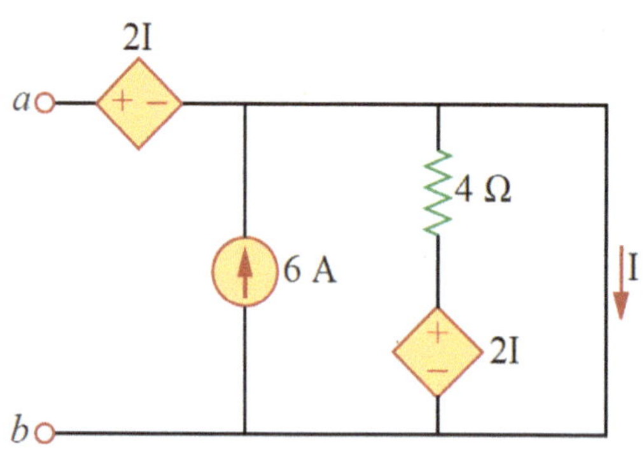

Fig. 5.29 The circuit of problem 5.28

Partially Solved Exercise

5.29. Calculate the Thevenin voltage and the resistance seen from terminal a-b in the circuit of Fig. 5.30.

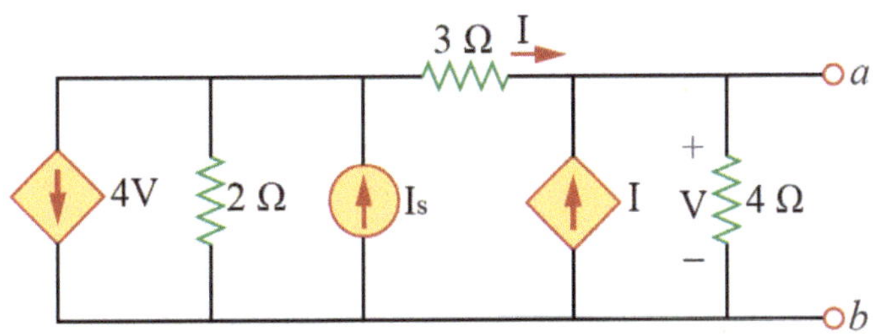

Fig. 5.30 The circuit of partially solved exercise 5.29

Solution

Step 1 (determining the method of solution)

Method 1: First, a test source (independent voltage or current source) with the voltage and current of V_t and I_t is connected to the input terminal of the circuit. Then, the circuit is analyzed by using a circuit analysis method to find a relation between the voltage and current of the test source in the following form.

$$V_t = \alpha I_t + \beta$$

Then, it is concluded that:

$$R_{Th} = \alpha$$

$$V_{Th} = \beta$$

As can be seen, in this method, the value of V_{Th} and R_{Th} are simultaneously calculated.

Method 2: In this method, the value of V_{Th} and R_{Th} are separately calculated. To calculate the value of V_{Th}, the terminal is assumed to be an open circuit branch. Then, the circuit is analyzed by using a circuit analysis method to find the value of V_{oc}. Next, we have:

$$V_{Th} = V_{oc}$$

Moreover, to calculate R_{Th}, which is the same as input resistance (R_{in}), first, all the independent sources (independent voltage and current sources) are turned off (each independent voltage source is replaced by short circuit branch and each independent current source is replaced by an open circuit branch). Then, if the circuit does not include any dependent source (dependent voltage or current source), the series–parallel rule is applied to calculate the input resistance of the circuit. However, if the circuit includes at least a dependent source, a test source (independent voltage or current source) with the voltage and current of V_t and I_t is connected to the input terminal of the circuit. Next, the circuit is analyzed by using a circuit analysis method to find the value of $\frac{V_t}{I_t}$. After that, we have:

$$R_{Th} = R_{in} = \frac{V_t}{I_t}$$

In this problem, the first method is applied because it is more straightforward.

Step 2 (solving the problem)

From the circuit of Fig. 5.31, it is noticed that:

$$V = V_t \quad (1)$$

KCL in the left-side node:

$$(\quad) + \frac{(\quad)}{(\quad)} + (\quad) + \frac{(\quad) - (\quad)}{(\quad)} = 0$$

$$\Rightarrow 4V + \frac{5}{6}e_1 - I_s - \frac{V_t}{3} = 0 \quad (2)$$

Using Ohm's law for the 3-Ω resistor to define I based on the node voltages:

$$I = \frac{(\quad) - (\quad)}{(\quad)} \quad (3)$$

KCL in the right-side node:

$$(\quad) + \frac{(\quad)}{(\quad)} + (\quad) + \frac{(\quad) - (\quad)}{(\quad)} = 0$$

$$\Rightarrow -I_t + \frac{7}{12}V_t - I - \frac{e_1}{3} = 0 \quad (4)$$

Solving (1) and (2):

$$4V_t + \frac{5}{6}e_1 - I_s - \frac{V_t}{3} = 0$$

$$\Rightarrow \frac{5}{6}e_1 = I_s - \frac{11}{3}V_t$$

$$\Rightarrow e_1 = \frac{6}{5}I_s - \frac{66}{15}V_t \quad (5)$$

Solving (3) and (4):

$$-I_t + \frac{7}{12}V_t - \frac{e_1 - V_t}{3} - \frac{e_1}{3} = 0$$

$$\Rightarrow -I_t + \frac{11}{12}V_t - \frac{2}{3}e_1 = 0 \quad (6)$$

Solving (5) and (6):

$$-I_t + \frac{11}{12}V_t - \frac{2}{3}\left(\frac{6}{5}I_s - \frac{66}{15}V_t\right) = 0$$

$$\Rightarrow -I_t + \frac{11}{12}V_t - \frac{4}{5}I_s + \frac{44}{15}V_t = 0$$

$$\Rightarrow (\quad)V_t = (\quad)I_t + (\quad)$$

$$\Rightarrow V_t = (\quad)I_t + (\quad)I_s$$

Therefore, the following results are concluded.

$$V_{Th} = \frac{16}{77} I_s$$

$$R_{Th} = \frac{20}{77}$$

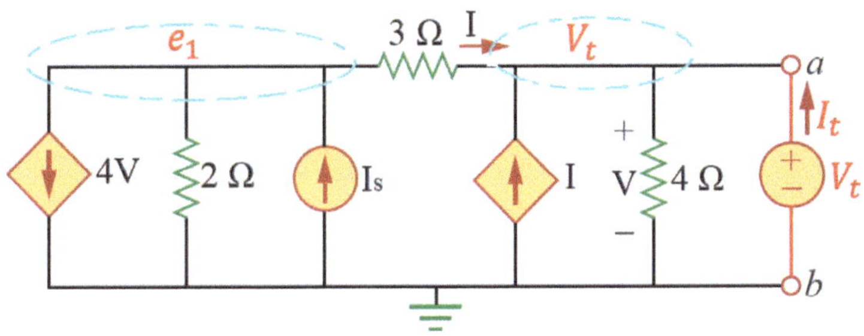

Fig. 5.31 The circuit of partially solved exercise 5.29

Problem

5.30. In the circuit of Fig. 5.32, determine the structure of the Thevenin equivalent circuit seen from terminal A-B.

Difficulty level ○ Easy ● Normal ○ Hard
Calculation amount ○ Small ● Normal ○ Large

1) An independent voltage source with the size of $\frac{5}{3} I_s$
2) An independent voltage source with the size of $\frac{5}{3} I_s$ series with a $\frac{4}{5}$ Ω resistor
3) An independent voltage source with the size of $\frac{3}{5} I_s$
4) An independent voltage source with the size of $\frac{5}{3} I_s$ series with a $\frac{4}{5}$ Ω resistor

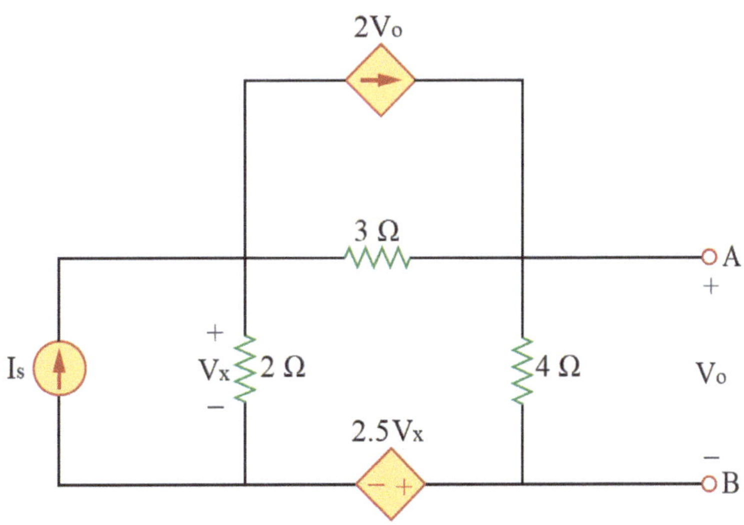

Fig. 5.32 The circuit of problem 5.30

Problem

5.31. Calculate the parameters of Thevenin equivalent circuit seen from terminal A-B in the circuit of Fig. 5.33.

Difficulty level ○ Easy ● Normal ○ Hard
Calculation amount ○ Small ○ Normal ● Large

1) $R_{Th} = -2\ \Omega,\ V_{Th} = -36\ I_s$
2) $R_{Th} = -2\ \Omega,\ V_{Th} = 36\ I_s$
3) $R_{Th} = 2\ \Omega,\ V_{Th} = -36\ I_s$
4) $R_{Th} = 2\ \Omega,\ V_{Th} = 36\ I_s$

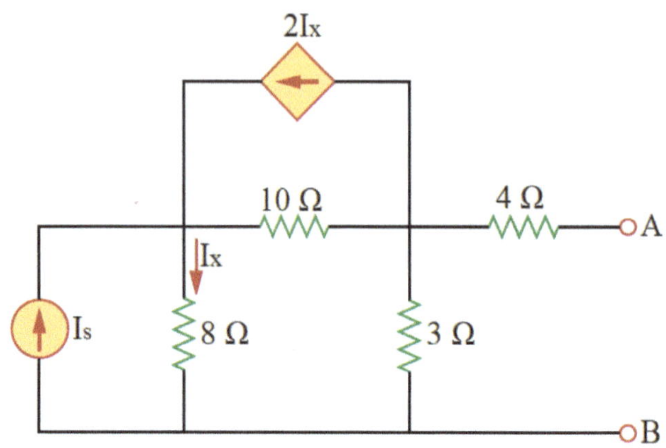

Fig. 5.33 The circuit of problem 5.31

Problem

5.32. In the circuit of Fig. 5.34, calculate the parameters of Thevenin equivalent circuit seen from terminal a-b.

Difficulty level ○ Easy ○ Normal ● Hard
Calculation amount ● Small ○ Normal ○ Large

1) $V_{Th} = \frac{5}{6}\ \text{E},\ R_{Th} = \frac{5}{6}\ \text{R}$
2) $V_{Th} = \frac{5}{6}\ \text{E},\ R_{Th} = 5\ \text{R}$
3) $V_{Th} = \frac{5}{6}\ \text{E},\ R_{Th} = 0$
4) There is no Thevenin equivalent circuit for this circuit.

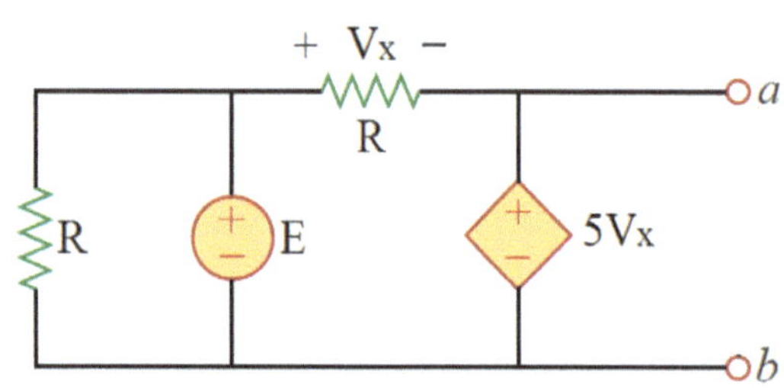

Fig. 5.34 The circuit of problem 5.32

Problem

5.33. In the circuit of Fig. 5.35, calculate the value of R_{Th} and V_{Th}, seen from terminal a-b, if $a \to \infty$.

Difficulty level ○ Easy ○ Normal ● Hard
Calculation amount ○ Small ● Normal ○ Large

1) $\lim_{a\to\infty} R_{Th} = 1\ \Omega,\ \lim_{a\to\infty} V_{Th} = 2\ \text{V}$
2) $\lim_{a\to\infty} R_{Th} = 0\ \Omega,\ \lim_{a\to\infty} V_{Th} = -2\ \text{V}$
3) $\lim_{a\to\infty} R_{Th} = 0\ \Omega,\ \lim_{a\to\infty} V_{Th} = 1\ \text{V}$
4) $\lim_{a\to\infty} R_{Th} = 1\ \Omega,\ \lim_{a\to\infty} V_{Th} = -1\ \text{V}$

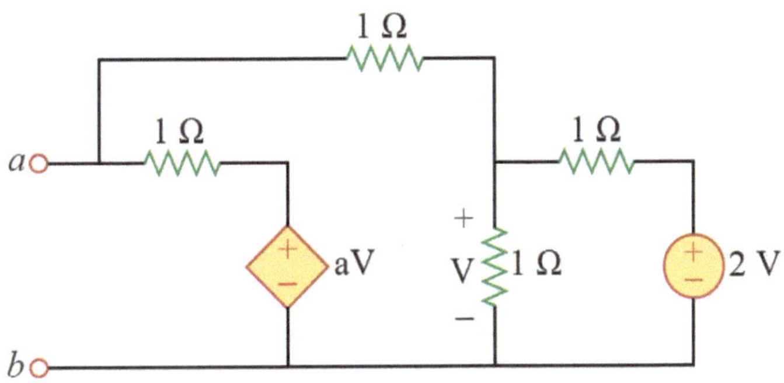

Fig. 5.35 The circuit of problem 5.33

Exercise

5.34. Calculate the Thevenin voltage and resistance seen from the terminal in the circuit of Fig. 5.36.

Final Answer

$V_{Th} = 2\ \text{V},\ R_{Th} = 0.5\ \Omega$

Note: Read the "educational support" part of the preface of the textbook.

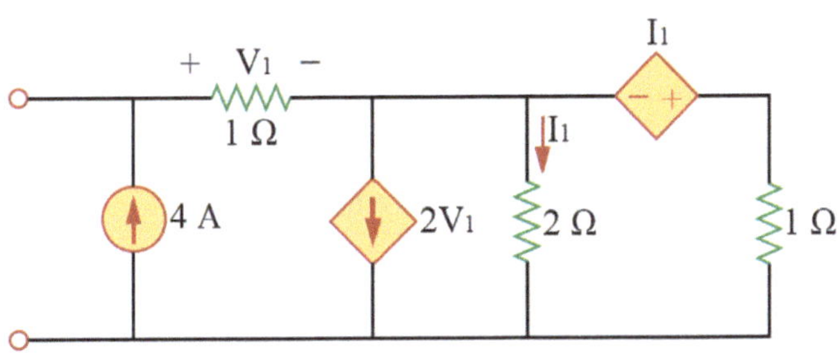

Fig. 5.36 The circuit of exercise 5.34

5.3 Norton Theorem

Problem

5.35. Calculate the Norton resistance and Norton current of a circuit if its Thevenin resistance and Thevenin voltage are 2Ω and $3V$, respectively.

Difficulty level ● Easy ○ Normal ○ Hard
Calculation amount ● Small ○ Normal ○ Large

1) $R_N = 2\ \Omega,\ I_N = \frac{3}{2}$ A
2) $R_N = 2\ \Omega,\ I_N = 6$ A
3) $R_N = 6\ \Omega,\ I_N = \frac{3}{2}$ A
4) Impossible to calculate

Problem

5.36. In the circuit of Fig. 5.37, calculate the Norton resistance seen from terminal a-b.

Difficulty level ○ Easy ● Normal ○ Hard
Calculation amount ○ Small ● Normal ○ Large

1) 30 Ω
2) 60 Ω
3) 90 Ω
4) 120 Ω

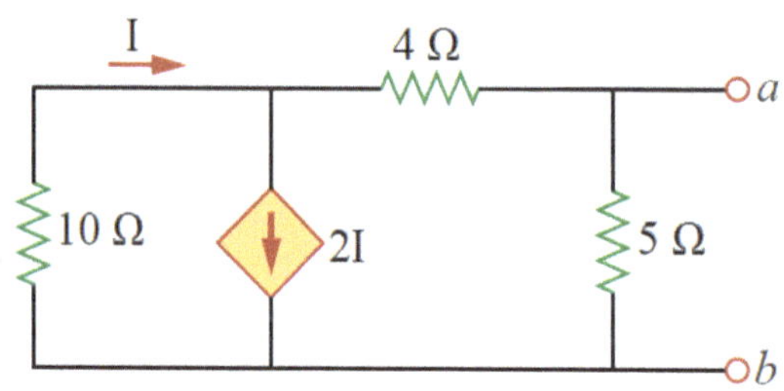

Fig. 5.37 The circuit of problem 5.36

Problem

5.37. In Problem 5.36, calculate the Norton current flowing through terminal a-b.

Difficulty level ● Easy ○ Normal ○ Hard
Calculation amount ● Small ○ Normal ○ Large

1) 30 A
2) 15 A
3) 0 A
4) ∞

Problem

5.38. Which one of the circuits shown below is the Norton equivalent circuit seen from terminal a-b of the circuit of Fig. 5.38?

Difficulty level ○ Easy ● Normal ○ Hard
Calculation amount ○ Small ● Normal ○ Large

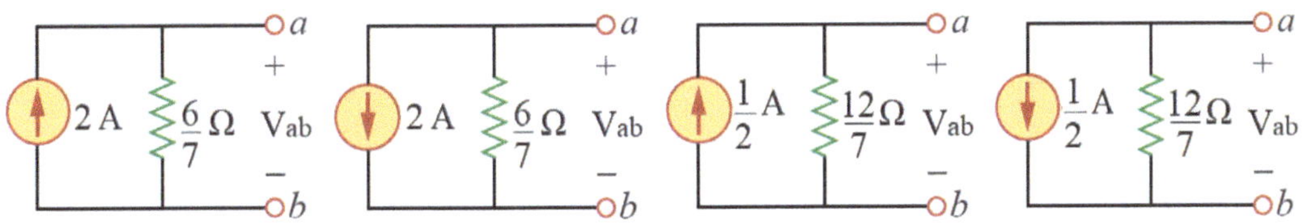

(1) (2) (3) (4)

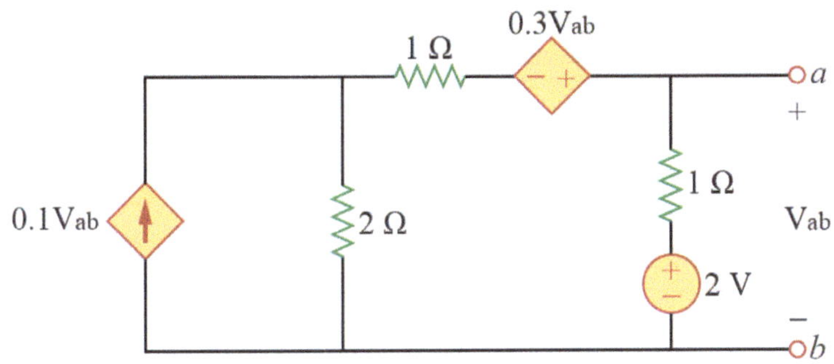

Fig. 5.38 The circuit of problem 5.38

Exercise

5.39. In Problem 5.38, calculate the parameters of Thevenin equivalent circuit.

Final Answer

$V_{Th} = \frac{12}{7}$ V, $R_{Th} = \frac{6}{7}$ Ω

Note: Read the "educational support" part of the preface of the textbook.

Problem

5.40. Calculate the parameters of Norton equivalent circuit seen from terminal a-b in the circuit of Fig. 5.39.

Difficulty level ○ Easy ○ Normal ● Hard
Calculation amount ○ Small ● Normal ○ Large

1) $I_N = 1$ A, $R_N = \frac{4}{3}$ Ω

2) $I_N = 1$ A, $R_N = \frac{7}{3}$ Ω

3) $I_N = \frac{3}{4}$ A, $R_N = \frac{4}{3}$ Ω

4) $I_N = \frac{3}{4}$ A, $R_N = \frac{7}{3}$ Ω

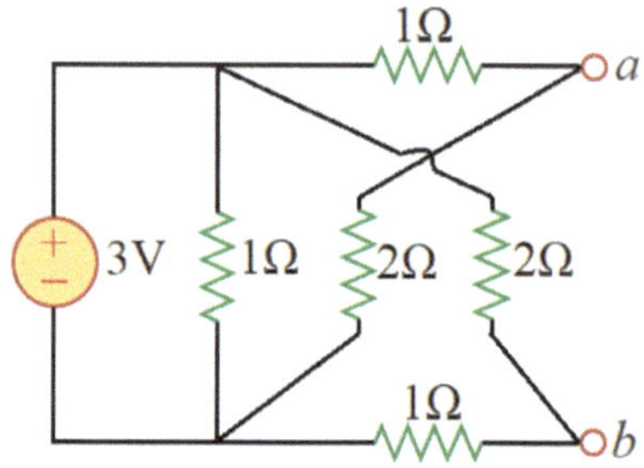

Fig. 5.39 The circuit of problem 5.40

Exercise

5.41. In the circuit of Fig. 5.40, calculate the parameters of Norton equivalent circuit seen by the 7-Ω resistor.

Final Answer

$R_N = \frac{40}{3}\ \Omega,\ I_N = \frac{1}{2}\ \text{A}$

Note: Read the "educational support" part of the preface of the textbook.

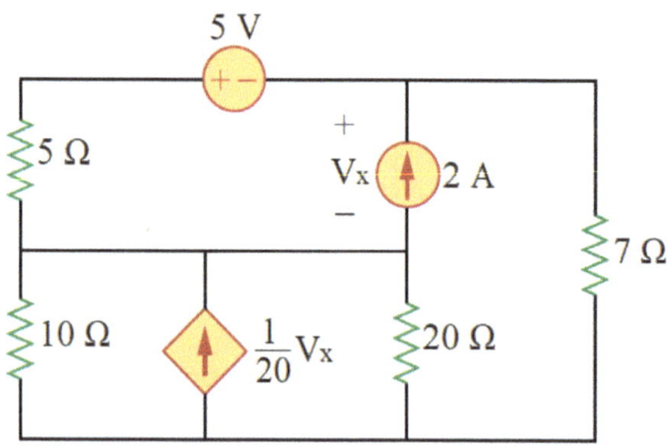

Fig. 5.40 The circuit of exercise 5.41

5.4 Source Transformation Theorem

Problem

5.42. In the circuit of Fig. 5.41, calculate the value of V_{AB} by using source transformation theorem.

Difficulty level ○ Easy ○ Normal ● Hard
Calculation amount ○ Small ○ Normal ● Large

1) 1 V
2) 2 V
3) 0.5 V
4) 0.25 V

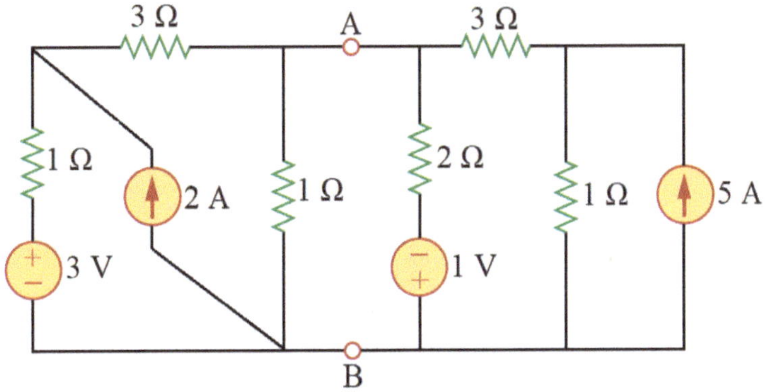

Fig. 5.41 The circuit of problem 5.42

Exercise

5.43. Apply source transformation to calculate the output voltage v_o in the circuit of Fig. 5.42.

Final Answer

$v_o = 2$ V

Note: *Read the "educational support" part of the preface of the textbook.*

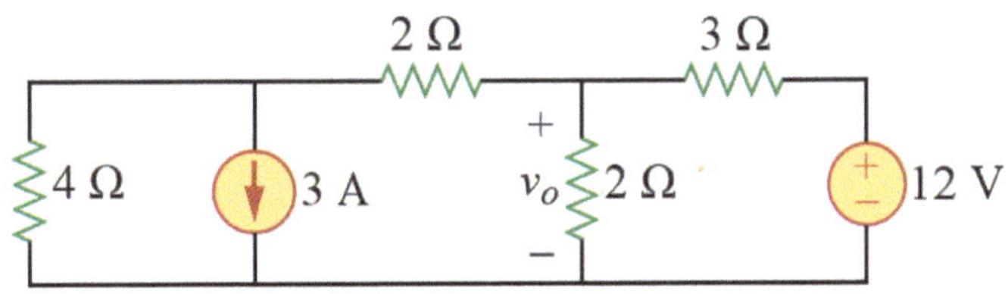

Fig. 5.42 The circuit of exercise 5.43

Problem

5.44. Apply source transformation to calculate the power absorbed by the 2-Ω resistor in the circuit of Fig. 5.43.

Difficulty level	○ Easy	● Normal	○ Hard
Calculation amount	○ Small	● Normal	○ Large

1) 0.08 W
2) 0.32 W
3) 0.64 W
4) 3.92 W

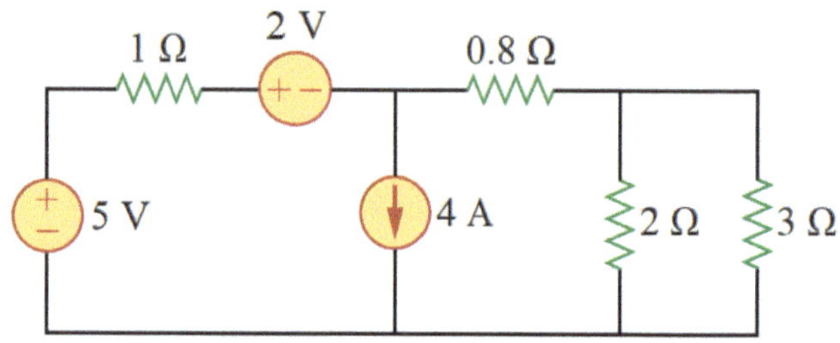

Fig. 5.43 The circuit of problem 5.44

Exercise

5.45. In Problem 5.44, calculate the power absorbed by the 0.8-Ω resistor.

Final Answer

$P = 0.089\ \text{W}$

Note: *Read the "educational support" part of the preface of the textbook.*

Exercise

5.46. Apply source transformation theorem to calculate the Thevenin equivalent circuit seen from terminal a-b in the circuit of Fig. 5.44.

Final Answer

$R_{Th} = \frac{R}{2},\ V_{Th} = \frac{E - IR}{4}$

Note: *Read the "educational support" part of the preface of the textbook.*

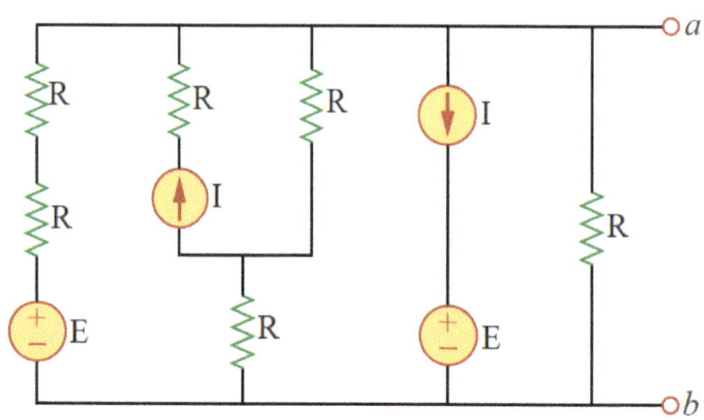

Fig. 5.44 The circuit of exercise 5.46

5.5 Superposition Theorem

Problem

5.47. Apply superposition theorem to calculate the output voltage v_o in the circuit of Fig. 5.45.

Difficulty level ● Easy ○ Normal ○ Hard
Calculation amount ● Small ○ Normal ○ Large

1) 2 V
2) 6 V
3) 3 V
4) 4 V

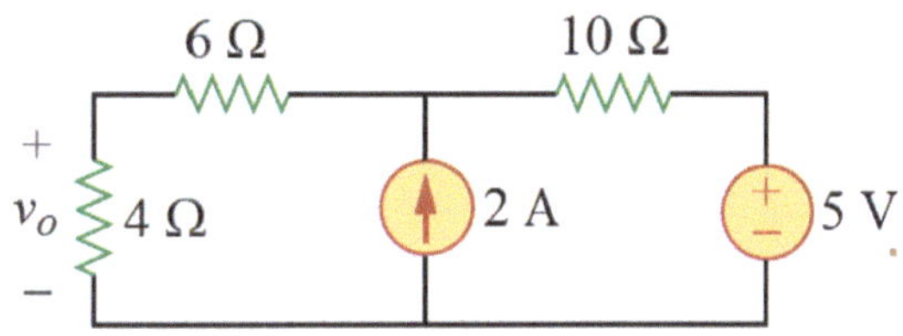

Fig. 5.45 The circuit of problem 5.47

Problem

5.48. Apply superposition theorem to calculate the output current I_o in the circuit of Fig. 5.46.

Difficulty level ● Easy ○ Normal ○ Hard
Calculation amount ● Small ○ Normal ○ Large

1) 4 A
2) 1 A
3) 2.5 A
4) 5 A

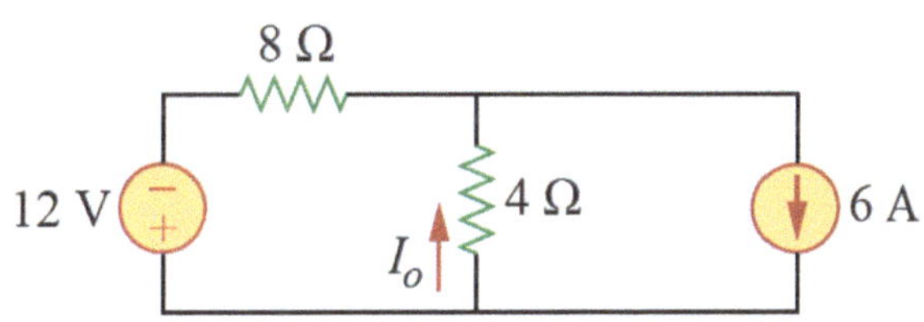

Fig. 5.46 The circuit of problem 5.48

Exercise

5.49. Apply superposition theorem to calculate the current I in the circuit of Fig. 5.47.

Final Answer

$I = 1.5$ A

Note: Read the "educational support" part of the preface of the textbook.

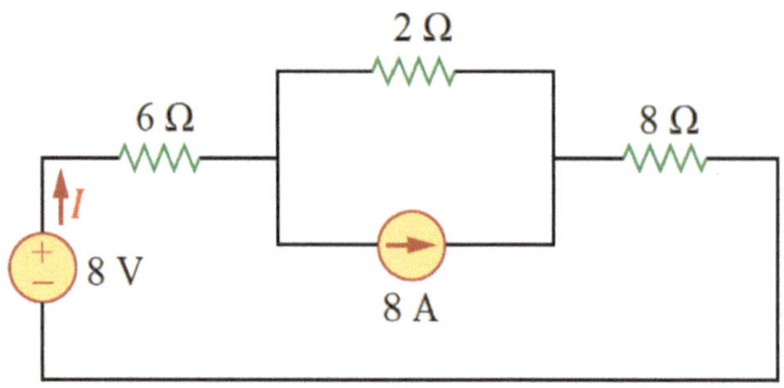

Fig. 5.47 The circuit of exercise 5.49

Partially Solved Exercise

5.50. In the circuit of Fig. 5.48, for what value of k, the voltage V, resulted only from the independent current source I_s, is equal to half value of the magnitude of independent current source. In other words, calculate the value of k so that the following relation is achieved.

$$V\Big|_{V_s=0} = \frac{I_s}{2}$$

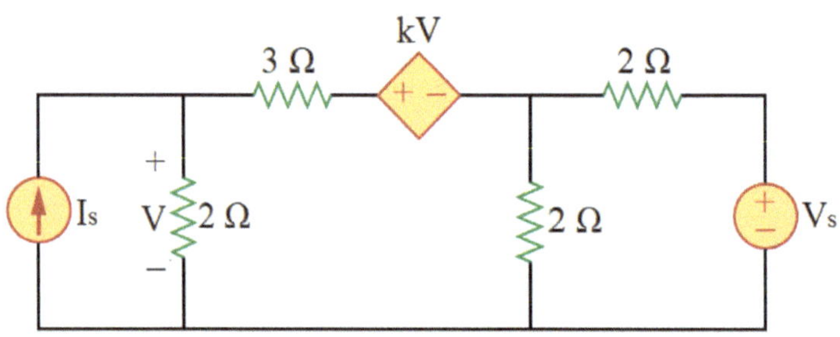

Fig. 5.48 The circuit of partially solved exercise 5.50

Solution

Based on superposition theorem, since we are interested in determining the effect of independent current source on the desirable voltage (V), we must turn off the other independent sources (herein, the independent voltage source is replaced by a short circuit branch) as is illustrated in Fig. 5.49.

In this problem, nodal analysis is the best approach to solve it.

Based on the given information:

$$V\Big|_{V_s=0} = \frac{I_s}{2} \quad (1)$$

KCL in the left-side node:

$$(\quad) + \frac{(\quad)}{(\quad)} + \frac{(\quad) - (\quad)}{(\quad)} = 0$$

$$\Rightarrow -I_s + \frac{5}{6}V - \frac{e_2}{3} = 0 \quad (2)$$

Defining the voltage of dependent voltage source based on the node voltages:

$$(\quad) - (\quad) = (\quad)$$

$$\Rightarrow e_3 = e_2 - kV \quad (3)$$

KCL in the supernode:

$$\frac{(\quad) - (\quad)}{(\quad)} + \frac{(\quad)}{(\quad)} + \frac{(\quad)}{(\quad)} = 0$$

$$\Rightarrow \frac{e_2}{3} - \frac{V}{3} + e_3 = 0 \quad (4)$$

Solving (1) and (2):

$$-I_s + \frac{5}{6} \times (\quad) - \frac{e_2}{3} = 0$$

$$\Rightarrow -\frac{7I_s}{12} - \frac{e_2}{3} = 0$$

$$\Rightarrow e_2 = -\frac{7}{4}I_s \quad (5)$$

Solving (1), (3) and (4):

$$\frac{e_2}{3} - \frac{\left(\frac{I_s}{2}\right)}{3} + e_2 - k\left(\frac{I_s}{2}\right) = 0$$

$$\Rightarrow \frac{4}{3}e_2 = \left(\frac{1}{6} + \frac{k}{2}\right)I_s$$

$$\Rightarrow e_2 = \left(\frac{1}{8} + \frac{3}{8}k\right)I_s \quad (6)$$

Solving (5) and (6):

$$(\quad) + (\quad)k = (\quad)$$

$$\Rightarrow (\quad)k = (\quad)$$

$$\Rightarrow k = -5$$

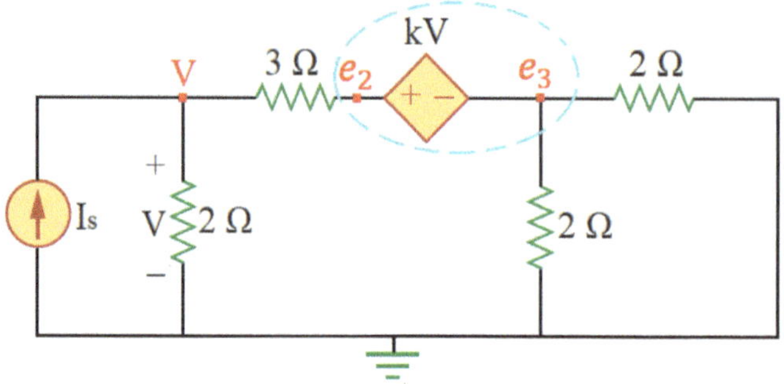

Fig. 5.49 The circuit of partially solved exercise 5.50

Exercise

5.51. Apply superposition theorem to calculate the output voltage v_o in the circuit of Fig. 5.50.

Final Answer

$v_o = -3$ V

Note: Read the "educational support" part of the preface of the textbook.

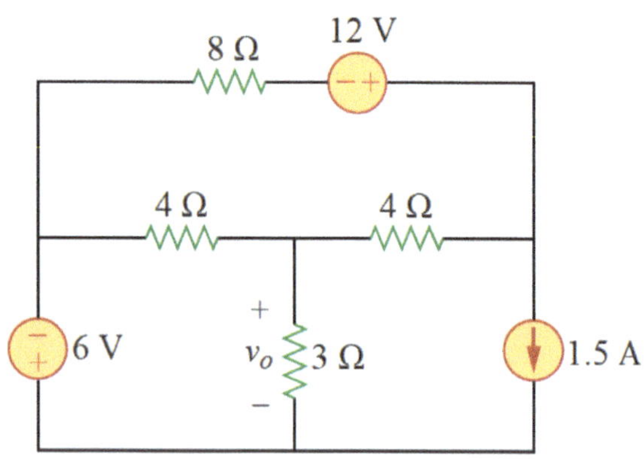

Fig. 5.50 The circuit of exercise 5.51

5.6 Maximum Power Transfer Theorem

Problem

5.52. Calculate the size of R_L so that it can absorb the maximum power in the circuit of Fig. 5.51. Moreover, calculate the maximum power transferable to the R_L.

Difficulty level ○ Easy ● Normal ○ Hard
Calculation amount ● Small ○ Normal ○ Large

1) $R_L = 5\ \Omega,\ P_{max} = 2.5$ W
2) $R_L = 10\ \Omega,\ P_{max} = 2.5$ W
3) $R_L = 7.5\ \Omega,\ P_{max} = 1.25$ W
4) $R_L = 2.5\ \Omega,\ P_{max} = 1.75$ W

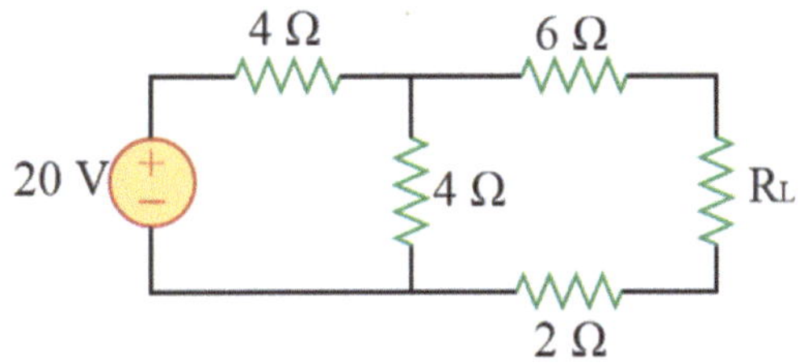

Fig. 5.51 The circuit of problem 5.52

Problem

5.53. Calculate the size of the load resistance (R_L) so that it can absorb the maximum power in the circuit of Fig. 5.52.

Difficulty level ○ Easy ● Normal ○ Hard
Calculation amount ● Small ○ Normal ○ Large

1) 5 Ω
2) 10 Ω
3) 15 Ω
4) 20 Ω

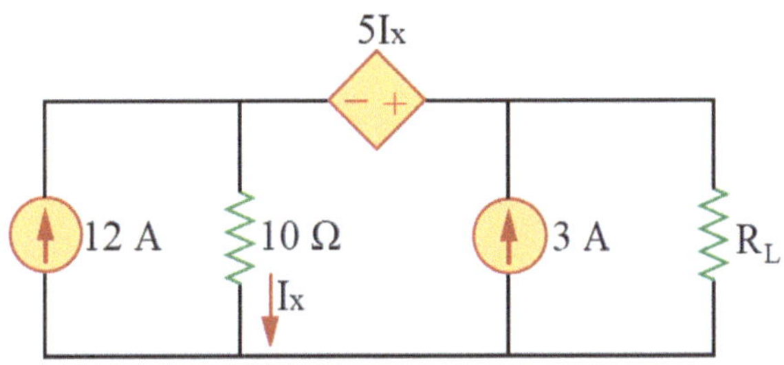

Fig. 5.52 The circuit of problem 5.53

Problem

5.54. Calculate the size of R_L so that it can absorb the maximum power in the circuit of Fig. 5.53.

Difficulty level ○ Easy ● Normal ○ Hard
Calculation amount ○ Small ● Normal ○ Large

1) 6 Ω
2) 18 Ω
3) 12 Ω
4) 8 Ω

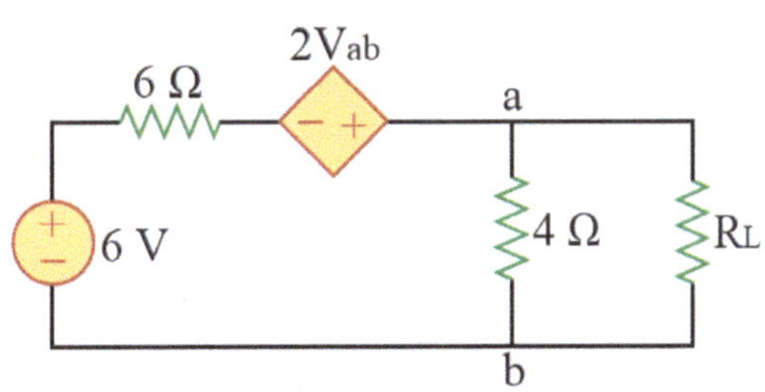

Fig. 5.53 The circuit of problem 5.54

Partially Solved Exercise

5.55. Calculate the size of R_L so that it can absorb the maximum power in the circuit of Fig. 5.54. In addition, calculate the amount of maximum power that can be transferred to the load.

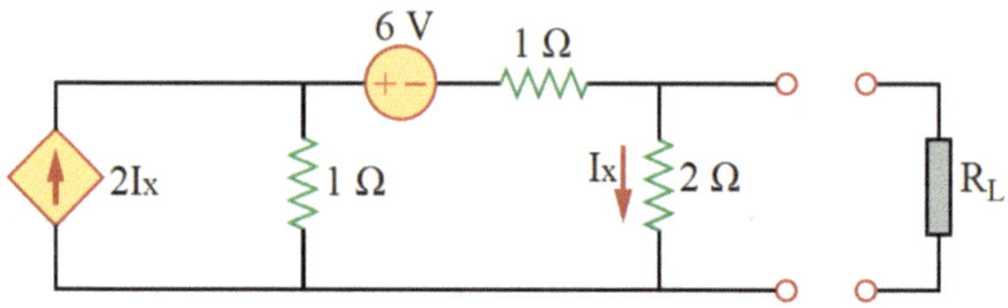

Fig. 5.54 The circuit of partially solved exercise 5.55

Solution

To absorb the maximum power, R_L must be equal to the Thevenin resistance (R_{Th}) seen by the R_L. Moreover, the maximum power transferable to the R_L can be calculated as follows.

$$P_{max} = \frac{V_{Th}^2}{4R_{Th}}$$

Therefore, the value of Thevenin resistance (R_{Th}) and Thevenin voltage (V_{Th}) must be calculated.

Step 1 (determining the method of solution)

Method 1: First, a test source (independent voltage or current source) with the voltage and current of V_t and I_t is connected to the input terminal of the circuit. Then, the circuit is analyzed by using a circuit analysis method to find a relation between the voltage and current of the test source in the following form.

$$V_t = \alpha I_t + \beta$$

Then, it is concluded that:

$$R_{Th} = \alpha$$

$$V_{Th} = \beta$$

As can be seen, in this method, the value of V_{Th} and R_{Th} are simultaneously calculated.

Method 2: In this method, the value of V_{Th} and R_{Th} are separately calculated. To calculate the value of V_{Th}, the terminal is assumed to be an open circuit branch. Then, the circuit is analyzed by using a circuit analysis method to find the value of V_{oc}. Next, we have:

$$V_{Th} = V_{oc}$$

Moreover, to calculate R_{Th}, which is the same as input resistance (R_{in}), first, all the independent sources (independent voltage and current sources) are turned off (each independent voltage source is replaced by short circuit branch and each independent current source is replaced by an open circuit branch). Then, if the circuit does not include any dependent source (dependent voltage or current source), the series–parallel rule is applied to calculate the input resistance of the circuit. However, if the circuit includes at least a dependent source, a test source (independent voltage or current source) with the voltage and current of V_t and I_t is connected to the input terminal of the circuit. Next, the circuit is analyzed by using a circuit analysis method to find the value of $\frac{V_t}{I_t}$. After that, we have:

$$R_{Th} = \frac{V_t}{I_t}$$

In this problem, the second method is applied because it is more straightforward.

Step 2 (solving the problem)

Calculating V_{Th}: KVL in the right-side mesh of the circuit of Fig. 5.55.1:

$$1(I_x - 2I_x) + (\quad) + (\quad) + (\quad) = 0$$

$$\Rightarrow (\quad)I_x + (\quad) = 0$$

$$\Rightarrow I_x = -3\ A \quad (1)$$

Applying Ohm's law for the 2-Ω resistor:

$$V_{oc} = 2I_x \quad (2)$$

Solving (1) and (2):

$$V_{oc} = (\quad) \times (\quad)$$

$$\Rightarrow V_{oc} = (\quad)$$

$$\Rightarrow V_{Th} = -6\ V$$

Calculating R_{Th}: Using Ohm's law for the 2-Ω resistor to define I_x based on the node voltage in the circuit of Fig. 5.55.2:

$$I_x = \frac{(\quad)}{(\quad)} \quad (3)$$

KCL in the left-side node:

$$(\quad) + \frac{(\quad)}{(\quad)} + \frac{(\quad) - (\quad)}{(\quad)} = 0$$

$$\Rightarrow -2I_x + 2e_1 - V_t = 0 \quad (4)$$

KCL in the right-side node:

$$(\quad) + \frac{(\quad)}{(\quad)} + \frac{(\quad) - (\quad)}{(\quad)} = 0$$

$$\Rightarrow -I_t + V_t - \frac{e_1}{2} = 0$$

$$\Rightarrow e_1 = -2I_t + 2V_t \quad (5)$$

Solving (3)-(5):

$$-2\left(\frac{V_t}{2}\right)+2(-2I_t+2V_t)-V_t=0$$

$$\Rightarrow (\quad)V_t+(\quad)I_t=0$$

$$\Rightarrow \frac{V_t}{I_t}=(\quad)$$

$$\Rightarrow R_{Th}=(\quad)\ \Omega$$

$$\Rightarrow R_L=2\ \Omega$$

Calculating P_{max}: The maximum power transferable to the R_L can be calculated as follows.

$$P_{max}=\frac{(\quad)^2}{(\quad)\times(\quad)}$$

$$\Rightarrow P_{max}=4.5\ \text{W}$$

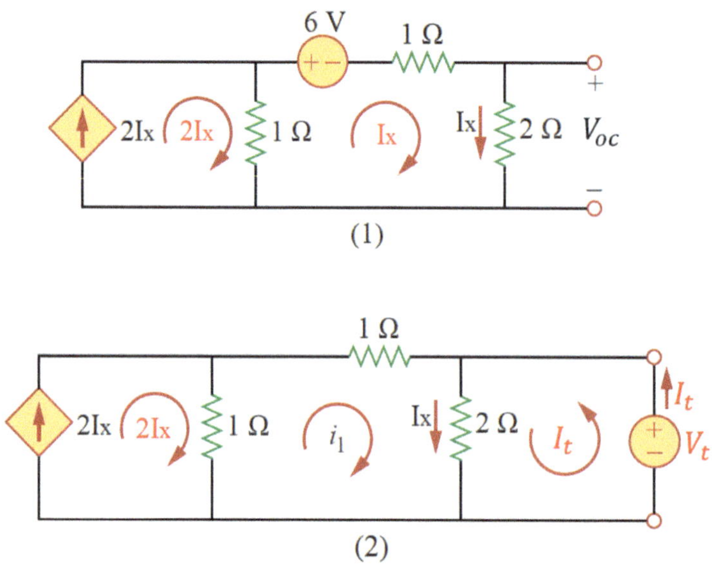

Fig. 5.55 The circuits of partially solved exercise 5.55

Exercise

5.56. Calculate the size of the load resistor so that it can absorb the maximum power in the circuit of Fig. 5.56.

Final Answer

$R_L=2\ \Omega$

Note: *Read the "educational support" part of the preface of the textbook.*

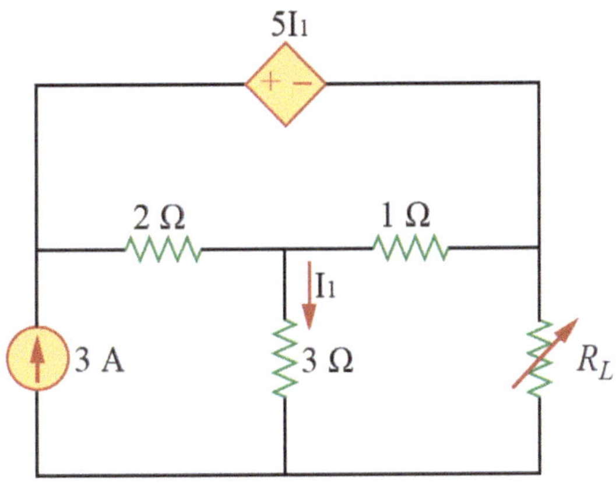

Fig. 5.56 The circuit of exercise 5.56

Exercise

5.57. In Problem 5.56, calculate the amount of maximum power that the load resistor can absorb.

Final Answer

$P_{max} = 4.5$ W

***Note:** Read the "educational support" part of the preface of the textbook.*

Exercise

5.58. The network shown in Fig. 5.57 is resistive, linear, and time-invariant. About 20% of the power of the voltage source is absorbed by the 5-Ω resistor. How much increase in the voltage of voltage source is required so that 30% of the power of voltage source is transferred to the resistor?

Final Answer

It is impossible to calculate the power absorbed by the resistor without knowing the configuration of network.

***Note:** Read the "educational support" part of the preface of the textbook.*

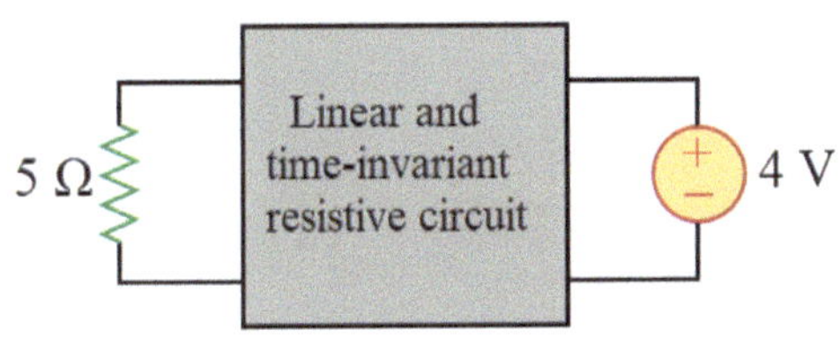

Fig. 5.57 The circuit of problem 5.58

References

1. Rahmani-Andebili, M., DC Electrical Circuit Analysis - Practice Problems, Methods, and Solutions (1st Ed.), Springer Nature, 2020.
2. Rahmani-Andebili, M., General Physics II - Practice Problems, Methods, and Solutions, Sringer Nature, 2025.
3. Rahmani-Andebili, M., General Physics I - Practice Problems, Methods, and Solutions, Springer Nature, 2025.
4. Rahmani-Andebili, M., Mathematics of Engineering and Science - Comprehensive Lessons, Springer Nature, 2024.
5. Rahmani-Andebili, M., Differential Equations – Practice Problems, Methods, and Solutions, Springer Nature, 2022.
6. Rahmani-Andebili, M., Calculus III – Practice Problems, Methods, and Solutions, Springer Nature, 2023.
7. Rahmani-Andebili, M., Calculus II – Practice Problems, Methods, and Solutions, Springer Nature, 2023.
8. Rahmani-Andebili, M., Calculus I (2nd Ed.) – Practice Problems, Methods, and Solutions, Springer Nature, 2023.
9. Rahmani-Andebili, M., Precalculus (2nd Ed.) – Practice Problems, Methods, and Solutions, Springer Nature, 2024.

Theorems of Electrical Circuits Analysis: Part B 6

Abstract

In this chapter, the problems of the fifth chapter are fully solved, in detail, step-by-step, and with different methods. Herein, a "Notes Box" has been added after the solution of each problem that includes all the formulas and important notes that are needed to solve the problem. Moreover, a "Traps Box" has been added that discloses the traps that have been designed in the choices to challenge students' minds and encourage them to deeply think about the problem solutions as well as discourage them from careless selection of choices. The "Traps Box", in fact, predicts the possible confusions and tricks that might be encountered by students in real tests.

6.1 Input Resistance

6.1. *Step 1 (turning off all the independent sources)*

If the circuit includes any independent voltage and current sources, they need to be turned off. In this regard, each independent voltage source must be replaced by a short-circuit branch and each independent current source is replaced by an open-circuit branch [1–9].

Step 2 (determining the type of problem)

Type 1: If the circuit does not include any dependent source (dependent voltage or current source), the series-parallel rule is applied to calculate the input resistance of the circuit.

Type 2: If the circuit includes at least a dependent source (dependent voltage or current source), a test source (independent voltage or current source) with the voltage and current of V_t and I_t needs to be connected to the input terminal of the circuit. Apparently, this problem is a type 2 problem, but, in fact, it is type 1 (simpler one) as is shown in the following.

Step 3 (determining the type of circuit analysis)

Heuristic approach is the best method to apply in this problem since it will result in the least calculation.

Step 4

As can be seen in Fig. 6.1.2, after turning off the independent current source (replacing it by an open circuit branch), we have:

$$I_x = 0$$

Consequently, the dependent current source is turned off and changed to an open circuit branch (see Fig. 6.1.3).

Now, the remaining circuit includes only two series resistors. Thus:

$$R_{in} = 1 + 1$$

S. Manias, *DC Electrical Circuits Analysis*, https://doi.org/10.1007/978-3-032-00253-2_6

$$R_{in} = 2\ \Omega$$

Choice (4) is the answer.

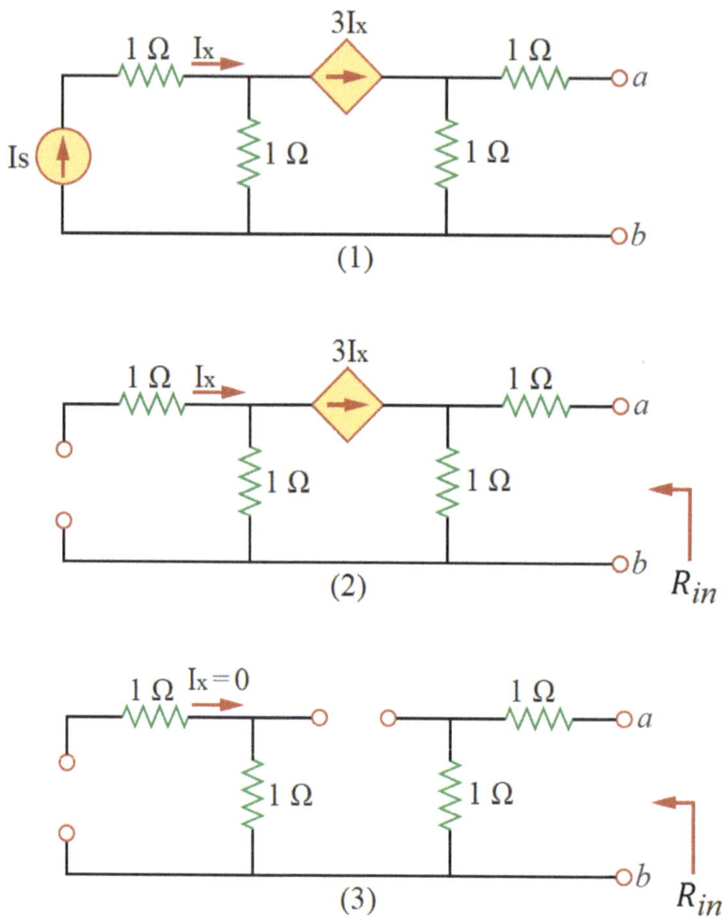

Fig. 6.1 The circuits of problem 6.1

6.2. *Step 1 (turning off all the independent sources)*

If the circuit includes any independent voltage and current sources, they need to be turned off. In this regard, each independent voltage source must be replaced by a short-circuit branch and each independent current source is replaced by an open-circuit branch.
Herein, the circuit does not include any independent source.

Step 2 (determining the type of problem)

Type 1: If the circuit does not include any dependent source (dependent voltage or current source), the series-parallel rule is applied to calculate the input resistance of the circuit.

Type 2: If the circuit includes at least a dependent source (dependent voltage or current source), a test source (independent voltage or current source) with the voltage and current of V_t and I_t needs to be connected to the input terminal of the circuit. This problem is a type 2 problem.

Step 3 (determining the type of circuit analysis)

Mesh analysis is the best method to apply in this problem since it will result in the least calculation.

Step 4 (formulating the circuit analysis equations)

Using Ohm's law for the 2-Ω resistor, to define V based on mesh current:

$$V = 2i_1 \tag{1}$$

KVL in the left-side mesh:

$$-V_t + 4I_t + 3(I_t - i_1) + kV = 0$$

$$\Rightarrow -V_t + 7I_t - 3i_1 + kV = 0 \tag{2}$$

KVL in the right-side mesh:

$$-kV + 3(i_1 - I_t) + 2i_1 = 0$$

$$\Rightarrow -kV + 5i_1 - 3I_t = 0 \tag{3}$$

Step 5 (solving the system of equations to calculate the value of $R_{in} = \frac{V_t}{I_t}$)

Solving (1) and (2):

$$-V_t + 7I_t + (2k - 3)i_1 = 0 \tag{4}$$

Solving (1) and (3):

$$(5 - 2k)i_1 - 3I_t = 0$$

$$\Rightarrow i_1 = \frac{3I_t}{5 - 2k} \tag{5}$$

Solving (4) and (5):

$$-V_t + 7I_t + (2k - 3)\frac{3I_t}{5 - 2k} = 0$$

$$\Rightarrow -V_t + I_t\left(7 + \frac{3(2k - 3)}{5 - 2k}\right) = 0$$

$$\Rightarrow -V_t + I_t\left(\frac{35 - 14k + 6k - 9}{5 - 2k}\right) = 0$$

$$\Rightarrow -V_t + I_t\left(\frac{26 - 8k}{5 - 2k}\right) = 0$$

$$\Rightarrow \frac{V_t}{I_t} = \frac{26 - 8k}{5 - 2k}$$

$$\Rightarrow \frac{V_t}{I_t} = \frac{8k - 26}{2k - 5}$$

$$\Rightarrow R_{ab} = \frac{8k - 26}{2k - 5}$$

Choice (4) is the answer (Fig. 6.2).

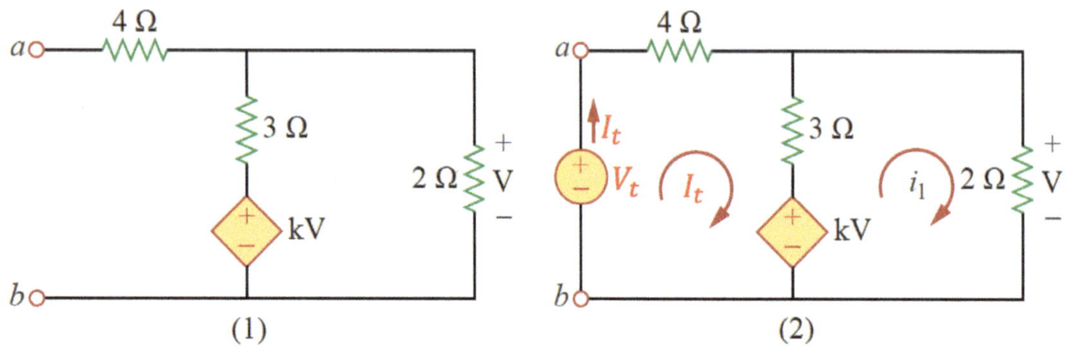

Fig. 6.2 The circuits of problem 6.2

6.5. *Step 1 (turning off all the independent sources)*

If the circuit includes any independent voltage and current sources, they need to be turned off. In this regard, each independent voltage source must be replaced by a short-circuit branch and each independent current source is replaced by an open-circuit branch. Herein, the circuit does not include any independent source.

Step 2 (determining the type of problem)

Type 1: If the circuit does not include any dependent source (dependent voltage or current source), the series-parallel rule is applied to calculate the input resistance of the circuit.

Type 2: If the circuit includes at least a dependent source (dependent voltage or current source), a test source (independent voltage or current source) with the voltage and current of V_t and I_t needs to be connected to the input terminal of the circuit. This problem is a type 2 problem.

Step 3 (determining the type of circuit analysis)

Nodal analysis is the best method to apply in this problem since it will result in the least calculation.

Step 4 (formulating the circuit analysis equations)

KCL in node 1:

$$\frac{e_1}{1} + \frac{e_1 - V_t}{1} + \alpha I + \frac{e_1 - V_t}{1} = 0$$

$$\Rightarrow 3e_1 - 2V_t + \alpha I = 0 \qquad (1)$$

KCL in the top node:

$$-I_t + \frac{V_t - e_1}{1} - \alpha I + \frac{V_t - e_1}{1} + \frac{V_t}{2} = 0$$

$$\Rightarrow -I_t + 2.5V_t - 2e_1 - \alpha I = 0 \tag{2}$$

Defining I based on the node voltages:

$$I = \frac{V_t - e_1}{1}$$

$$\Rightarrow I = V_t - e_1 \tag{3}$$

Step 5 (solving the system of equations to calculate the value of $R_{in} = \frac{V_t}{I_t}$)

Solving (1) and (3):

$$3e_1 - 2V_t + \alpha(V_t - e_1) = 0$$

$$\Rightarrow (3 - \alpha)e_1 + (\alpha - 2)V_t = 0$$

$$\Rightarrow e_1 = \frac{\alpha - 2}{\alpha - 3}V_t \tag{4}$$

Solving (2) and (3):

$$-I_t + 2.5V_t - 2e_1 - \alpha(V_t - e_1) = 0$$

$$\Rightarrow -I_t + V_t(2.5 - \alpha) + e_1(\alpha - 2) = 0 \tag{5}$$

Solving (4) and (5):

$$-I_t + V_t(2.5 - \alpha) + \left(\frac{\alpha - 2}{\alpha - 3}V_t\right)(\alpha - 2) = 0$$

$$\Rightarrow -I_t + V_t\left(2.5 - \alpha + \frac{\alpha^2 - 4\alpha + 4}{\alpha - 3}\right) = 0$$

$$\Rightarrow \frac{V_t}{I_t} = \left(2.5 - \alpha + \frac{\alpha^2 - 4\alpha + 4}{\alpha - 3}\right)^{-1}$$

$$\Rightarrow \frac{V_t}{I_t} = \left(\frac{2.5\alpha - \alpha^2 - 7.5 + 3\alpha + \alpha^2 - 4\alpha + 4}{\alpha - 3}\right)^{-1}$$

$$\Rightarrow \frac{V_t}{I_t} = \frac{\alpha - 3}{1.5\alpha - 3.5}$$

$$\Rightarrow R_{in} = \frac{2(\alpha - 3)}{3\alpha - 7}$$

Choice (1) is the answer (Fig. 6.3).

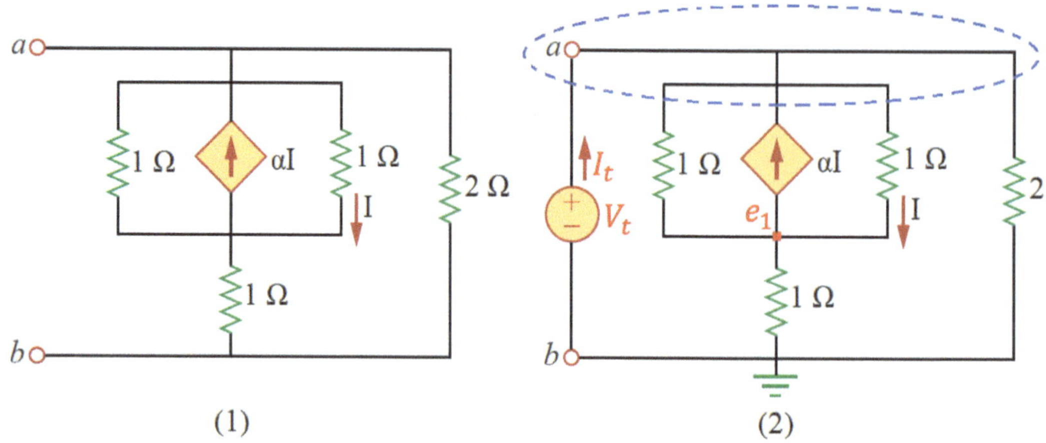

Fig. 6.3 The circuits of problem 6.5

6.8. *Step 1 (turning off all the independent sources)*

If the circuit includes any independent voltage and current sources, they need to be turned off. In this regard, each independent voltage source must be replaced by a short-circuit branch and each independent current source is replaced by an open-circuit branch.
Herein, the circuit does not include any independent source.

Step 2 (determining the type of problem)

Type 1: If the circuit does not include any dependent source (dependent voltage or current source), the series-parallel rule is applied to calculate the input resistance of the circuit.

Type 2: If the circuit includes at least a dependent source (dependent voltage or current source), a test source (independent voltage or current source) with the voltage and current of V_t and I_t needs to be connected to the input terminal of the circuit. This problem is a type 2 problem.

Step 3 (determining the type of circuit analysis)

Nodal analysis is the best method to apply in this problem since it will result in the least calculation.

Step 4 (formulating the circuit analysis equations)

KCL in the node:

$$-I_x + \frac{V_x}{1} - 2I_x + \frac{V_x - 2V_x}{1} = 0$$

$$\Rightarrow -3I_x = 0$$

$$\Rightarrow I_x = 0$$

Step 5 (solving the system of equations to calculate the value of $R_{in} = \frac{V_t}{I_t}$*)*

$$\Rightarrow \frac{V_t}{I_x} = \infty$$

$$\Rightarrow R_{in} = \infty$$

Choice (3) is the answer (Fig. 6.4).

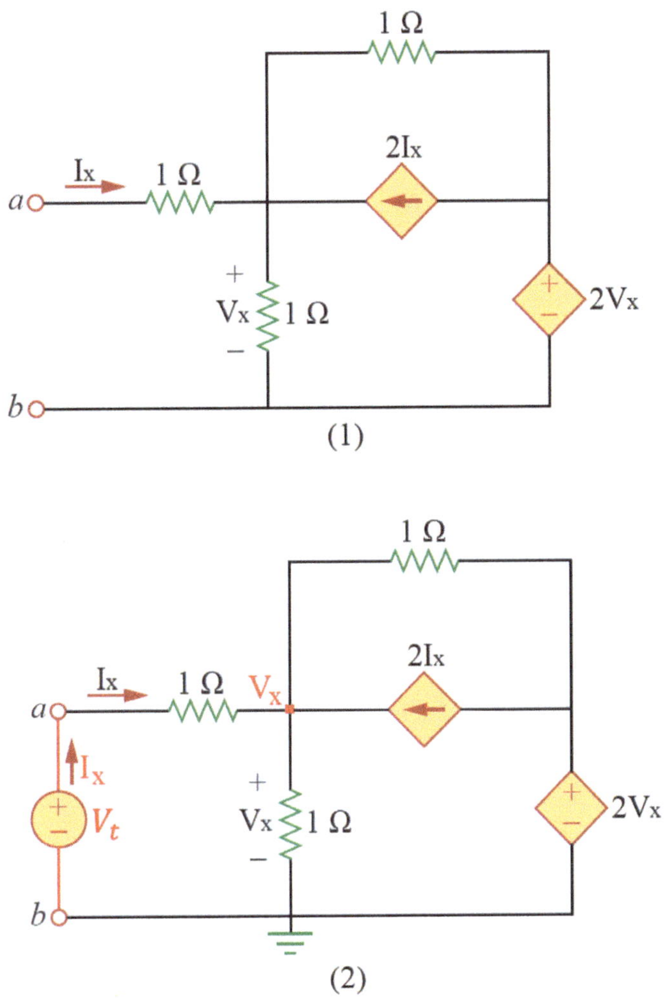

Fig. 6.4 The circuits of problem 6.8

6.9. *Step 1 (turning off all the independent sources)*

If the circuit includes any independent voltage and current sources, they need to be turned off. In this regard, each independent voltage source must be replaced by a short-circuit branch and each independent current source is replaced by an open-circuit branch.
Herein, the circuit does not include any independent source.

Step 2 (determining the type of problem)

Type 1: If the circuit does not include any dependent source (dependent voltage or current source), the series-parallel rule is applied to calculate the input resistance of the circuit.

Type 2: If the circuit includes at least a dependent source (dependent voltage or current source), a test source (independent voltage or current source) with the voltage and current of V_t and I_t needs to be connected to the input terminal of the circuit. This problem is a type 2 problem.

Step 3 (determining the type of circuit analysis)

Nodal analysis is the best method to apply in this problem since it will result in the least calculation. Herein, the node voltages are defined based on the available variables (V_t and V_2) to decrease the number of variables.

Step 4 (formulating the circuit analysis equations)

KCL in the left-side node:

$$-I_t + I_1 = 0$$

$$\Rightarrow I_1 = I_t \tag{1}$$

KCL in the right-side node:

$$\alpha I_1 + \frac{V_2}{R} + \frac{V_2}{R} = 0$$

$$\Rightarrow \alpha I_1 + \frac{2V_2}{R} = 0$$

$$\Rightarrow V_2 = -\frac{R\alpha I_1}{2} \tag{2}$$

Defining I_1 based on the node voltages:

$$I_1 = \frac{V_t - (-\mu V_2)}{R} \tag{3}$$

Step 5 (solving the system of equations to calculate the value of $R_{in} = \frac{V_t}{I_t}$)

Solving (1) and (2):

$$V_2 = -\frac{R\alpha I_t}{2} \tag{4}$$

Solving (1) and (3):

$$I_t = \frac{V_t + \mu V_2}{R} \tag{5}$$

Solving (4) and (5):

$$I_t = \frac{V_t + \mu\left(-\frac{R\alpha I_t}{2}\right)}{R}$$

$$\Rightarrow V_t - \frac{\mu R\alpha I_t}{2} = RI_t$$

$$\Rightarrow V_t = I_t R\left(1 + \frac{\mu\alpha}{2}\right)$$

$$\Rightarrow \frac{V_t}{I_t} = R\left(1 + \frac{\mu\alpha}{2}\right)$$

$$\Rightarrow R_{in} = R\left(1 + \frac{\mu\alpha}{2}\right)$$

Choice (3) is the answer (Fig. 6.5).

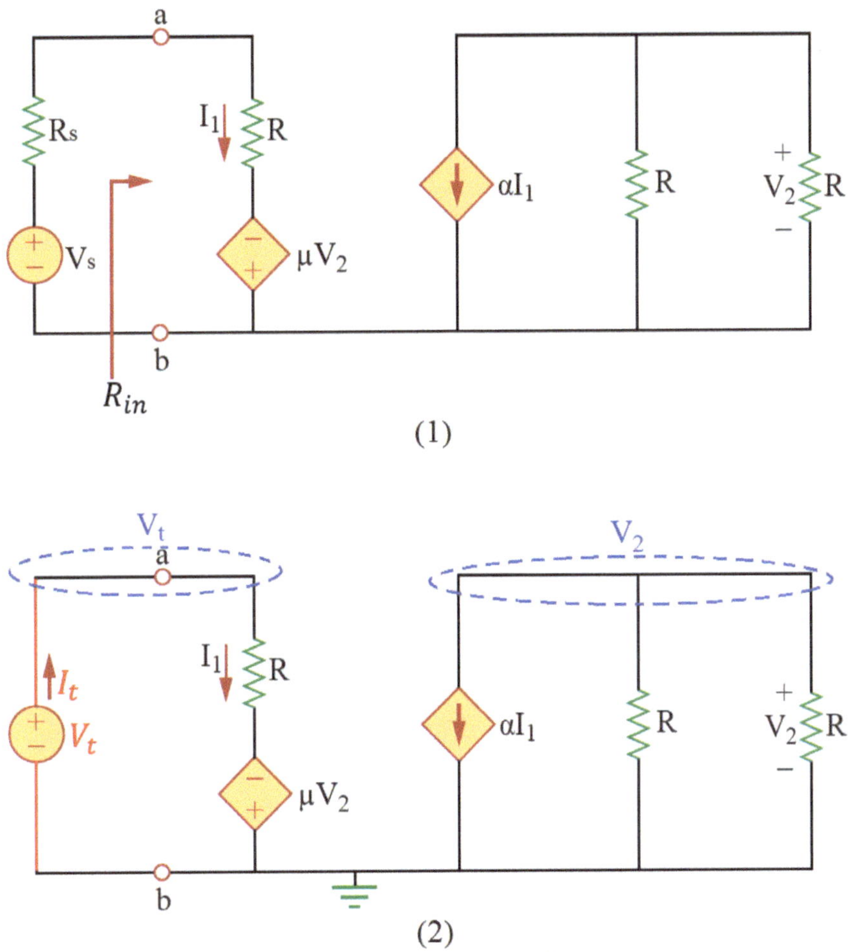

Fig. 6.5 The circuits of problem 6.9

6.10. *Step 1 (turning off all the independent sources)*

If the circuit includes any independent voltage and current sources, they need to be turned off. In this regard, each independent voltage source must be replaced by a short-circuit branch and each independent current source is replaced by an open-circuit branch.
Herein, the circuit does not include any independent source.

Step 2 (determining the type of problem)

Type 1: If the circuit does not include any dependent source (dependent voltage or current source), the series-parallel rule is applied to calculate the input resistance of the circuit.

Type 2: If the circuit includes at least a dependent source (dependent voltage or current source), a test source (independent voltage or current source) with the voltage and current of V_t and I_t needs to be connected to the input terminal of the circuit. This problem is a type 2 problem.

Step 3 (determining the type of circuit analysis)

Nodal analysis is the best method to apply in this problem since it will result in the least calculation.

Step 4 (formulating the circuit analysis equations)

KCL in the left-side node:

$$-I_t + I + \frac{V_t - e_1}{1} = 0$$

$$\Rightarrow -I_t + I + V_t - e_1 = 0 \tag{1}$$

KCL in node 1:

$$-I + \frac{e_1 - V_t}{1} + \frac{e_1 - V_t}{1} = 0$$

$$\Rightarrow -I + 2e_1 - 2V_t = 0 \tag{2}$$

KCL in the right-side node:

$$-I + \frac{V_t - e_1}{1} + \frac{V_t}{R} = 0$$

$$\Rightarrow -I - e_1 + \left(1 + \frac{1}{R}\right)V_t = 0 \tag{3}$$

Step 5 (solving the system of equations to calculate the value of $R_{in} = \frac{V_t}{I_t}$)

Solving (1) and (2) to eliminate I:

$$-I_t + (2e_1 - 2V_t) + V_t - e_1 = 0$$

$$\Rightarrow -I_t + e_1 - V_t = 0$$

$$\Rightarrow e_1 = I_t + V_t \tag{4}$$

Solving (2) and (3) to eliminate I:

$$-(2e_1 - 2V_t) - e_1 + \left(1 + \frac{1}{R}\right)V_t = 0$$

$$\Rightarrow -3e_1 + \left(3 + \frac{1}{R}\right)V_t = 0$$

$$\Rightarrow e_1 = \left(1 + \frac{1}{3R}\right)V_t \tag{5}$$

Solving (4) and (5) to eliminate e_1:

$$I_t + V_t = \left(1 + \frac{1}{3R}\right)V_t$$

$$\Rightarrow I_t = \frac{1}{3R}V_t$$

$$\Rightarrow \frac{V_t}{I_t} = 3\ \text{R}$$

$$\Rightarrow R_{in} = 3\ \text{R}$$

Choice (3) is the answer (Fig. 6.6).

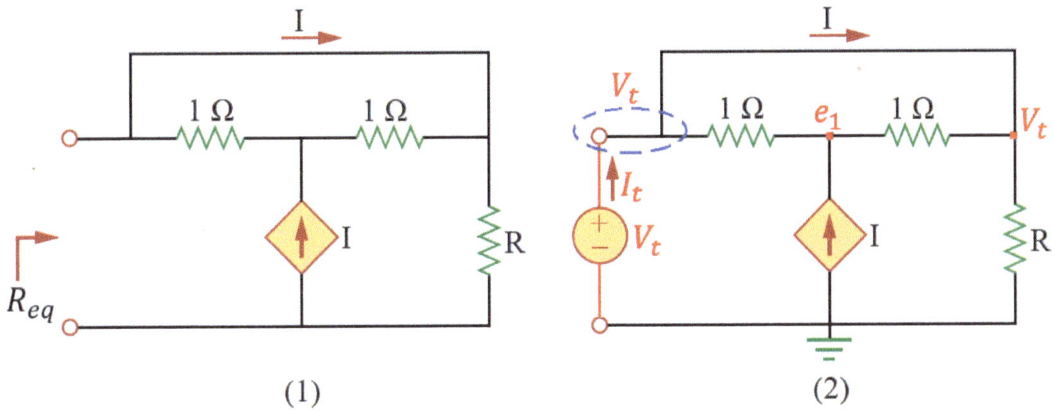

Fig. 6.6 The circuits of problem 6.10

6.12. *Step 1 (turning off all the independent sources)*

If the circuit includes any independent voltage and current sources, they need to be turned off. In this regard, each independent voltage source must be replaced by a short-circuit branch and each independent current source is replaced by an open-circuit branch.

Herein, the circuit does not include any independent source.

Step 2 (determining the type of problem)

Type 1: If the circuit does not include any dependent source (dependent voltage or current source), the series-parallel rule is applied to calculate the input resistance of the circuit.

Type 2: If the circuit includes at least a dependent source (dependent voltage or current source), a test source (independent voltage or current source) with the voltage and current of V_t and I_t needs to be connected to the input terminal of the circuit. This problem is a type 2 problem.

Step 3 (determining the type of circuit analysis)

Mesh analysis is the best method to apply in this problem since it will result in the least calculation.

Step 4 (formulating the circuit analysis equations)

To avoid being involved with the unknown voltages of independent current sources, we need to apply the supermesh shown in Fig. 6.7.

Defining the current of the left-side dependent current source based on the mesh currents:

$$13I' = i_1 - i_2$$

$$\Rightarrow i_1 = 13I' + i_2 \quad (1)$$

Defining the current of the right-side dependent current source based on the mesh currents:

$$11I' = -i_2 - I_t$$

$$\Rightarrow i_2 = -I_t - 11I' \quad (2)$$

KVL in the supermesh:

$$Ri_1 + 3(i_2 + I') + 7(-I_t + I') + V_t = 0$$

$$\Rightarrow Ri_1 + 3i_2 + 10I' - 7I_t + V_t = 0 \quad (3)$$

KVL in the top mesh:

$$3(i_2 + I') + 7(-I_t + I') + 3I' = 0$$

$$\Rightarrow 3i_2 + 13I' - 7I_t = 0 \quad (4)$$

Step 5 (solving the system of equations to calculate the value of $R_{in} = \frac{V_t}{I_t}$)

Solving (1), (2) and (3):

$$R(13I' + (-I_t - 11I')) + 3(-I_t - 11I') + 10I' - 7I_t + V_t = 0$$

$$\Rightarrow (2R - 23)I' + (-R - 10)I_t + V_t = 0 \quad (5)$$

Solving (1), (2), and (4):

$$3(-I_t - 11I') + 13I' - 7I_t = 0$$

$$\Rightarrow -20I' - 10I_t = 0$$

$$\Rightarrow I' = -0.5I_t \quad (6)$$

Solving (5) and (6):

$$(2R - 23)(-0.5I_t) + (-R - 10)I_t + V_t = 0$$

$$\Rightarrow (-2R + 1.5)I_t + V_t = 0$$

$$\Rightarrow V_t = (2R - 1.5)I_t$$

$$\Rightarrow \frac{V_t}{I_t} = 2R - 1.5$$

$$\Rightarrow R_{AB} = (2R - 1.5)\ \Omega$$

Choice (2) is the answer.

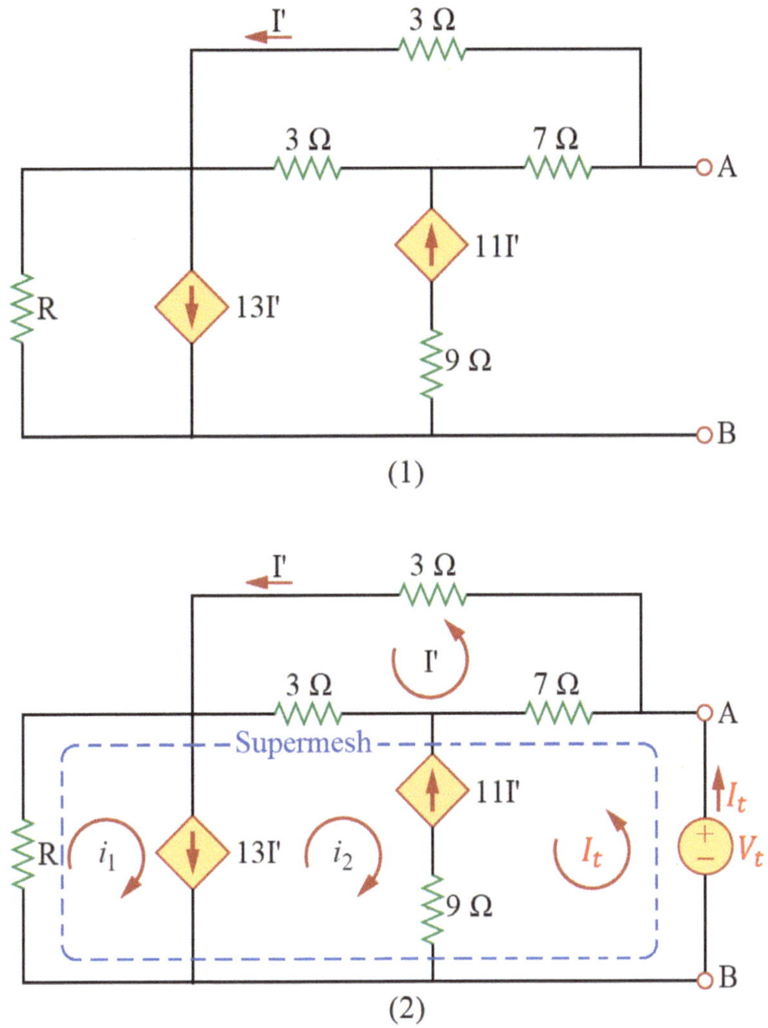

Fig. 6.7 The circuits of problem 6.12

6.13. *Step 1 (turning off all the independent sources)*

If the circuit includes any independent voltage and current sources, they need to be turned off. In this regard, each independent voltage source must be replaced by a short-circuit branch and each independent current source is replaced by an open-circuit branch.

Herein, the circuit does not include any independent source.

Step 2 (determining the type of problem)

Type 1: If the circuit does not include any dependent source (dependent voltage or current source), the series-parallel rule is applied to calculate the input resistance of the circuit.

Type 2: If the circuit includes at least a dependent source (dependent voltage or current source), a test source (independent voltage or current source) with the voltage and current of V_t and I_t needs to be connected to the input terminal of the circuit. This problem is a type 2 problem.

Step 3 (determining the type of circuit analysis)

Nodal analysis is the best method to apply in this problem since it will result in the least calculation.

Step 4 (formulating the circuit analysis equations)

KCL in the left-side node:

$$-I_t + I_x + \frac{V_t - e_1}{R} = 0$$

$$\Rightarrow -I_t + I_x + \frac{V_t}{R} - \frac{e_1}{R} = 0 \tag{1}$$

KCL in node 1:

$$\beta I_x + \frac{e_1 - V_t}{R} + \frac{e_1 - V_t}{R} = 0$$

$$\Rightarrow \beta I_x + \frac{2}{R}e_1 - \frac{2}{R}V_t = 0 \tag{2}$$

KCL in the right-side node:

$$-I_x + \frac{V_t - e_1}{R} + \frac{V_t}{R_L} = 0$$

$$\Rightarrow -I_x - \frac{1}{R}e_1 + \left(\frac{1}{R} + \frac{1}{R_L}\right)V_t = 0 \tag{3}$$

Step 5 (solving the system of equations to calculate the value of $R_{in} = \frac{V_t}{I_t}$)

Solving (1) and (2) to eliminate I_x:

$$-I_t + \left(-\frac{2}{R\beta}e_1 + \frac{2}{R\beta}V_t\right) + \frac{V_t}{R} - \frac{e_1}{R} = 0$$

$$\Rightarrow -I_t + \left(\frac{2}{R\beta} + \frac{1}{R}\right)V_t + \left(-\frac{2}{R\beta} - \frac{1}{R}\right)e_1 = 0$$

$$\Rightarrow e_1 = \frac{I_t - \left(\frac{2}{R\beta} + \frac{1}{R}\right)V_t}{-\left(\frac{2}{R\beta} + \frac{1}{R}\right)}$$

$$\Rightarrow e_1 = \frac{R\beta I_t - (2+\beta)V_t}{-(2+\beta)}$$

$$\Rightarrow e_1 = -\frac{R\beta}{2+\beta}I_t + V_t \tag{4}$$

Solving (2) and (3) to eliminate I_x:

$$\Rightarrow \beta\left(-\frac{1}{R}e_1 + \left(\frac{1}{R}+\frac{1}{R_L}\right)V_t\right) + \frac{2}{R}e_1 - \frac{2}{R}V_t = 0$$

$$\Rightarrow \left(\frac{2-\beta}{R}\right)e_1 + \left(\frac{\beta-2}{R}+\frac{\beta}{R_L}\right)V_t = 0 \tag{5}$$

Solving (4) and (5) to eliminate e_1:

$$\left(\frac{2-\beta}{R}\right)\left(-\frac{R\beta}{2+\beta}I_t + V_t\right) + \left(\frac{\beta-2}{R}+\frac{\beta}{R_L}\right)V_t = 0$$

$$\Rightarrow \left(-\frac{2-\beta}{R}\times\frac{R\beta}{2+\beta}\right)I_t + \left(\frac{2-\beta}{R}+\frac{\beta-2}{R}+\frac{\beta}{R_L}\right)V_t = 0$$

$$\Rightarrow \frac{V_t}{I_t} = \frac{\frac{2-\beta}{R}\times\frac{R\beta}{2+\beta}}{\frac{2-\beta}{R}+\frac{\beta-2}{R}+\frac{\beta}{R_L}}$$

$$\Rightarrow \frac{V_t}{I_t} = \frac{\frac{\beta(2-\beta)}{2+\beta}}{\frac{\beta}{R_L}}$$

$$\Rightarrow \frac{V_t}{I_t} = \frac{2-\beta}{2+\beta}R_L$$

$$\Rightarrow R_{eq} = \frac{2-\beta}{2+\beta}R_L$$

Choice (3) is the answer (Fig. 6.8).

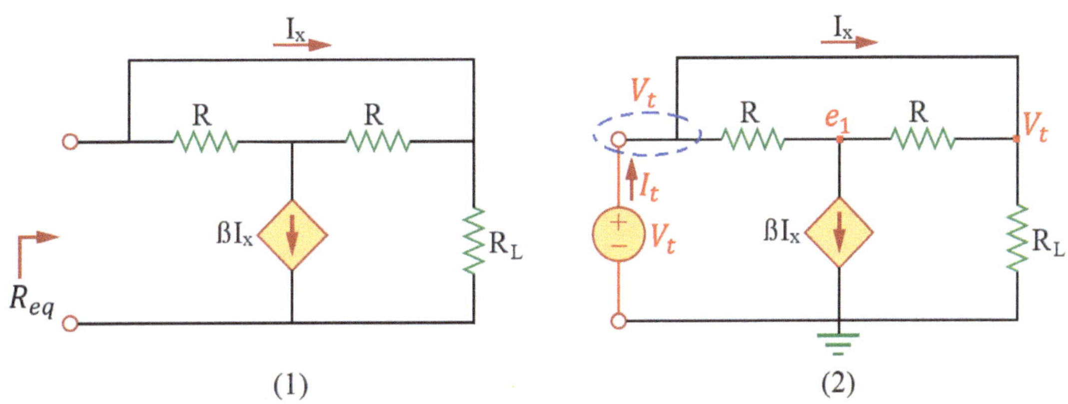

Fig. 6.8 The circuits of problem 6.13

6.14. *Step 1 (turning off all the independent sources)*

If the circuit includes any independent voltage and current sources, they need to be turned off. In this regard, each independent voltage source must be replaced by a short-circuit branch and each independent current source is replaced by an open-circuit branch.

Step 2 (determining the type of problem)

Type 1: If the circuit does not include any dependent source (dependent voltage or current source), the series-parallel rule is applied to calculate the input resistance of the circuit.

Type 2: If the circuit includes at least a dependent source (dependent voltage or current source), a test source (independent voltage or current source) with the voltage and current of V_t and I_t needs to be connected to the input terminal of the circuit. This problem is a type 2 problem.

Step 3 (determining the type of circuit analysis)

Mesh analysis is the best method to apply in this problem since it will result in the least calculation. To minimize the number of variables, the current of the test voltage source (I_t) and the current of the independent current source ($3V$) are considered as the currents of mesh 1 and mesh 3, respectively.

Step 4 (formulating the circuit analysis equations)

Defining V_1 based on the mesh current:

$$V_1 = 4 \times (-3V) = -12 \text{ V} \tag{1}$$

KVL in mesh 1:

$$-V_t + 1(I_t - 3V) + 1 \times I_t - \frac{7}{6}V_1 = 0$$

$$\Rightarrow -V_t + 2I_t - 3V - \frac{7}{6}V_1 = 0 \tag{2}$$

KVL in mesh 2:

$$-1 \times I_t + 4(-3V) + V = 0$$

$$\Rightarrow -I_t - 11V = 0$$

$$\Rightarrow V = -\frac{1}{11}I_t \tag{3}$$

Step 5 (solving the system of equations to calculate the value of $R_{in} = \frac{V_t}{I_t}$)

Solving (1), (2), and (3):

$$-V_t + 2I_t - 3\left(-\frac{1}{11}I_t\right) - \frac{7}{6}\left(-12\left(-\frac{1}{11}I_t\right)\right) = 0$$

$$\Rightarrow -V_t + I_t\left(2 + \frac{3}{11} - \frac{7}{6} \times \frac{12}{11}\right) = 0$$

$$\Rightarrow -V_t + I_t(1) = 0$$

$$\Rightarrow \frac{V_t}{I_t} = 1$$

$$\Rightarrow R_{eq} = 1\ \Omega$$

Choice (2) is the answer (Fig. 6.9).

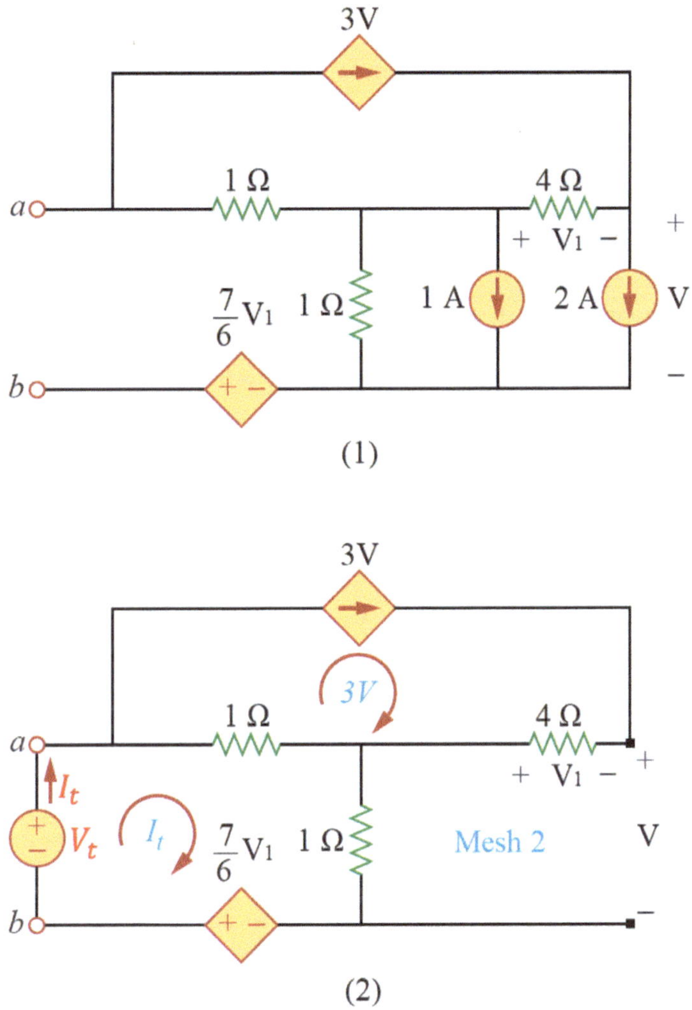

Fig. 6.9 The circuits of problem 6.14

6.15. *Step 1 (turning off all the independent sources)*

If the circuit includes any independent voltage and current sources, they need to be turned off. In this regard, each independent voltage source must be replaced by a short-circuit branch and each independent current source is replaced by an open-circuit branch.

Herein, the circuit does not include any independent source.

Step 2 (determining the type of problem)

Type 1: If the circuit does not include any dependent source (dependent voltage or current source), the series-parallel rule is applied to calculate the input resistance of the circuit.

Type 2: If the circuit includes at least a dependent source (dependent voltage or current source), a test source (independent voltage or current source) with the voltage and current of V_t and I_t needs to be connected to the input terminal of the circuit. This problem is a type 2 problem.

Step 3 (determining the type of circuit analysis)

Nodal analysis is the best method to apply in this problem since it will result in the least calculation.

Step 4 (formulating the circuit analysis equations)

Using Ohm's law for the vertical 1-Ω resistor to define I based on the node voltages:

$$I = \frac{e_2}{1}$$

$$\Rightarrow e_2 = I \quad (1)$$

Defining the voltage of the dependent voltage source based on the node voltages:

$$mI = e_2 - e_1 \quad (2)$$

KCL in the supernode:

$$\frac{e_1 - V_t}{1} + \frac{e_2}{1} + \frac{e_2}{0.5} = 0$$

$$\Rightarrow e_1 - V_t + 3e_2 = 0 \quad (3)$$

KCL in the left-side node:

$$-I_t + \frac{V_t}{0.5} + \frac{V_t - e_1}{1} = 0$$

$$\Rightarrow -I_t + 3V_t - e_1 = 0 \quad (4)$$

Step 5 (solving the system of equations to calculate the value of $R_{in} = \frac{V_t}{I_t}$*)*

Solving (1) and (2):

$$me_2 = e_2 - e_1$$

$$\Rightarrow e_2 = \frac{e_1}{1 - m} \quad (5)$$

Solving (3) and (5):

$$e_1 - V_t + \frac{3e_1}{1 - m} = 0$$

$$\Rightarrow e_1 = \frac{V_t}{1 + \frac{3}{1 - m}}$$

$$\Rightarrow e_1 = \frac{1-m}{4-m} V_t \quad (6)$$

Solving (4) and (6):

$$-I_t + 3V_t - \frac{1-m}{4-m} V_t = 0$$

$$\Rightarrow \left(3 - \frac{1-m}{4-m}\right) V_t = I_t$$

$$\Rightarrow \left(\frac{12 - 3m - 1 + m}{4-m}\right) V_t = I_t$$

$$\Rightarrow \frac{V_t}{I_t} = \frac{4-m}{11-2m}$$

$$\Rightarrow R_{in} = \frac{4-m}{11-2m}$$

Choice (3) is the answer (Fig. 6.10).

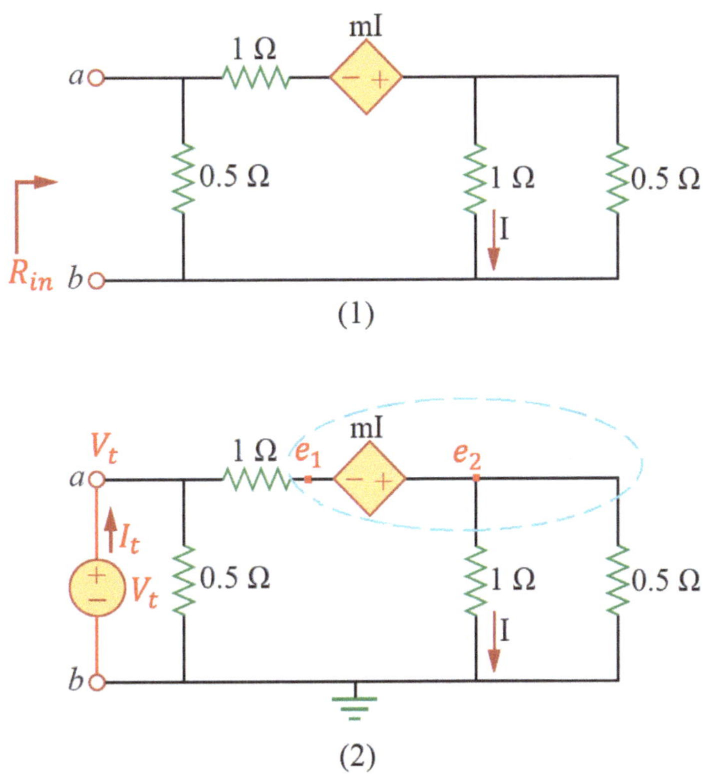

Fig. 6.10 The circuits of problem 6.15

6.17. *Step 1 (turning off all the independent sources)*

If the circuit includes any independent voltage and current sources, they need to be turned off. In this regard, each independent voltage source must be replaced by a short-circuit branch and each independent current source is replaced by an open-circuit branch.

Step 2 (determining the type of problem)

Type 1: If the circuit does not include any dependent source (dependent voltage or current source), the series-parallel rule is applied to calculate the input resistance of the circuit.

Type 2: If the circuit includes at least a dependent source (dependent voltage or current source), a test source (independent voltage or current source) with the voltage and current of V_t and I_t needs to be connected to the input terminal of the circuit. This problem is a type 2 problem.

Step 3 (determining the type of circuit analysis)

Nodal analysis is the best method to apply in this problem since it will result in the least calculation.

Step 4 (formulating the circuit analysis equations)

From the circuit, it is seen that:

$$V_1 = V_t \qquad (1)$$

Defining the voltage of the dependent voltage source based on the node voltages:

$$\alpha V_1 = V_t - e_1 \qquad (2)$$

KCL in node a:

$$3V_2 - I_t + \frac{V_t}{R} + \frac{e_1 - V_2}{1} = 0$$

$$\Rightarrow 2V_2 - I_t + \frac{V_t}{R} + e_1 = 0 \qquad (3)$$

KCL in node 2:

$$\frac{V_2}{1} - 3V_2 + \frac{V_2 - e_1}{1} = 0$$

$$\Rightarrow e_1 = -V_2 \qquad (4)$$

Step 5 (solving the system of equations to calculate the value of $R_{in} = \frac{V_t}{I_t}$*)*

Solving (1), (2) and (4):

$$\alpha V_t = V_t + V_2$$

$$\Rightarrow V_2 = (\alpha - 1)V_t \qquad (5)$$

Solving (3), (4) and (5):

$$2(\alpha-1)V_t - I_t + \frac{V_t}{R} - (\alpha-1)V_t = 0$$

$$\Rightarrow \left(2\alpha - 2 + \frac{1}{R} - \alpha + 1\right)V_t = I_t$$

$$\Rightarrow \frac{(\alpha-1)R+1}{R}V_t = I_t$$

$$\Rightarrow \frac{V_t}{I_t} = \frac{R}{(\alpha-1)R+1}$$

$$\Rightarrow R_{in} = \frac{R}{(\alpha-1)R+1}$$

Choice (3) is the answer (Fig. 6.11).

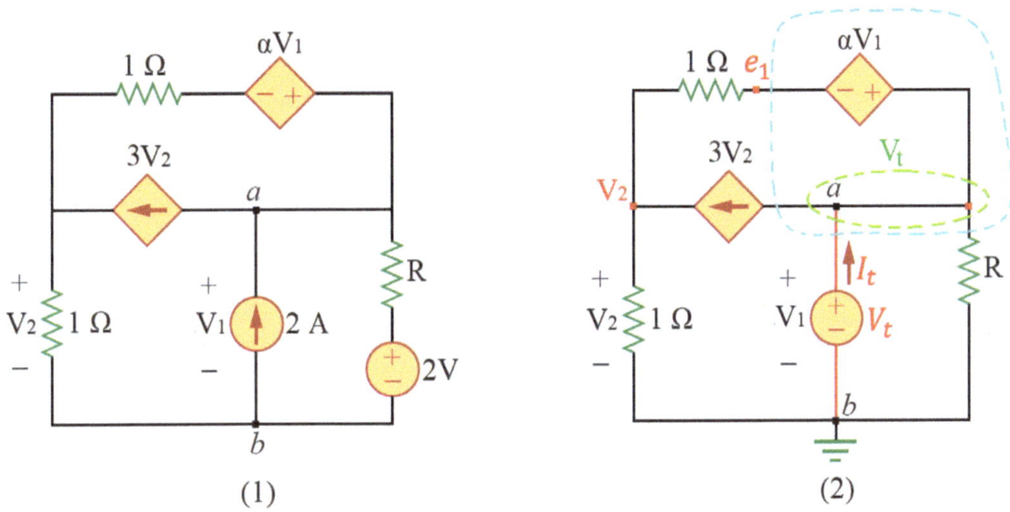

Fig. 6.11 The circuits of problem 6.17

6.19. *Step 1 (turning off all the independent sources)*

If the circuit includes any independent voltage and current sources, they need to be turned off. In this regard, each independent voltage source must be replaced by a short-circuit branch and each independent current source is replaced by an open-circuit branch.

Herein, the circuit does not include any independent source.

Step 2 (determining the type of problem)

Type 1: If the circuit does not include any dependent source (dependent voltage or current source), the series-parallel rule is applied to calculate the input resistance of the circuit.

Type 2: If the circuit includes at least a dependent source (dependent voltage or current source), a test source (independent voltage or current source) with the voltage and current of V_t and I_t needs to be connected to the input terminal of the circuit. This problem is a type 2 problem.

Step 3 (determining the type of circuit analysis)

Nodal analysis is applied to solve the problem although it does not matter which method of analysis is used in this problem because the circuit includes three meshes and three independent nodes.

Step 4 (formulating the circuit analysis equations)

KCL in node 2:

$$\frac{e_2}{4R}+\frac{e_2-0}{2R}-\alpha I=0$$

$$\Rightarrow \frac{3e_2}{4R}-\alpha I=0$$

$$\Rightarrow I=\frac{3}{4\alpha R}e_2 \tag{1}$$

KCL in node "a":

$$-I_t+\frac{V_t-0}{R}+\alpha I=0$$

$$\Rightarrow -I_t+\frac{V_t}{R}+\alpha I=0 \tag{2}$$

KCL in node 1:

$$I+\frac{0-V_t}{R}+\frac{0-e_2}{2R}=0$$

$$\Rightarrow I-\frac{V_t}{R}-\frac{e_2}{2R}=0 \tag{3}$$

Step 5 (solving the system of equations to calculate the value of $R_{in}=\frac{V_t}{I_t}$)

Solving (1) and (2):

$$-I_t+\frac{V_t}{R}+\frac{3}{4R}e_2=0 \tag{4}$$

Solving (1) and (3):

$$\frac{3}{4\alpha R}e_2-\frac{V_t}{R}-\frac{e_2}{2R}=0$$

$$\Rightarrow \left(\frac{3}{4\alpha R}-\frac{1}{2R}\right)e_2=\frac{V_t}{R}$$

$$\Rightarrow e_2=\frac{\frac{V_t}{R}}{\frac{3-2\alpha}{4\alpha R}}$$

$$\Rightarrow e_2=\frac{4\alpha}{3-2\alpha}V_t \tag{5}$$

Solving (4) and (5):

$$-I_t + \frac{V_t}{R} + \frac{3}{4R}\left(\frac{4\alpha}{3-2\alpha}V_t\right) = 0$$

$$\Rightarrow \left(\frac{1}{R} + \frac{3\alpha}{R(3-2\alpha)}\right)V_t = I_t$$

$$\Rightarrow \left(\frac{3-2\alpha+3\alpha}{R(3-2\alpha)}\right)V_t = I_t$$

$$\Rightarrow \left(\frac{3+\alpha}{R(3-2\alpha)}\right)V_t = I_t$$

$$\Rightarrow \frac{V_t}{I_t} = \frac{R(3-2\alpha)}{3+\alpha}$$

$$\Rightarrow R_{in} = \frac{R(3-2\alpha)}{3+\alpha}$$

Choice (4) is the answer (Fig. 6.12).

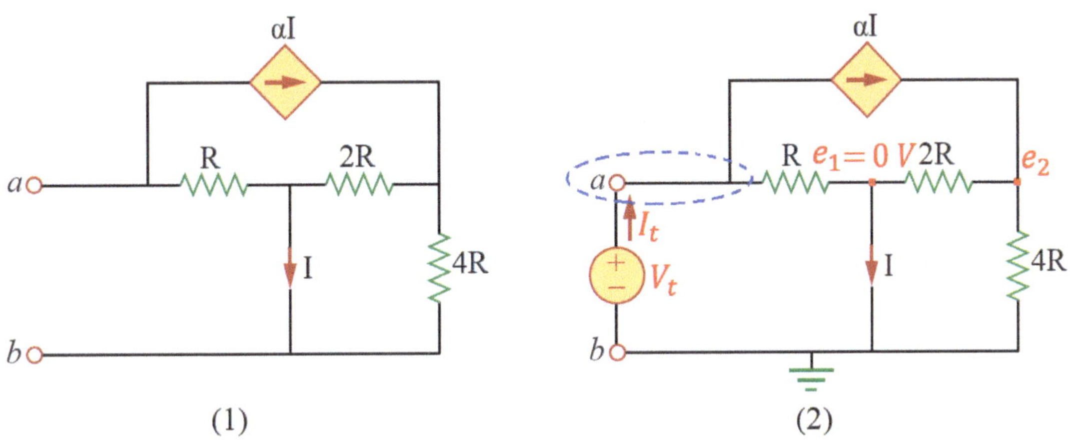

Fig. 6.12 The circuits of problem 6.19

6.2 Thevenin Theorem

6.21. *Step 1 (determining the method of solution)*

Method 1: First, a test source (independent voltage or current source) with the voltage and current of V_t and I_t is connected to the input terminal of the circuit. Then, the circuit is analyzed by using a circuit analysis method to find a relation between the voltage and current of the test source in the following form.

$$V_t = \alpha I_t + \beta$$

Then, it is concluded that:

$$R_{Th} = \alpha V_{Th} = \beta$$

As can be seen, in this method, the value of V_{Th} and R_{Th} are simultaneously calculated.

Method 2: In this method, the value of V_{Th} and R_{Th} are separately calculated. To calculate the value of V_{Th}, the terminal is assumed to be an open circuit branch. Then, the circuit is analyzed by using a circuit analysis method to find the value of V_{oc}. Next, we have:

$$V_{Th} = V_{oc}$$

Moreover, to calculate R_{Th}, which is the same as input resistance (R_{in}), first, all the independent sources (independent voltage and current sources) are turned off (each independent voltage source is replaced by short circuit branch and each independent current source is replaced by an open circuit branch). Then, if the circuit does not include any dependent source (dependent voltage or current source), the series-parallel rule is applied to calculate the input resistance of the circuit. However, if the circuit includes at least a dependent source, a test source (independent voltage or current source) with the voltage and current of V_t and I_t is connected to the input terminal of the circuit. Next, the circuit is analyzed by using a circuit analysis method to find the value of $\frac{V_t}{I_t}$. After that, we have:

$$R_{Th} = \frac{V_t}{I_t}$$

In this problem, the second method is applied because it is more straightforward.

Step 2 (solving the problem)

Calculating V_{Th}: As can be noticed from Fig. 6.13.2, $I = 1A$. Thus, by writing a KVL in the indicated mesh, we have:

$$-V_{oc} + 10 + 2 \times 1 + 3 \times 0 = 0$$

$$\Rightarrow V_{oc} = 12 \text{ V}$$

$$\Rightarrow V_{Th} = 12 \text{ V}$$

Calculating R_{Th}: As can be seen in Fig. 6.13.3, the input resistance of the circuit is a series connection of 2-Ω and 3-Ω resistors. Therefore:

$$R_{in} = 2 + 3$$

$$\Rightarrow R_{Th} = 5\ \Omega$$

Choice (3) is the answer.

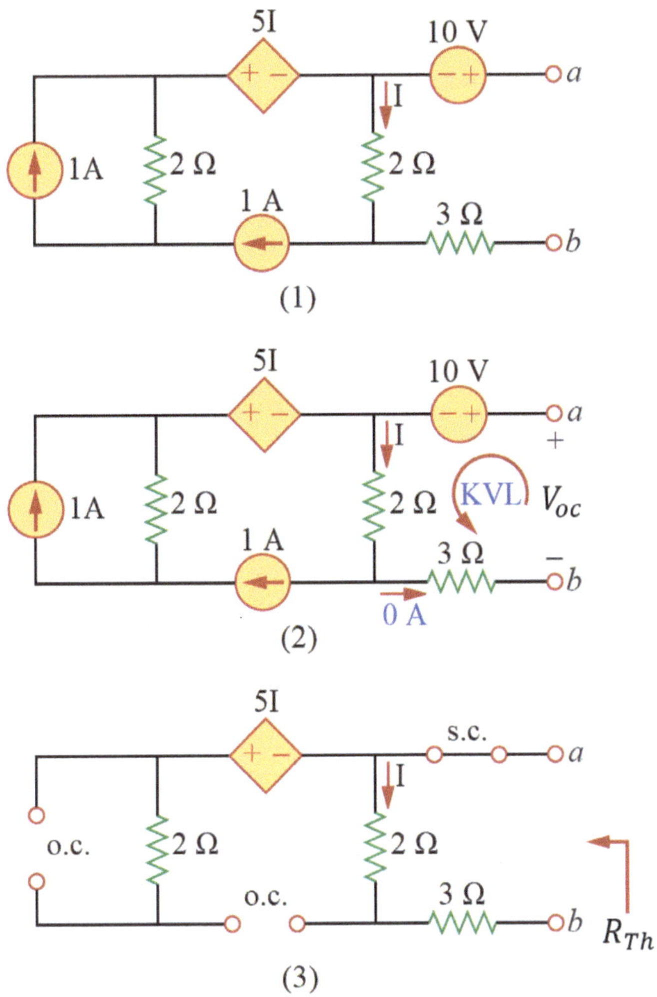

Fig. 6.13 The circuits of problem 6.21

6.22. *Step 1 (determining the method of solution)*

Method 1: First, a test source (independent voltage or current source) with the voltage and current of V_t and I_t is connected to the input terminal of the circuit. Then, the circuit is analyzed by using a circuit analysis method to find a relation between the voltage and current of the test source in the following form.

$$V_t = \alpha I_t + \beta$$

Then, it is concluded that:

$$R_{Th} = \alpha$$

$$V_{Th} = \beta$$

As can be seen, in this method, the value of V_{Th} and R_{Th} are simultaneously calculated.

Method 2: In this method, the value of V_{Th} and R_{Th} are separately calculated. To calculate the value of V_{Th}, the terminal is assumed to be an open circuit branch. Then, the circuit is analyzed by using a circuit analysis method to find the value of V_{oc}. Next, we have:

$$V_{Th} = V_{oc}$$

Moreover, to calculate R_{Th}, which is the same as input resistance (R_{in}), first, all the independent sources (independent voltage and current sources) are turned off (each independent voltage source is replaced by short circuit branch and each independent current source is replaced by an open circuit branch). Then, if the circuit does not include any dependent source (dependent voltage or current source), the series-parallel rule is applied to calculate the input resistance of the circuit. However, if the circuit includes at least a dependent source, a test source (independent voltage or current source) with the voltage and current of V_t and I_t is connected to the input terminal of the circuit. Next, the circuit is analyzed by using a circuit analysis method to find the value of $\frac{V_t}{I_t}$. After that, we have:

$$R_{Th} = \frac{V_t}{I_t}$$

In this problem, the first method is applied because it is more straightforward.

Step 2 (solving the problem)

Using Ohm's law for the 2-Ω resistor to define V_1 based on the mesh currents:

$$V_1 = -2i_1 \quad (1)$$

KVL in the supermesh:

$$-20 + 2i_1 + 3V_1 = 0 \quad (2)$$

KVL in the right-side mesh:

$$-V_t + 5I_t + 3V_1 = 0 \quad (3)$$

Solving (1) and (2):

$$-20 - V_1 + 3V_1 = 0$$

$$\Rightarrow V_1 = 10 \quad (4)$$

Solving (3) and (4):

$$-V_t + 5I_t + 3 \times 10 = 0$$

$$\Rightarrow V_t = 5I_t + 30$$

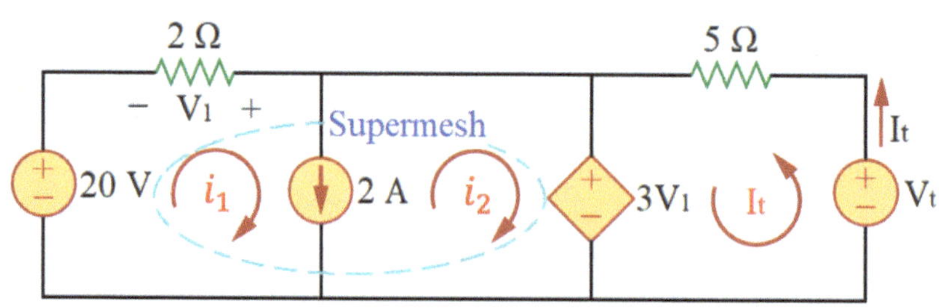

Fig. 6.14 The circuit of problem 6.22

Choice (2) is the answer (Fig. 6.14).

6.24. To calculate the value of V_{Th}, the terminal is assumed to be an open circuit branch. Then, the circuit is analyzed by using a circuit analysis method to find the value of V_{oc}. Next, we have:

$$V_{Th} = V_{oc}$$

As can be noticed from the circuit, we have:

$$V_{oc} = -3\ \text{V} \tag{1}$$

Applying KCL in the supernode:

$$I_x - 1 = 0$$

$$\Rightarrow I_x = 1$$

Using Ohm's law for the horizontal 1-Ω resistor:

$$V = -1 \times 1$$

$$\Rightarrow V = -1\ \text{V} \tag{2}$$

Solving (1) and (2):

$$V_{oc} = -3 \times (-1)$$

$$\Rightarrow V_{oc} = 3\ \text{V}$$

$$\Rightarrow V_{Th} = 3\ \text{V}$$

Choice (1) is the answer (Fig. 6.15).

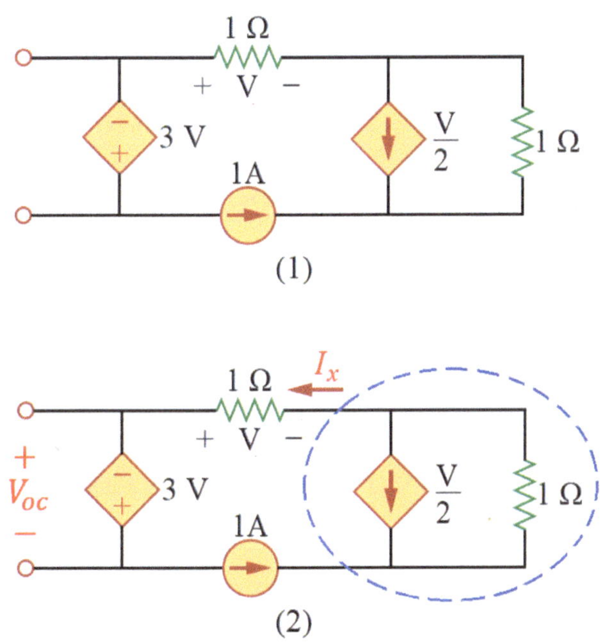

Fig. 6.15 The circuits of problem 6.24

6.26. *Step 1 (determining the method of solution)*

Method 1: First, a test source (independent voltage or current source) with the voltage and current of V_t and I_t is connected to the input terminal of the circuit. Then, the circuit is analyzed by using a circuit analysis method to find a relation between the voltage and current of the test source in the following form.

$$V_t = \alpha I_t + \beta$$

Then, it is concluded that:

$$R_{Th} = \alpha$$

$$V_{Th} = \beta$$

As can be seen, in this method, the value of V_{Th} and R_{Th} are simultaneously calculated.

Method 2: In this method, the value of V_{Th} and R_{Th} are separately calculated. To calculate the value of V_{Th}, the terminal is assumed to be an open circuit branch. Then, the circuit is analyzed by using a circuit analysis method to find the value of V_{oc}. Next, we have:

$$V_{Th} = V_{oc}$$

Moreover, to calculate R_{Th}, which is the same as input resistance (R_{in}), first, all the independent sources (independent voltage and current sources) are turned off (each independent voltage source is replaced by short circuit branch and each independent current source is replaced by an open circuit branch). Then, if the circuit does not include any dependent source (dependent voltage or current source), the series-parallel rule is applied to calculate the input resistance of the circuit. However, if the circuit includes at least a dependent source, a test source (independent voltage or current source) with the voltage and current of V_t and I_t is connected to the input terminal of the circuit. Next, the circuit is analyzed by using a circuit analysis method to find the value of $\frac{V_t}{I_t}$. After that, we have:

$$R_{Th} = R_{in} = \frac{V_t}{I_t}$$

In this problem, the second method is applied because it is more straightforward.

Step 2 (solving the problem)

Calculating R_{Th}: From the circuit shown in Fig. 6.16.2, it is noticed that:

$$I_2 = I_t \qquad (1)$$

Defining the current of the dependent current source based on the mesh currents:

$$3I_2 = i_1 - I_t \qquad (2)$$

Using Ohm's law for the bottom 1-Ω resistor to define V_1 based on mesh currents:

$$V_1 = 1 \times i_1$$

$$\Rightarrow V_1 = i_1 \qquad (3)$$

KVL in the supermesh including both meshes:

$$-V_t + 5V_1 + I_2 + V_1 = 0$$

$$\Rightarrow -V_t + 6V_1 + I_2 = 0 \quad (4)$$

Now, the equations can be solved as follows.
Solving (1)-(3):

$$3I_t = V_1 - I_t$$

$$\Rightarrow V_1 = 4\,I_t \quad (5)$$

Solving (1), (4), and (5):

$$-V_t + 6 \times 4I_t + I_t = 0$$

$$\Rightarrow V_t = 25\,I_t$$

$$\Rightarrow \frac{V_t}{I_t} = 25$$

$$\Rightarrow R_{Th} = 25\ \Omega$$

Calculating V_{Th}: Writing KVL in the left-side mesh of the circuit shown in Fig. 6.16.3:

$$-V_{oc} + 5V_1 + I_2 \times 1 + V_1 = 0$$

$$\Rightarrow V_{oc} = 6V_1 + I_2 \quad (6)$$

KCL in the top node:

$$-1 + I_2 = 0$$

$$\Rightarrow I_2 = 1\ \text{A} \quad (7)$$

KCL in the node with the node voltage of V_1:

$$-I_2 - 3I_2 + \frac{V_1}{1} = 0$$

$$\Rightarrow V_1 = 4I_2 \quad (8)$$

Solving (6)–(8):

$$V_{oc} = 6 \times 4 \times 1 + 1$$

$$\Rightarrow V_{oc} = 25\ \text{V}$$

$$\Rightarrow V_{Th} = 25\ \text{V}$$

Choice (4) is the answer.

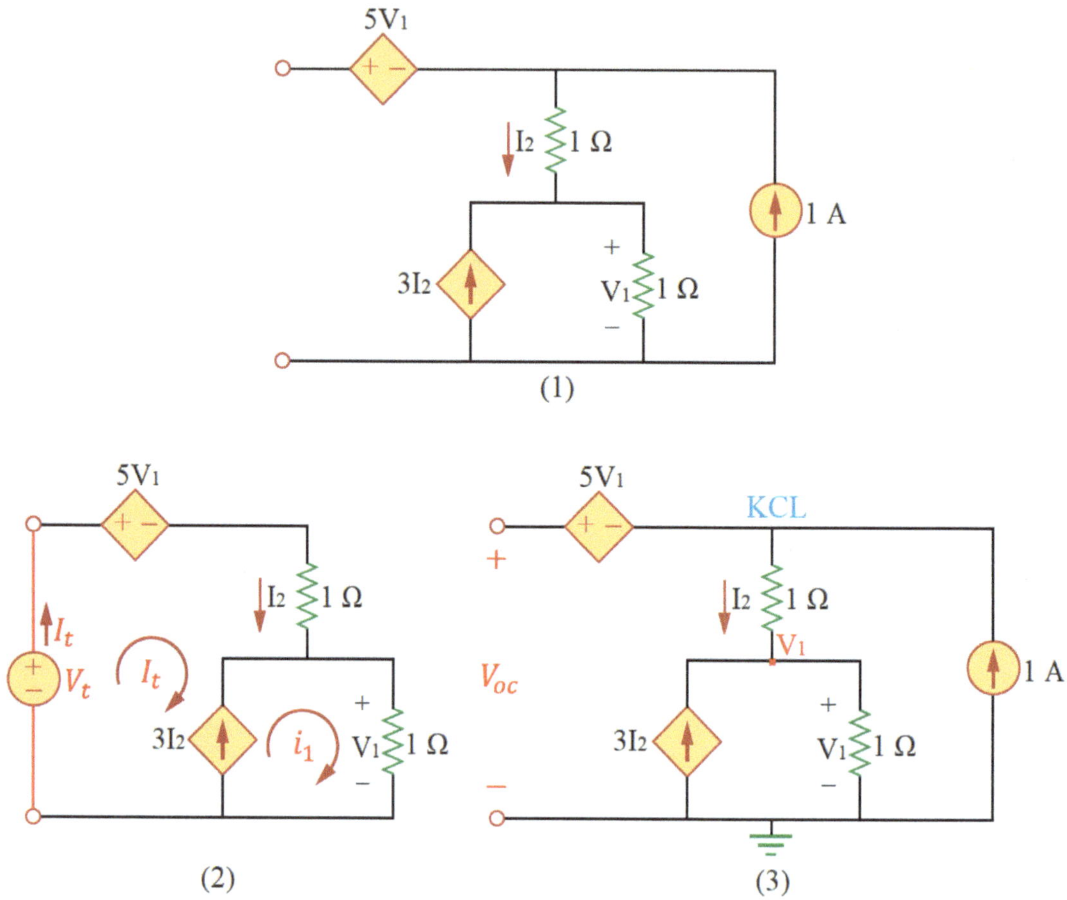

Fig. 6.16 The circuits of problem 6.26

6.28. *Step 1 (determining the method of solution)*

Method 1: First, a test source (independent voltage or current source) with the voltage and current of V_t and I_t is connected to the input terminal of the circuit. Then, the circuit is analyzed by using a circuit analysis method to find a relation between the voltage and current of the test source in the following form.

$$V_t = \alpha I_t + \beta$$

Then, it is concluded that:

$$R_{Th} = \alpha$$

$$V_{Th} = \beta$$

As can be seen, in this method, the value of V_{Th} and R_{Th} are simultaneously calculated.

Method 2: In this method, the value of V_{Th} and R_{Th} are separately calculated. To calculate the value of V_{Th}, the terminal is assumed to be an open circuit branch. Then, the circuit is analyzed by using a circuit analysis method to find the value of V_{oc}. Next, we have:

$$V_{Th} = V_{oc}$$

Moreover, to calculate R_{Th}, which is the same as input resistance (R_{in}), first, all the independent sources (independent voltage and current sources) are turned off (each independent voltage source is replaced by short circuit branch and each independent current source is replaced by an open circuit branch). Then, if the circuit does not include any dependent source (dependent voltage or current source), the series-parallel rule is applied to calculate the input resistance of the circuit. However, if the circuit includes at least a dependent source, a test source (independent voltage or current source) with the voltage and current of V_t and I_t is connected to the input terminal of the circuit. Next, the circuit is analyzed by using a circuit analysis method to find the value of $\frac{V_t}{I_t}$. After that, we have:

$$R_{Th} = R_{in} = \frac{V_t}{I_t}$$

In this problem, the second method is applied because it is more straightforward.

Step 2 (solving the problem)

Calculating V_{Th}: As can be seen in the circuit of Fig. 6.17.2, the node with the node voltage e_1 has been grounded. Thus:

$$e_1 = 0 \text{ V} \tag{1}$$

Defining the voltage of horizontal dependent voltage source based on the node voltages:

$$2I = V_{oc} - e_1$$

$$\Rightarrow I = \frac{V_{oc} - e_1}{2} \tag{2}$$

KCL in the supernode:

$$-6 + \frac{e_1 - 2I}{4} + I = 0$$

$$\Rightarrow -6 + \frac{e_1}{4} + \frac{1}{2}I = 0 \tag{3}$$

Solving (1)-(3):

$$-6 + \frac{0}{4} + \frac{1}{2} \times \frac{V_{oc} - 0}{2} = 0$$

$$\Rightarrow V_{oc} = 24$$

$$\Rightarrow V_{Th} = 24 \text{ V}$$

Calculating R_{Th}: Defining the voltage of the horizontal dependent voltage source based on the node voltages in the circuit of Fig. 6.17.3:

$$2I = V_t - e_1$$

$$\Rightarrow I = \frac{V_t - e_1}{2} \tag{4}$$

KCL in the supernode:

$$-I_t+\frac{e_1-2I}{4}+I=0$$

$$\Rightarrow -I_t+\frac{e_1}{4}+\frac{1}{2}I=0 \qquad (5)$$

Solving (1), (4), and (5):

$$-I_t+\frac{0}{4}+\frac{1}{2}\times\frac{V_t-0}{2}=0$$

$$\Rightarrow \frac{V_t}{I_t}=4$$

$$\Rightarrow R_{Th}=4\ \Omega$$

Choice (4) is the answer.

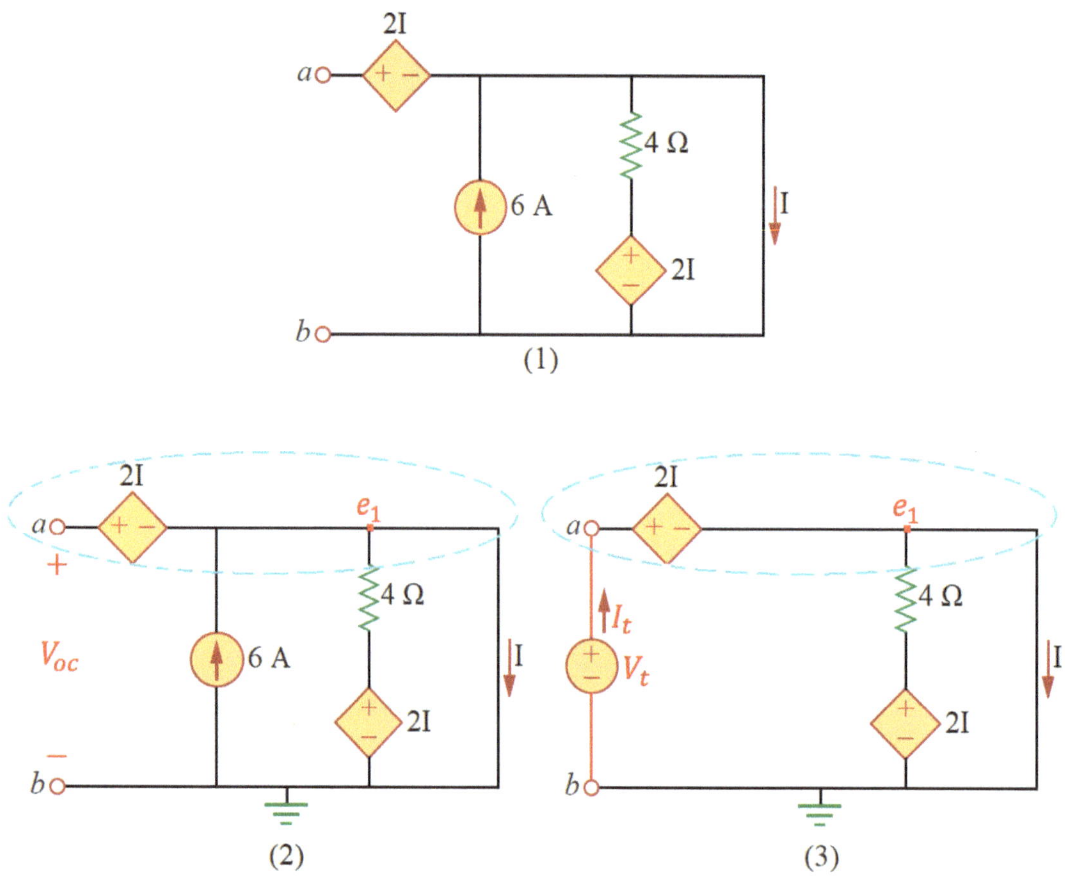

Fig. 6.17 The circuits of problem 6.28

6.30. *Step 1 (determining the method of solution)*

Method 1: First, a test source (independent voltage or current source) with the voltage and current of V_t and I_t is connected to the input terminal of the circuit. Then, the circuit is analyzed by using a circuit analysis method to find a relation between the voltage and current of the test source in the following form.

$$V_t = \alpha I_t + \beta$$

Then, it is concluded that:

$$R_{Th} = \alpha$$

$$V_{Th} = \beta$$

As can be seen, in this method, the value of V_{Th} and R_{Th} are simultaneously calculated.

Method 2: In this method, the value of V_{Th} and R_{Th} are separately calculated. To calculate the value of V_{Th}, the terminal is assumed to be an open circuit branch. Then, the circuit is analyzed by using a circuit analysis method to find the value of V_{oc}. Next, we have:

$$V_{Th} = V_{oc}$$

Moreover, to calculate R_{Th}, which is the same as input resistance (R_{in}), first, all the independent sources (independent voltage and current sources) are turned off (each independent voltage source is replaced by short circuit branch and each independent current source is replaced by an open circuit branch). Then, if the circuit does not include any dependent source (dependent voltage or current source), the series-parallel rule is applied to calculate the input resistance of the circuit. However, if the circuit includes at least a dependent source, a test source (independent voltage or current source) with the voltage and current of V_t and I_t is connected to the input terminal of the circuit. Next, the circuit is analyzed by using a circuit analysis method to find the value of $\frac{V_t}{I_t}$. After that, we have:

$$R_{Th} = R_{in} = \frac{V_t}{I_t}$$

In this problem, the second method is applied because it is more straightforward.

Step 2 (solving the problem)

Calculating R_{Th}: As can be noticed from the circuit of Fig. 6.18.2, we have:

$$V_o = V_t \tag{1}$$

Using Ohm's law for the 2-Ω resistor to define V_x based on the mesh current:

$$V_x = -2i_1 \tag{2}$$

KVL in mesh 1:

$$2.5V_x + 2i_1 + 3(i_1 - 2V_o) + 4(i_1 + I_t) = 0$$

$$\Rightarrow 2.5V_x + 9i_1 - 6V_o + 4I_t = 0 \tag{3}$$

KVL in the right-side mesh:

$$-V_t + 4(i_1 + I_t) = 0 \tag{4}$$

Solving (1)-(3):

$$2.5(-2i_1) + 9i_1 - 6V_t + 4I_t = 0$$

$$i_1 = 1.5V_t - I_t \tag{5}$$

Solving (4) and (5):

$$-V_t + 4(1.5V_t - I_t + I_t) = 0$$

$$\Rightarrow 5V_t = 0$$

$$\Rightarrow V_t = 0$$

$$\Rightarrow \frac{V_t}{I_t} = 0$$

$$\Rightarrow R_{Th} = 0 \tag{6}$$

Calculating V_{Th}: Using Ohm's law for the 4-Ω resistor to define V_o based on the mesh current in the circuit of Fig. 6.18.3:

$$V_o = 4i_1 \tag{7}$$

Using Ohm's law for the 2-Ω resistor to define V_x based on the mesh currents:

$$V_x = 2(I_s - i_1) \tag{8}$$

KVL in mesh 1:

$$2.5V_x + 2(i_1 - I_s) + 3(i_1 - 2V_o) + 4i_1 = 0$$

$$\Rightarrow 2.5V_x + 9i_1 - 2I_s - 6V_o = 0 \tag{9}$$

Solving (7)-(9):

$$2.5 \times 2\left(I_s - \frac{V_o}{4}\right) + 9 \times \frac{V_o}{4} - 2I_s - 6V_o = 0$$

$$\Rightarrow 3I_s - 5V_o = 0$$

$$\Rightarrow V_o = \frac{3}{5}I_s$$

$$\Rightarrow V_{oc} = \frac{3}{5}I_s$$

$$\Rightarrow V_{Th} = \frac{3}{5}I_s \tag{10}$$

Therefore, from (6) and (10), it is concluded that the Thevenin equivalent circuit is a single independent voltage source with the size of $\frac{3}{5}I_s$. Choice (3) is the answer.

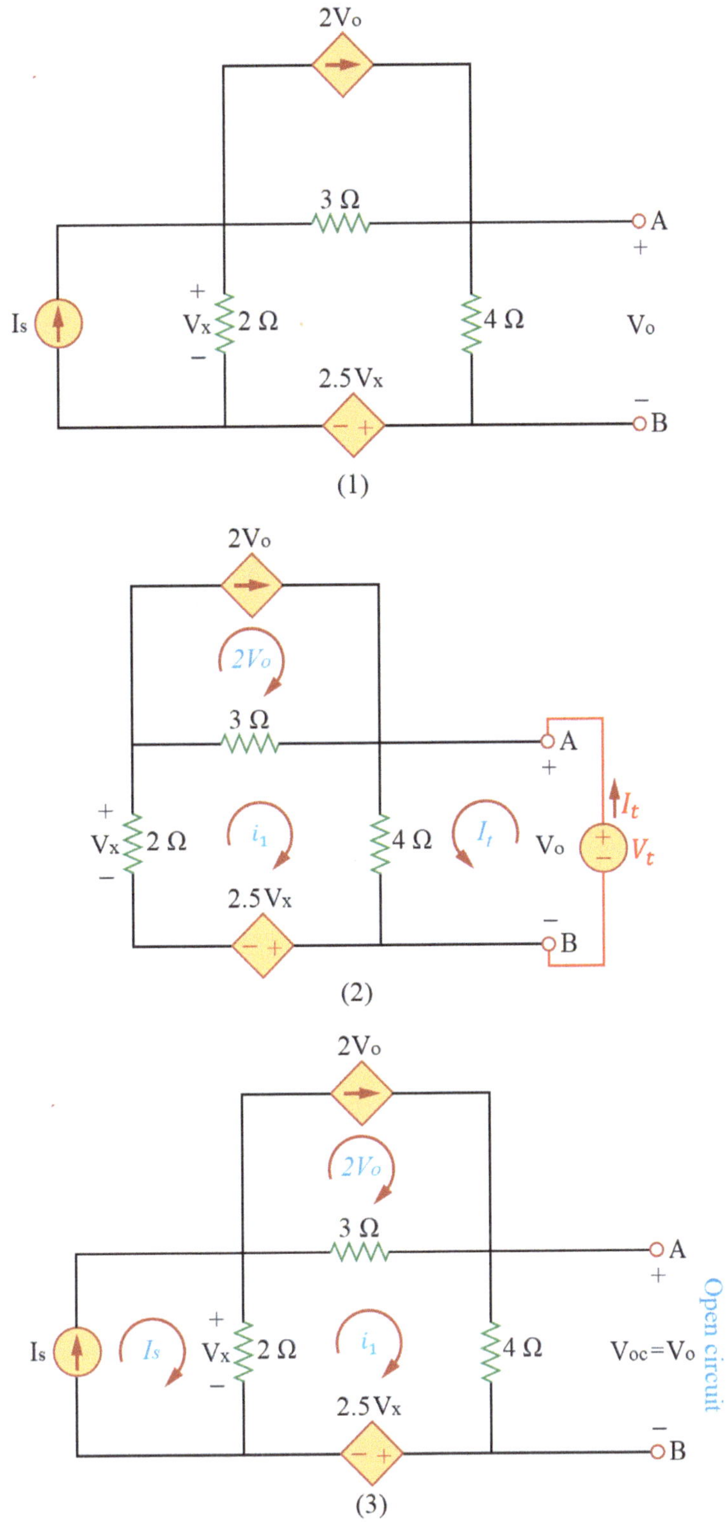

Fig. 6.18 The circuits of problem 6.30

6.31. *Step 1 (determining the method of solution)*

Method 1: First, a test source (independent voltage or current source) with the voltage and current of V_t and I_t is connected to the input terminal of the circuit. Then, the circuit is analyzed by using a circuit analysis method to find a relation between the voltage and current of the test source in the following form.

$$V_t = \alpha I_t + \beta$$

Then, it is concluded that:

$$R_{Th} = \alpha$$

$$V_{Th} = \beta$$

As can be seen, in this method, the value of V_{Th} and R_{Th} are simultaneously calculated.

Method 2: In this method, the value of V_{Th} and R_{Th} are separately calculated. To calculate the value of V_{Th}, the terminal is assumed to be an open circuit branch. Then, the circuit is analyzed by using a circuit analysis method to find the value of V_{oc}. Next, we have:

$$V_{Th} = V_{oc}$$

Moreover, to calculate R_{Th}, which is the same as input resistance (R_{in}), first, all the independent sources (independent voltage and current sources) are turned off (each independent voltage source is replaced by short circuit branch and each independent current source is replaced by an open circuit branch). Then, if the circuit does not include any dependent source (dependent voltage or current source), the series-parallel rule is applied to calculate the input resistance of the circuit. However, if the circuit includes at least a dependent source, a test source (independent voltage or current source) with the voltage and current of V_t and I_t is connected to the input terminal of the circuit. Next, the circuit is analyzed by using a circuit analysis method to find the value of $\frac{V_t}{I_t}$. After that, we have:

$$R_{Th} = R_{in} = \frac{V_t}{I_t}$$

In this problem, the second method is applied because it is more straightforward.

Step 2 (solving the problem)

Calculating R_{Th}: Defining I_x based on the node voltages in the circuit of Fig. 6.19.2:

$$I_x = \frac{e_1}{8} \tag{1}$$

KCL in node 1:

$$\frac{e_1}{8} - 2I_x + \frac{e_1 - e_2}{10} = 0$$

$$\Rightarrow \frac{9}{40}e_1 - 2I_x - \frac{1}{10}e_2 = 0 \tag{2}$$

KCL in node 2:

$$\frac{e_2 - e_1}{10} + \frac{e_2}{3} + \frac{e_2 - V_t}{4} + 2I_x = 0$$

$$\Rightarrow -\frac{1}{10}e_1 + \frac{41}{60}e_2 + \frac{1}{4}V_t + 2I_x = 0 \quad (3)$$

Defining I_t based on the node voltages:

$$I_t = \frac{V_t - e_2}{4}$$

$$\Rightarrow e_2 = V_t - 4I_t \quad (4)$$

Solving (1) and (2):

$$\frac{9}{40}e_1 - 2 \times \frac{e_1}{8} - \frac{1}{10}e_2 = 0$$

$$\Rightarrow -\frac{1}{40}e_1 - \frac{1}{10}e_2 = 0$$

$$\Rightarrow e_1 = -4e_2 \quad (5)$$

Solving (1), (3), and (5):

$$-\frac{1}{10} \times (-4e_2) + \frac{41}{60}e_2 + \frac{1}{4}V_t + 2 \times \frac{(-4e_2)}{8} = 0$$

$$\Rightarrow \frac{1}{12}e_2 - \frac{V_t}{4} = 0$$

$$\Rightarrow e_2 = 3V_t \quad (6)$$

Solving (4) and (6):

$$3V_t = V_t - 4I_t$$

$$\Rightarrow V_t = -2I_t$$

$$\Rightarrow \frac{V_t}{I_t} = -2$$

$$R_{Th} = -2\ \Omega$$

Calculating V_{Th}: KCL in node 1 in the circuit of Fig. 6.19.3:

$$-I_s + \frac{e_1}{8} - 2I_x + \frac{e_1 - e_2}{10} = 0$$

$$\Rightarrow -I_s + \frac{9}{40}e_1 - 2I_x - \frac{1}{10}e_2 = 0 \quad (7)$$

KCL in node 2:

$$\frac{e_2 - e_1}{10} + \frac{e_2}{3} + 2I_x = 0$$

$$\Rightarrow -\frac{1}{10}e_1 + \frac{13}{30}e_2 + 2I_x = 0 \quad (8)$$

Solving (1) and (7):

$$-I_s + \frac{9}{40}e_1 - 2 \times \frac{e_1}{8} - \frac{1}{10}e_2 = 0$$

$$\Rightarrow -I_s - \frac{1}{40}e_1 - \frac{1}{10}e_2 = 0 \quad (9)$$

Solving (1) and (8):

$$-\frac{1}{10}e_1 + \frac{13}{30}e_2 + 2 \times \frac{e_1}{8} = 0$$

$$\Rightarrow \frac{3}{20}e_1 + \frac{13}{30}e_2 = 0$$

$$\Rightarrow e_1 = -\frac{26}{9}e_2 \quad (10)$$

Solving (9) and (10):

$$-I_s - \frac{1}{40}\left(-\frac{26}{9}e_2\right) - \frac{1}{10}e_2 = 0$$

$$\Rightarrow -\frac{1}{36}e_2 = I_s$$

$$\Rightarrow e_2 = -36I_s$$

Since there is no voltage drop on the 4-Ω resistor (the terminal is open circuit), we have:

$$\Rightarrow V_{oc} = -36I_s$$

$$\Rightarrow V_{Th} = -36\ I_s$$

Choice (1) is the answer.

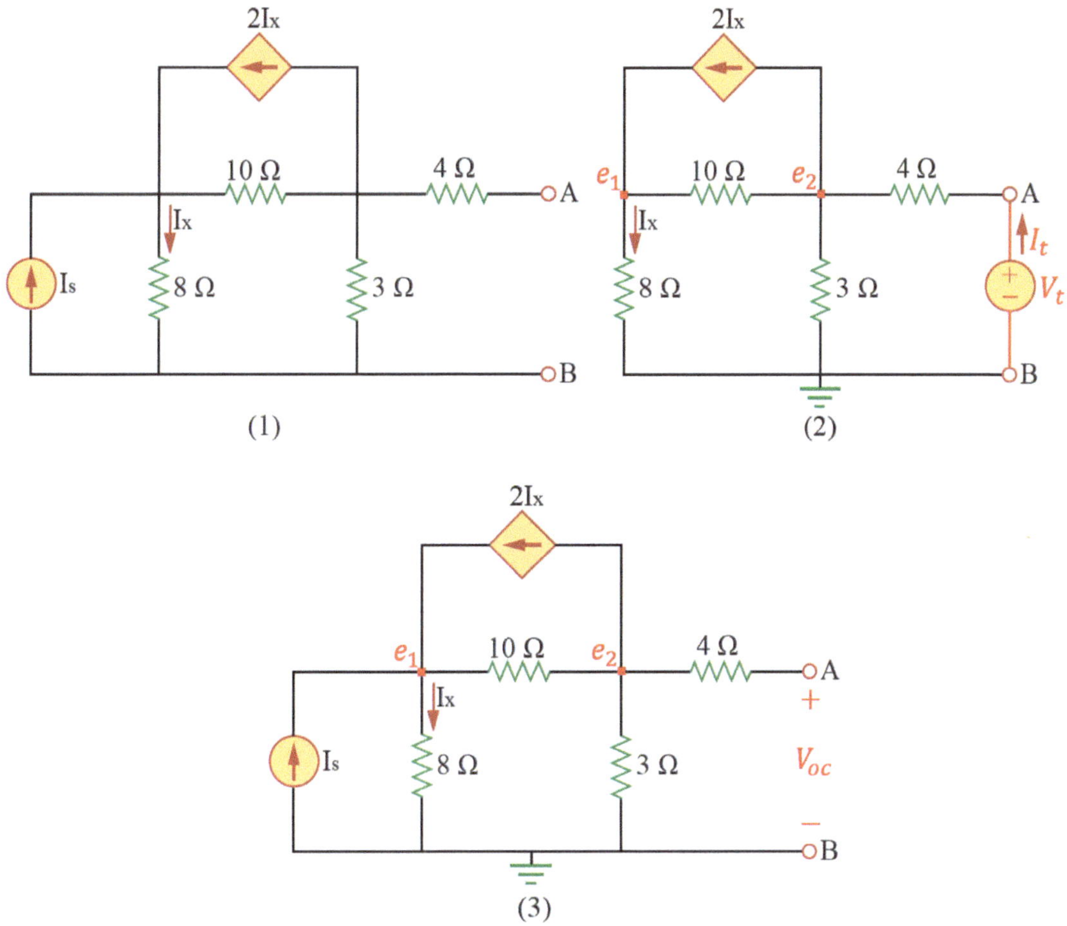

Fig. 6.19 The circuits of problem 6.31

6.32. *Step 1 (determining the method of solution)*

Method 1: First, a test source (independent voltage or current source) with the voltage and current of V_t and I_t is connected to the input terminal of the circuit. Then, the circuit is analyzed by using a circuit analysis method to find a relation between the voltage and current of the test source in the following form.

$$V_t = \alpha I_t + \beta$$

Then, it is concluded that:

$$R_{Th} = \alpha$$

$$V_{Th} = \beta$$

As can be seen, in this method, the value of V_{Th} and R_{Th} are simultaneously calculated.

Method 2: In this method, the value of V_{Th} and R_{Th} are separately calculated. To calculate the value of V_{Th}, the terminal is assumed to be an open circuit branch. Then, the circuit is analyzed by using a circuit analysis method to find the value of V_{oc}. Next, we have:

$$V_{Th} = V_{oc}$$

Moreover, to calculate R_{Th}, which is the same as input resistance (R_{in}), first, all the independent sources (independent voltage and current sources) are turned off (each independent voltage source is replaced by short circuit branch and each independent current source is replaced by an open circuit branch). Then, if the circuit does not include any dependent source (dependent voltage or current source), the series-parallel rule is applied to calculate the input resistance of the circuit. However, if the circuit includes at least a dependent source, a test source (independent voltage or current source) with the voltage and current of V_t and I_t is connected to the input terminal of the circuit. Next, the circuit is analyzed by using a circuit analysis method to find the value of $\frac{V_t}{I_t}$. After that, we have:

$$R_{Th} = R_{in} = \frac{V_t}{I_t}$$

In this problem, the second method is applied because it is more straightforward.

Step 2 (solving the problem)

Calculating V_{Th}: KVL in the indicated loop of the circuit of Fig. 6.20.2:

$$-5V_x - V_x + E = 0$$

$$\Rightarrow V_x = \frac{1}{6}\text{E} \tag{1}$$

As can be noticed from the circuit, we have:

$$V_{oc} = 5V_x \tag{2}$$

Solving (1) and (2):

$$V_{oc} = \frac{5}{6}\text{E}$$

$$\Rightarrow V_{Th} = \frac{5}{6}\text{E}$$

Calculating R_{Th}: As can be seen in Fig. 6.20.3, after turning off the independent voltage source (replacing it by a short circuit branch), its parallel resistor is eliminated.

KVL in the loop of the circuit of Fig. 6.20.4:

$$-5V_x - V_x = 0$$

$$\Rightarrow V_x = 0 \tag{3}$$

As can be noticed from the circuit, we have:

$$V_t = 5V_x \tag{4}$$

Solving (3) and (4):

$$V_t = 0$$

$$\Rightarrow \frac{V_t}{I_t} = 0$$

$$\Rightarrow R_{Th} = 0$$

Choice (3) is the answer.

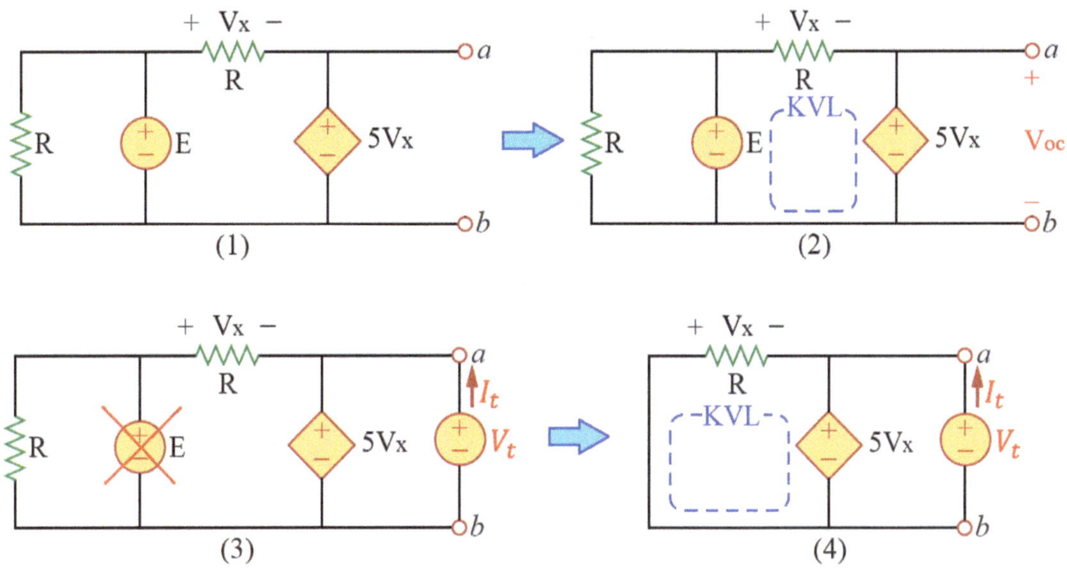

Fig. 6.20 The circuits of problem 6.32

6.33. *Step 1 (determining the method of solution)*

Method 1: First, a test source (independent voltage or current source) with the voltage and current of V_t and I_t is connected to the input terminal of the circuit. Then, the circuit is analyzed by using a circuit analysis method to find a relation between the voltage and current of the test source in the following form.

$$V_t = \alpha I_t + \beta$$

Then, it is concluded that:

$$R_{Th} = \alpha V_{Th} = \beta$$

As can be seen, in this method, the value of V_{Th} and R_{Th} are simultaneously calculated.

Method 2: In this method, the value of V_{Th} and R_{Th} are separately calculated. To calculate the value of V_{Th}, the terminal is assumed to be an open circuit branch. Then, the circuit is analyzed by using a circuit analysis method to find the value of V_{oc}. Next, we have:

$$V_{Th} = V_{oc}$$

Moreover, to calculate R_{Th}, which is the same as input resistance (R_{in}), first, all the independent sources (independent voltage and current sources) are turned off (each independent voltage source is replaced by short circuit branch and each independent current source is replaced by an open circuit branch). Then, if the circuit does not include any dependent source (dependent voltage or current source), the series-parallel rule is applied to calculate the input resistance of the circuit. However, if the circuit includes at least a dependent source, a test source (independent voltage or current source) with the voltage and current of V_t and I_t is connected to the input terminal of the circuit. Next, the circuit is analyzed by using a circuit analysis method to find the value of $\frac{V_t}{I_t}$. After that, we have:

$$R_{Th} = R_{in} = \frac{V_t}{I_t}$$

In this problem, the first method is applied because it is more straightforward.

Step 2 (solving the problem)

KCL in the node with the node voltage of V (see Fig. 6.21.2):

$$\frac{V}{1} + \frac{V-2}{1} + \frac{V-V_t}{1} = 0$$

$$\Rightarrow 3V - 2 - V_t = 0$$

$$\Rightarrow V = \frac{2+V_t}{3} \tag{1}$$

KCL in node "a":

$$-I_t + \frac{V_t - aV}{1} + \frac{V_t - V}{1} = 0$$

$$\Rightarrow -I_t + 2V_t - (a+1)V = 0 \tag{2}$$

Solving (1) and (2):

$$-I_t + 2V_t - (a+1)\frac{2+V_t}{3} = 0$$

$$\Rightarrow -I_t + \left(\frac{5-a}{3}\right)V_t - \frac{2a+2}{3} = 0$$

$$\Rightarrow V_t = \frac{3}{5-a}I_t + \frac{2a+2}{5-a}$$

Therefore, the Thevenin resistance and voltage of the circuit are as follows.

$$R_{Th} = \frac{3}{5-a}$$

$$V_{Th} = \frac{2a+2}{5-a}$$

Now, the condition $a \to \infty$ is applied as follows.

$$\lim_{a\to\infty} R_{Th} = \lim_{a\to\infty} \frac{3}{5-a}$$

$$\Rightarrow \lim_{a\to\infty} R_{Th} = 0\ \Omega$$

$$\lim_{a\to\infty} V_{Th} = \lim_{a\to\infty} \frac{2a+2}{5-a}$$

$$\Rightarrow \lim_{a\to\infty} V_{Th} = -2\ \mathrm{V}$$

Choice (2) is the answer.

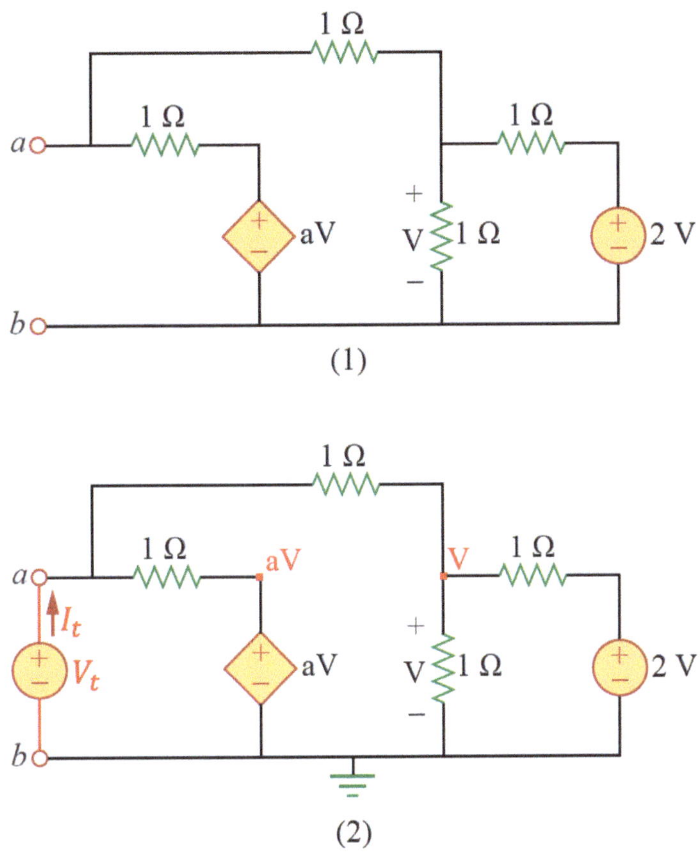

Fig. 6.21 The circuits of problem 6.33

6.3 Norton Theorem

6.35. Based on the information given in the problem, we have:

$$R_{Th} = 2\ \Omega$$

$$V_{Th} = 3\ \mathrm{V}$$

As we know, the relations below are available between the Norton and Thevenin parameters of a circuit.

$$R_N = R_{Th}$$

$$I_N = \frac{V_{Th}}{R_{Th}}$$

Therefore:

$$R_N = 2\ \Omega$$

$$I_N = \frac{3}{2}\ \text{A}$$

Choice (1) is the answer.

6.36. To calculate R_N, which is the same as input resistance (R_{in}), first, all the independent sources (independent voltage and current sources) are turned off (each independent voltage source is replaced by short circuit branch and each independent current source is replaced by an open circuit branch). Then, if the circuit does not include any dependent source (dependent voltage or current source), the series-parallel rule is applied to calculate the input resistance of the circuit. However, if the circuit includes at least a dependent source, a test source (independent voltage or current source) with the voltage and current of V_t and I_t is connected to the input terminal of the circuit. Next, the circuit is analyzed by using a circuit analysis method to find the value of $\frac{V_t}{I_t}$. After that, we have:

$$R_N = \frac{V_t}{I_t}$$

Defining I based on the node voltages:

$$I = -\frac{e_1}{10} \qquad (1)$$

KCL in the left-side node:

$$\frac{e_1}{10} + 2I + \frac{e_1 - V_t}{4} = 0$$

$$\Rightarrow \frac{7}{20}e_1 + 2I - \frac{V_t}{4} = 0 \qquad (2)$$

KCL in the right-side node:

$$-I_t + \frac{V_t}{5} + \frac{V_t - e_1}{4} = 0$$

$$\Rightarrow -I_t + \frac{9}{20}V_t - \frac{1}{4}e_1 = 0 \qquad (3)$$

Solving (1) and (2):

$$\frac{7}{20}e_1 + 2\left(-\frac{e_1}{10}\right) - \frac{V_t}{4} = 0$$

$$\Rightarrow \frac{3}{20}e_1 - \frac{V_t}{4} = 0$$

$$\Rightarrow e_1 = \frac{5}{3}V_t \qquad (4)$$

Solving (3) and (4):

$$-I_t + \frac{9}{20}V_t - \frac{1}{4}\left(\frac{5}{3}V_t\right) = 0$$

$$\Rightarrow -I_t + \frac{1}{30}V_t = 0$$

$$\Rightarrow \frac{V_t}{I_t} = 30$$

$$\Rightarrow R_N = 30\ \Omega$$

Choice (1) is the answer (Fig. 6.22).

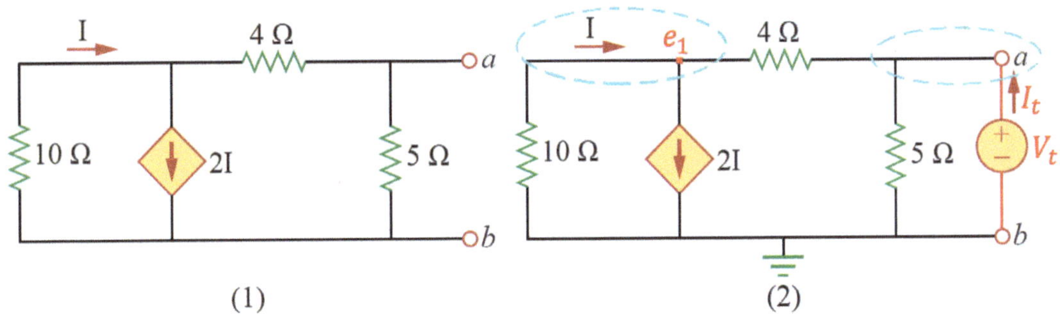

Fig. 6.22 The circuits of problem 6.35

6.37. Since the circuit does not have any independent source, it is concluded that:

$$I_N = 0$$

Choice (3) is the answer.

6.38. *Step 1 (determining the method of solution)*.

Method 1: First, a test source (independent voltage or current source) with the voltage and current of V_t and I_t is connected to the input terminal of the circuit. Then, the circuit is analyzed by using a circuit analysis method to find a relation between the voltage and current of the test source in the following form.

$$I_t = \frac{V_t}{\alpha} - \beta$$

Then, it is concluded that:

$$R_N = \alpha$$

$$I_N = \beta$$

As can be seen, in this method, the value of I_N and R_N are simultaneously calculated.

Method 2: In this method, the value of I_N and R_N are separately calculated. To calculate the value of I_N, the terminal is assumed to be a short circuit branch. Then, the circuit is analyzed by using a circuit analysis method to find the value of I_{sc}. Next, we have:

$$I_N = I_{sc}$$

Moreover, to calculate R_N, which is the same as input resistance (R_{in}), first, all the independent sources (independent voltage and current sources) are turned off (each independent voltage source is replaced by short circuit branch and each independent current source is replaced by an open circuit branch). Then, if the circuit does not include any dependent source (dependent voltage or current source), the series-parallel rule is applied to calculate the input resistance of the circuit. However, if the circuit includes at least a dependent source, a test source (independent voltage or current source) with the voltage and current of V_t and I_t is connected to the input terminal of the circuit. Next, the circuit is analyzed by using a circuit analysis method to find the value of $\frac{V_t}{I_t}$. After that, we have:

$$R_N = \frac{V_t}{I_t}$$

Method 3: In this method, the parameters of Thevenin equivalent circuit (V_{Th} and R_{Th}) are calculated, and then the parameters of Norton equivalent circuit (I_N and R_N) are calculated by using the following relations.

$$R_N = R_{Th}$$

$$I_N = \frac{V_{Th}}{R_{Th}}$$

In this problem, the second method is applied because it is more straightforward.

Step 2 (solving the problem)

Calculating R_N: In Fig. 6.23.2, it noticed that:

$$V_{ab} = V_t \quad (1)$$

KVL in mesh 2:

$$2(i_2 - 0.1V_{ab}) + i_2 - 0.3V_{ab} + (i_2 + I_t) = 0$$

$$\Rightarrow 4i_2 - 0.5V_{ab} + I_t = 0 \quad (2)$$

KVL in the right-side mesh:

$$-V_t + (i_2 + I_t) = 0$$

$$\Rightarrow i_2 = V_t - I_t \quad (3)$$

Solving (1)-(3):

$$4(V_t - I_t) - 0.5V_t + I_t = 0$$

$$\Rightarrow 3.5V_t - 3I_t = 0$$

$$\Rightarrow \frac{V_t}{I_t} = \frac{6}{7}$$

$$\Rightarrow R_N = \frac{6}{7}\ \Omega$$

Calculating I_N: As can be seen in Fig. 6.23.3, since the a-b terminal is short-circuited, $V_{ab} = 0$. Therefore, the current of the dependent current source and the voltage of the dependent voltage source are zero. Hence, the dependent current and voltage sources must be replaced by an open circuit branch and a short circuit branch, respectively, as is shown in Fig. 6.23.3. As is illustrated in Fig. 6.23.4, the left-side branch, including the 2-Ω and 1-Ω resistors, is eliminated due to the short circuit branch. Thus, by using Ohm's law, we have:

$$I_{sc} = \frac{2}{1} = 2\ \text{A}$$

$$\Rightarrow I_N = 2\ \text{A}$$

Therefore, the equivalent Norton circuit is a $\frac{6}{7} - \Omega$ resistor in parallel to a $2 - A$ independent current source (with the upward current direction), as is shown in Fig. 6.23.5. Choice (1) is the answer.

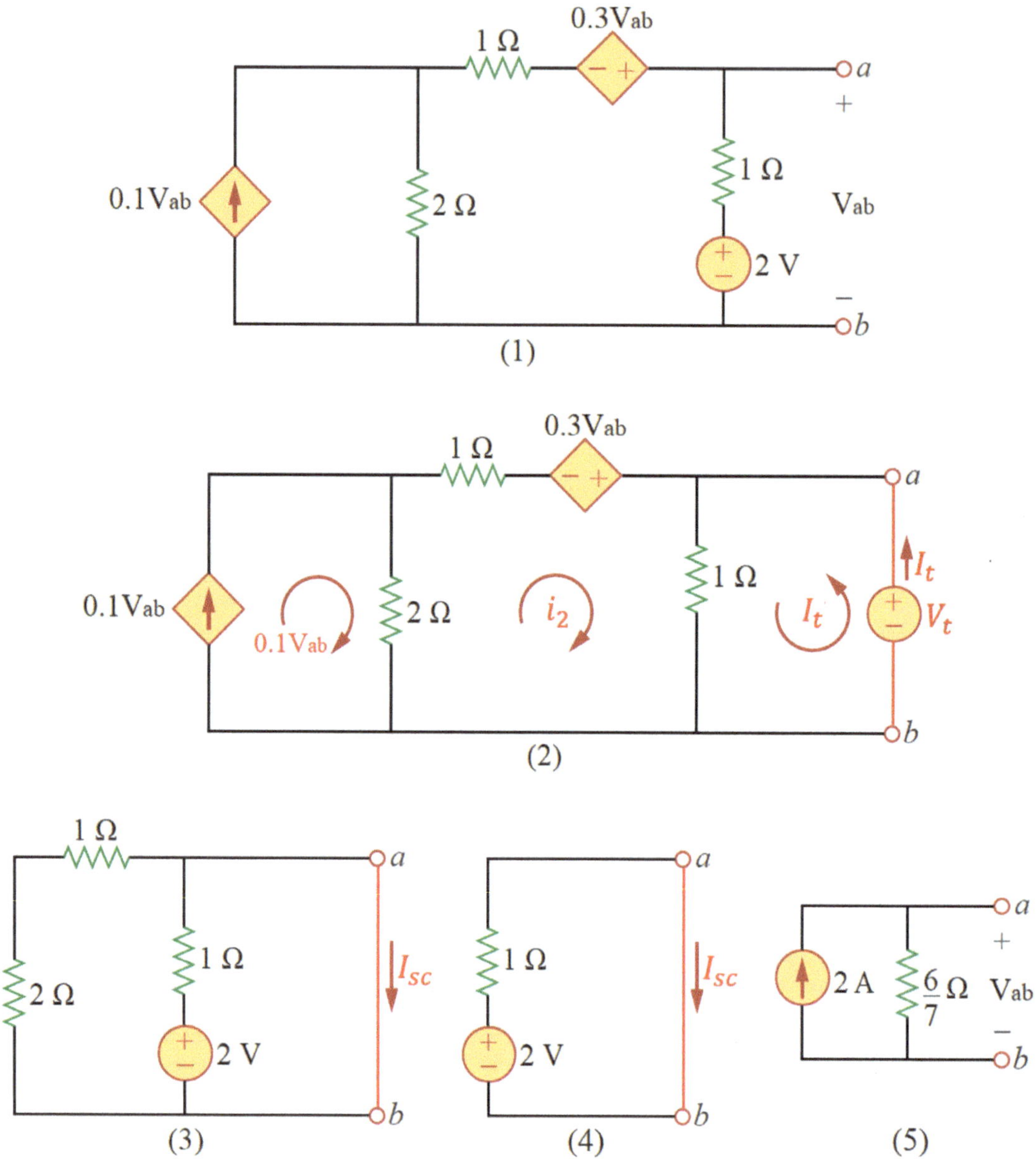

Fig. 6.23 The circuits of problem 6.38

6.40. *Step 1 (determining the method of solution)*

Method 1: First, a test source (independent voltage or current source) with the voltage and current of V_t and I_t is connected to the input terminal of the circuit. Then, the circuit is analyzed by using a circuit analysis method to find a relation between the voltage and current of the test source in the following form.

$$I_t = \frac{V_t}{\alpha} - \beta$$

Then, it is concluded that:

$$R_N = \alpha$$

$$I_N = \beta$$

As can be seen, in this method, the value of I_N and R_N are simultaneously calculated.

Method 2: In this method, the value of I_N and R_N are separately calculated. To calculate the value of I_N, the terminal is assumed to be a short circuit branch. Then, the circuit is analyzed by using a circuit analysis method to find the value of I_{sc}. Next, we have:

$$I_N = I_{sc}$$

Moreover, to calculate R_N, which is the same as input resistance (R_{in}), first, all the independent sources (independent voltage and current sources) are turned off (each independent voltage source is replaced by short circuit branch and each independent current source is replaced by an open circuit branch). Then, if the circuit does not include any dependent source (dependent voltage or current source), the series-parallel rule is applied to calculate the input resistance of the circuit. However, if the circuit includes at least a dependent source, a test source (independent voltage or current source) with the voltage and current of V_t and I_t is connected to the input terminal of the circuit. Next, the circuit is analyzed by using a circuit analysis method to find the value of $\frac{V_t}{I_t}$. After that, we have:

$$R_N = R_{in} = \frac{V_t}{I_t}$$

Method 3: In this method, the parameters of Thevenin equivalent circuit (V_{Th} and R_{Th}) are calculated, and then the parameters of Norton equivalent circuit (I_N and R_N) are calculated by using the following relations.

$$R_N = R_{Th}$$

$$I_N = \frac{V_{Th}}{R_{Th}}$$

In this problem, the third method is applied because it is more straightforward.

Step 2 (solving the problem)

Calculating R_N: As can be seen in Fig. 6.24.2, the 1-Ω resistor in the vertical branch is eliminated since its two ends are short-circuited. Then, the circuit of Fig. 6.24.3 is simplified and converted to the circuit of Fig. 6.24.4.

$$R_N = (1||2) + (1||2)$$

$$\Rightarrow R_N = \frac{1 \times 2}{1 + 2} + \frac{1 \times 2}{1 + 2}$$

$$\Rightarrow R_N = \frac{4}{3}\ \Omega \qquad (1)$$

Calculating I_N: As can be seen in Fig. 6.24.5, to find the Thevenin voltage of the circuit, we can simply use the voltage division rule as follows.

$$V_a = \frac{2}{2 + 1} \times 3 = 2\ \text{V} \qquad (2)$$

$$V_b = \frac{1}{1 + 2} \times 3 = 1\ \text{V} \qquad (3)$$

$$V_{oc} = V_a - V_b \qquad (4)$$

Solving (2)-(4):

$$V_{oc} = 2 - 1 = 1 \text{ V}$$

$$\Rightarrow V_{Th} = 1 \text{ V} \tag{5}$$

Solving (1) and (5):

$$I_N = \frac{1}{\frac{4}{3}}$$

$$\Rightarrow I_N = \frac{3}{4} \text{ A}$$

Choice (3) is the answer.

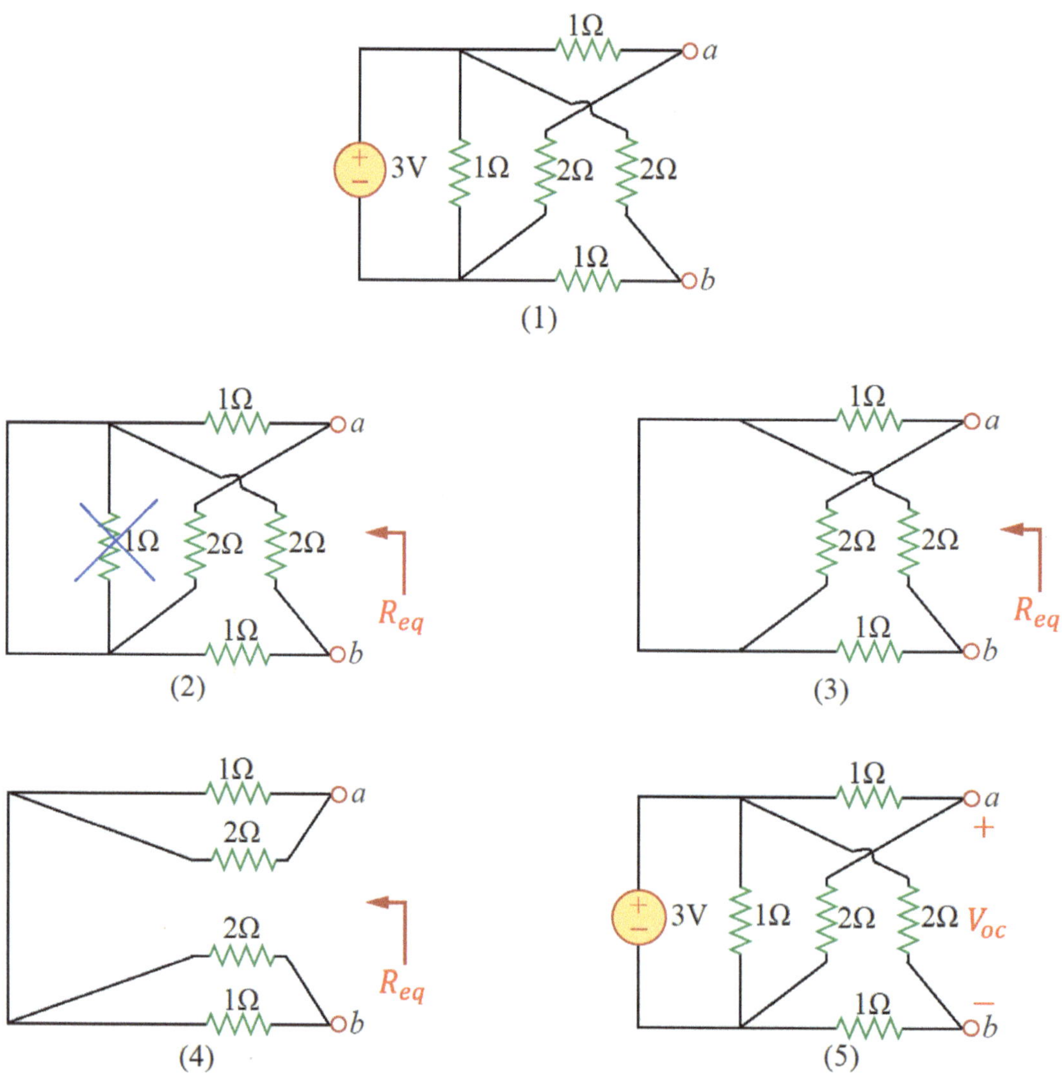

Fig. 6.24 The circuits of problem 6.40

6.4 Source Transformation Theorem

6.42. Source transformation theorem has been recommended to be used to solve this problem. It should be noted that mesh and nodal analyses are the worst methods to solve this problem since the circuit includes many nodes and meshes.

Source transformation from Fig. 6.25.1 to Fig. 6.25.2: In the left-side dashed box, source transformation theorem is applied as follows.

$$I = \frac{V}{R} = \frac{3}{1} = 3\ \text{A}(\text{Upward})$$

In the right-side dashed box, source transformation theorem is used as follows.

$$V = IR = 5 \times 1 = 5\ \text{V}$$

Circuit simplification Form Fig. 6.25.2 to Fig. 6.25.3: In the left-side dashed box of Fig. 6.25.2, the parallel current sources are combined first as follows.

$$I = 3 + 2 = 5\,\text{A}(\text{Upward})$$

Then, in the right-side dashed box of Fig. 6.25.2, the series resistors are combined as follows.

$$R = 3 + 1 = 4\ \Omega$$

Source transformation from Fig. 6.25.3 to Fig. 6.25.4: In the left-side dashed box of Fig. 6.25.3, source transformation theorem is applied as follows.

$$V = IR = 5 \times 1 = 5\ \text{V}$$

In the right-side dashed boxes of Fig. 6.25.3, source transformation theorem is applied twice as follows.

$$I_1 = \frac{V_1}{R_1} = \frac{1}{2} = 0.5\ \text{A} \quad (\text{Downward})$$

$$I_2 = \frac{V_2}{R_2} = \frac{5}{4} = 1.25\ \text{A} \quad (\text{Upward})$$

Circuit simplification Form Fig. 6.25.4 to Fig. 6.25.5: In the left-side dashed box of Fig. 6.25.4, the resistors are combined as follows.

$$R = 1 + 3 = 4\ \Omega$$

In the right-side dashed boxes of Fig. 6.25.4, the parallel resistors and the parallel current sources are combined, respectively, as follows.

$$R = 2||4 = \frac{2 \times 4}{2 + 4} = \frac{4}{3}\ \Omega$$

$$I = 1.25 - 0.5 = 0.75\ \text{A} \quad (\text{Upward})$$

Source transformation from Fig. 6.25.5 to Fig. 6.25.6: In the dashed box of Fig. 6.25.5, source transformation theorem is applied as follows.

$$I = \frac{V}{R} = \frac{5}{4} = 1.25\ \text{A} \quad (\text{Upward})$$

Circuit simplification Form Fig. 6.25.6 to Fig. 6.25.7: In the dashed box of Fig. 6.25.6, the parallel resistors and parallel current sources are combined, respectively, as follows.

$$R = (4||1)||\frac{4}{3} = \left(\frac{4\times 1}{4+1}\right)\bigg|\bigg|\frac{4}{3} = \frac{4}{5}\bigg|\bigg|\frac{4}{3} = \frac{\frac{4}{5}\times\frac{4}{3}}{\frac{4}{5}+\frac{4}{3}} = \frac{\frac{16}{15}}{\frac{12+20}{15}} = \frac{16}{32} = 0.5\ \Omega$$

$$I = 1.25 + 0.75 = 2\ \text{A} \quad \text{(Upward)}$$

Applying Ohm's law in Fig. 6.25.7:

$$V_{ab} = 2\times 0.5$$

$$\Rightarrow V_{ab} = 1\ \text{V}$$

Choice (1) is the answer.

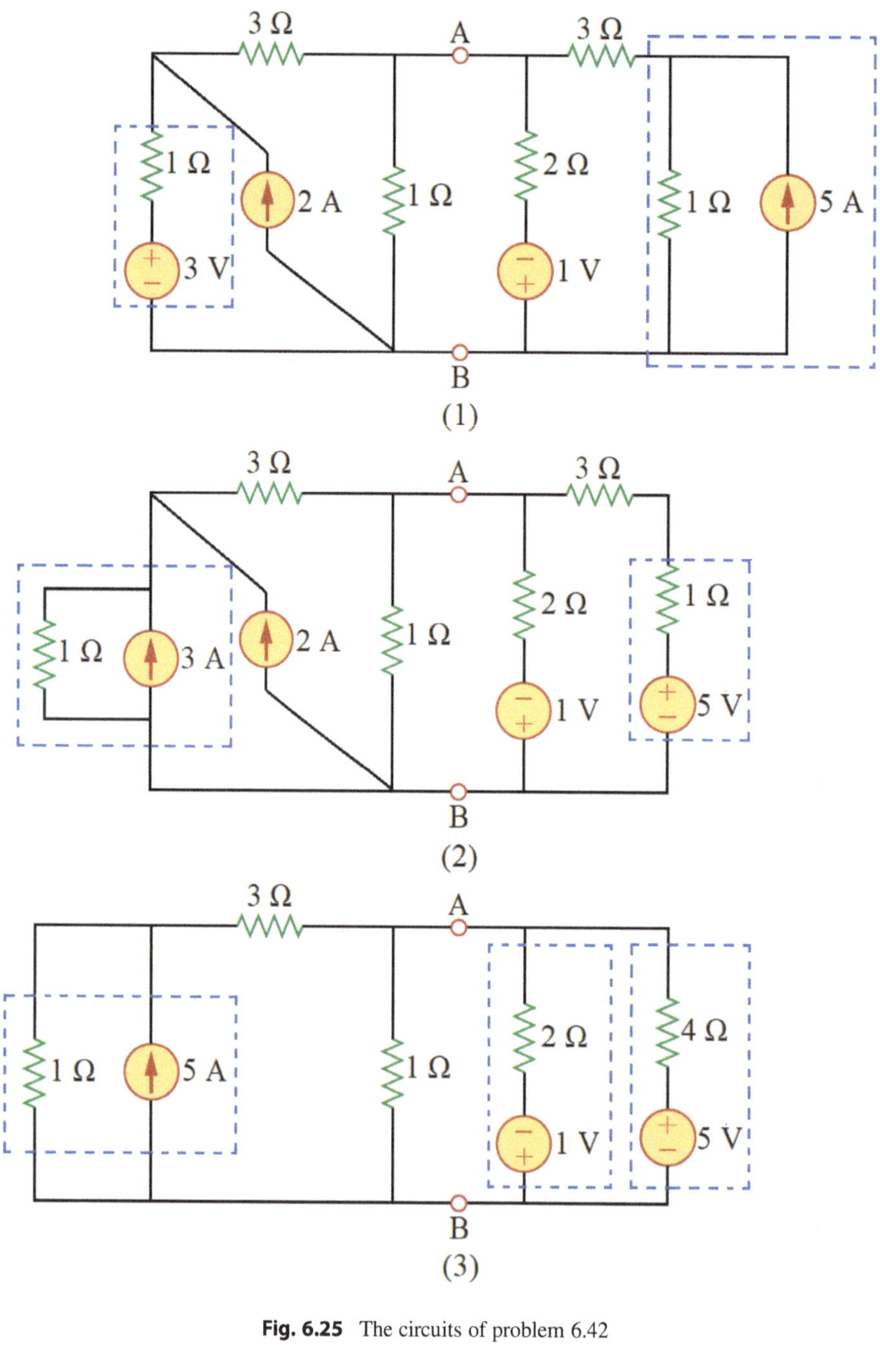

Fig. 6.25 The circuits of problem 6.42

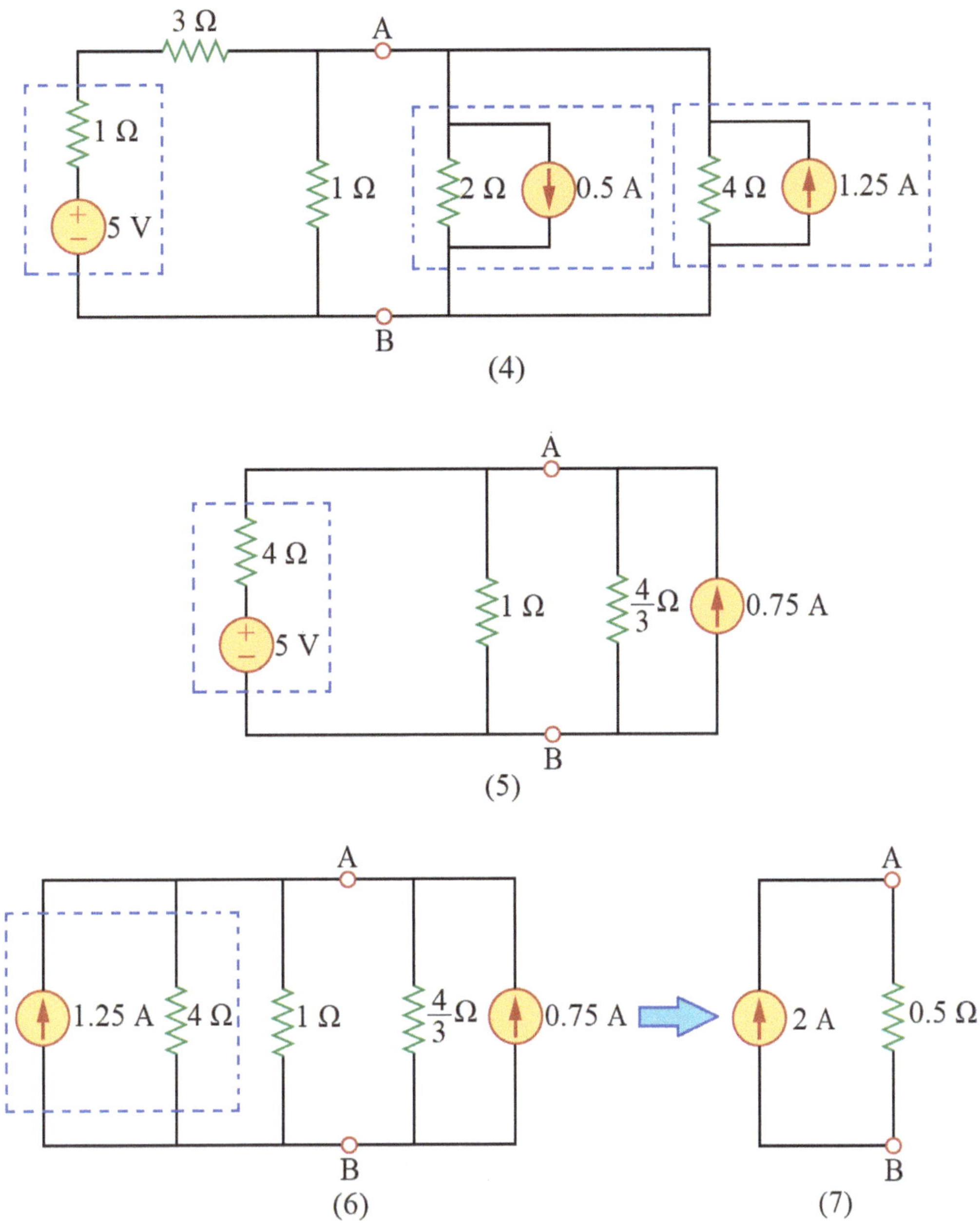

Fig. 6.25 (continued)

6.44. ***Circuit simplification Form Fig. 6.26.1 to Fig. 6.26.2***: In Fig. 6.26.1, the two series voltage sources are combined considering their polarities as follows.

$$V = 5 - 2 = 3 \text{ V}$$

Source transformation from Fig. 6.26.2 to Fig. 6.26.3: In the dashed box of Fig. 6.26.2, source transformation theorem is applied as follows.

$$I = \frac{V}{R} = \frac{3}{1} = 3 \text{ A} \quad \text{(Upward)}$$

Circuit simplification Form Fig. 6.26.3 to Fig. 6.26.4: In Fig. 6.26.3, the two parallel current sources are combined as follows.

$$I = 4 - 3 = 1 \text{ A} \quad \text{(Downward)}$$

Source transformation from Fig. 6.26.4 to Fig. 6.26.5: In the dashed box of Fig. 6.26.4, source transformation theorem is applied as follows.

$$V = IR = 1 \times 1 = 1 \text{ V}$$

Circuit simplification Form Fig. 6.26.5 to Fig. 6.26.6: In Fig. 6.26.5, the two series resistors are combined as follows.

$$R = 1 + 0.8 = 1.8 \ \Omega$$

Applying voltage division rule:

$$V_{2\Omega} = \frac{2||3}{2||3 + 1.8} \times (-1) = -\frac{1.2}{1.2 + 1.8}$$

$$\Rightarrow V_{2\Omega} = -0.4 \text{ V}$$

Calculating the power of 2-Ω resistor:

$$P = \frac{V^2}{R}$$

$$\Rightarrow P = \frac{(-0.4)^2}{2}$$

$$\Rightarrow P = 0.08 \text{ W}$$

Choice (1) is the answer.

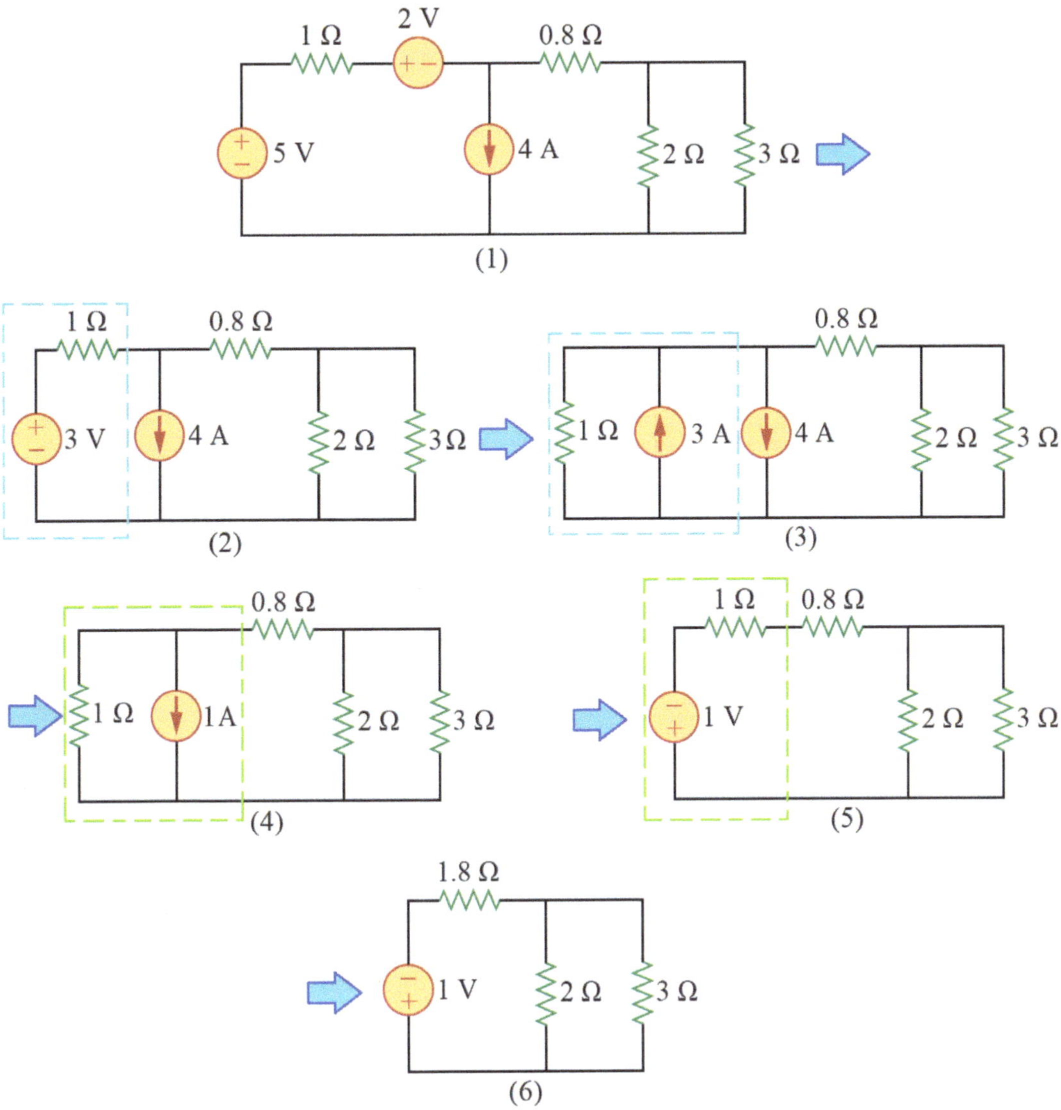

Fig. 6.26 The circuits of problem 6.44

Notes

The power consumed by a resistor can be calculated by using one of the following relations.

$$P = RI^2$$

$$P = VI$$

$$P = \frac{V^2}{R}$$

6.5 Superposition Theorem

6.47. The circuit includes two independent sources. Thus, the circuit needs to be analyzed twice as follows.

Step 1 (analyzing the circuit for the first independent source)

As can be seen in Fig. 6.27.2, the independent voltage source is kept but the independent current source is turned off (replaced by an open circuit branch). Now, by applying voltage division rule, we have:

$$v_{o1} = \frac{4}{4+6+10} \times 10 = 2 \text{ V}$$

Step 2 (analyzing the circuit for the second independent source)

As is shown in Fig. 6.27.3, the independent current source is kept but the independent voltage source is turned off (replaced by a short circuit branch). Now, by applying current division rule, we have:

$$i_{o2} = \frac{10}{4+6+10} \times 2 = 1 \text{ A}$$

Applying Ohm's law for the 4-Ω resistor:

$$v_{o2} = 4 \times 1 = 4 \text{ V}$$

Step 3 (calculating the total contribution)

Based on superposition theorem, the output voltage can be calculated as follows.

$$v_o = v_{o1} + v_{o2}$$

$$\Rightarrow v_o = 2 + 4$$

$$\Rightarrow v_o = 6 \text{ V}$$

Choice (2) is the answer.

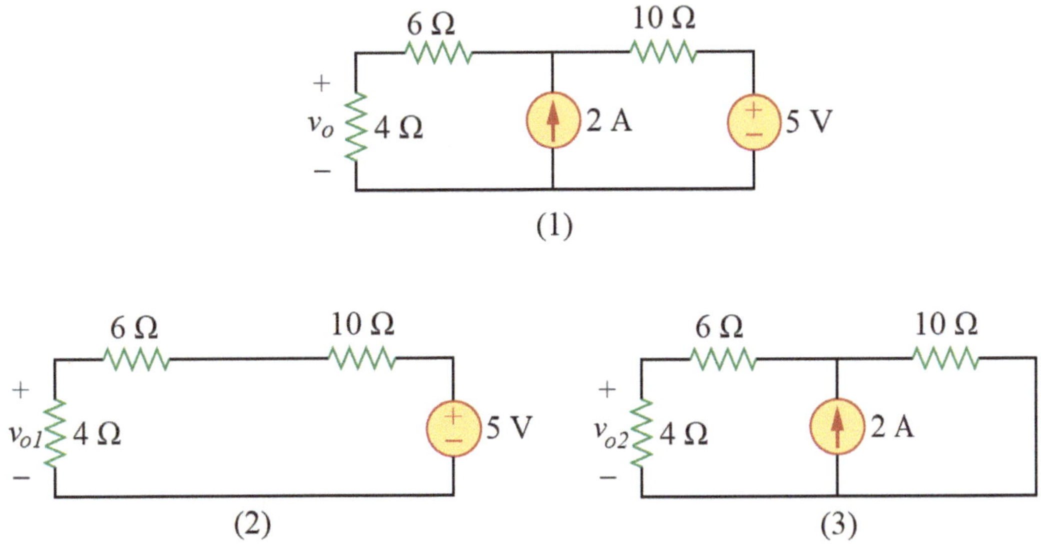

Fig. 6.27 The circuits of problem 6.47

6.48. The circuit includes two independent sources. Thus, the circuit needs to be analyzed twice as follows.

Step 1 (analyzing the circuit for the first independent source)

As can be seen in Fig. 6.28.2, the independent current source is kept but the independent voltage source is turned off (replaced by a short circuit branch). Now, by applying current division rule, we have:

$$I_{o1} = \frac{8}{8+4} \times 6 = 4 \text{ A}$$

Step 2 (analyzing the circuit for the second independent source)

As is shown in Fig. 6.28.3, the independent voltage source is kept but the independent current source is turned off (replaced by an open circuit branch). Now, by applying Ohm's law, we have:

$$i_{o2} = \frac{12}{8+4} = 1 \text{ A}$$

Step 3 (calculating the total contribution)

Based on superposition theorem, the output current can be calculated as follows.

$$I_o = I_{o1} + I_{o2}$$

$$\Rightarrow I_o = 4 + 1$$

$$\Rightarrow I_o = 5 \text{ A}$$

Choice (4) is the answer.

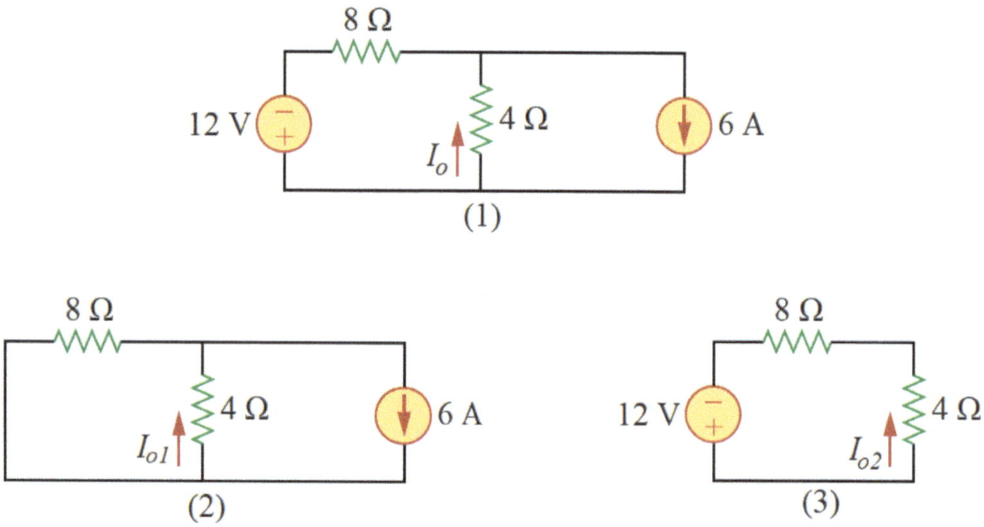

Fig. 6.28 The circuits of problem 6.48

6.6 Maximum Power Transfer Theorem

6.52. To absorb the maximum power, R_L must be equal to the Thevenin resistance (R_{Th}) seen by the R_L. In other words:

$$R_L = R_{Th}$$

Moreover, the maximum power transferable to the R_L can be calculated as follows.

$$P_{max} = \frac{V_{Th}^2}{4R_{Th}}$$

Therefore, the value of Thevenin resistance (R_{Th}) and Thevenin voltage (V_{Th}) must be calculated.

Step 1 (determining the method of solution)

Method 1: First, a test source (independent voltage or current source) with the voltage and current of V_t and I_t is connected to the input terminal of the circuit. Then, the circuit is analyzed by using a circuit analysis method to find a relation between the voltage and current of the test source in the following form.

$$V_t = \alpha I_t + \beta$$

Then, it is concluded that:

$$R_{Th} = \alpha$$

$$V_{Th} = \beta$$

As can be seen, in this method, the value of V_{Th} and R_{Th} are simultaneously calculated.

Method 2: In this method, the value of V_{Th} and R_{Th} are separately calculated. To calculate the value of V_{Th}, the terminal is assumed to be an open circuit branch. Then, the circuit is analyzed by using a circuit analysis method to find the value of V_{oc}. Next, we have:

$$V_{Th} = V_{oc}$$

Moreover, to calculate R_{Th}, which is the same as input resistance (R_{in}), first, all the independent sources (independent voltage and current sources) are turned off (each independent voltage source is replaced by short circuit branch and each independent current source is replaced by an open circuit branch). Then, if the circuit does not include any dependent source (dependent voltage or current source), the series-parallel rule is applied to calculate the input resistance of the circuit. However, if the circuit includes at least a dependent source, a test source (independent voltage or current source) with the voltage and current of V_t and I_t is connected to the input terminal of the circuit. Next, the circuit is analyzed by using a circuit analysis method to find the value of $\frac{V_t}{I_t}$. After that, we have:

$$R_{Th} = \frac{V_t}{I_t}$$

In this problem, the second method is applied because it is more straightforward.

Step 2 (solving the problem)

Calculating R_{Th}: As can be seen in Fig. 6.29.1, since there is no dependent source in the circuit, we can use series-parallel rule.

$$R_{Th} = 6 + (4||4) + 2$$

$$\Rightarrow R_{Th} = 10\ \Omega$$

$$\Rightarrow R_L = 10\ \Omega$$

Calculating V_{Th}: As can be seen in Fig. 6.29.3, since there is no current flowing through the 6-Ω and 2-Ω resistors (no voltage drop), the V_{oc} is equal to the voltage of the vertical 4-Ω resistor. Therefore, by using voltage division formula, we have:

$$V_{oc} = \frac{4}{4+4} \times 20 = 10\ \text{V}$$

$$\Rightarrow V_{Th} = 10\ \text{V}$$

Calculating P_{max}: The maximum power transferable to the R_L can be calculated as follows.

$$P_{max} = \frac{10^2}{4 \times 10}$$

$$\Rightarrow P_{max} = 2.5\ \text{W}$$

Choice (2) is the answer.

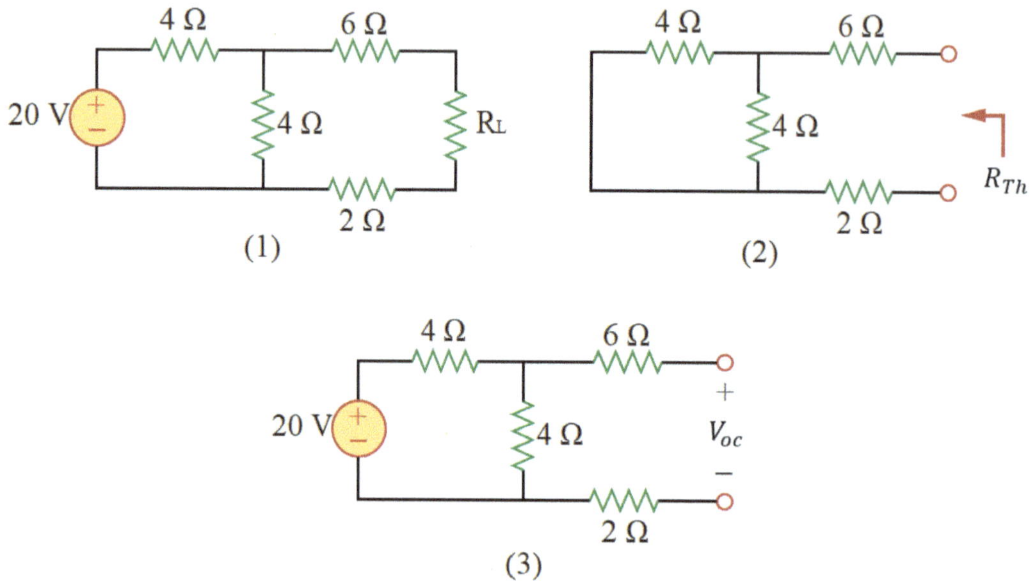

Fig. 6.29 The circuits of problem 6.52

6.53. To absorb the maximum power, R_L must be equal to the Thevenin resistance (R_{Th}) seen by the R_L. In other words:

$$R_L = R_{Th}$$

Moreover, to calculate R_{Th}, which is the same as input resistance (R_{in}), first, all the independent sources (independent voltage and current sources) are turned off (each independent voltage source is replaced by short circuit branch and each independent current source is replaced by an open circuit branch). Then, if the circuit does not include any dependent source (dependent voltage or current source), the series-parallel rule is applied to calculate the input resistance of the circuit. However, if the circuit includes at least a dependent source, a test source (independent voltage or current source) with the voltage and current of V_t and I_t is connected to the input terminal of the circuit. Next, the circuit is analyzed by using a circuit analysis method to find the value of $\frac{V_t}{I_t}$. After that, we have:

$$R_{Th} = R_{in} = \frac{V_t}{I_t}$$

Since the circuit includes at least one dependent source, we must apply a test source at the terminal.
As can be noticed from the circuit of Fig. 6.30.2, we have:

$$I_x = I_t \qquad (1)$$

Applying KVL in the loop:

$$-V_t + 5I_x + 10I_x = 0$$

$$\Rightarrow -V_t + 15I_x = 0 \qquad (2)$$

Solving (1) and (2):

$$-V_t + 15\,I_t = 0$$

$$\Rightarrow V_t = 15\,I_t$$

$$\Rightarrow \frac{V_t}{I_t} = 15$$

$$\Rightarrow R_{Th} = 15\ \Omega$$

$$\Rightarrow R_L = 15\ \Omega$$

Choice (3) is the answer.

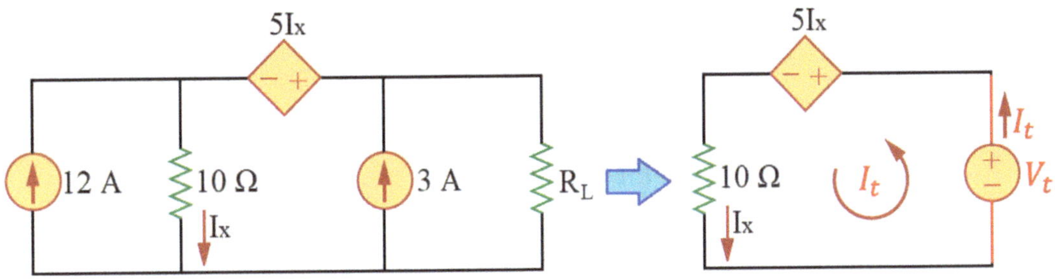

Fig. 6.30 The circuits of problem 6.53

6.54. To absorb the maximum power, R_L must be equal to the Thevenin resistance (R_{Th}) seen by the R_L. In other words:

$$R_L = R_{Th}$$

Moreover, to calculate R_{Th}, which is the same as input resistance (R_{in}), first, all the independent sources (independent voltage and current sources) are turned off (each independent voltage source is replaced by short circuit branch and each independent current source is replaced by an open circuit branch). Then, if the circuit does not include any dependent source (dependent voltage or current source), the series-parallel rule is applied to calculate the input resistance of the circuit. However, if the circuit includes at least a dependent source, a test source (independent voltage or current source) with the voltage and current of V_t and I_t is connected to the input terminal of the circuit. Next, the circuit is analyzed by using a circuit analysis method to find the value of $\frac{V_t}{I_t}$. After that, we have:

$$R_{Th} = R_{in} = \frac{V_t}{I_t}$$

In this problem, mesh analysis is applied to solve the problem.
As can be noticed from the circuit of Fig. 6.31.2, we have:

$$V_{ab} = V_t \tag{1}$$

KVL in the right-side mesh:

$$-V_t + 4(I_t + I_1) = 0 \quad (2)$$

KVL in the left-side mesh:

$$6I_1 - 2V_{ab} + 4(I_t + I_1) = 0$$

$$\Rightarrow 10I_1 - 2V_{ab} + 4I_t = 0$$

$$\Rightarrow I_1 = 0.2V_{ab} - 0.4I_t \quad (3)$$

Solving (1)-(3):

$$-V_t + 4(I_t + (0.2V_t - 0.4I_t)) = 0$$

$$\Rightarrow -0.2V_t + 2.4I_t = 0$$

$$\Rightarrow \frac{V_t}{I_t} = 12\ \Omega$$

$$\Rightarrow R_{Th} = 12\ \Omega$$

$$\Rightarrow R_L = 12\ \Omega$$

Choice (3) is the answer.

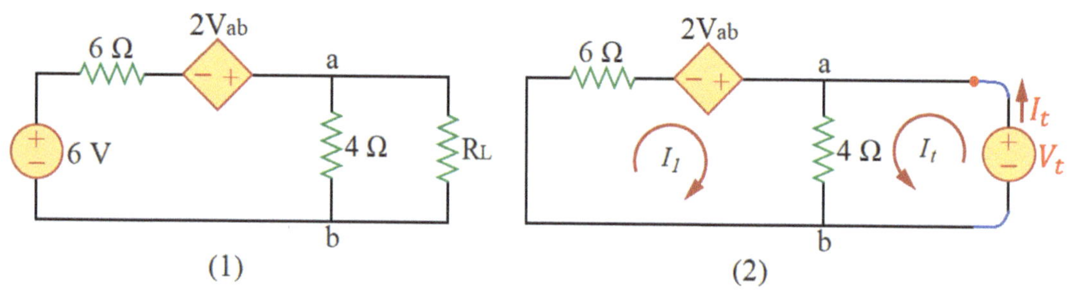

Fig. 6.31 The circuits of problem 6.54

References

1. Rahmani-Andebili, M., DC Electrical Circuit Analysis - Practice Problems, Methods, and Solutions (1st Ed.), Springer Nature, 2020.
2. Rahmani-Andebili, M., General Physics II – Practice Problems, Methods, and Solutions, Sringer Nature, 2025.
3. Rahmani-Andebili, M., General Physics I – Practice Problems, Methods, and Solutions, Springer Nature, 2025.
4. Rahmani-Andebili, M., Mathematics of Engineering and Science - Comprehensive Lessons, Springer Nature, 2024.
5. Rahmani-Andebili, M., Differential Equations – Practice Problems, Methods, and Solutions, Springer Nature, 2022.
6. Rahmani-Andebili, M., Calculus III – Practice Problems, Methods, and Solutions, Springer Nature, 2023.
7. Rahmani-Andebili, M., Calculus II – Practice Problems, Methods, and Solutions, Springer Nature, 2023.
8. Rahmani-Andebili, M., Calculus I (2nd Ed.) – Practice Problems, Methods, and Solutions, Springer Nature, 2023.
9. Rahmani-Andebili, M., Precalculus (2nd Ed.) – Practice Problems, Methods, and Solutions, Springer Nature, 2024.

Energy Storage Elements and Circuit Analysis Under Steady-State Condition: Part A

7

Abstract

In this chapter, the basic and advanced problems of energy storage elements and circuit analysis under steady-state condition are studied. The subjects include equivalent capacitance of series, parallel, and series-parallel connections, equivalent inductance of series, parallel, and series-parallel connections, and circuit analysis under steady-state condition. Herein, different types of problems and exercises are presented that are categorized as follows.

- ***Problems with detailed solution***: They have been designed to teach students the subjects in detail. Moreover, they have been categorized in different levels based on their difficulty levels (easy, normal, and hard) and calculation amounts (small, normal, and large).
- ***Partially solved exercises***: They have been designed to encourage students to practice problems while guiding them through the problem-solving procedure and hinting the required formulas.
- ***Exercises with final answer***: They have been designed to encourage students to practice more by themselves while hinting them by the final answer as well as to help instructors to give tests or quizzes.

7.1 Equivalent Capacitance of Series, Parallel, and Series-Parallel Connections

Problem

7.1. In the circuit shown in Fig. 7.1, the capacitors are connected in series. Calculate the equivalent capacitance of the circuit [1–9].

Difficulty level	● Easy	○ Normal	○ Hard
Calculation amount	● Small	○ Normal	○ Large

1) 11 F
2) 1 F
3) 4 F
4) 2 F

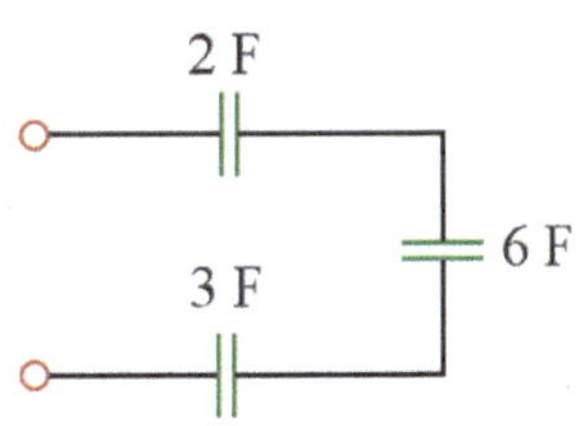

Fig. 7.1 The circuit of problem 7.1

S. Manias, *DC Electrical Circuits Analysis*, https://doi.org/10.1007/978-3-032-00253-2_7

Problem

7.2. In the circuit shown in Fig. 7.2, the capacitors are connected in parallel. Calculate the equivalent capacitance of the circuit.

Difficulty level ● Easy ○ Normal ○ Hard

Calculation amount ● Small ○ Normal ○ Large

1) 11 F
2) 1 F
3) 4 F
4) 2 F

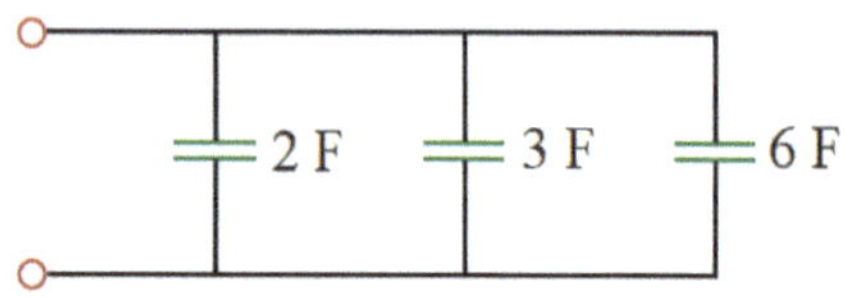

Fig. 7.2 The circuit of problem 7.2

Problem

7.3. In the circuit shown in Fig. 7.3, the capacitors are connected in series–parallel from. Calculate the equivalent capacitance of the circuit.

Difficulty level ● Easy ○ Normal ○ Hard

Calculation amount ● Small ○ Normal ○ Large

1) 4 F
1) 2.5 F
1) 2 F
1) 1 F

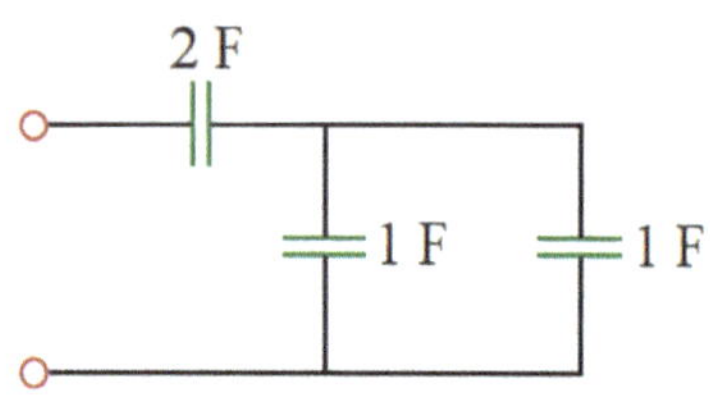

Fig. 7.3 The circuit of problem 7.3

Partially Solved Exercise

7.4. Calculate the Equivalent Capacitance of the circuit shown in Fig. 7.4.

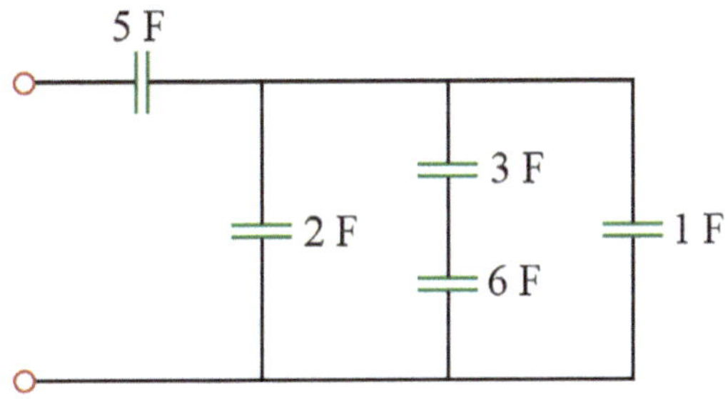

Fig. 7.4 The circuit of partially solved exercise 7.4

Solution

First, the series connection of the middle branch should be simplified as follows.

$$\frac{1}{C_{3,6}} = \frac{1}{C_3} + \frac{1}{C_6}$$

Or:

$$C_{3,6} = \frac{C_3 C_6}{C_3 + C_6}$$

$$\Rightarrow C_{3,6} = \frac{(\quad) \times (\quad)}{(\quad) + (\quad)} = 2\ F$$

Then, the three parallel branches can be combined as follows.

$$C_{2,2,1} = C_2 + C_{3,6} + C_1 = (\quad) + (\quad) + (\quad)$$

$$\Rightarrow C_{2,2,1} = 5\ F$$

Finally, two $5F$ capacitors are connected in series. Thus:

$$C_{eq} = \frac{C_5}{2} = \frac{(\quad)}{2}$$

$$\Rightarrow C_{eq} = 2.5\ \text{F}$$

Exercise

7.5. Calculate the Equivalent Capacitance of the circuit shown in Fig. 7.5.

Final Answer

$$C_{eq} = \frac{88}{81}\ \text{F}$$

Note: *Read the "educational support" part of the preface of the textbook.*

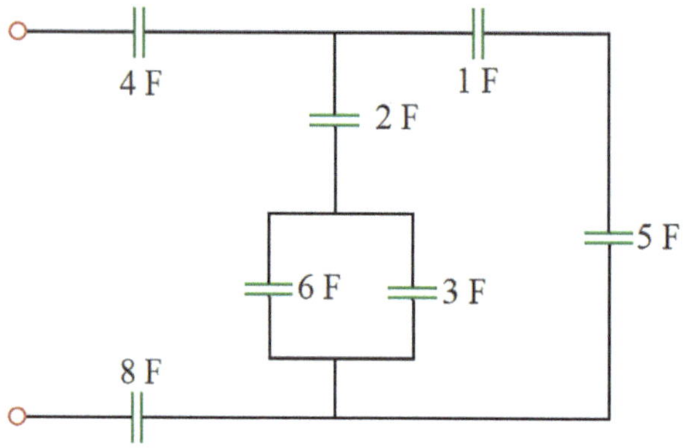

Fig. 7.5 The circuit of exercise 7.5

Exercise

7.6. Calculate the equivalent capacitance seen from the terminal of the circuit shown in Fig. 7.6. The circuit includes an infinite number of capacitors connected in series and the capacitance of each of them is 1 F.

Final Answer

$C_{eq} = 0$ F

Note: *Read the "educational support" part of the preface of the textbook.*

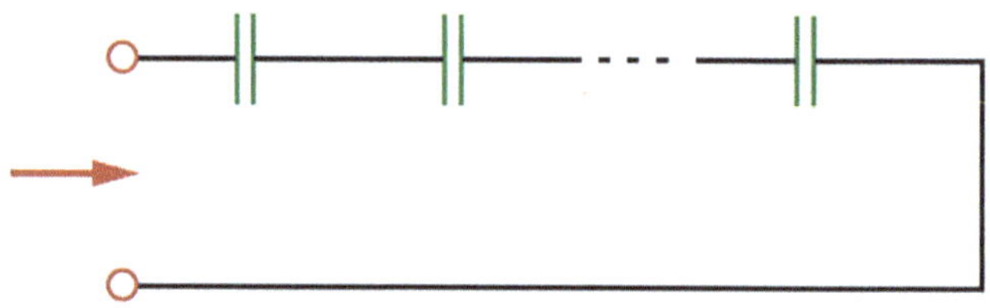

Fig. 7.6 The circuit of exercise 7.6

Exercise

7.7. Calculate the equivalent capacitance seen from the terminal of the circuit shown in Fig. 7.7. The circuit includes an infinite number of capacitors connected in parallel and the capacitance of each of them is 1 F.

Final Answer

$C_{eq} = \infty$ F

Note: *Read the "educational support" part of the preface of the textbook.*

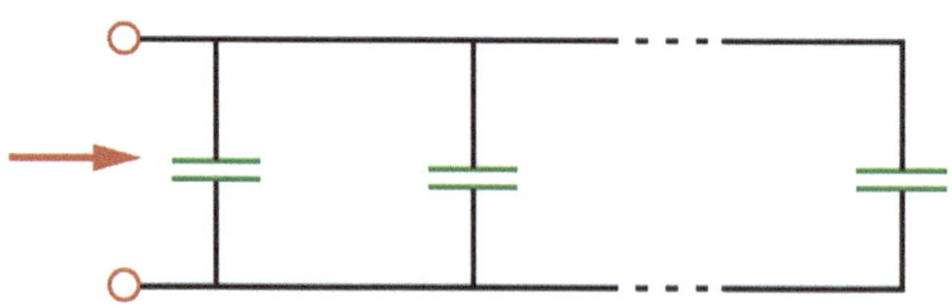

Fig. 7.7 The circuit of exercise 7.7

Problem

7.8. The capacitive network shown in Fig. 7.8 is infinitely repeated from its right side. Calculate the equivalent capacitance of the network.

Difficulty level ○ Easy ○ Normal ● Hard
Calculation amount ○ Small ● Normal ○ Large

1) 0.23 F
2) 0.36 F
3) 1.36 F
4) 0.36 F or 1.36 F

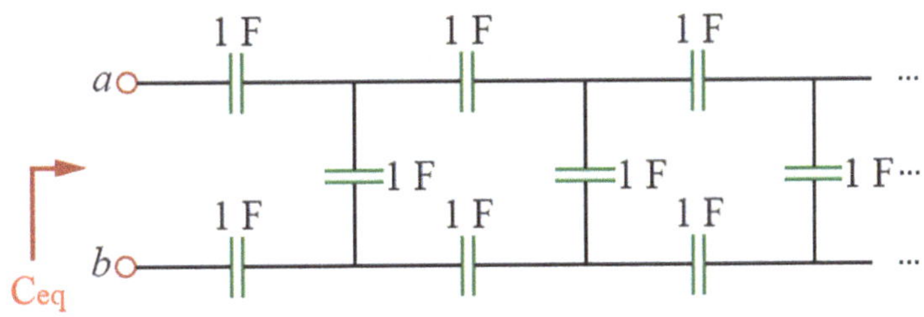

Fig. 7.8 The circuit of problem 7.8

7.2 Equivalent Inductance of Series, Parallel, and Series-Parallel Connections

Problem

7.9. In the circuit shown in Fig. 7.9, the inductors are connected in series. Calculate the equivalent inductance of the circuit.

Difficulty level ● Easy ○ Normal ○ Hard
Calculation amount ● Small ○ Normal ○ Large

1) 11 H
2) 1 H
3) 4 H
4) 2 H

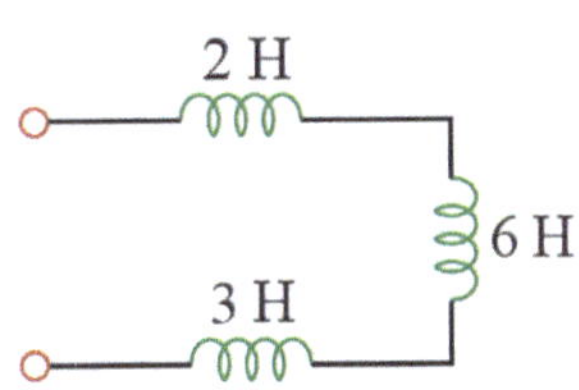

Fig. 7.9 The circuit of problem 7.9

Problem

7.10. In the circuit shown in Fig. 7.10, the inductors are connected in parallel. Calculate the equivalent inductance of the circuit.

Difficulty level ● Easy ○ Normal ○ Hard
Calculation amount ● Small ○ Normal ○ Large

1) 11 H
2) 1 H
3) 4 H
4) 2 H

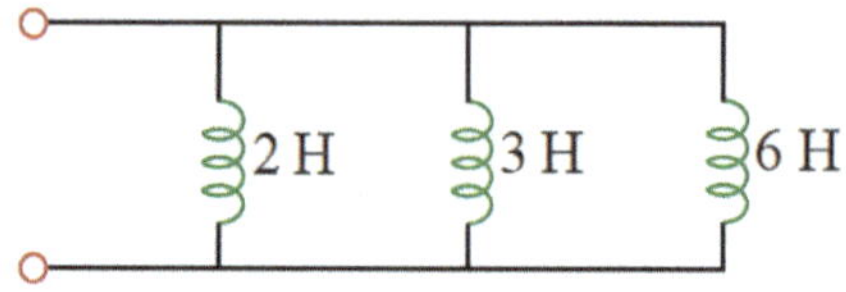

Fig. 7.10 The circuit of problem 7.10

Problem

7.11. In the circuit shown in Fig. 7.11, the inductors are connected in series–parallel from. Calculate the equivalent inductance of the circuit.

Difficulty level ● Easy ○ Normal ○ Hard
Calculation amount ● Small ○ Normal ○ Large

1) 1 H
2) 2 H
3) 4 H
4) 2.5 H

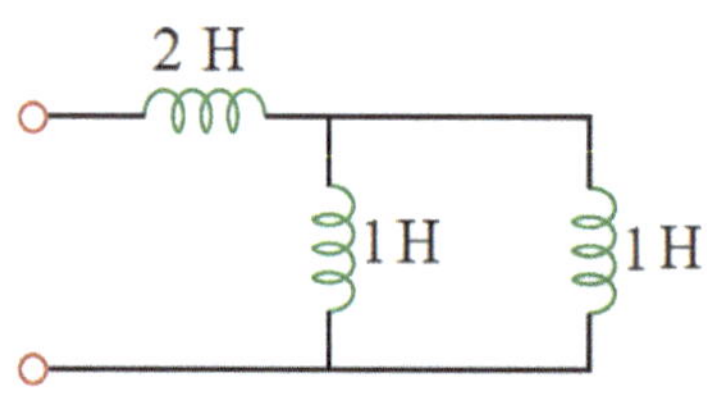

Fig. 7.11 The circuit of problem 7.11

Partially Solved Exercise

7.12. Calculate the Equivalent Inductance of the circuit shown in Fig. 7.12.

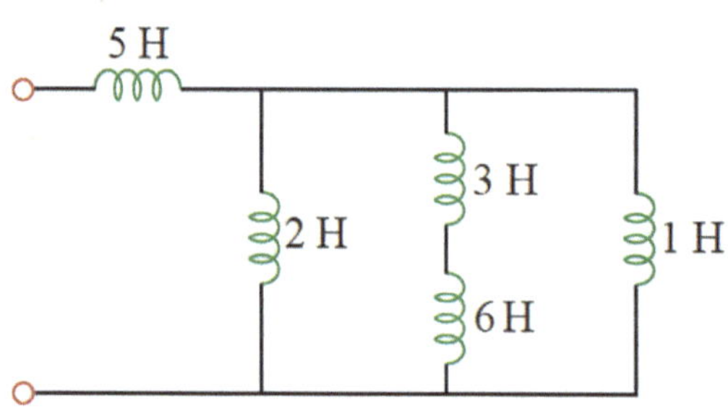

Fig. 7.12 The circuit of partially solved exercise 7.12

Solution

First, the series connection of the middle branch should be simplified as follows.

$$L_{3,6} = L_3 + L_6 = (\quad) + (\quad) = 9 \text{ H}$$

Then, the three parallel branches can be combined as follows.

$$\frac{1}{L_{2,9,1}} = \frac{1}{L_2} + \frac{1}{L_9} + \frac{1}{L_1}$$

$$\Rightarrow \frac{1}{L_{2,9,1}} = \frac{1}{(\quad)} + \frac{1}{(\quad)} + \frac{1}{(\quad)}$$

$$\Rightarrow \frac{1}{L_{2,9,1}} = \frac{(\quad) + (\quad) + (\quad)}{18}$$

$$\Rightarrow L_{2,9,1} = \frac{18}{29} \text{ H}$$

Finally, there are only two inductors connected in series. Thus:

$$L_{eq} = (\quad) + (\quad) = \frac{(\quad) + (\quad)}{(\quad)}$$

$$\Rightarrow L_{eq} = \frac{163}{29} \text{ H}$$

Exercise

7.13. Calculate the Equivalent Inductance of the circuit shown in Fig. 7.13.

Final Answer

$L_{eq} = 14.4$ F

Note: *Read the "educational support" part of the preface of the textbook.*

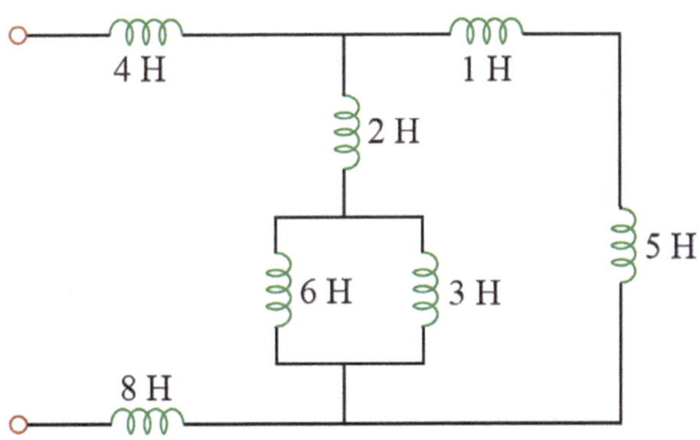

Fig. 7.13 The circuit of exercise 7.13

Exercise

7.14. Calculate the equivalent inductance seen from the terminal of the circuit shown in Fig. 7.14. The circuit includes an infinite number of inductors connected in series and the inductance of each of them is 1 H.

Final Answer

$L_{eq} = \infty$

Note: *Read the "educational support" part of the preface of the textbook.*

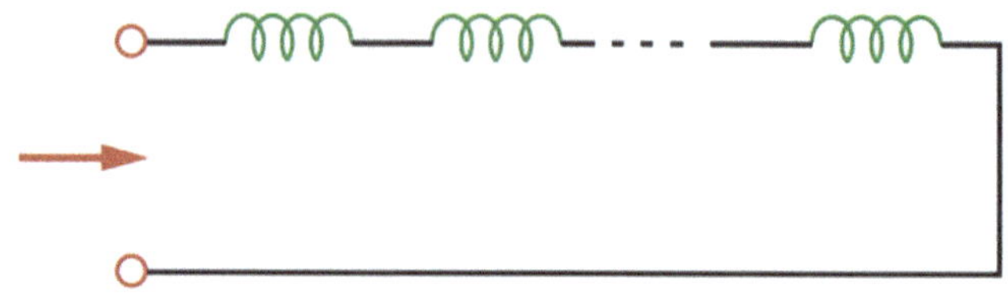

Fig. 7.14 The circuit of exercise 7.14

Exercise

7.15 Calculate the equivalent inductance seen from the terminal of the circuit shown in Fig. 7.15. The circuit includes an infinite number of inductors connected in parallel and the inductance of each of them is 1 H.

Final Answer

$L_{eq} = 0$ H

Note: *Read the "educational support" part of the preface of the textbook.*

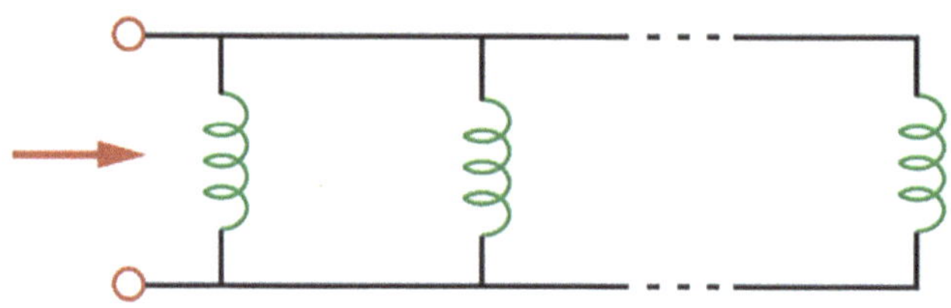

Fig. 7.15 The circuit of exercise 7.15

7.3 Circuit Analysis Under Steady-State Condition

Problem

7.16. In the circuit of Fig. 7.16, calculate the value of output voltage V_o in the steady state condition.

Difficulty level ○ Easy ● Normal ○ Hard

Calculation amount ○ Small ● Normal ○ Large

1) 6 V
2) 7 V
3) 8 V
4) 10 V

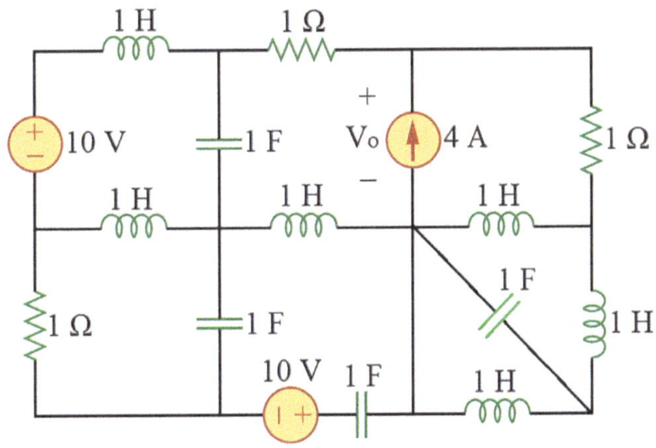

Fig. 7.16 The circuit of problem 7.16

Problem

7.17. In the circuit of Fig. 7.17, calculate the current of inductor and voltage of capacitor in the steady state condition.

Difficulty level ○ Easy ● Normal ○ Hard

Calculation amount ● Small ○ Normal ○ Large

1) $I_L = 0$ A, $V_C = 80$ V
2) $I_L = 0$ A, $V_C = 20$ V
3) $I_L = 2$ A, $V_C = 80$ V
4) $I_L = 2$ A, $V_C = 20$ V

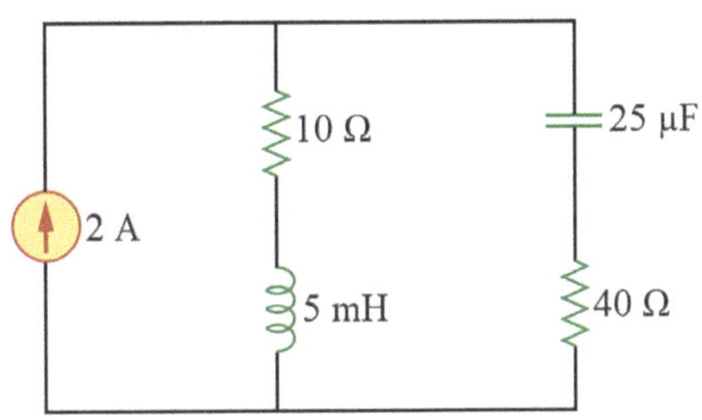

Fig. 7.17 The circuit of problem 7.17

Problem

7.18. In Problem 7.17, calculate the amount of energy stored in the inductor and capacitor in the steady state condition.

Difficulty level ● Easy ○ Normal ○ Hard

Calculation amount ● Small ○ Normal ○ Large

1) $W_L = 10$ J, $W_C = 5$ J
2) $W_L = 10$ mJ, $W_C = 5$ mJ
3) $W_L = 5$ mJ, $W_C = 10$ mJ
4) $W_L = 2.5$ J, $W_C = 5$ J

Partially Solved Exercise

7.19. In the circuit of Fig. 7.18, calculate the amount of energy stored in the circuit in a steady state condition.

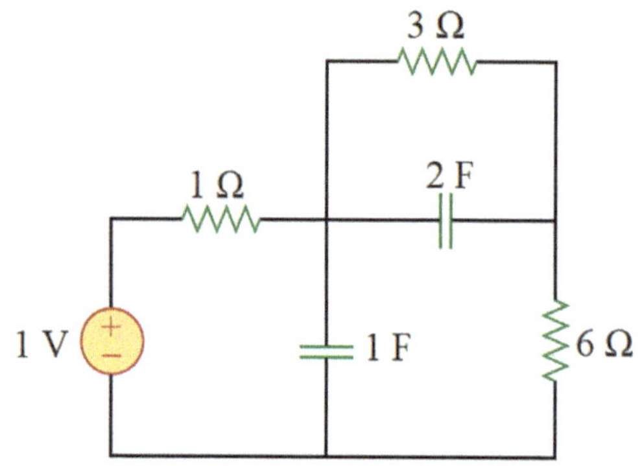

Fig. 7.18 The circuit of partially solved exercise 7.19

Solution

As we know, in a steady state condition, a capacitor behaves like an open circuit branch and an inductor acts like a short circuit branch. Figure 7.19 shows the circuit in the steady state condition.

Applying voltage division to calculate V_{C1}:

$$V_{C1} = \frac{(\quad)+(\quad)}{(\quad)+(\quad)+(\quad)} \times (\quad) = 0.9 \text{ V}$$

Applying voltage division to calculate V_{C2}:

$$V_{C1} = \frac{(\quad)}{(\quad)+(\quad)+(\quad)} \times 1 = 0.3 \text{ V}$$

Moreover, the amount of energy stored in a capacitor in the steady state condition can be calculated as follows.

$$W_C = \frac{1}{2} C V_C^2$$

Thus:

$$W_{C1} = \frac{1}{2} \times (\quad) \times (\quad)^2$$

$$\Rightarrow W_{C1} = 0.405 \text{ J}$$

$$W_{C2} = \frac{1}{2} \times (\quad) \times (\quad)^2$$

$$\Rightarrow W_{C2} = 0.09 \text{ J}$$

Hence, the total energy stored in the circuit is as follows.

$$W = W_{C1} + W_{C2}$$

$$\Rightarrow W = 0.414 \text{ J}$$

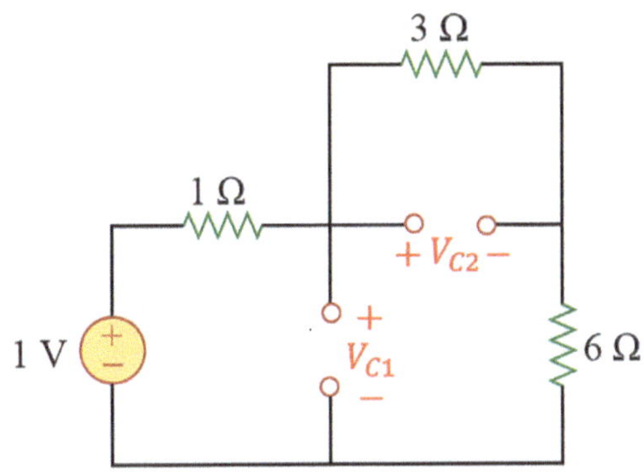

Fig. 7.19 The circuit of partially solved exercise 7.19

Exercise

7.20 In the circuit of Fig. 7.20, calculate the total amount of energy stored in the circuit in the steady state condition.

Final Answer

$W = 0.25$ J

Note: Read the "educational support" part of the preface of the textbook.

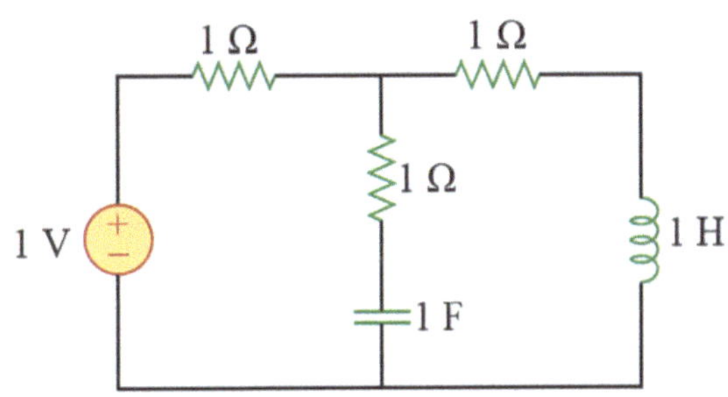

Fig. 7.20 The circuit of exercise 7.20

Problem

7.21. In the circuit of Fig. 7.21, calculate the total amount of energy stored in the circuit in the steady state condition.

Difficulty level ○ Easy ● Normal ○ Hard

Calculation amount ● Small ○ Normal ○ Large

1) 0.25 J
2) 0.125 J
3) 0.5 J
4) 1 J

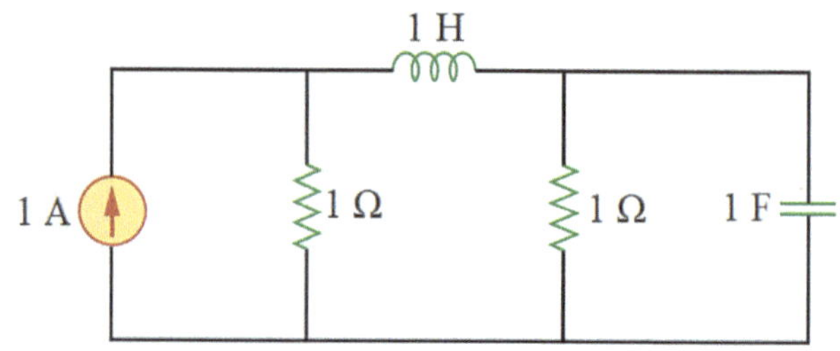

Fig. 7.21 The circuit of problem 7.21

Problem

7.22. In the circuit of Fig. 7.22, calculate the absolute value of charge stored in the capacitor in a steady state condition.

Difficulty level ○ Easy ● Normal ○ Hard

Calculation amount ○ Small ● Normal ○ Large

1) 2.21 μC
2) 2.01 μC
3) 1.28 mC
4) 0.48 mC

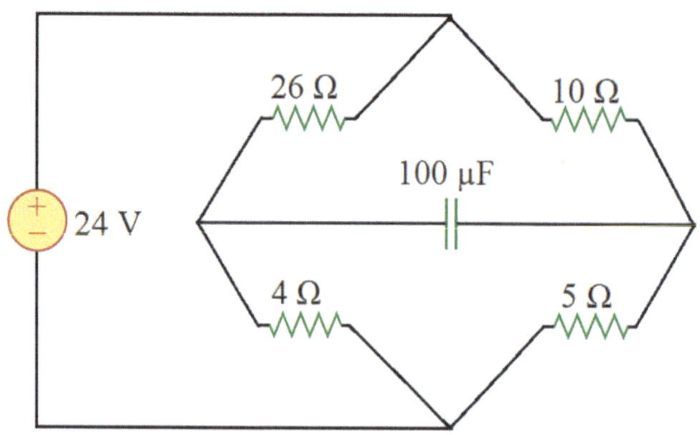

Fig. 7.22 The circuit of problem 7.22

References

1. Rahmani-Andebili, M., DC Electrical Circuit Analysis - Practice Problems, Methods, and Solutions (1st Ed.), Springer Nature, 2020.
2. Rahmani-Andebili, M., General Physics II - Practice Problems, Methods, and Solutions, Sringer Nature, 2025.
3. Rahmani-Andebili, M., General Physics I - Practice Problems, Methods, and Solutions, Springer Nature, 2025.
4. Rahmani-Andebili, M., Mathematics of Engineering and Science - Comprehensive Lessons, Springer Nature, 2024.
5. Rahmani-Andebili, M., Differential Equations – Practice Problems, Methods, and Solutions, Springer Nature, 2022.
6. Rahmani-Andebili, M., Calculus III – Practice Problems, Methods, and Solutions, Springer Nature, 2023.
7. Rahmani-Andebili, M., Calculus II – Practice Problems, Methods, and Solutions, Springer Nature, 2023.
8. Rahmani-Andebili, M., Calculus I (2nd Ed.) – Practice Problems, Methods, and Solutions, Springer Nature, 2023.
9. Rahmani-Andebili, M., Precalculus (2nd Ed.) – Practice Problems, Methods, and Solutions, Springer Nature, 2024.

Energy Storage Elements and Circuit Analysis Under Steady-State Condition: Part B

8

Abstract

In this chapter, the problems of the seventh chapter are fully solved, in detail, step-by-step, and with different methods. Herein, a "Notes Box" has been added after the solution of each problem that includes all the formulas and important notes that are needed to solve the problem. Moreover, a "Traps Box" has been added that discloses the traps that have been designed in the choices to challenge students' minds and encourage them to deeply think about the problem solutions as well as discourage them from careless selection of choices. The "Traps Box", in fact, predicts the possible confusions and tricks that might be encountered by students in real tests.

8.1 Equivalent Capacitance of Series, Parallel, and Series-Parallel Connections

8.1. The circuit shown in Fig. 8.1 is the series connection of three capacitors. Therefore, the equivalent capacitance of the circuit can be calculated as follows [1–9]

$$\frac{1}{C_{eq}} = \frac{1}{C_1} + \frac{1}{C_2} + \frac{1}{C_3}$$

$$\Rightarrow \frac{1}{C_{eq}} = \frac{1}{2} + \frac{1}{6} + \frac{1}{3}$$

$$\Rightarrow \frac{1}{C_{eq}} = \frac{3+1+2}{6} = \frac{6}{6}$$

$$\Rightarrow C_{eq} = 1\ \text{F}$$

Choice (2) is the answer.

Traps

- You would choose Choice (1) if you had mistakenly used the formula concerned with parallel capacitors.

S. Manias, *DC Electrical Circuits Analysis*, https://doi.org/10.1007/978-3-032-00253-2_8

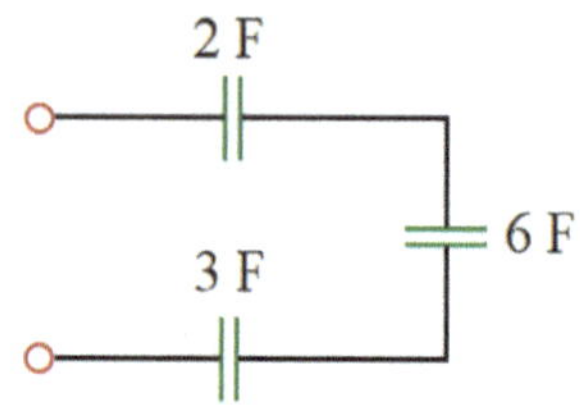

Fig. 8.1 The circuit of problem 8.1

Notes

The equivalent capacitance of n series capacitors can be calculated as follows.

$$\frac{1}{C_{eq}} = \frac{1}{C_1} + \ldots + \frac{1}{C_n}$$

If $C_1 = \ldots = C_n = C$, we have:

$$C_{eq} = \frac{C}{n}$$

If $n = 2$, we have:

$$\frac{1}{C_{eq}} = \frac{1}{C_1} + \frac{1}{C_2} \Rightarrow C_{eq} = \frac{C_1 C_2}{C_1 + C_2}$$

8.2. The circuit shown in Fig. 8.2 is the parallel connection of three capacitors. Therefore, the equivalent capacitance of the circuit can be calculated as follows.

$$C_{eq} = C_1 + C_2 + C_3$$

$$\Rightarrow C_{eq} = 2 + 3 + 6$$

$$\Rightarrow C_{eq} = 11 \text{ F}$$

Choice (1) is the answer.

Traps

- You would choose Choice (2) if you had mistakenly used the formula concerned with series capacitors.

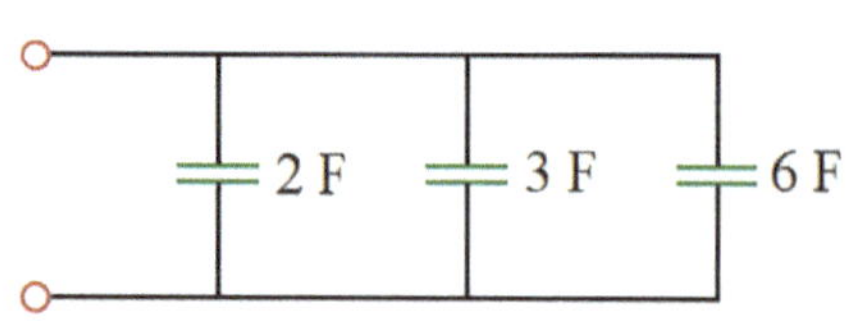

Fig. 8.2 The circuit of problem 8.2

Notes

The equivalent capacitance of n parallel capacitors can be calculated as follows.

$$C_{eq} = C_1 + \ldots + C_n$$

If $C_1 = \ldots = C_n = C$, we have:

$$C_{eq} = nC$$

8.3. The circuit shown in Fig. 8.3 is the series-parallel connection of three capacitors. Therefore, the equivalent capacitance of the circuit can be calculated as follows.

$$\frac{1}{C_{eq}} = \frac{1}{C_1} + \frac{1}{C_2 + C_3}$$

$$\Rightarrow \frac{1}{C_{eq}} = \frac{1}{2} + \frac{1}{1+1}$$

$$\Rightarrow \frac{1}{C_{eq}} = \frac{1}{2} + \frac{1}{2}$$

$$\Rightarrow C_{eq} = 1 \text{ F}$$

Choice (4) is the answer.

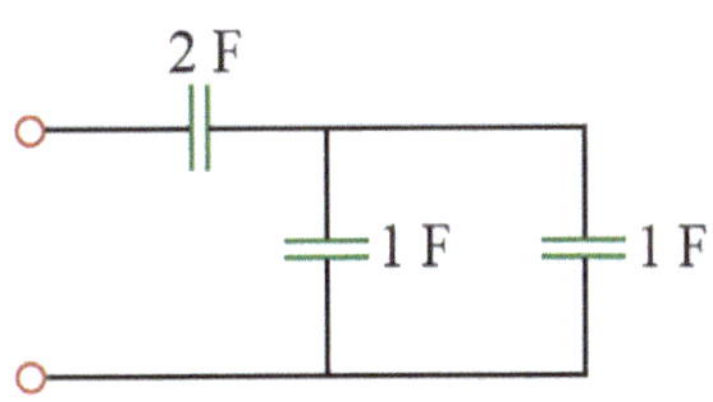

Fig. 8.3 The circuit of problem 8.3

Notes

The equivalent capacitance of n parallel capacitors can be calculated as follows.

$$C_{eq} = C_1 + \ldots + C_n$$

If $C_1 = \ldots = C_n = C$, we have:

$$C_{eq} = nC$$

The equivalent capacitance of n series capacitors can be calculated as follows.

$$\frac{1}{C_{eq}} = \frac{1}{C_1} + \ldots + \frac{1}{C_n}$$

If $C_1 = \ldots = C_n = C$, we have:

$$C_{eq} = \frac{C}{n}$$

If $n = 2$, we have:

$$\frac{1}{C_{eq}} = \frac{1}{C_1} + \frac{1}{C_2} \Rightarrow C_{eq} = \frac{C_1 C_2}{C_1 + C_2}$$

8.8. As can be seen in the circuit of Fig. 8.4.2, the equivalent capacitance of the circuit seen by terminal $a' - b'$ is approximately equal to the one seen by terminal $a - b$ since the size of the circuit is infinite. In other words:

$$C_{eq'} \approx C_{eq} \tag{1}$$

Now, we can calculate the total capacitance of the circuit by using the circuit of Fig. 8.4.3 as follows.

$$\frac{1}{C_{eq}} = \frac{1}{1} + \frac{1}{C_{eq'} + 1} + \frac{1}{1}$$

$$\Rightarrow \frac{1}{C_{eq}} = 2 + \frac{1}{C_{eq'} + 1}$$

$$\Rightarrow \frac{1}{C_{eq}} = \frac{2C_{eq'} + 3}{C_{eq'} + 1}$$

$$\Rightarrow C_{eq} = \frac{C_{eq'} + 1}{2C_{eq'} + 3} \tag{2}$$

Solving (1) and (2):

$$C_{eq} = \frac{C_{eq} + 1}{2C_{eq} + 3}$$

$$\Rightarrow 2C_{eq}^2 + 3C_{eq} = C_{eq} + 1$$

$$\Rightarrow 2C_{eq}^2 + 2C_{eq} - 1 = 0$$

$$\Rightarrow C_{eq} = \frac{-2 \pm \sqrt{2^2 - 4 \times 2 \times (-1)}}{2 \times 2}$$

$$\Rightarrow C_{eq} = \frac{-2 \pm \sqrt{12}}{4} = -1.36, 0.36$$

$C_{eq} = -1.36F$ is not acceptable. Therefore:

$$C_{eq} = 0.36 \text{ F}$$

Choice (2) is the answer.

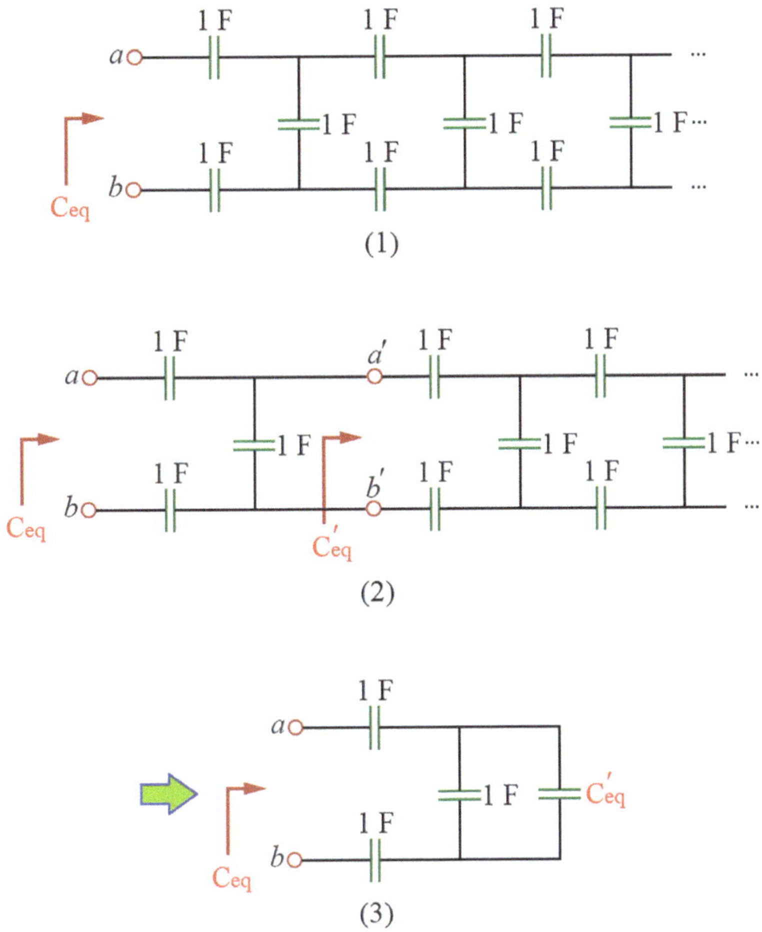

Fig. 8.4 The circuits of problem 8.8

Notes

In this problem, the relation below has been used.

$$ax^2 + bx + c = 0 \Rightarrow x = \frac{-b \pm \sqrt{b^2 - 4ac}}{2a}$$

8.2 Equivalent Inductance of Series, Parallel, and Series-Parallel Connections

8.9. The equivalent inductance of the series inductors can be calculated as follows.

$$L_{eq} = 2 + 6 + 3$$

$$\Rightarrow L_{eq} = 11 \text{ H}$$

Choice (1) is the answer.

Traps

- You would choose Choice (2) if you had mistakenly used the formula concerned with parallel inductors.

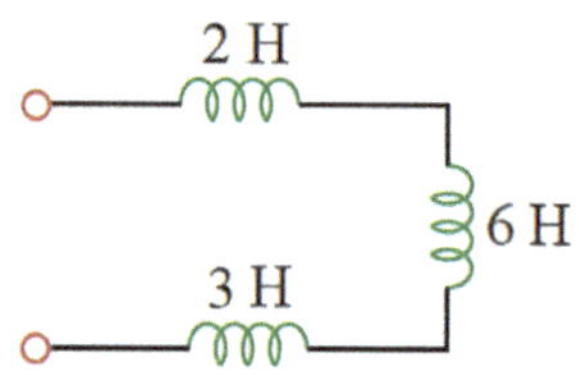

Fig. 8.5 The circuit of problem 8.9

Notes

The equivalent inductance of n series inductors can be calculated as follows.

$$L_{eq} = L_1 + \ldots + L_n$$

If $L_1 = \ldots = L_n = L$, we have:

$$L_{eq} = nL$$

8.10. The equivalent inductance of the parallel inductors can be calculated as follows.

$$\frac{1}{L_{eq}} = \frac{1}{2} + \frac{1}{3} + \frac{1}{6}$$

$$\Rightarrow \frac{1}{L_{eq}} = \frac{3 + 2 + 1}{6}$$

$$\Rightarrow L_{eq} = 1 \text{ H}$$

Choice (2) is the answer.

Traps

- You would choose Choice (1) if you had mistakenly used the formula concerned with series inductors.

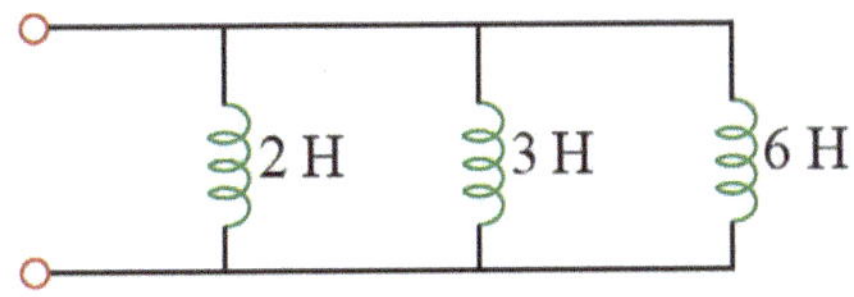

Fig. 8.6 The circuit of problem 8.10

Notes

The equivalent inductance of n parallel inductors can be calculated as follows.

$$\frac{1}{L_{eq}} = \frac{1}{L_1} + \ldots + \frac{1}{L_n}$$

If $L_1 = \ldots = L_n = L$, we have:

$$L_{eq} = \frac{L}{n}$$

If $n = 2$, we have:

$$\frac{1}{L_{eq}} = \frac{1}{L_1} + \frac{1}{L_2} \Rightarrow L_{eq} = \frac{L_1 L_2}{L_1 + L_2}$$

8.11 In the circuit, the inductors are connected in series-parallel from. Therefore:

$$L_{eq} = 2 + 1||1$$

$$\Rightarrow L_{eq} = 2 + \frac{1 \times 1}{1 + 1}$$

$$\Rightarrow L_{eq} = 2.5 \text{ H}$$

Choice (4) is the answer.

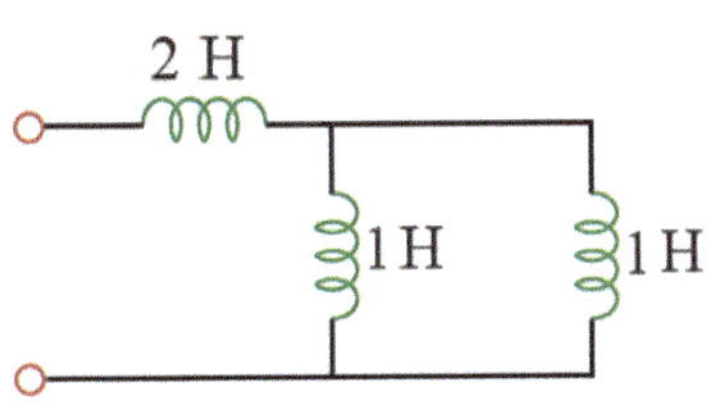

Fig. 8.7 The circuit of problem 8.11

Notes

The equivalent inductance of n series inductors can be calculated as follows.

$$L_{eq} = L_1 + \ldots + L_n$$

If $L_1 = \ldots = L_n = L$, we have:

$$L_{eq} = nL$$

The equivalent inductance of n parallel inductors can be calculated as follows.

$$\frac{1}{L_{eq}} = \frac{1}{L_1} + \ldots + \frac{1}{L_n}$$

If $L_1 = \ldots = L_n = L$, we have:

$$L_{eq} = \frac{L}{n}$$

If $n = 2$, we have:

$$\frac{1}{L_{eq}} = \frac{1}{L_1} + \frac{1}{L_2} \Rightarrow L_{eq} = \frac{L_1 L_2}{L_1 + L_2}$$

8.3 Circuit Analysis Under Steady-State Condition

8.16. To analyze a circuit in a steady state condition, we need to replace the capacitors and inductors with the open circuit and short circuit branches, respectively. Hence, for this problem, the circuit becomes like the one shown in Fig. 8.8.2.

By simplifying the circuit of Fig. 8.8.2, the circuit shown in Fig. 8.8.3 is achieved. Now, by applying superposition theorem, the problem can be solved as follows.

Applying voltage division rule in Fig. 8.8.4:

$$V_{o1} = \frac{1}{1+1} \times 10 = 5 \text{ V}$$

Applying current division rule in Fig. 8.8.5:

$$I_{o2} = \frac{1}{1+1} \times 4 = 2 \text{ A}$$

Applying Ohm's law in Fig. 8.8.5:

$$V_{o2} = 2 \times 1 = 2 \text{ V}$$

In overall, we have:

$$V_o = V_{o1} + V_{o2}$$

$$\Rightarrow V_o = 5 + 2$$

$$\Rightarrow V_o = 7 \text{ V}$$

Choice (2) is the answer.

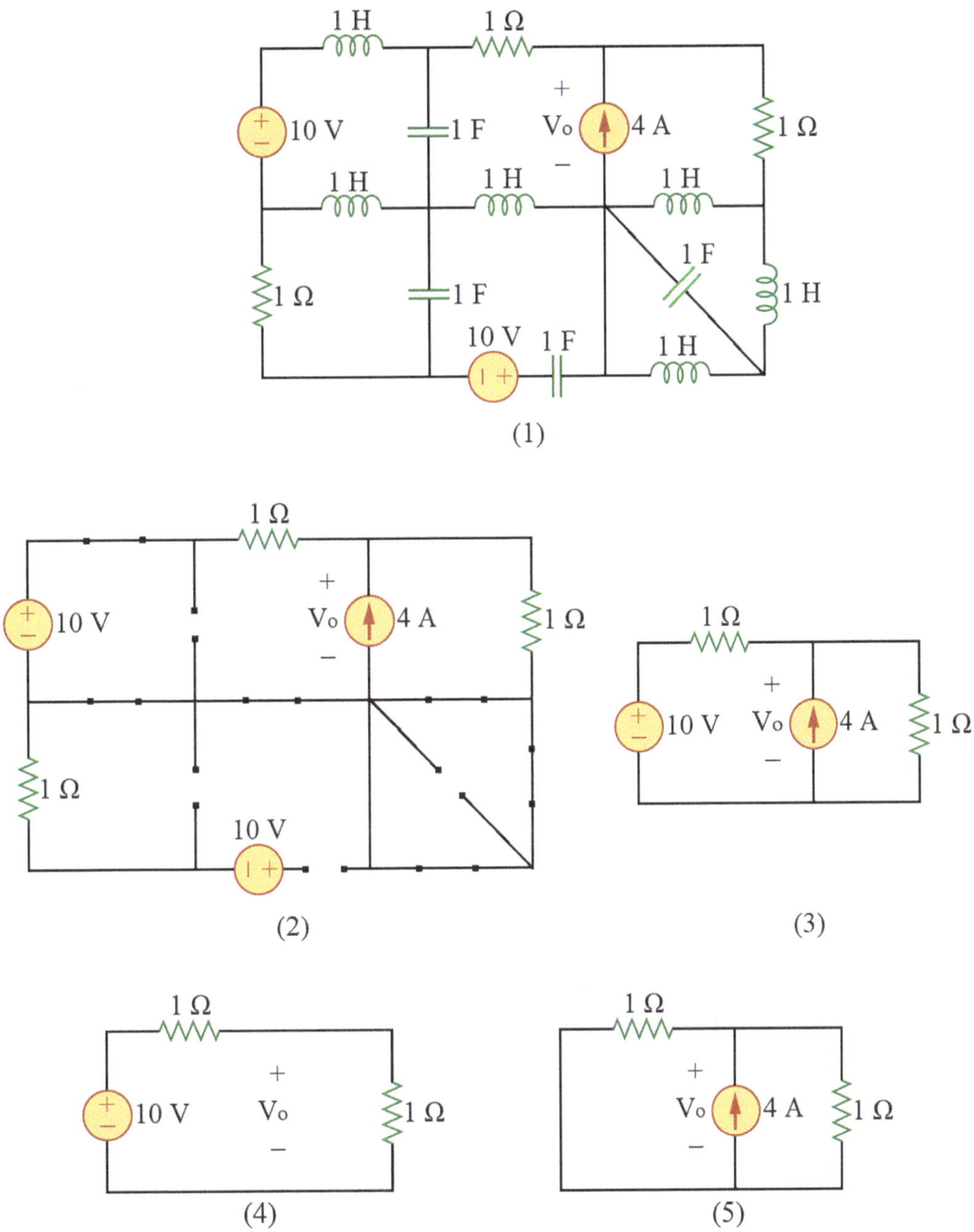

Fig. 8.8 The circuits of problem 8.16

8.17 As we know, in a steady state condition, a capacitor behaves like an open circuit branch and an inductor acts like a short circuit branch.

Hence, the following result is concluded from the circuit of Fig. 8.9.2.

$$I_L = 2 \text{ A}$$

Using Ohm's law for the 10-Ω resistor:

$$V_C = 10 \times 2$$

$$\Rightarrow V_C = 20 \text{ V}$$

Choice (4) is the answer.

Traps

- You would choose Choice (1) if you had mistakenly assumed that, in steady state conditions, capacitors and inductors behave like short circuit and open circuit branches, respectively.

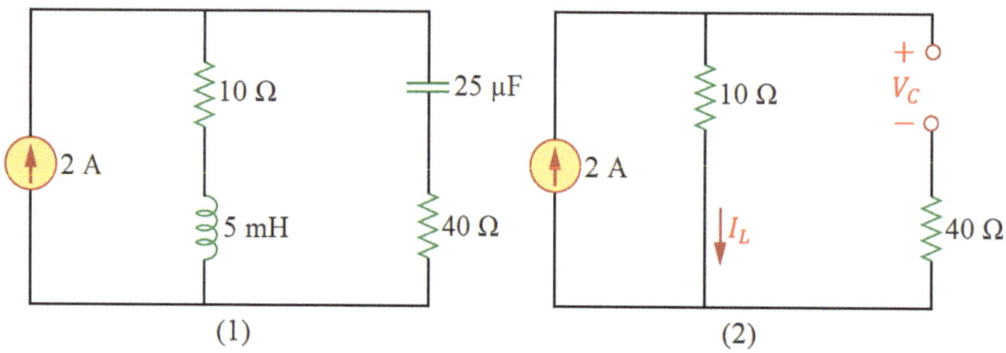

Fig. 8.9 The circuits of problem 8.17

8.18. As we know, the amount of energy stored in an inductor and capacitor in a steady state condition can be calculated as follows.

$$W_L = \frac{1}{2} L I_L^2$$

$$W_C = \frac{1}{2} C V_C^2$$

Therefore:

$$W_L = \frac{1}{2} \times 5 \times 10^{-3} \times 2^2 = 10^{-2} \text{ J}$$

$$\Rightarrow W_L = 10 \text{ mJ}$$

$$W_C = \frac{1}{2} \times 25 \times 10^{-6} \times 20^2 = 5 \times 10^{-3} \text{ J}$$

$$\Rightarrow W_C = 5 \text{ mJ}$$

Choice (2) is the answer.

8.21. As we know, in a steady state condition, a capacitor behaves like an open circuit branch and an inductor acts like a short circuit branch.

Hence, by applying current division rule, we have:

$$I_L = \frac{1}{1+1} \times 1 = 0.5 \text{ A}$$

Applying Ohm's law:

$$V_C = 1 \times 0.5 = 0.5 \text{ V}$$

Moreover, the amount of energy stored in an inductor and capacitor in a steady state condition can be calculated as follows.

$$W_L = \frac{1}{2} L I_L^2$$

$$W_C = \frac{1}{2} C V_C^2$$

Therefore:

$$W_L = \frac{1}{2} \times 1 \times 0.5^2$$

$$\Rightarrow W_L = 0.125 \text{ J}$$

$$W_C = \frac{1}{2} \times 1 \times 0.5^2$$

$$\Rightarrow W_C = 0.125 \text{ J}$$

Therefore, the total energy stored in the circuit is as follows.

$$W = W_L + W_C$$

$$\Rightarrow W = 0.25 \text{ J}$$

Choice (1) is the answer.

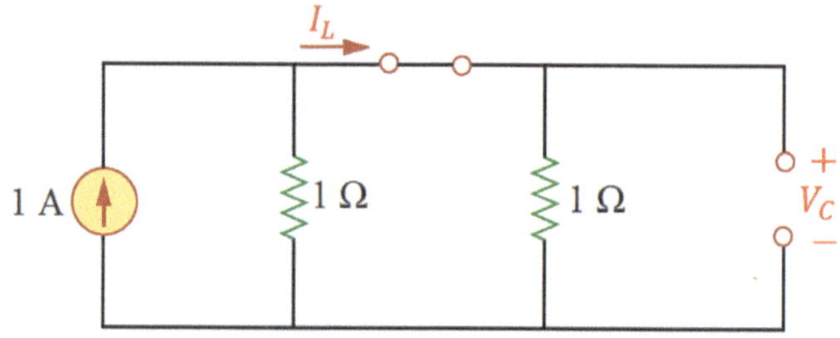

Fig. 8.10 The circuit of problem 8.21

8.22. The circuit is in a steady state condition. Therefore, the capacitor can be modeled by an open circuit branch, as is shown in Fig. 8.11.2.

Using voltage division rule for points "*a*" and "*b*":

$$V_C(\infty) = V_a - V_b$$

$$\Rightarrow V_C(\infty) = \frac{4}{4+26} \times 24 - \frac{5}{5+10} \times 24$$

$$\Rightarrow V_C(\infty) = \left(\frac{4}{30} - \frac{5}{15}\right) \times 24$$

$$\Rightarrow V_C(\infty) = -\frac{3}{15} \times 24$$

$$\Rightarrow V_C(\infty) = -4.8 \text{ V}$$

Using voltage-charge relation of capacitors:

$$Q_C(\infty) = V_C(\infty)C$$

$$\Rightarrow Q_C(\infty) = -4.8 \times 100 \times 10^{-6}$$

$$\Rightarrow Q_C(\infty) = -0.48 \text{ mC}$$

Therefore, the absolute value of charge stored in the capacitor is as follows.

$$|Q_C(\infty)| = 0.48 \text{ mC}$$

Choice (4) is the answer.

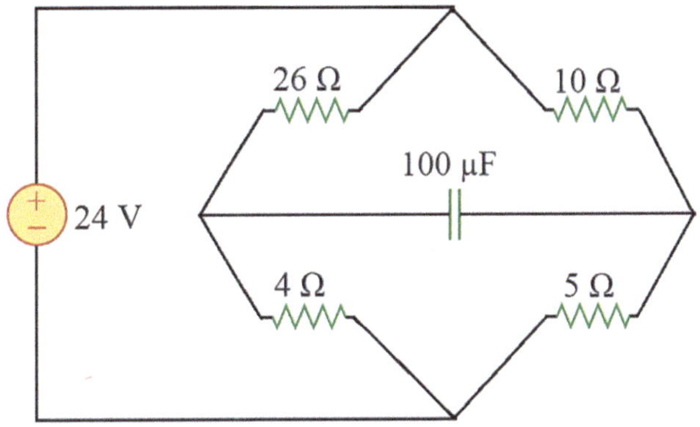

(1)

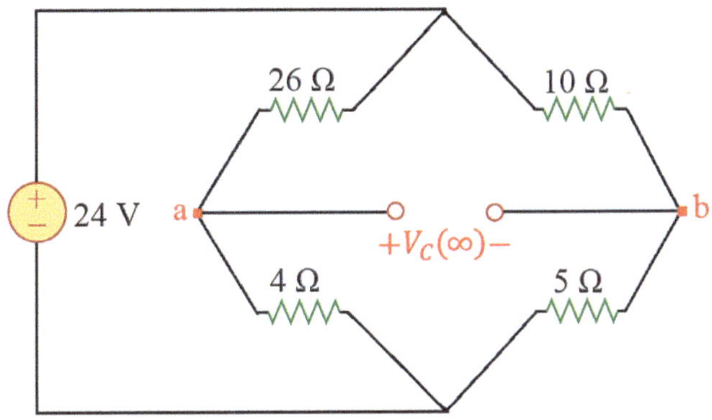

(2)

Fig. 8.11 The circuits of problem 8.22

References

1. Rahmani-Andebili, M., DC Electrical Circuit Analysis - Practice Problems, Methods, and Solutions (1st Ed.), Springer Nature, 2020.
2. Rahmani-Andebili, M., General Physics II - Practice Problems, Methods, and Solutions, Sringer Nature, 2025.
3. Rahmani-Andebili, M., General Physics I - Practice Problems, Methods, and Solutions, Springer Nature, 2025.
4. Rahmani-Andebili, M., Mathematics of Engineering and Science - Comprehensive Lessons, Springer Nature, 2024.
5. Rahmani-Andebili, M., Differential Equations – Practice Problems, Methods, and Solutions, Springer Nature, 2022.
6. Rahmani-Andebili, M., Calculus III – Practice Problems, Methods, and Solutions, Springer Nature, 2023.
7. Rahmani-Andebili, M., Calculus II – Practice Problems, Methods, and Solutions, Springer Nature, 2023.
8. Rahmani-Andebili, M., Calculus I (2nd Ed.) – Practice Problems, Methods, and Solutions, Springer Nature, 2023.
9. Rahmani-Andebili, M., Precalculus (2nd Ed.) – Practice Problems, Methods, and Solutions, Springer Nature, 2024.

9 Analysis of First-Order Circuits: Part A

Abstract

In this chapter, the basic and advanced problems related to the analysis of first-order circuits are studied. The subjects include time constant of first-order RC circuits, source-free response of an RC circuit, step response of an RC circuit, complete response of an RC circuit, time constant of first-order RL circuits, source-free response of an RL circuit, step response of an RL circuit, heuristic methods for first-order circuits, first-order circuits including ramp and singularity functions, differential equation of first-order circuits, and energy stored in first-order circuits. Herein, different types of problems and exercises are presented that are categorized as follows.

- ***Problems with detailed solution***: They have been designed to teach students the subjects in detail. Moreover, they have been categorized in different levels based on their difficulty levels (easy, normal, and hard) and calculation amounts (small, normal, and large).
- ***Partially solved exercises***: They have been designed to encourage students to practice problems while guiding them through the problem-solving procedure and hinting the required formulas.
- ***Exercises with final answer***: They have been designed to encourage students to practice more by themselves while hinting them by the final answer as well as to help instructors to give tests or quizzes.

9.1 Time Constant of First-Order RC Circuits

Problem

9.1. Calculate the time constant of the circuit shown in Fig. 9.1 [1–9].

Difficulty level ○ Easy ● Normal ○ Hard

Calculation amount ○ Small ● Normal ○ Large

1) 14.4 sec
2) 15.5 sec
3) 31.5 sec
4) 33.5 sec

S. Manias, *DC Electrical Circuits Analysis*, https://doi.org/10.1007/978-3-032-00253-2_9

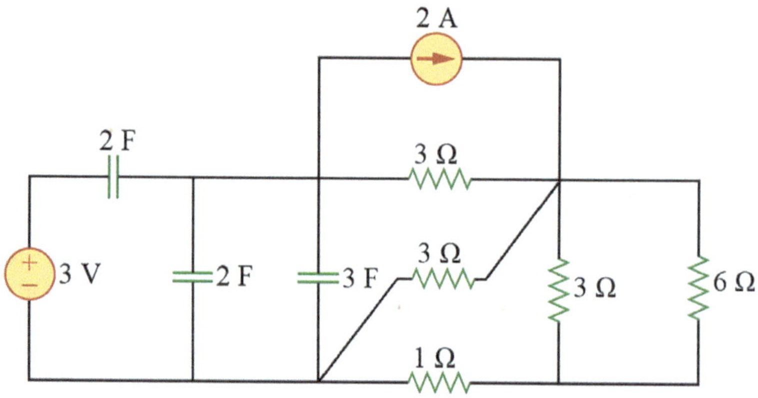

Fig. 9.1 The circuit of problem 9.1

Partially Solved Exercise

9.2. Calculate the time constant of the first-order RC circuit shown in Fig. 9.2.

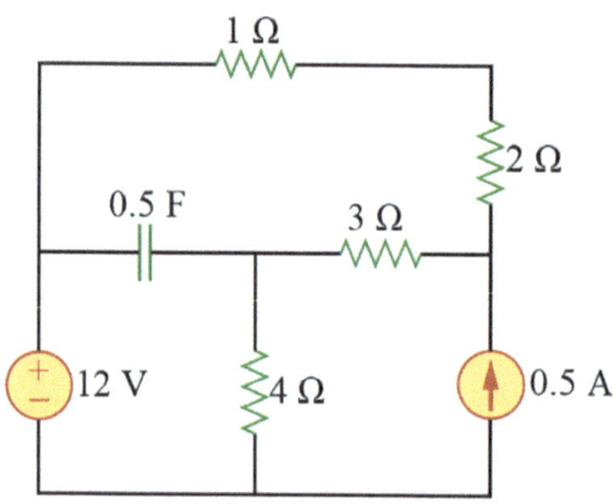

Fig. 9.2 The circuit of partially solved exercise 9.2

Solution

As we know, the time constant of a first-order RC circuit can be calculated as follows.

$$\tau = R_{Th}C_{eq}$$

Herein, C_{eq} is the equivalent capacitor and R_{Th} is the Thevenin resistance seen by the equivalent capacitor. From previous chapters, we know that R_{Th} can be calculated as follows.

First, all the independent sources (independent voltage and current sources) are turned off (each independent voltage source is replaced by a short circuit branch and each independent current source is replaced by an open circuit branch). Then, if the circuit does not include any dependent source (dependent voltage or current source), the series-parallel rule is applied to calculate the input resistance of the circuit. However, if the circuit includes at least a dependent source, a test source (independent voltage or current source) with the voltage and current of V_t and I_t is connected to the input terminal of the circuit. Next, the circuit is analyzed by using a circuit analysis method to find the value of $\frac{V_t}{I_t}$. After that, we have:

$$R_{Th} = \frac{V_t}{I_t}$$

Therefore, by using the circuit of Fig. 9.3, we can write as follows.

$$R_{Th} = (1+2+3)||4$$

$$\Rightarrow R_{Th} = (\quad)||(\quad)$$

$$\Rightarrow R_{Th} = \frac{(\quad)\times(\quad)}{(\quad)+(\quad)}$$

$$\Rightarrow R_{Th} = 2.4\ \Omega$$

The time constant of the RC circuit can be calculated as follows.

$$\tau = (\quad)\times(\quad)$$

$$\Rightarrow \tau = 1.2\ \text{sec}$$

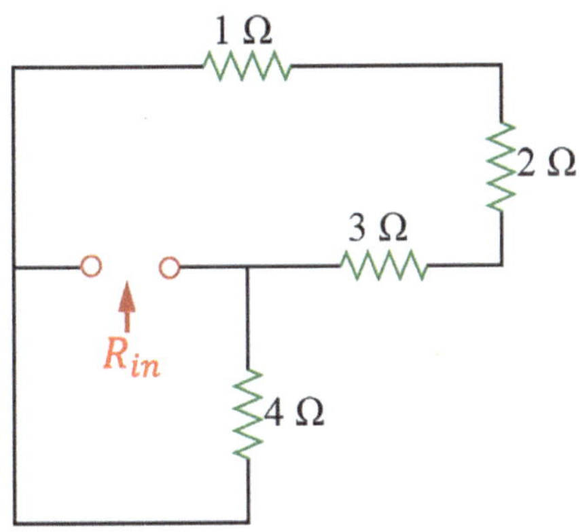

Fig. 9.3 The circuit of partially solved exercise 9.2

Problem

9.3. Calculate the time constant of the circuit of Fig. 9.4.

Difficulty level ● Easy ○ Normal ○ Hard

Calculation amount ● Small ○ Normal ○ Large

1) $\frac{R_1R_2C}{R_1+R_2}$
2) $(R_1+R_2)C$
3) $\frac{R_1}{R_2}C$
4) $\frac{R_2}{R_1}C$

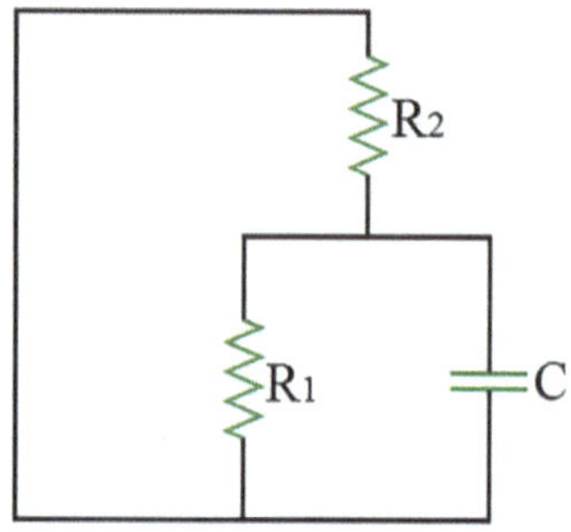

Fig. 9.4 The circuit of problem 9.3

Problem

9.4. Calculate the time constant of the circuit illustrated in Fig. 9.5.

Difficulty level ○ Easy ○ Normal ● Hard

Calculation amount ○ Small ● Normal ○ Large

1) 1.5 sec
2) 2 sec
3) 3.5 sec
4) 5 sec

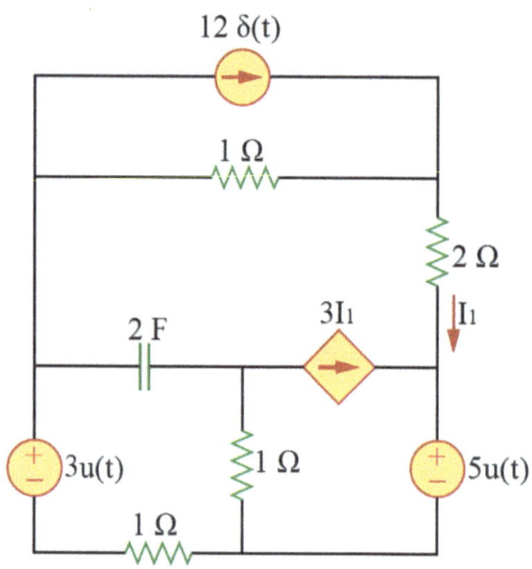

Fig. 9.5 The circuit of problem 9.4

Partially Solved Exercise

9.5. In the circuit of Fig. 9.6, determine the range of k so that the time constant of the circuit is below 8 seconds.

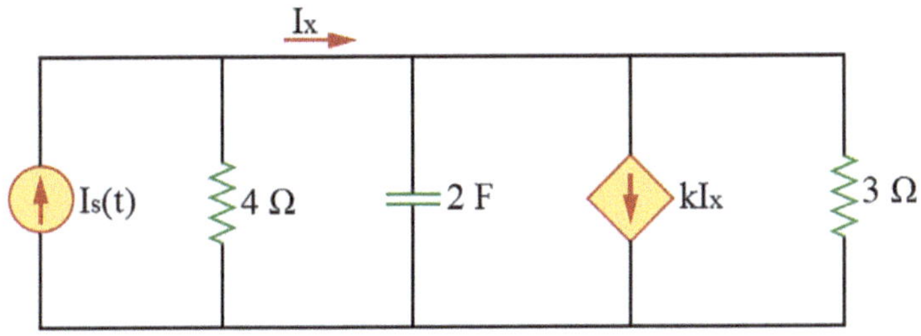

Fig. 9.6 The circuit of partially solved exercise 9.5

Solution

As we know, the time constant of a first-order RC circuit can be calculated as follows.

$$\tau = R_{Th} C_{eq}$$

Herein, C_{eq} is the equivalent capacitor and R_{Th} is the Thevenin resistance seen by the equivalent capacitor. From previous chapters, we know that R_{Th} can be calculated as follows.

First, all the independent sources (independent voltage and current sources) are turned off (each independent voltage source is replaced by a short circuit branch and each independent current source is replaced by an open circuit branch). Then, if the circuit does not include any dependent source (dependent voltage or current source), the series-parallel rule is applied to calculate the input resistance of the circuit. However, if the circuit includes at least a dependent source, a test source (independent voltage or current source) with the voltage and current of V_t and I_t is connected to the input terminal of the circuit. Next, the circuit is analyzed by using a circuit analysis method to find the value of $\frac{V_t}{I_t}$. After that, we have:

$$R_{Th} = \frac{V_t}{I_t}$$

Using Ohm's law for the 4-Ω resistor:

$$I_x = (\qquad) \quad (1)$$

KCL in the indicated node:

$$(\qquad) + (\qquad) + (\qquad) + (\qquad) = 0$$

$$\Rightarrow -I_t + (k-1)I_x + \frac{V_t}{3} = 0 \quad (2)$$

Solving (1) and (2):

$$-I_t + (k-1)\left(-\frac{V_t}{4}\right) + \frac{V_t}{3} = 0$$

$$\Rightarrow (\qquad\qquad) I_t + (\qquad\qquad\qquad\qquad) V_t = 0$$

$$\Rightarrow (\qquad\qquad) I_t + (\qquad\qquad\qquad\qquad) V_t = 0$$

$$\Rightarrow \frac{V_t}{I_t} = (\qquad)$$

$$\Rightarrow R_{Th} = \frac{12}{7-3k}$$

The time constant of the RC circuit can be calculated as follows (Fig. 9.7).

$$\tau = (\qquad\qquad) \times (\qquad\qquad)$$

$$\Rightarrow \tau = \frac{24}{7-3k} \quad (3)$$

Based on the information given in the problem, we have:

$$\tau < 8sec \quad (4)$$

Solving (3) and (4):

$$\frac{24}{7-3k} < 8$$

$$\Rightarrow (\qquad\qquad) < (\qquad\qquad)$$

$$\Rightarrow (\quad)k < (\quad)$$

$$\Rightarrow k < \frac{4}{3}$$

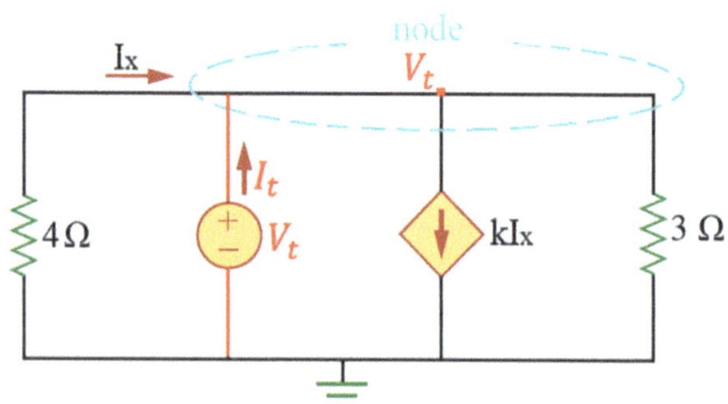

Fig. 9.7 The circuit of partially solved exercise 9.5

9.2 Source-Free Response of an RC Circuit

Problem

9.6. In the circuit of Fig. 9.8, $V_C(0^+) = 10V$. Calculate $V_C(t)$ for $t \geq 0$.

Difficulty level ● Easy ○ Normal ○ Hard

Calculation amount ● Small ○ Normal ○ Large

1) $10e^{-\frac{t}{36}}$ V

2) $e^{-\frac{t}{36}}$ V

3) $10e^{-\frac{t}{24}}$ V

4) $e^{-\frac{t}{24}}$ V

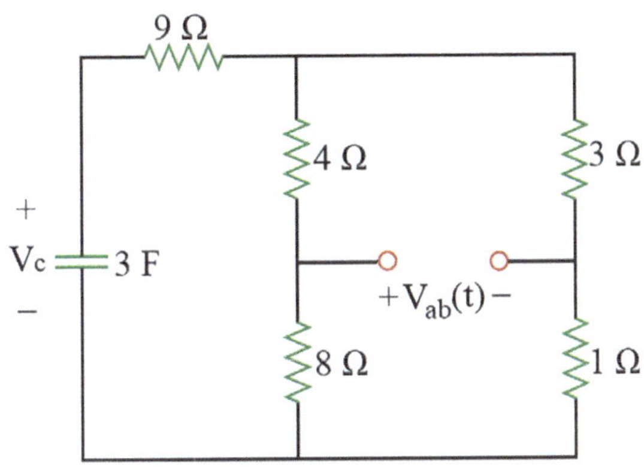

Fig. 9.8 The circuit of problem 9.6

Problem

9.7. In Problem 9.6, calculate $V_{ab}(t)$ for $t \geq 0$.

Difficulty level ○ Easy ● Normal ○ Hard

Calculation amount ○ Small ● Normal ○ Large

1) $2.1e^{-\frac{t}{36}}$ V

2) $1.04e^{-\frac{t}{36}}$ V

3) $1.04e^{-\frac{t}{24}}$ V

4) $2.1e^{-\frac{t}{24}}$ V

Partially Solved Exercise

9.8. The switch shown in the circuit of Fig. 9.9 is closed at $t = 0$ after a long time. Calculate the voltage of the capacitor for $t \geq 0$.

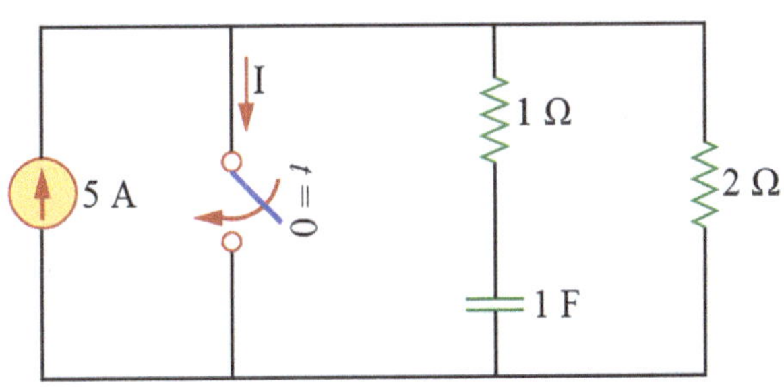

Fig. 9.9 The circuit of partially solved exercise 9.8

Solution

Since the switch is closed at $t = 0$ and remove the power source (independent current source), the type of response of the circuit is source-free.
The general form of source-free response of an RC circuit is as follows.

$$V_C(t) = V_C(0^+)e^{-\frac{t}{\tau}}$$

Herein, $V_C(0^+)$ is the initial voltage of the capacitor and $\tau = R_{Th}C$ is the time constant of the circuit.

Calculating the initial voltage of the capacitor: Fig. 9.10.2 shows the circuit before closing the switch. Since the switch was open for a long time, the circuit has reached its steady state condition, and the capacitor is like an open circuit.
The voltage of the capacitor (before the switching) can be calculated by using Ohm's law as follows.

$$V_C(0^-) = (\quad) \times (\quad)$$

$$\Rightarrow V_C(0^-) = 10 \text{ V}$$

Due to voltage continuity of capacitors in lack of a power source with an infinite size (e.g. impulse power source), it is concluded that:

$$V_C(0^+) = (\quad) \text{ V}$$

Calculating the time constant of the circuit: Fig. 9.10.3 illustrates the circuit after closing the switch. Moreover, Fig. 9.10.4 shows the simplified circuit of Fig. 9.10.3. As can be seen, the 2-Ω resistor has been eliminated by the short-circuited branch of the switch. Thus, the Thevenin resistance of the circuit is as follows.

$$R_{Th} = 1 \ \Omega$$

Hence, the time constant of the RC circuit is calculated as follows.

$$\tau = (\quad) \times (\quad) = 1 \text{ sec}$$

Calculating the source-free response of the RC circuit:

$$V_C(t) = 10e^{-t} \text{ V}$$

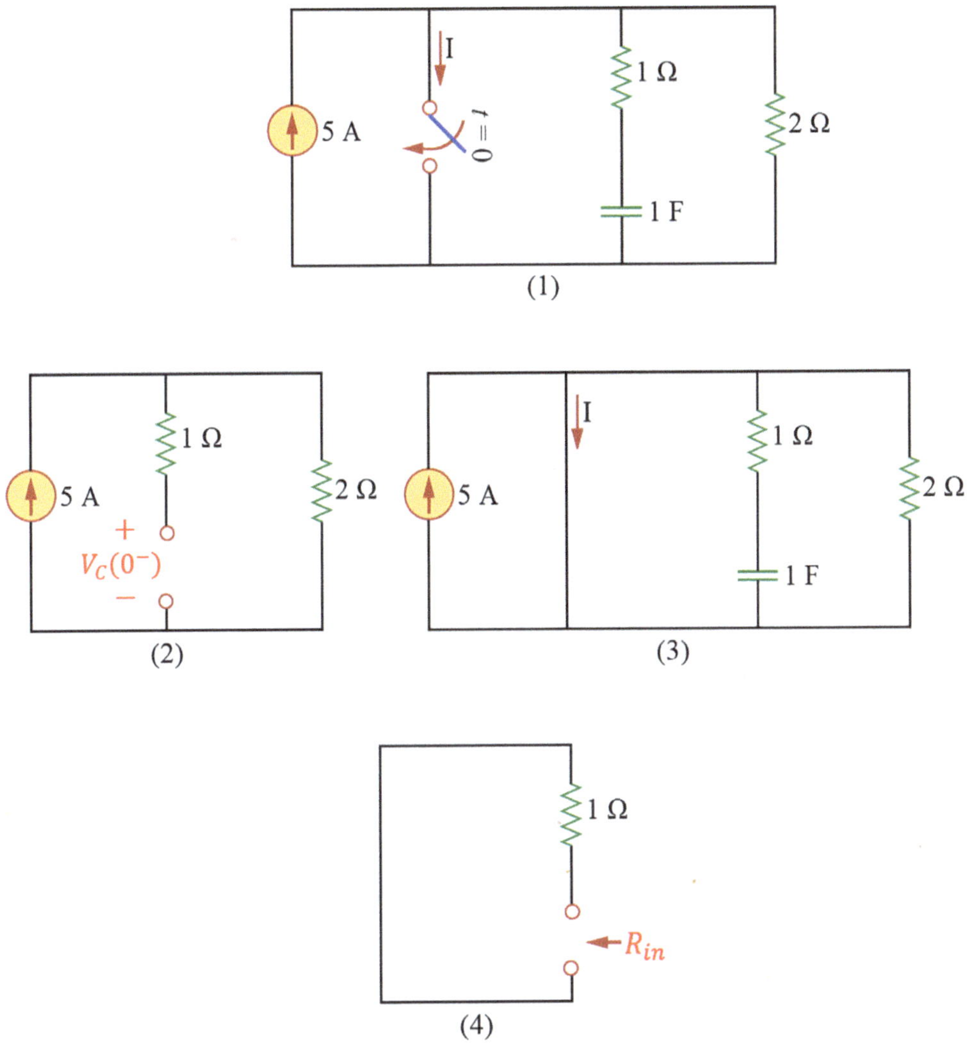

Fig. 9.10 The circuits of partially solved exercise 9.8

Problem

9.9. In Problem 9.8, calculate the current of the switch at $t = 1$ sec.

Difficulty level ○ Easy ○ Normal ● Hard

Calculation amount ○ Small ● Normal ○ Large

1) $5(1 - e^{-1})$ A
2) $5(1 + e^{-1})$ A
3) $5(1 - 2e^{-1})$ A
4) $5(1 + 2e^{-1})$ A

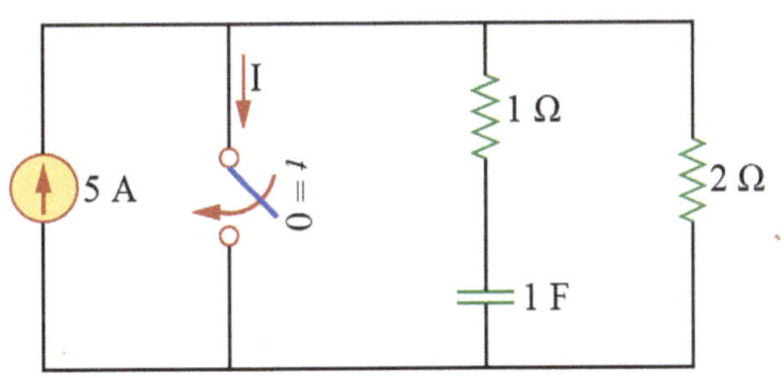

Fig. 9.11 The circuit of problem 9.9

Problem

9.10. In the circuit of Fig. 9.12, calculate the voltage of the capacitor for $t \geq 0$. The primary voltage of the capacitor is $2V$ at $t = 0^-$.

Difficulty level ○ Easy ● Normal ○ Hard
Calculation amount ○ Small ● Normal ○ Large

1) e^{-5t} V
2) $2e^{-5t}$ V
3) $e^{-\frac{t}{5}}$ V
4) $2e^{-\frac{t}{5}}$ V

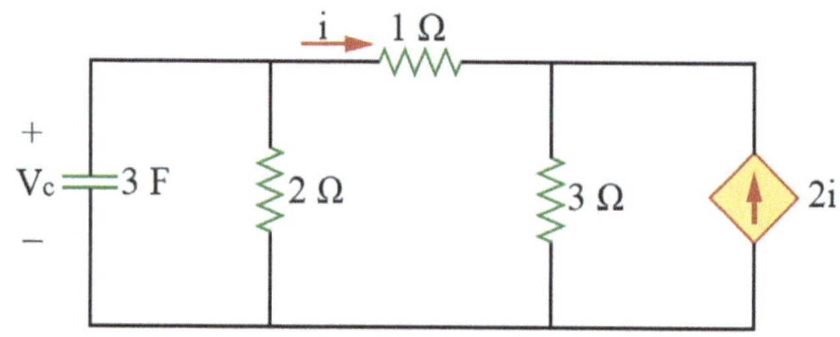

Fig. 9.12 The circuit of problem 9.10

Problem

9.11. In Problem 9.10, calculate the time that will take for the initial voltage of the capacitor to be halved.

Difficulty level ○ Easy ● Normal ○ Hard
Calculation amount ○ Small ● Normal ○ Large

1) $5 \ln 2$ sec
2) $4 \ln 2$ sec
3) $\frac{4}{3} \ln 2$ sec
4) $\frac{3}{4} \ln 2$ sec

9.3 Step Response of an RC Circuit

Problem

9.12. In the circuit of Fig. 9.13, the independent current source is a unit step function, and the capacitor does not have any primary voltage. Calculate the voltage across the current source for $t \geq 0$.

Difficulty level ○ Easy ○ Normal ● Hard
Calculation amount ● Small ○ Normal ○ Large

1) $0.5\left(1 - e^{-0.5t}\right)u(t)$ V
2) $0.5\left(1 - e^{-0.25t}\right)u(t)$ V
3) $2\left(1 - e^{-0.5t}\right)u(t)$ V
4) $2\left(1 - e^{-0.25t}\right)u(t)$ V

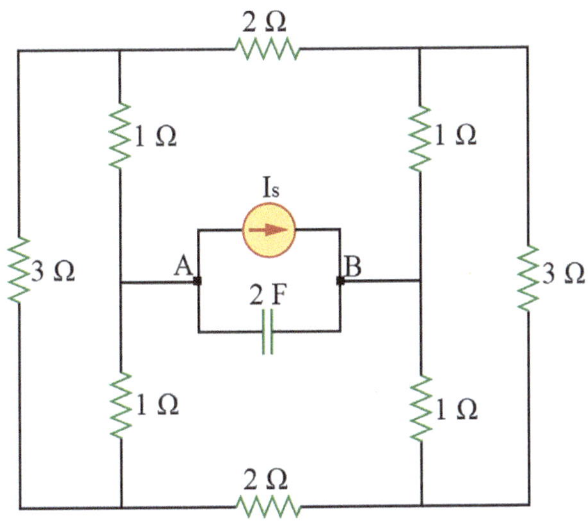

Fig. 9.13 The circuit of problem 9.12

Partially Solved Exercise

9.13. In the circuit of Fig. 9.14, the initial voltage of the capacitor is zero. Calculate its voltage for $t \geq 0$.

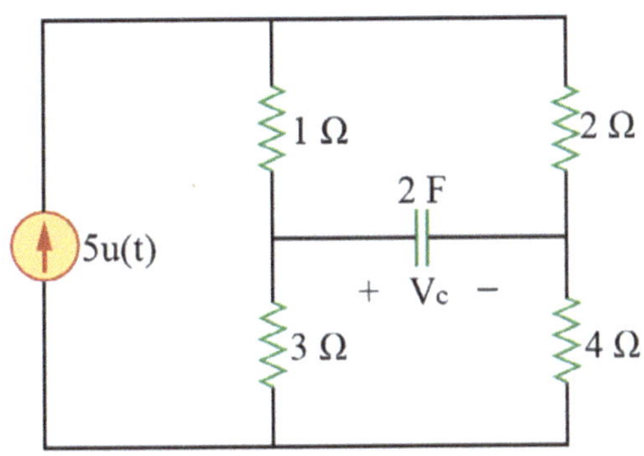

Fig. 9.14 The circuit of partially solved exercise 9.13

Solution

Since the initial voltage of the capacitor is zero and the power source (independent current source) is a step function, the response of the circuit is step response.
The general form of step response of an RC circuit is as follows.

$$V_C(t) = V_C(\infty)\left(1 - e^{-\frac{t}{\tau}}\right)$$

Herein, $V_C(\infty)$ is the voltage of the capacitor after a long time and $\tau = R_{Th}C$ is the time constant of the circuit.

Calculating the time constant of the circuit: Fig. 9.15.1 is used to calculate the Thevenin resistance of the circuit as follows.

$$R_{Th} = (1+2)||(3+4)$$

$$\Rightarrow R_{Th} = (\quad)||(\quad)$$

$$\Rightarrow R_{Th} = \frac{(\quad)(\quad)}{(\quad)+(\quad)}$$

$$\Rightarrow R_{Th} = \frac{21}{10}\ \Omega$$

The time constant of the circuit:

$$\tau = (\quad) \times (\quad)$$

$$\Rightarrow \tau = \frac{21}{5}\ \text{sec}$$

Calculating the final voltage of the capacitor: Applying current division rule in the circuit of Fig. 9.15.2:

$$I_{1,3} = \frac{(\quad)+(\quad)}{(\quad)+(\quad)+(\quad)+(\quad)} \times (\quad) = 3\ \text{A}$$

$$I_{2,4} = \frac{(\quad)+(\quad)}{(\quad)+(\quad)+(\quad)+(\quad)} \times (\quad) = 2\ \text{A}$$

Applying Ohm's law in the circuit of Fig. 9.15.2:

$$V_C(\infty) = V_C^+ - V_C^-$$

$$\Rightarrow V_C(\infty) = 3I_{1,3} - 4I_{2,4}$$

$$\Rightarrow V_C(\infty) = (\quad)(\quad) - (\quad)(\quad)$$

$$V_C(\infty) = 1\ \text{V}$$

Calculating the step response of the RC circuit:

$$V_C(t) = 1 - e^{-\frac{5}{21}t}\ \text{V}$$

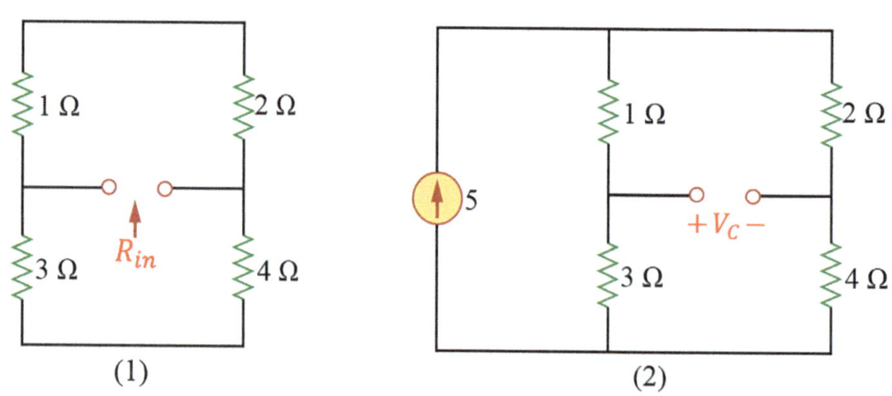

Fig. 9.15 The circuits of partially solved exercise 9.13

Problem

9.14. In the circuit of Fig. 9.16, the primary voltage of the capacitor is zero. Calculate the voltage of the capacitor at $t = 1ms$.

Difficulty level ○ Easy ● Normal ○ Hard

Calculation amount ○ Small ● Normal ○ Large

1) $2\left(1+e^{-1}\right)$ V
2) $2\left(1-e^{-1}\right)$ V
3) $3\left(1+e^{-1}\right)$ V
4) $3\left(1-e^{-1}\right)$ V

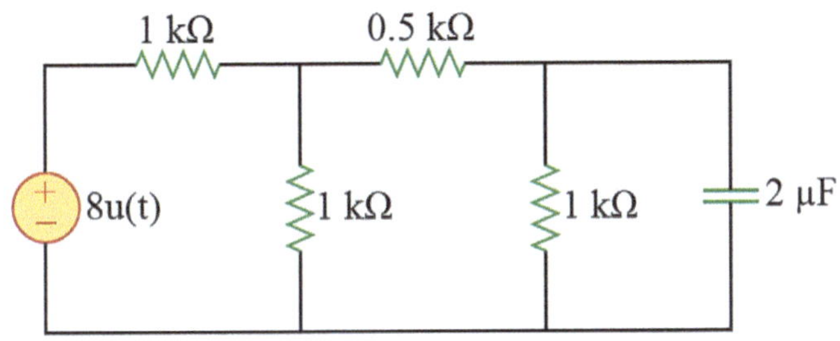

Fig. 9.16 The circuit of problem 9.14

Problem

9.15. In the circuit of Fig. 9.17, at $t = 0$, the position of switch is changed from “A” to “B”. Calculate the time (t_0) that the voltage of capacitor reaches 50% of its final value.

Difficulty level ○ Easy ● Normal ○ Hard

Calculation amount ○ Small ● Normal ○ Large

1) $t_0 = 0.5 \ln 2$
2) $t_0 = \ln 2$
3) $t_0 = 1.5 \ln 2$
4) $t_0 = 2 \ln 2$

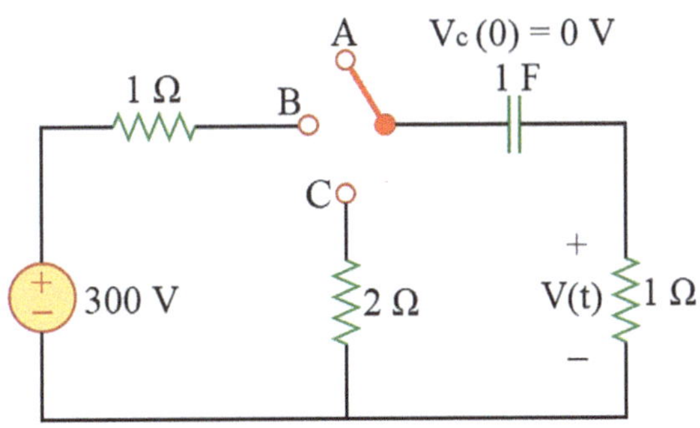

Fig. 9.17 The circuit of problem 9.15

Problem

9.16. In Problem 9.15, the switch is moved to position "C" at $t_0 = 0$. Calculate $V(t)$ after this switching operation.

Difficulty level ○ Easy ○ Normal ● Hard

Calculation amount ○ Small ● Normal ○ Large

1) $-50e^{-(t-2\ln 2)/2}$ V
2) $-150e^{-(t-\ln 2)/3}$ V
3) $-150e^{-(t-\ln 2)/2}$ V
4) $-50e^{-(t-2\ln 2)/3}$ V

Partially Solved Exercise

9.17. In the circuit of Fig. 9.18, calculate the voltage of the capacitor after the switching operation at $t_1 = R_1C$. The primary voltage of the capacitor is zero.

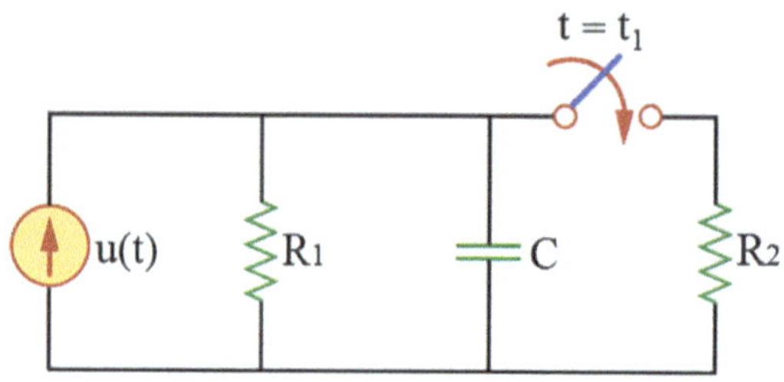

Fig. 9.18 The circuit of partially solved exercise 9.17

Solution

Since the initial voltage of the capacitor is zero and the power source (independent current source) is a step function, the response of the circuit is step response.

The general form of step response of an RC circuit is as follows.

$$V_C(t) = V_C(\infty)\left(1 - e^{-\frac{t}{\tau}}\right)$$

Herein, $V_C(\infty)$ is the voltage of the capacitor after a long time and $\tau = R_{Th}C$ is the time constant of the circuit.

Calculating the time constant of the circuit: The time constant of the circuit before the switching is as follows. See Fig. 9.19.1.

$$\tau = R_1C$$

Calculating the final voltage of the capacitor: Since after a long time, the capacitor will be an open circuit, its voltage can be calculated by using Ohm's law as follows. See Fig. 9.19.2.

$$V_C(\infty) = (\qquad) \times (\qquad) = R_1$$

Calculating the step response of the RC circuit:

$$V_C(t) = R_1\left(1 - e^{-\frac{t}{R_1 C}}\right)$$

The voltage of the capacitor after the switching operation at $t_1 = R_1 C$ can be calculated as follows.

$$V_C(t_1 = R_1 C) = R_1\left(1 - e^{-\frac{(\quad)}{R_1 C}}\right)$$

$$\Rightarrow V_C(t_1 = R_1 C) = R_1\left(1 - e^{-(\quad)}\right)$$

$$\Rightarrow V_C(t_1 = R_1 C) = R_1\left(1 - \frac{1}{e}\right)$$

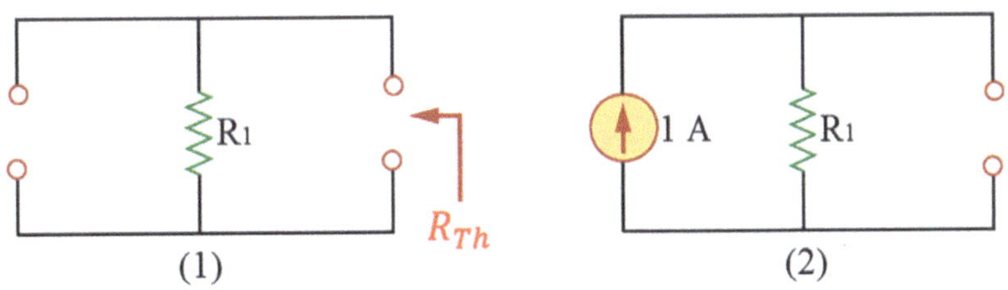

Fig. 9.19 The circuits of partially solved exercise 9.17

Problem

9.18. The voltage pulse shown in Fig. 9.20.1 is applied to the circuit of Fig. 9.20.2. Calculate the current of the circuit.

Difficulty level ○ Easy ○ Normal ● Hard

Calculation amount ○ Small ● Normal ○ Large

1) $-\left(10e^{-(t-1)} - 10e^{-t}\right)u(t)$
2) $-\left(10e^{-(t-1)} - 10e^{-t}\right)u(t-1)$
3) $\left(10e^{-(t-1)} - 10e^{-t}\right)u(t-1)$
4) $\left(10e^{-(t-1)} - 10e^{-t}\right)u(t)$

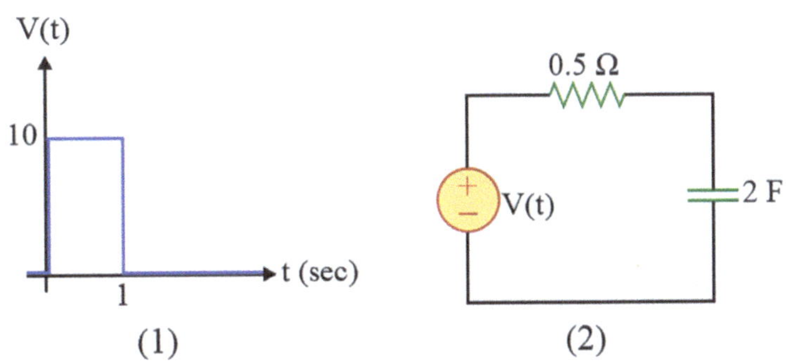

Fig. 9.20 The circuit of problem 9.18

Exercise

9.19. The circuit of Fig. 9.21 is in zero state before $t = 0$. Calculate the voltage of the capacitor for $t > 0$.

Final Answer

$V_C(t) = 0$

***Note:** Read the "educational support" part of the preface of the textbook.*

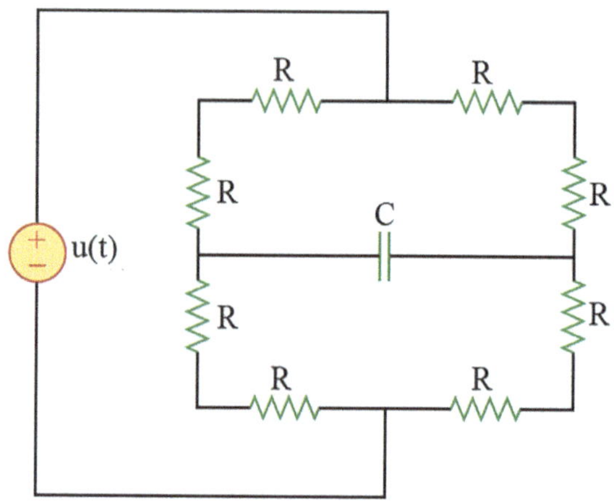

Fig. 9.21 The circuit of exercise 9.19

Problem

9.20. The circuit of Fig. 9.22 is in zero state before $t = 0$. Calculate the current of capacitor $I_C(t)$ for $t > 0$.

Difficulty level ○ Easy ○ Normal ● Hard

Calculation amount ○ Small ○ Normal ● Large

1) $-\frac{1}{4R}e^{-\frac{t}{RC}}u(t)$

2) $-\frac{1}{2R}e^{-\frac{t}{2RC}}u(t)$

3) $-\frac{1}{4R}e^{-\frac{t}{2RC}}u(t)$

4) $\frac{1}{2R}e^{-\frac{t}{RC}}u(t)$

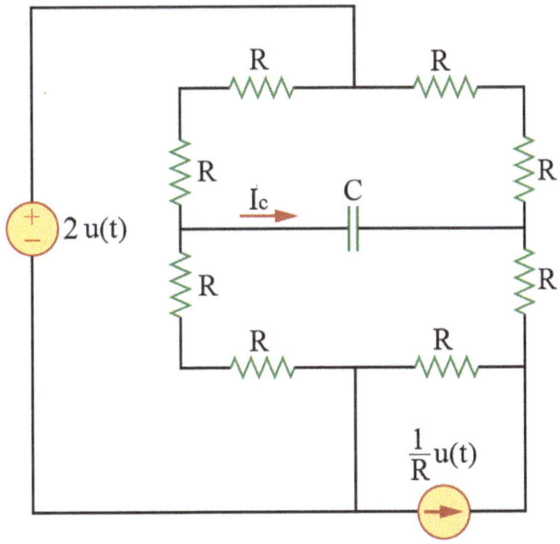

Fig. 9.22 The circuit of problem 9.20

Problem

9.21. The network under study is shown in Fig. 9.23. The response of the circuit is as follows.

$$V(t) = \frac{4}{3}\left(1 - e^{-1.5t}\right)u(t)\ \text{V}$$

If the 2-Ω resistor of the network is replaced by a 0.5-F capacitor, calculate the output voltage.

Difficulty level ○ Easy ○ Normal ● Hard
Calculation amount ○ Small ● Normal ○ Large

1) $2\left(1 - e^{-0.25t}\right)u(t)$ V
2) $2\left(1 - e^{-0.5t}\right)u(t)$ V
3) $4\left(1 - e^{-0.5t}\right)u(t)$ V
4) $4\left(1 - e^{-0.25t}\right)u(t)$ V

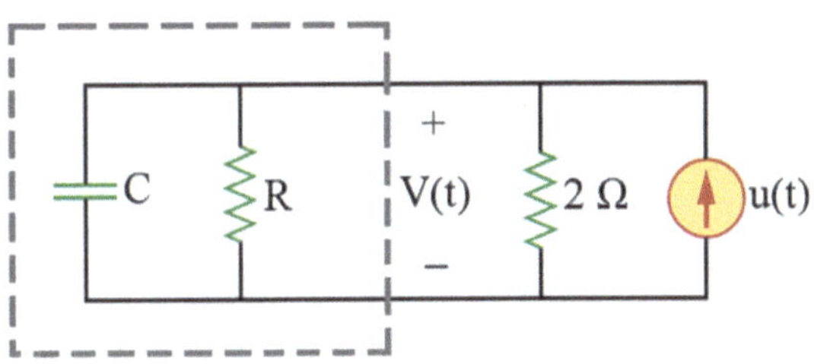

Fig. 9.23 The circuit of problem 9.21

Exercise

9.22. The network under study is shown in Fig. 9.24. The response of the circuit is as follows.

$$V(t) = \left(1 - e^{-t}\right)u(t)\ \text{V}$$

Calculate the size of the resistor and capacitor.

Final Answer

$R = \frac{1}{3}\ \Omega,\ C = 3\ \text{F}$

Note: Read the "educational support" part of the preface of the textbook.

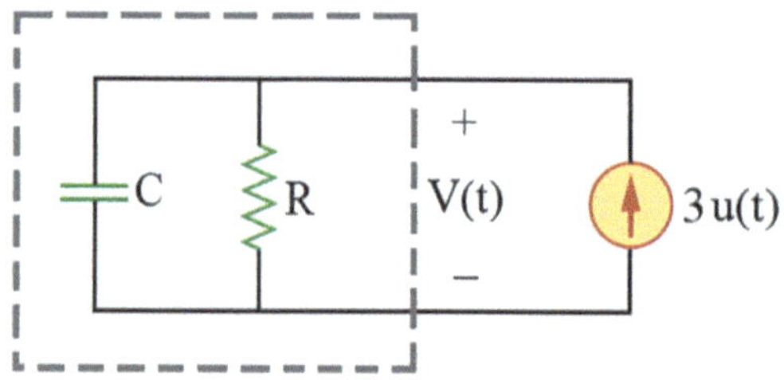

Fig. 9.24 The circuit of exercise 9.22

Problem

9.23. In the circuit of Fig. 9.25, switch S1 is closed at $t = 0$. Calculate the voltage of the capacitor for $t > 0$.

Difficulty level ○ Easy ○ Normal ● Hard

Calculation amount ○ Small ● Normal ○ Large

1) $9\left(1 - e^{-\frac{2}{7}t}\right)$ V

2) $12\left(1 - e^{-\frac{7}{2}t}\right)$ V

3) $12\left(1 - e^{-\frac{2}{7}t}\right)$ V

4) $9\left(1 - e^{-\frac{7}{2}t}\right)$ V

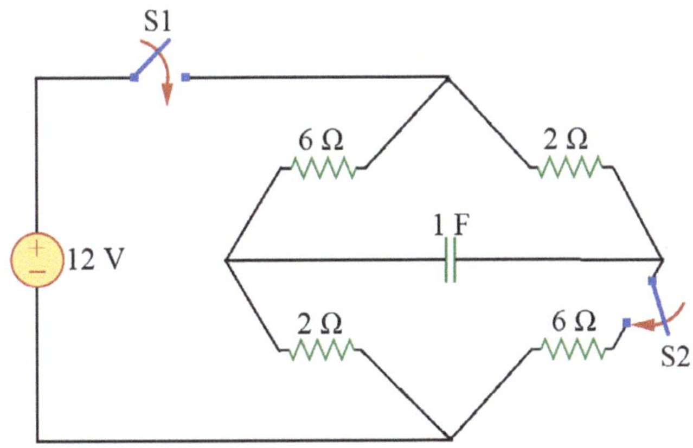

Fig. 9.25 The circuit of problem 9.23

9.4 Complete Response of an RC Circuit

Problem

9.24. In Problem 9.17, calculate the resistance of R_2 so that the voltage of the capacitor remains constant due to the switching operation at $t_1 = R_1C$.

Difficulty level ○ Easy ○ Normal ● Hard
Calculation amount ○ Small ● Normal ○ Large

1) $\frac{R_1}{e-1}$
2) $\frac{R_1}{e}$
3) eR_1
4) $(e-1)R_1$

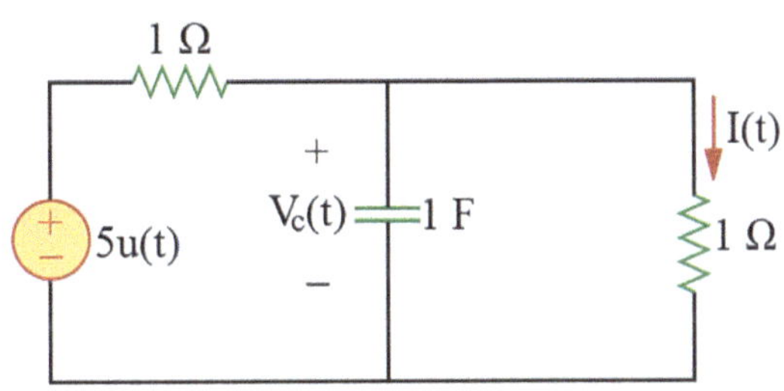

Fig. 9.26 The circuit of partially solved exercise 9.25

Partially Solved Exercise

9.25. Calculate $I(t)$ in the circuit of Fig. 9.26 while assuming that $V_C(0^-) = -2$ V.

Solution

Since the capacitor includes an initial voltage as well as the power source (independent voltage source) is available in the circuit and is a step function, the response of the circuit is complete.

The general form of complete response of an RC circuit is as follows.

$$V_C(t) = V_C(\infty) + (V_C(0^+) - V_C(\infty))e^{-\frac{t}{\tau}}$$

Herein, $V_C(0^+)$ is the initial voltage of the capacitor, $V_C(\infty)$ is the voltage of the capacitor after a long time, and $\tau = R_{Th}C$ is the time constant of the circuit.

Calculating the initial voltage of the capacitor: Based on the information given in the problem, we have:

$$V_C(0^-) = -2 \text{ V} \quad (1)$$

Due to voltage continuity of capacitors in lack of a power source with an infinite size (e.g. impulse power source), it is concluded that:

$$V_C(0^+) = (\quad) \text{ V} \quad (2)$$

Calculating the time constant of the circuit: By using Fig. 9.27.2, the time constant of the circuit can be calculated as follows.

$$R_{Th} = (\quad)||(\quad) = 0.5\ \Omega \quad (3)$$

$$\tau = (\quad) \times (\quad) = 0.5 \text{ sec} \quad (4)$$

Calculating the final voltage of the capacitor: By using Fig. 9.27.3 and voltage division rule, the final voltage of the capacitor can be calculated as follows.

$$V_C(\infty) = \frac{(\quad)}{(\quad) + (\quad)} \times (\quad) = 2.5 \text{ V} \quad (5)$$

Calculating the complete response of the RC circuit:

$$V_C(t) = (\quad) + ((\quad) + (\quad))e^{-\frac{t}{(\quad)}}$$

$$\Rightarrow V_C(t) = 2.5 - 4.5e^{-2t} \quad (6)$$

Now, by applying KCL in the indicated node of the circuit of Fig. 9.27.4, we have:

$$\frac{(\quad) - (\quad)}{(\quad)} + I_C(t) + I(t) = 0$$

$$\Rightarrow V_C(t) - 5 + 1 \times \frac{d}{dt} V_C(t) + I(t) = 0$$

$$\Rightarrow I(t) = -V_C(t) + 5 - \frac{d}{dt} V_C(t) \quad (7)$$

Solving (6) and (7):

$$I(t) = (\quad) + (\quad) + \frac{d}{dt}(\quad)$$

$$\Rightarrow I(t) = 2.5 - 4.5e^{-2t} \text{ A}$$

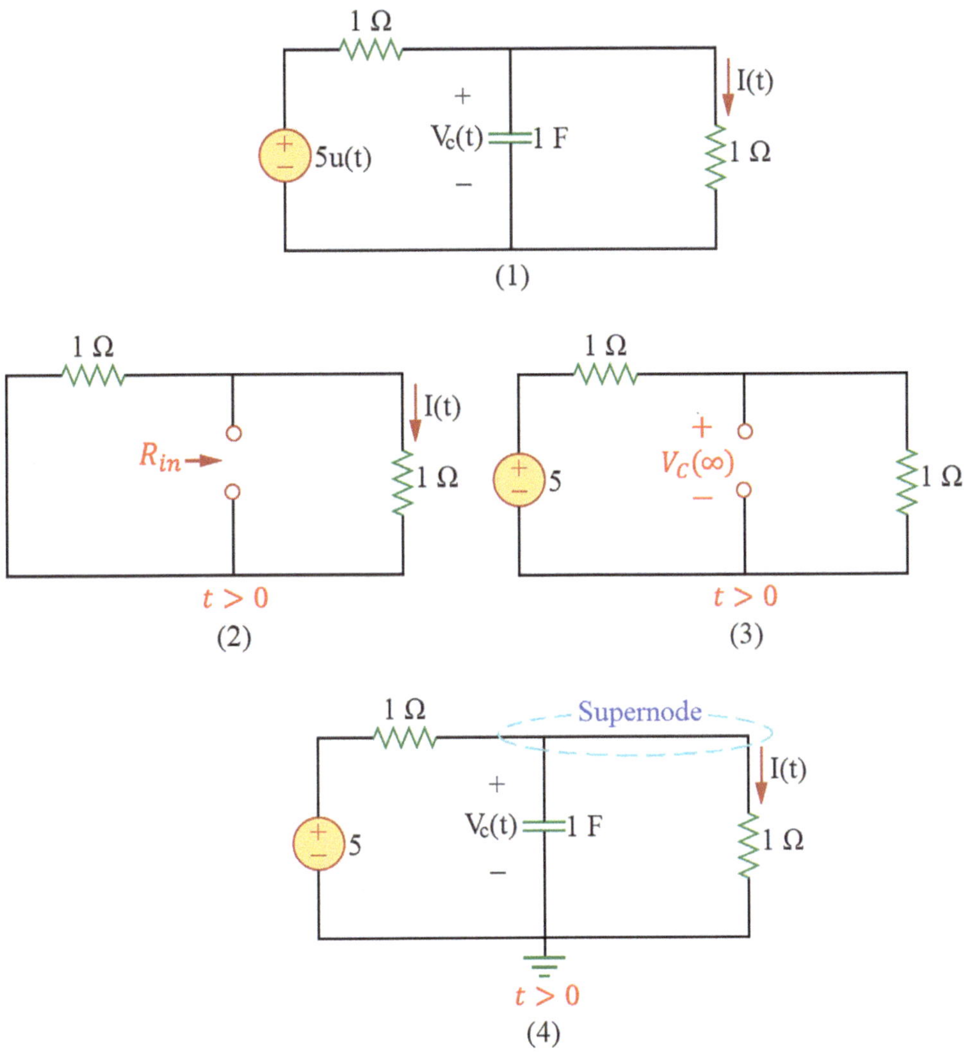

Fig. 9.27 The circuits of partially solved exercise 9.25

Problem

9.26. In the circuit of Fig. 9.28, we have $V_C(0^-) = 4V$. Find a relation between V_s and I_s so that the capacitor is not charged nor discharged for $t \geq 0$.

Difficulty level ○ Easy ○ Normal ● Hard
Calculation amount ○ Small ● Normal ○ Large

1) $V_s = -2I_s$
2) $V_s + 2I_s > 4$
3) $V_s + 2I_s > 12$
4) $V_s + 2I_s = 12$

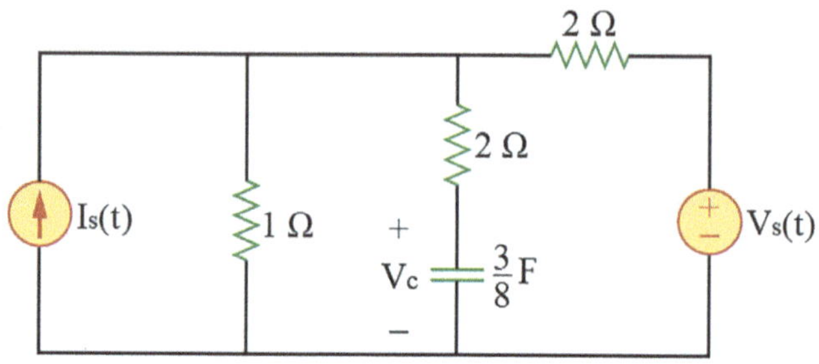

Fig. 9.28 The circuit of problem 9.26

Problem

9.27. From Problem 9.29, we know that switch S1 is closed at $t = 0$. After that, when the voltage of switch S2 reaches 9 V, this switch is also closed. Calculate $V_C(t)$ after the second switching operation (Fig. 9.29).

Difficulty level ○ Easy ○ Normal ● Hard

Calculation amount ○ Small ○ Normal ● Large

1) $6 + 1.8e^{-\frac{t-2.2}{3}}$ V

2) $6 - 2.25e^{-\frac{t-1.89}{3}}$ V

3) $6 + 1.8e^{-\frac{t-2.2}{0.75}}$ V

4) $6 + 2.25e^{-\frac{t-1.89}{0.75}}$ V

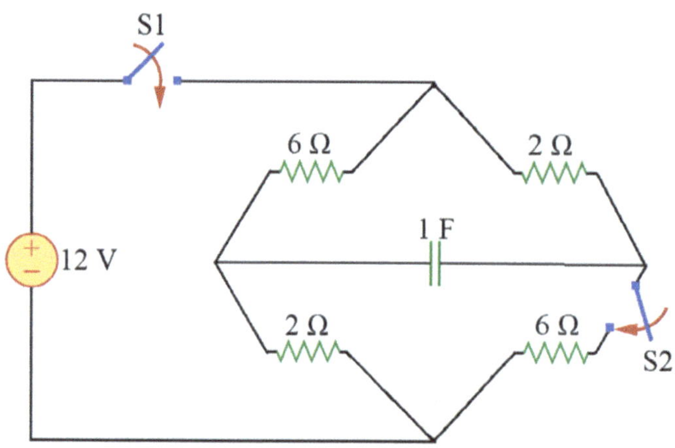

Fig. 9.29 The circuit of problem 9.27

9.5 Time Constant of First-Order RL Circuits

Problem

9.28. Calculate the time constant of the circuit of Fig. 9.30.

Difficulty level ○ Easy ● Normal ○ Hard

Calculation amount ○ Small ● Normal ○ Large

1) 2 sec
2) 1 sec
3) 0.05 sec
4) 0.01 sec

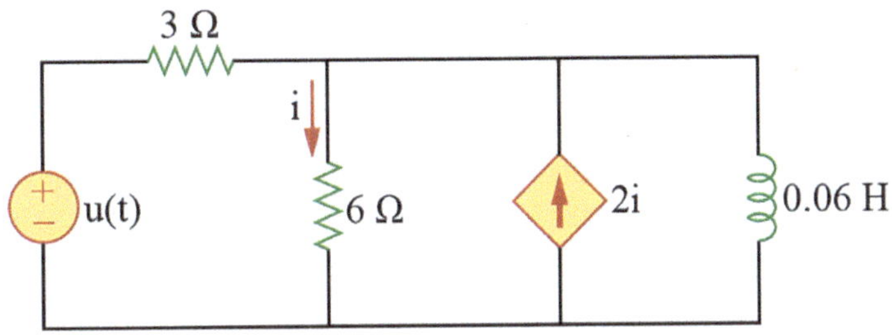

Fig. 9.30 The circuit of problem 9.28

Exercise

9.29. Calculate the time constant of the RL circuit illustrated in Fig. 9.31.

Final Answer

$\tau = 1$ sec

Note: Read the "educational support" part of the preface of the textbook.

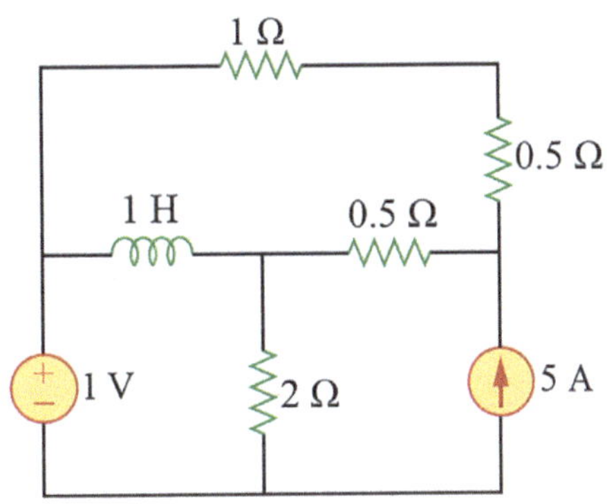

Fig. 9.31 The circuit of exercise 9.29

Exercise

9.30. Calculate the time constant of the RL circuit illustrated in Fig. 9.32.

Final Answer

$$\tau = \frac{L(R_1 + R_2)}{R_1 R_2}$$

Note: Read the "educational support" part of the preface of the textbook.

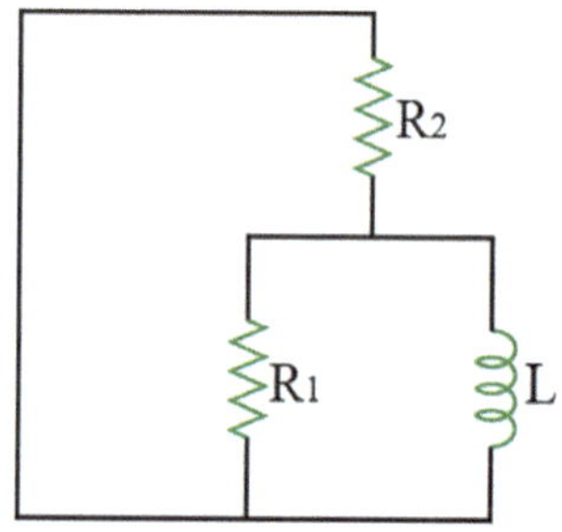

Fig. 9.32 The circuit of exercise 9.30

Partially Solved Exercise

9.31. In the circuit of Fig. 9.33, calculate the resistance of R so that the time constant of the circuit remains unchanged if the capacitor C is replaced by an inductor L with the same size ($L = C$).

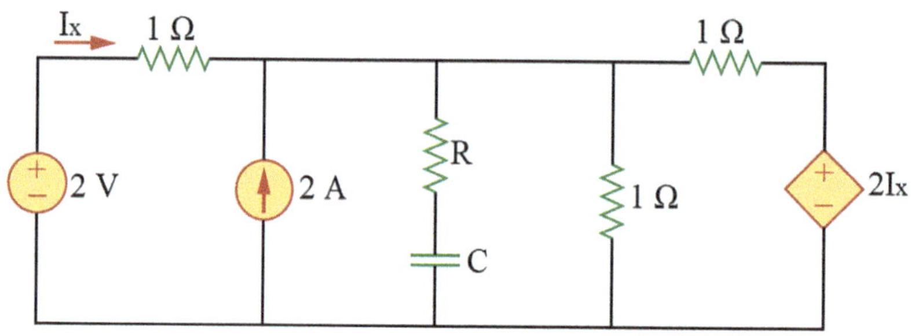

Fig. 9.33 The circuit of partially solved exercise 9.31

Solution

As we know, the time constant of a first-order RC and RL circuit can be calculated as follows.

$$\tau_{RC} = R_{Th} C_{eq} \quad (1)$$

$$\tau_{RL} = \frac{L_{eq}}{R_{Th}} \quad (2)$$

Herein, C_{eq} is the equivalent capacitor, L_{eq} is the equivalent inductor, and R_{Th} is the Thevenin resistance seen by the equivalent capacitor or equivalent inductor. From previous chapters, we know that R_{Th} can be calculated as follows.

First, all the independent sources (independent voltage and current sources) are turned off (each independent voltage source is replaced by a short circuit branch and each independent current source is replaced by an open circuit branch). Then, if the circuit does not include any dependent source (dependent voltage or current source), the series-parallel rule is applied to calculate the input resistance of the circuit. However, if the circuit includes at least a dependent source, a test source (independent voltage or current source) with the voltage and current of V_t and I_t is connected to the input terminal of the circuit. Next, the circuit is analyzed by using a circuit analysis method to find the value of $\frac{V_t}{I_t}$. After that, we have:

$$R_{Th} = \frac{V_t}{I_t}$$

Based on the information given in the problem, we have:

$$\tau_{RC} = \tau_{RL} \quad (3)$$

$$L = C \quad (4)$$

Solving (1)–(3):

$$R_{Th}C = \frac{L}{R_{Th}}$$

$$\Rightarrow R_{Th}^2 C = L \quad (5)$$

Solving (4) and (5):

$$R_{Th}^2 = (\quad)$$

$$\Rightarrow R_{Th} = \pm 1 \quad (6)$$

The Thevenin resistance seen by the capacitor can be calculated as follows. See Fig. 9.34.2. Defining I_x based on the appropriate node voltages:

$$I_x = \frac{(\quad)}{(\quad)}$$

$$\Rightarrow I_x = -e_1 \quad (7)$$

KCL for the node of the test source:

$$(\quad) + \frac{(\quad) - (\quad)}{(\quad)} = 0$$

$$\Rightarrow e_1 = V_t - RI_t \quad (8)$$

KCL in node 1:

$$(\quad) + (\quad) + (\quad) + (\quad) = 0$$

$$\Rightarrow -I_t + 3e_1 - 2I_x = 0 \quad (9)$$

Solving (7) and (9):

$$(\quad) + (\quad) + (\quad) = 0$$

$$\Rightarrow -I_t + 5e_1 = 0 \quad (10)$$

Solving (8) and (10):

$$(\quad) + (\quad) = 0$$

$$\Rightarrow -(\quad)I_t + (\quad)V_t = 0$$

$$\Rightarrow \frac{V_t}{I_t} = \frac{1 + 5R}{5}$$

$$\Rightarrow \frac{V_t}{I_t} = R + \frac{1}{5}$$

$$\Rightarrow R_{Th} = R + \frac{1}{5} \quad (11)$$

Solving (6) and (11):

$$R + \frac{1}{5} = \pm 1$$

$$\Rightarrow R = \frac{4}{5}, -\frac{6}{5}$$

The negative value of R is not acceptable. Thus:

$$R = 0.8\ \Omega$$

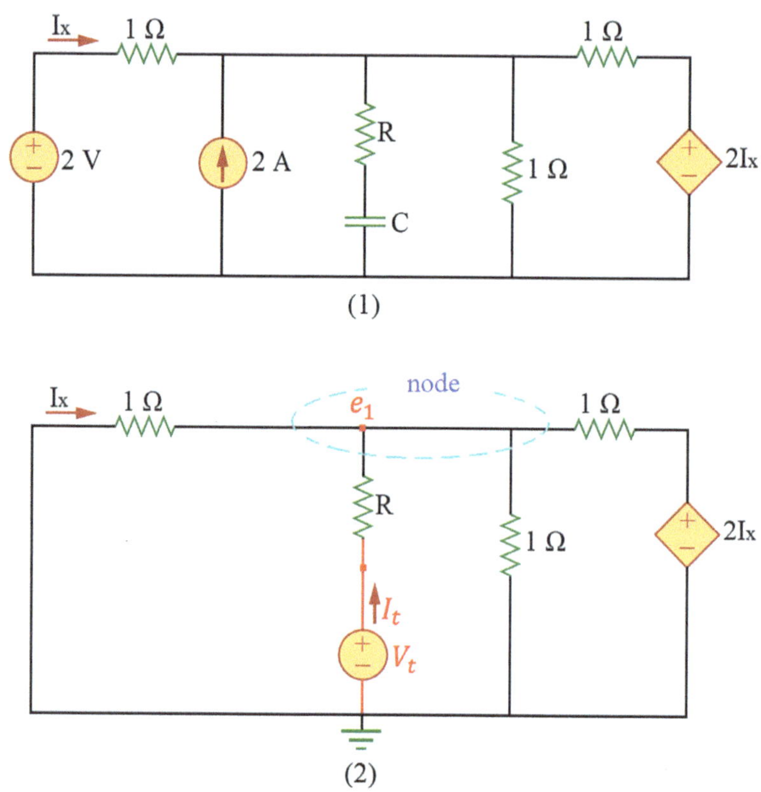

Fig. 9.34 The circuits of partially solved exercise 9.31

9.6 Source-Free Response of an RL Circuit

Problem

9.32. In the circuit of Fig. 9.35, $I_L(0) = 1A$. Calculate the current of the inductor for $t > 0$.

Difficulty level ● Easy ○ Normal ○ Hard
Calculation amount ● Small ○ Normal ○ Large

1) $e^{-3.4t}$ A
2) $0.2e^{-3.4t}$ A
3) $2e^{-3.4t}$ A
4) $0.4e^{-3.4t}$ A

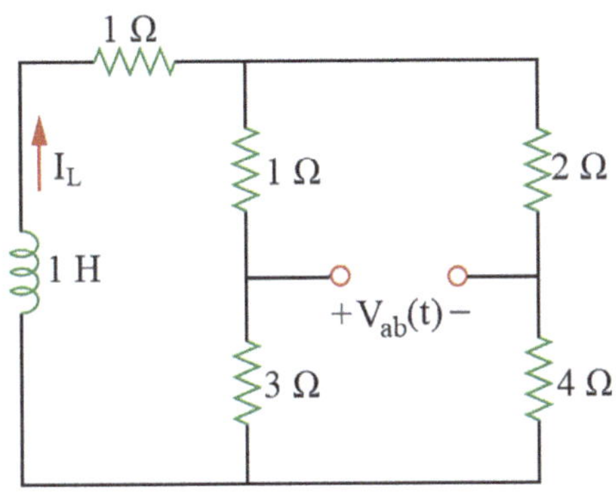

Fig. 9.35 The circuit of problem 9.32

Problem

9.33. In Problem 9.32, calculate the voltage $V_{ab}(t)$ for $t > 0$.

Difficulty level ○ Easy ● Normal ○ Hard
Calculation amount ● Small ○ Normal ○ Large

1) $-1.2e^{-3.4t}$ V
2) $0.2e^{-3.4t}$ V
3) $-0.2e^{-3.4t}$ V
4) $1.2e^{-3.4t}$ V

Partially Solved Exercise

9.34. The switch is closed at $t = 0$ in the circuit of Fig. 9.36. Calculate the current of the switch $I(t)$ for $t \geq 0$.

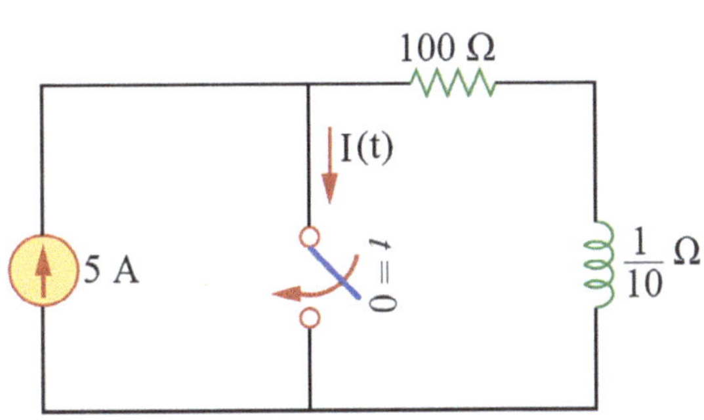

Fig. 9.36 The circuit of partially solved exercise 9.34

Solution

Since the power source (independent current source) is removed from the circuit due to a switching operation, the type of response of the circuit is source-free.
The general form of source-free response of an RL circuit is as follows.

$$I_L(t) = I_L(0^+)e^{-\frac{t}{\tau}}$$

Herein, $I_L(0^+)$ is the initial current of the inductor and $\tau = \frac{L}{R_{Th}}$ is the time constant of the circuit.

Calculating the initial current of the inductor: Before closing the switch at $t = 0$, the circuit is like the one shown in Fig. 9.37.2. As can be seen, the inductor has been modeled by a short circuit branch since the circuit is in its steady state condition. Therefore:

$$I_L(0^-) = 5 \text{ A}$$

After the switching operation, the circuit is changed to the one illustrated in Fig. 9.37.3. Due to current continuity of inductors in lack of a power source with an infinite size (e.g. impulse power source), it is concluded that:

$$I_L(0^+) = (\quad) \text{ A}$$

Calculating the time constant of the circuit: The circuit of Fig. 9.37.4 can be used to calculate the time constant of the circuit as follows.

$$\tau = \frac{(\quad)}{(\quad)} = \frac{1}{1000} \text{ sec}$$

Calculating the source-free response of the RL circuit:

$$I_L(t) = 5e^{-1000t}$$

Now, by applying KCL in the indicated node of the circuit of Fig. 9.37.3, we have:

$$I_L(t) + (\quad) + (\quad) = 0$$

$$\Rightarrow I(t) = (\quad) + (\quad)$$

$$\Rightarrow I(t) = 5\left(1 - e^{-1000t}\right) \text{ A}$$

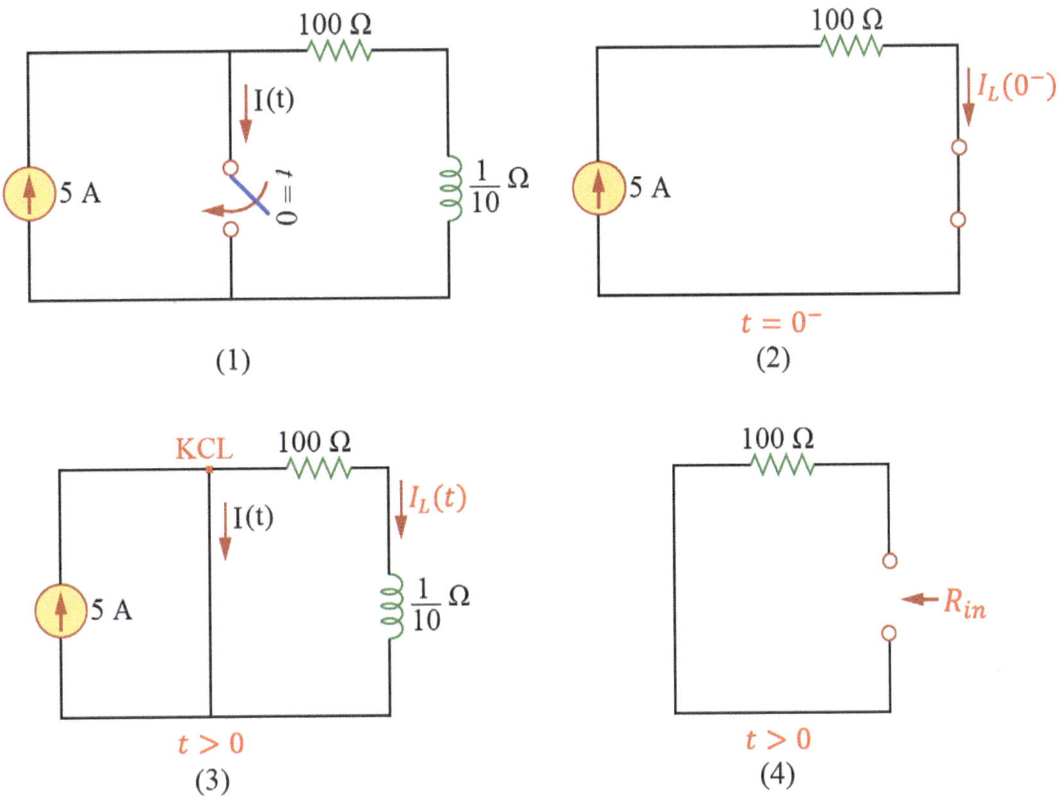

Fig. 9.37 The circuits of partially solved exercise 9.34

Problem

9.35. In the circuit of Fig. 9.38, the switch has been open for a long time, and then it is closed at $t = 0$. Calculate the short circuit current at $t = 4$ sec while assuming that $R_2 = 0.5\ \Omega, R_1 = 0.25\ \Omega, I_s = 6\ \text{A}, \text{L} = 0.5\ \text{H}$.

Difficulty level ○ Easy ○ Normal ● Hard

Calculation amount ○ Small ● Normal ○ Large

1) 8 A
2) 0 A
3) 4 A
4) 2 A

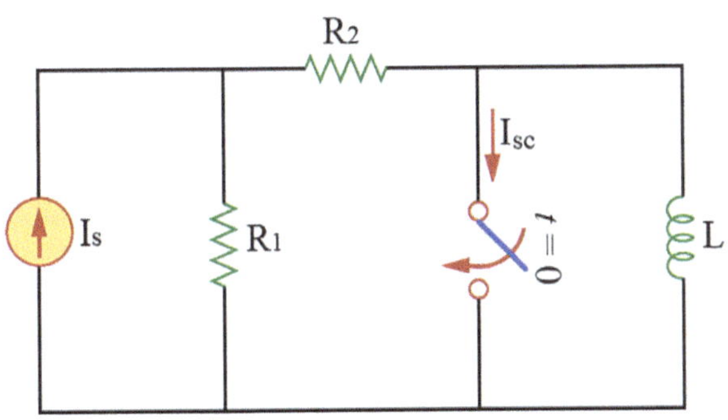

Fig. 9.38 The circuit of problem 9.35

Problem

9.36. In the circuit of Fig. 9.39, the switch has been closed for a long time, and then it is opened at $t = 0$. Calculate the current $I(t)$ for $t > 0$.

Difficulty level ○ Easy ● Normal ○ Hard
Calculation amount ○ Small ● Normal ○ Large

1) $7.5e^{-3t}$ A
2) $5e^{-2t}$ A
3) $7.5e^{-t}$ A
4) $5e^{-0.5t}$ A

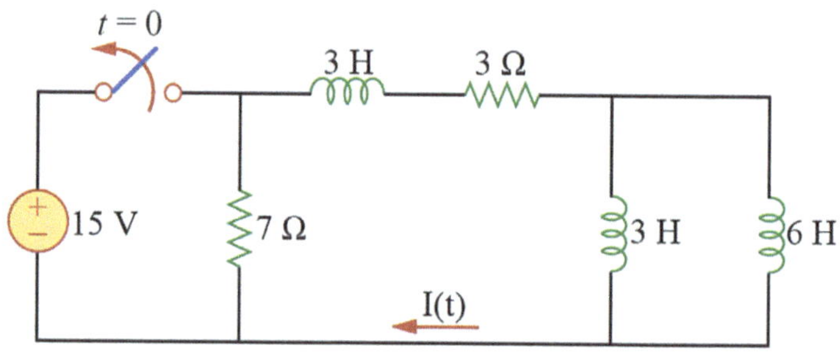

Fig. 9.39 The circuit of problem 9.36

Problem

9.37. In the circuit of Fig. 9.40, the switch has been in the closed position for a long time, but it is opened at $t = 0$. Calculate the voltage $V(t)$ for $t > 0$.

Difficulty level ○ Easy ○ Normal ● Hard
Calculation amount ○ Small ● Normal ○ Large

1) $12e^{-\frac{4}{9}t}$ V
2) $6e^{-\frac{9}{4}t}$ V
3) $-12e^{-\frac{4}{9}t}$ V
4) $-7.5e^{-\frac{9}{4}t}$ V

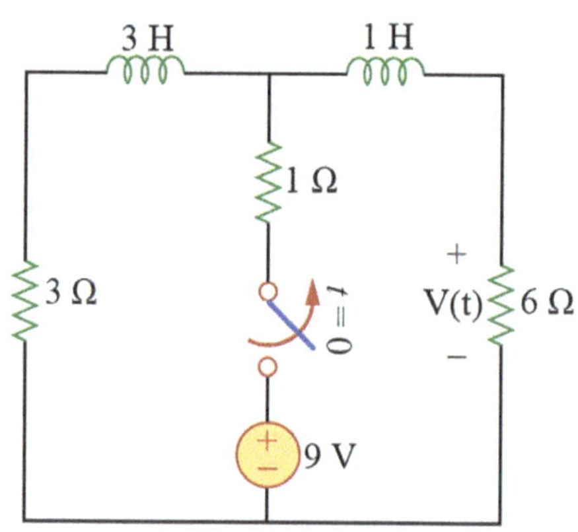

Fig. 9.40 The circuit of problem 9.37

9.7 Step Response of an RL Circuit

Problem

9.38. Calculate the current of the inductor $I_L(t)$ for $t \geq 0$ in the circuit of Fig. 9.41.

Difficulty level ○ Easy ○ Normal ● Hard

Calculation amount ○ Small ● Normal ○ Large

1) $7.5e^{-t}$ V
2) $2.5\left(1 - e^{-t}\right)$ V
3) $\left(7.5 - 5e^{-t}\right)$ V
4) $\left(7.5 - 2.5e^{-t}\right)$ V

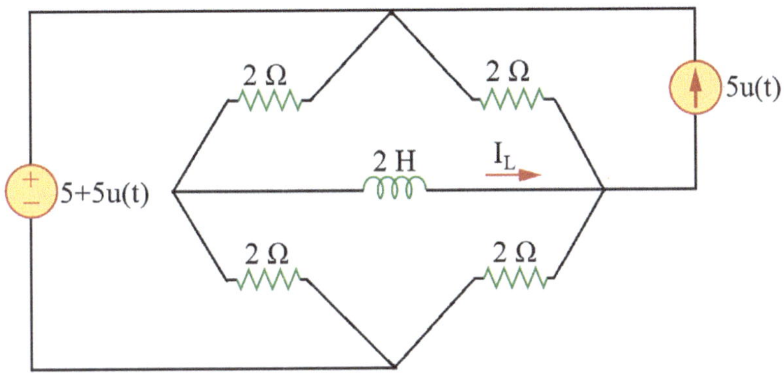

Fig. 9.41 The circuit of problem 9.38

Exercise

9.39. The circuit of Fig. 9.42 is in zero state before $t = 0$. Calculate the current of the inductor for $t > 0$.

Final Answer

$I_L(t) = 0$

Note: Read the "educational support" part of the preface of the textbook.

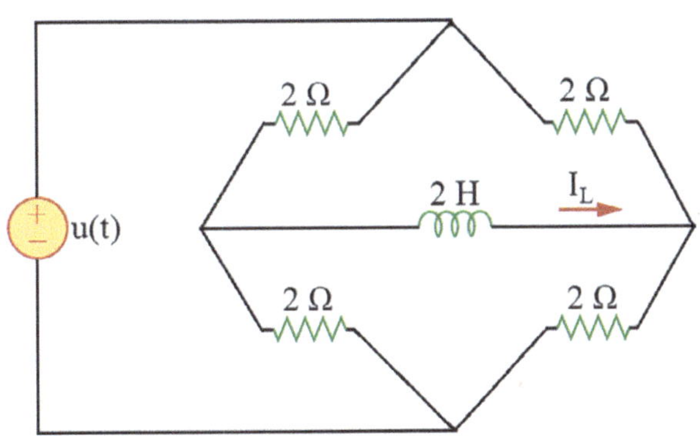

Fig. 9.42 The circuit of exercise 9.39

Problem

9.40. In the resistive circuit of Fig. 9.43, if a 1-F capacitor is connected to terminal A-B, $V(t) = 2\left(1 - e^{-\frac{t}{2}}\right)u(t)$ V is achieved. Calculate the voltage $V(t)$ if a 3-H inductor is connected to the terminal.

Difficulty level ○ Easy ○ Normal ● Hard

Calculation amount ○ Small ○ Normal ● Large

1) $2e^{-\frac{3}{2}t}u(t)$ V
2) $2e^{-\frac{2}{3}t}u(t)$ V
3) $\left(1 - 2e^{-\frac{2}{3}t}\right)u(t)$ V
4) $\left(1 - 2e^{-\frac{3}{2}t}\right)u(t)$ V

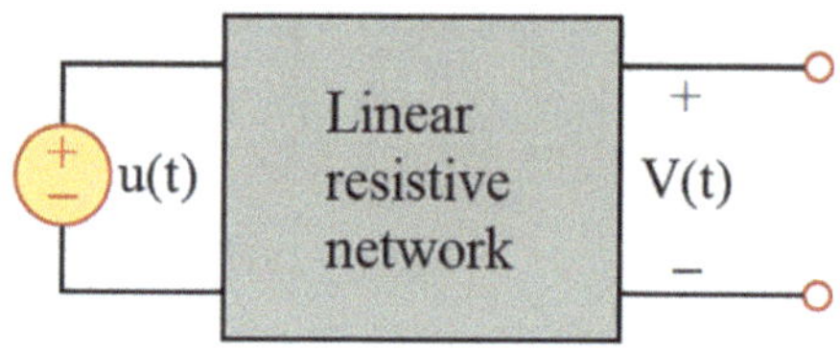

Fig. 9.43 The circuit of problem 9.40

Partially Solved Exercise

9.41. In the resistive circuit of Fig. 9.44, if a 1-F capacitor is connected to terminal A-B, $V(t) = \frac{1}{4}\left(1 - e^{-3t}\right)u(t)$ V is achieved. Calculate the voltage $V(t)$ if a 2-H inductor is connected to the terminal.

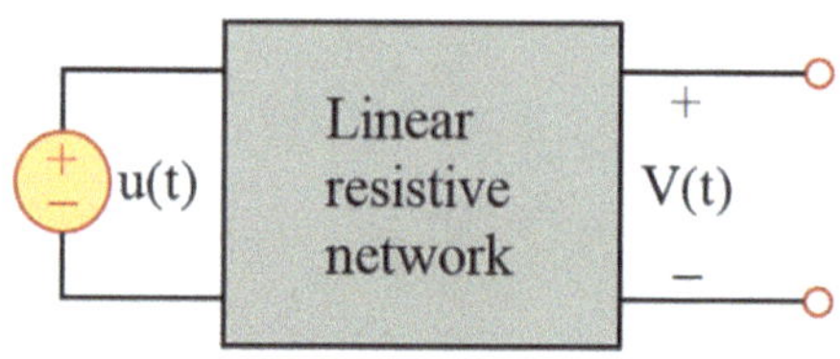

Fig. 9.44 The circuit of partially solved exercise 9.41

Solution

Since the network does not have any power source and the power source (independent voltage source) is a step function, the response of the circuit is step response.

Part 1: The general form of step response of an RC circuit is as follows.

$$V_C(t) = V_C(\infty)\left(1 - e^{-\frac{t}{\tau}}\right) \quad (1)$$

Herein, $V_C(\infty)$ is the voltage of the capacitor after a long time and the time constant of the circuit is as follows.

$$\tau = R_{Th}C \quad (2)$$

Calculating the time constant of the circuit: Based on the information given in the problem, we have:

$$V(t) = \quad (3)$$

$$C = \quad (4)$$

Comparing (1) and (3):

$$\tau = \frac{1}{3}\ \text{sec} \quad (5)$$

Solving (2), (4), and (5):

$$R_{Th} \times (\quad) = (\quad)$$

$$\Rightarrow R_{Th} = \frac{1}{3}\ \Omega \quad (6)$$

Calculating the final voltage of the capacitor: Fig. 9.45.1 shows the Thevenin circuit of the network. Since the capacitor will behave like an open circuit after a long time, the final voltage of the capacitor will be equal to the Thevenin voltage of the circuit, as can be noticed from the circuits of Fig. 9.45.1. Therefore, by using (3), we have:

$$\Rightarrow V_{Th} = \lim_{t \to \infty} \frac{1}{4}\left(1 - e^{-3t}\right)u(t)$$

$$\Rightarrow V_{Th} = \frac{1}{4}\ \text{V} \quad (7)$$

Part 2: The general form of step response of an RL circuit is as follows.

$$I_L(t) = I_L(\infty)\left(1 - e^{-\frac{t}{\tau}}\right) \quad (8)$$

Herein, $I_L(\infty)$ is the current of the inductor after a long time and the time constant of the circuit is calculated as follows.

$$\tau = \frac{L}{R_{Th}} \quad (9)$$

Calculating the time constant of the circuit: Based on the information given in the problem, we have:

$$L = 2\ H \quad (10)$$

Solving (6), (9), and (10):

$$\tau = \frac{(\quad)}{(\quad)} = 6\ \text{sec} \quad (11)$$

Calculating the final current of the inductor: The inductor will behave like a short circuit after a long time. Therefore, its final current can be calculated by using Ohm's law as follows (See Fig. 9.45.2).

$$I_L(\infty) = \frac{V_{Th}}{R_{Th}} = \frac{(\quad)}{(\quad)}$$

$$\Rightarrow I_L(\infty) = \frac{3}{4}\text{ A} \quad (12)$$

***Calculating the step response of the RL circuit*:**

$$I_L(t) = \frac{3}{4}\left(1 - e^{-\frac{t}{6}}\right) \quad (13)$$

Using the current-voltage relation of inductors:

$$V(t) = L\frac{d}{dt}I_L(t)$$

$$\Rightarrow V(t) = 2 \times \frac{d}{dt}\left(\frac{3}{4}\left(1 - e^{-\frac{t}{6}}\right)\right)$$

$$\Rightarrow V(t) = 2 \times (\quad) \times (\quad)e^{-\frac{t}{6}}$$

$$\Rightarrow V(t) = \frac{1}{4}e^{-\frac{t}{6}}\text{V}, \ t > 0 \quad (14)$$

Since the result is credible for $t > 0$, unit step function ($u(t)$) can be applied in (14) as follows.

$$V(t) = \frac{1}{4}e^{-\frac{t}{6}}u(t)\text{ V}$$

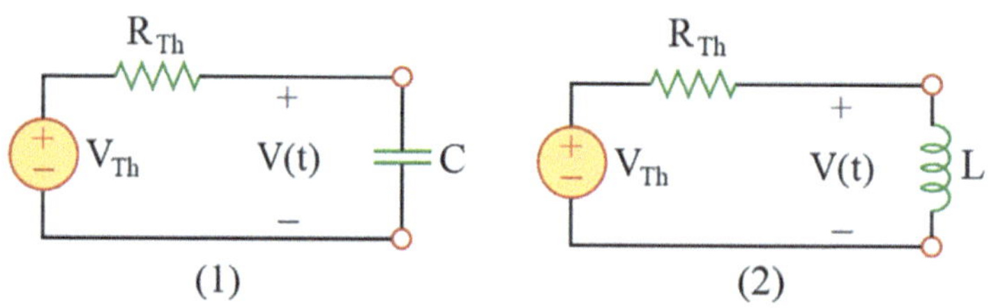

Fig. 9.45 The circuits of partially solved exercise 9.41

Exercise

9.42. In the circuit of Fig. 9.46, there is no energy stored in the inductor for $t < 0$. The switch is closed at $t = 1$ sec. Calculate the current flowing through the short-circuit branch at $t = 4$ sec.

Final Answer

$I_{sc}(t = 4) = 0.45$ A

Note: *Read the "educational support" part of the preface of the textbook.*

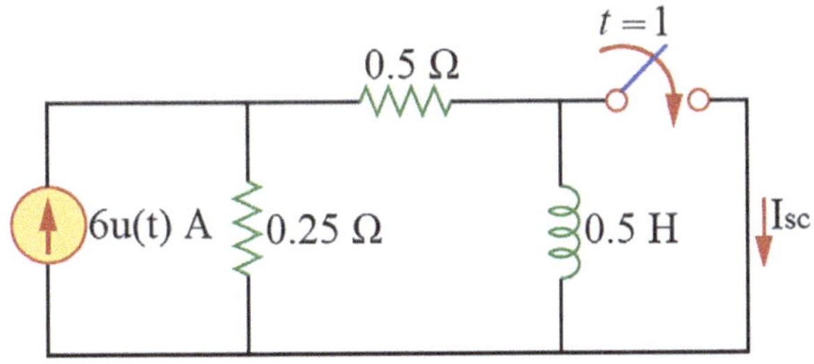

Fig. 9.46 The circuit of exercise 9.42

9.8 Heuristic Methods for First-Order Circuits

Problem

9.43. In the circuit of Fig. 9.47, the switch has been in the open position for a long time. Calculate the current of the capacitor exactly after closing the switch ($I_c(0^+)$).

Difficulty level ○ Easy ○ Normal ● Hard
Calculation amount ● Small ○ Normal ○ Large

1) − 6 A
2) 6 A
3) − 4 A
4) 4 A

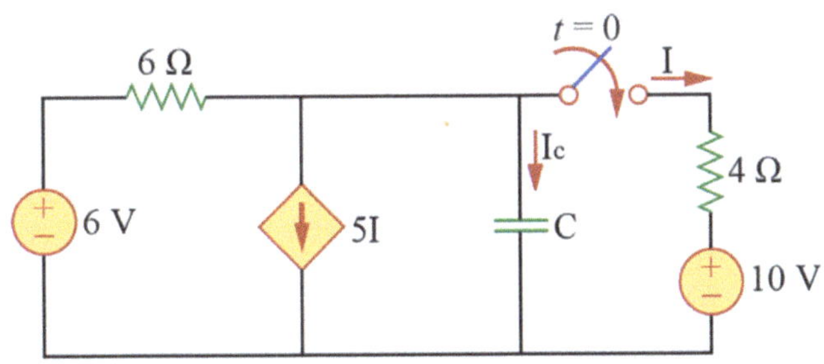

Fig. 9.47 The circuit of problem 9.43

Problem

9.44. In the circuit of Fig. 9.48, the switch is opened at $t = 0$. Calculate the value of the following term.

$$\left.\frac{dV_C}{dt}\right|_{t=0^+}$$

Difficulty level ○ Easy ○ Normal ● Hard
Calculation amount ○ Small ● Normal ○ Large

1) 0
2) − 3 E
3) − 6 E
4) − 3αE

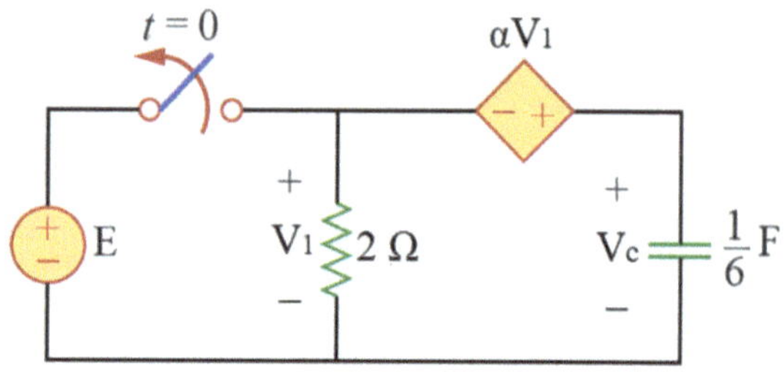

Fig. 9.48 The circuit of problem 9.44

Exercise

9.45. In the circuit of Fig. 9.49, the switch is opened at $t = 0$. Calculate the value of the term below.

$$\left.\frac{dV_C}{dt}\right|_{t=0^+}$$

Final Answer

$$\left.\frac{dV_C}{dt}\right|_{t=0^+} = -2\ \text{V/sec}$$

Note: *Read the "educational support" part of the preface of the textbook.*

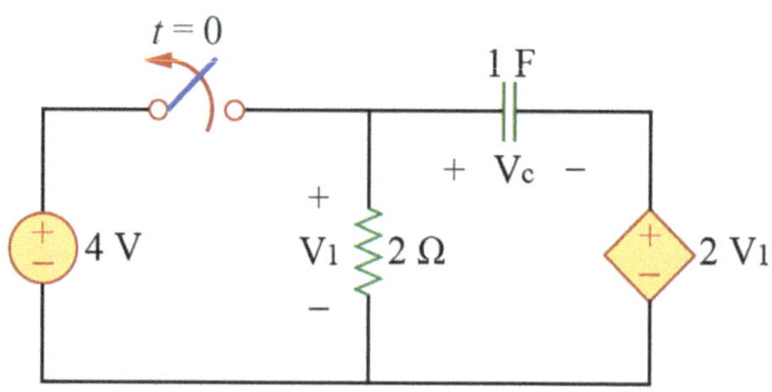

Fig. 9.49 The circuit of exercise 9.45

9.9 First-Order Circuits Including Ramp and Singularity Functions

Problem

9.46. Calculate the output voltage $V_o(t)$ in the circuit of Fig. 9.50 for $t \geq 0$ while assuming that $V_C(0^-) = 1$ V.

Difficulty level ○ Easy ○ Normal ● Hard

Calculation amount ○ Small ● Normal ○ Large

1) $1 + e^{-t}$ V
2) $1 - e^{-2t}$ V
3) $e^{-t} - e^{-2t}$ V
4) 0 V

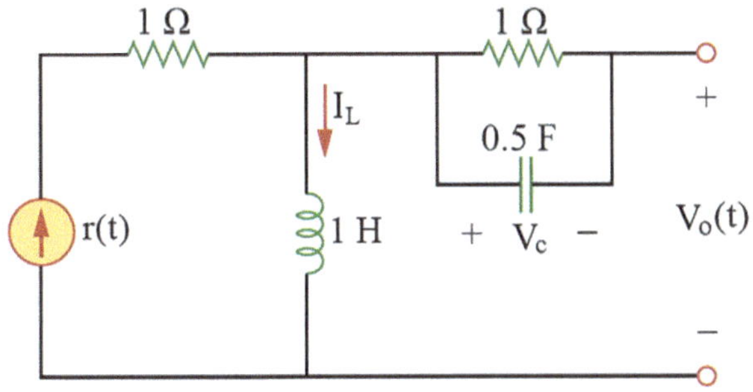

Fig. 9.50 The circuit of problem 9.46

Problem

9.47. In the circuit of Fig. 9.51, calculate the voltage of the capacitor at $t = 0^+$ while assuming that $V_C(0^-) = 1$ V.

Difficulty level ○ Easy ○ Normal ● Hard
Calculation amount ○ Small ● Normal ○ Large

1) 6 V
2) 2 V
3) 5 V
4) 4 V

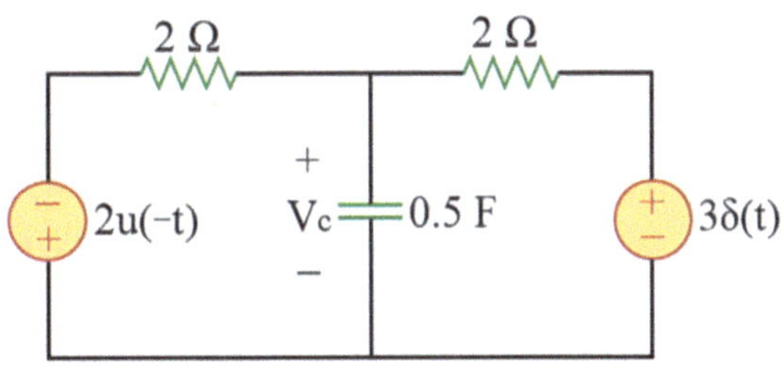

Fig. 9.51 The circuit of problem 9.47

9.10 Differential Equation of First-Order Circuits

Problem

9.48. Determine the differential equation of $V_C(t)$ in the circuit of Fig. 9.52 while assuming that the capacitor has an non-zero initial voltage.

Difficulty level ○ Easy ● Normal ○ Hard
Calculation amount ○ Small ● Normal ○ Large

1) $\frac{dV_C(t)}{dt} + 0.5\,V_C(t) = 0$
2) $\frac{dV_C(t)}{dt} + 0.5\,V_C(t) = I_s(t)$
3) $\frac{dV_C(t)}{dt} + V_C(t) = 0$
4) $\frac{dV_C(t)}{dt} + V_C(t) = I_s(t)$

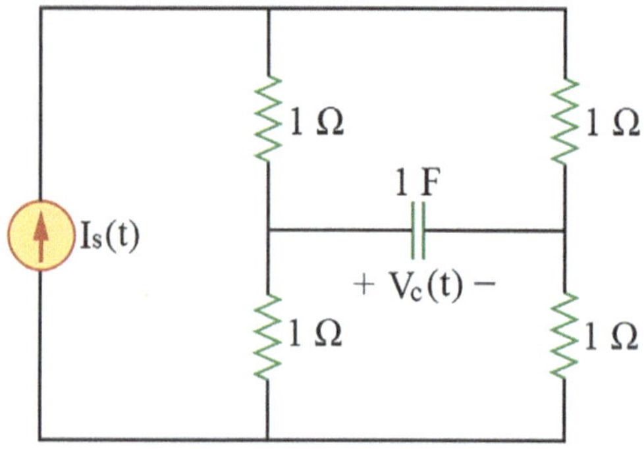

Fig. 9.52 The circuit of problem 9.48

Problem

9.49. The short-term voltage signal shown in Fig. 9.53 is part of a sine wave. This signal is applied to the circuit illustrated in Fig. 9.53. Calculate the output voltage at $t = 2.2$ s while assuming that $I_L(0^-) = 0$.

Difficulty level ○ Easy ○ Normal ● Hard

Calculation amount ○ Small ○ Normal ● Large

1) 0 V
2) 0.96 V
3) 0.58 V
4) 1.48 V

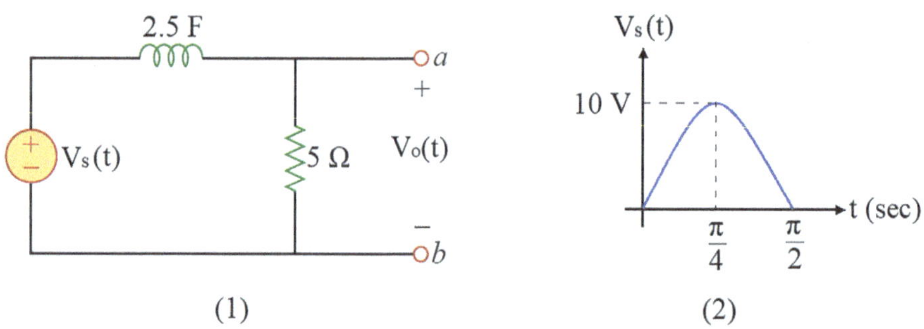

Fig. 9.53 The circuit of problem 9.49

9.11 Energy Stored in First-Order Circuits

Problem

9.50. In the circuit of Fig. 9.54, calculate the time that the voltage of capacitor changes from its primary value of 4 V at $t = 0$ sec to 1 V.

Difficulty level ○ Easy ○ Normal ● Hard

Calculation amount ○ Small ● Normal ○ Large

1) 0.91 sec
2) 0.45 sec
3) 1.82 sec
4) 1.00 sec

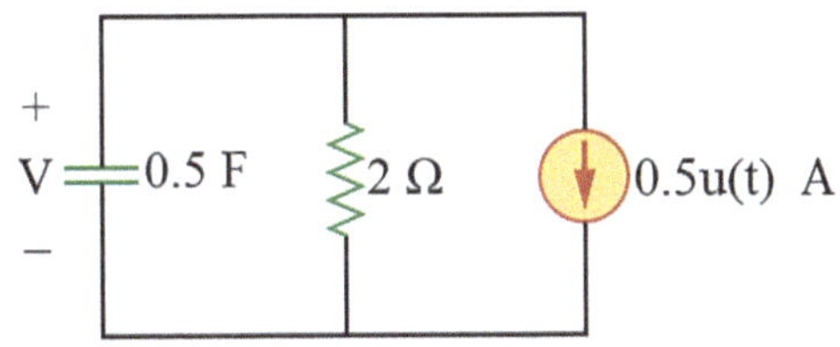

Fig. 9.54 The circuit of problem 9.50

Problem

9.51. In Problem 9.50, calculate the amount of energy that is generated or consumed by the current source during the time that the voltage of capacitor changes from its primary value of 4 V to 1 V.

Difficulty level ○ Easy ○ Normal ● Hard
Calculation amount ○ Small ● Normal ○ Large

1) 2.71 J, generated
2) 1.04 J, generated
3) 2.71 J, consumed
4) 1.04 J, consumed

Problem

9.52. In the circuit of Fig. 9.55, $V_C(0^-) = 4$ V. Determine the size of the resistor R so that the energy of the capacitor remains constant for $t \geq 0$.

Difficulty level ○ Easy ○ Normal ● Hard
Calculation amount ○ Small ● Normal ○ Large

1) $\frac{3}{4}$ Ω
2) $\frac{3}{2}$ Ω
3) $\frac{2}{3}$ Ω
4) $\frac{4}{3}$ Ω

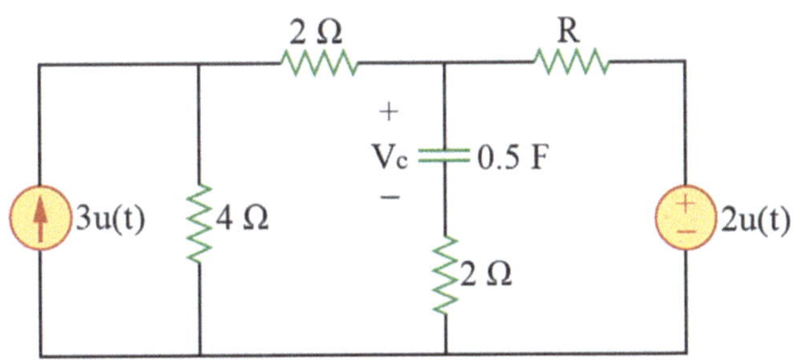

Fig. 9.55 The circuit of problem 9.52

References

1. Rahmani-Andebili, M., DC Electrical Circuit Analysis - Practice Problems, Methods, and Solutions (1st Ed.), Springer Nature, 2020.
2. Rahmani-Andebili, M., General Physics II - Practice Problems, Methods, and Solutions, Sringer Nature, 2025.
3. Rahmani-Andebili, M., General Physics I - Practice Problems, Methods, and Solutions, Springer Nature, 2025.
4. Rahmani-Andebili, M., Mathematics of Engineering and Science - Comprehensive Lessons, Springer Nature, 2024.
5. Rahmani-Andebili, M., Differential Equations – Practice Problems, Methods, and Solutions, Springer Nature, 2022.
6. Rahmani-Andebili, M., Calculus III – Practice Problems, Methods, and Solutions, Springer Nature, 2023.
7. Rahmani-Andebili, M., Calculus II – Practice Problems, Methods, and Solutions, Springer Nature, 2023.
8. Rahmani-Andebili, M., Calculus I (2nd Ed.) – Practice Problems, Methods, and Solutions, Springer Nature, 2023.
9. Rahmani-Andebili, M., Precalculus (2nd Ed.) – Practice Problems, Methods, and Solutions, Springer Nature, 2024.

Analysis of First-Order Circuits: Part B

10

Abstract

In this chapter, the problems of the ninth chapter are fully solved, in detail, step-by-step, and with different methods. Herein, a "Notes Box" has been added after the solution of each problem that includes all the formulas and important notes that are needed to solve the problem. Moreover, a "Traps Box" has been added that discloses the traps that have been designed in the choices to challenge students' minds and encourage them to deeply think about the problem solutions as well as discourage them from careless selection of choices. The "Traps Box", in fact, predicts the possible confusions and tricks that might be encountered by students in real tests.

10.1 Time Constant of First-Order RC Circuits

10.1. The time constant (τ) of a first-order RC circuit can be calculated as follows [1–9].

$$\tau = R_{Th} C_{eq}$$

Herein, C_{eq} is the equivalent capacitor and R_{Th} is the Thevenin resistance seen by the equivalent capacitor. From previous chapters, we know that R_{Th} can be calculated as follows.

First, all the independent sources (independent voltage and current sources) are turned off (each independent voltage source is replaced by a short circuit branch and each independent current source is replaced by an open circuit branch). Then, if the circuit does not include any dependent source (dependent voltage or current source), the series-parallel rule is applied to calculate the input resistance of the circuit. However, if the circuit includes at least a dependent source, a test source (independent voltage or current source) with the voltage and current of V_t and I_t is connected to the input terminal of the circuit. Next, the circuit is analyzed by using a circuit analysis method to find the value of $\frac{V_t}{I_t}$. After that, we have:

$$R_{Th} = \frac{V_t}{I_t}$$

S. Manias, *DC Electrical Circuits Analysis*, https://doi.org/10.1007/978-3-032-00253-2_10

In this problem, although the circuit includes three energy saving components (capacitors), the circuit is a first-order circuit. As can be seen in Fig. 10.1.2, the capacitors have a parallel connection. Thus,

$$C_{eq} = 2 + 2 + 3 = 7 \text{ F}$$

Calculating the Thevenin resistance seen by the equivalent capacitor (see Fig. 10.1.3):

$$R_{Th} = (6||3 + 1)||3 + 3$$

$$\Rightarrow R_{Th} = \left(\frac{6 \times 3}{6 + 3} + 1\right)||3 + 3$$

$$\Rightarrow R_{Th} = 3||3 + 3$$

$$\Rightarrow R_{Th} = \frac{3}{2} + 3$$

$$\Rightarrow R_{Th} = 4.5 \ \Omega$$

Hence:

$$\tau = 7 \times 4.5$$

$$\Rightarrow \tau = 31.5 \text{ sec}$$

Choice (3) is the answer.

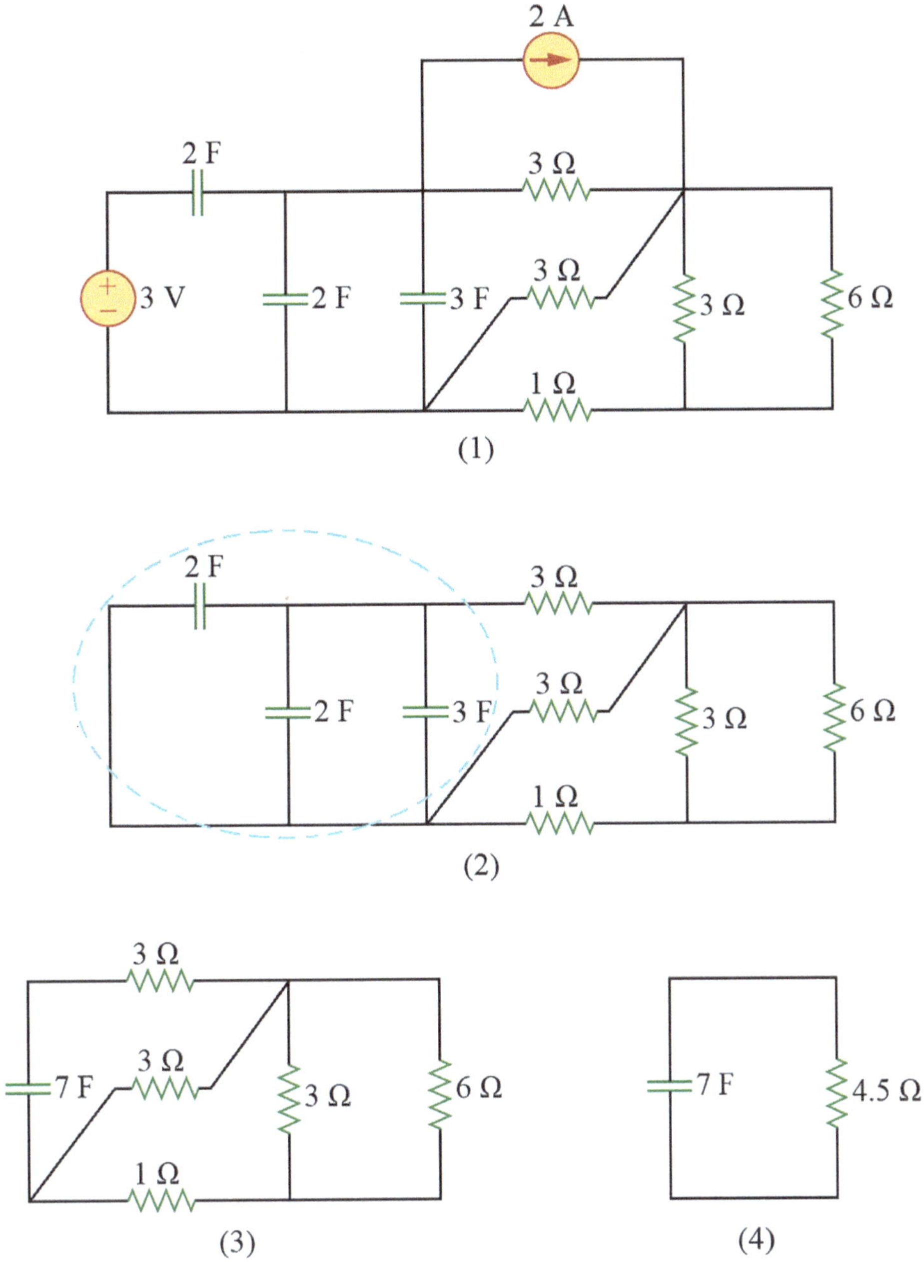

Fig. 10.1 The circuits of problem 10.1

10.3. As we know, the time constant of a first-order RC circuit can be calculated as follows.

$$\tau = R_{Th} C_{eq}$$

Herein, C_{eq} is the equivalent capacitor and R_{Th} is the Thevenin resistance seen by the equivalent capacitor. From previous chapters, we know that R_{Th} can be calculated as follows.

First, all the independent sources (independent voltage and current sources) are turned off (each independent voltage source is replaced by a short circuit branch and each independent current source is replaced by an open circuit branch). Then, if the circuit does not include any dependent source (dependent voltage or current source), the series-parallel rule is applied to calculate the input resistance of the circuit. However, if the circuit includes at least a dependent source, a test source (independent voltage or current source) with the voltage and current of V_t and I_t is connected to the input terminal of the circuit. Next, the circuit is analyzed by using a circuit analysis method to find the value of $\frac{V_t}{I_t}$. After that, we have:

$$R_{Th} = \frac{V_t}{I_t}$$

As can be noticed from Fig. 10.2.2, the resistors are in parallel. Therefore:

$$R_{Th} = R_1 || R_2$$

$$\Rightarrow R_{Th} = \frac{R_1 R_2}{R_1 + R_2}$$

Thus, the time constant of the circuit is as follows.

$$\tau = \frac{R_1 R_2}{R_1 + R_2} \times C$$

$$\Rightarrow \tau = \frac{R_1 R_2 C}{R_1 + R_2}$$

Choice (1) is the answer.

Traps

- You would choose Choice (2) if you had mistakenly assumed that the resistors are connected in series.

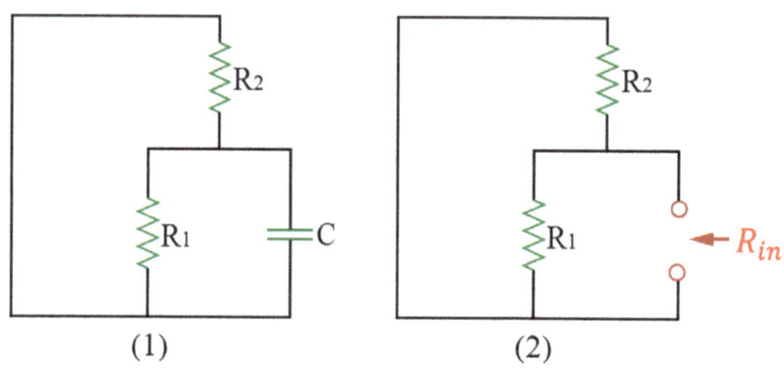

Fig. 10.2 The circuits of problem 10.3

10.4. As we know, the time constant of a first-order RC circuit can be calculated as follows.

$$\tau = R_{Th} C_{eq}$$

Herein, C_{eq} is the equivalent capacitor and R_{Th} is the Thevenin resistance seen by the equivalent capacitor. From previous chapters, we know that R_{Th} can be calculated as follows.

First, all the independent sources (independent voltage and current sources) are turned off (each independent voltage source is replaced by a short circuit branch and each independent current source is replaced by an open circuit branch). Then, if the circuit does not include any dependent source (dependent voltage or current source), the series-parallel rule is applied to calculate the input resistance of the circuit. However, if the circuit includes at least a dependent source, a test source (independent voltage or current source) with the voltage and current of V_t and I_t is connected to the input terminal of the circuit. Next, the circuit is analyzed by using a circuit analysis method to find the value of $\frac{V_t}{I_t}$. After that, we have:

$$R_{Th} = \frac{V_t}{I_t}$$

As is shown in Fig. 10.3.2, all the independent sources have been turned off, and a test source with the voltage and current of V_t and I_t has been connected to the terminal. Herein, nodal analysis is applied.

Defining I_1 based on the node voltage:

$$I_1 = -\frac{e_2}{3} \quad (1)$$

KCL in node 1:

$$-I_t + \frac{V_t - e_2}{1} + 3I_1 = 0 \quad (2)$$

KCL in node 2:

$$\frac{e_2}{1} + \frac{e_2 - V_t}{1} - 3I_1 - I_1 = 0$$

$$\Rightarrow 2e_2 - V_t - 4I_1 = 0 \quad (3)$$

Solving (1) and (2):

$$-I_t + \frac{V_t - e_2}{1} - e_2 = 0$$

$$\Rightarrow -I_t + V_t - 2e_2 = 0 \quad (4)$$

Solving (1) and (3):

$$2e_2 - V_t + \frac{4}{3}e_2 = 0$$

$$\Rightarrow e_2 = \frac{3}{10}V_t \quad (5)$$

Solving (4) and (5):

$$-I_t + V_t - \frac{3}{5}V_t = 0$$

$$\Rightarrow -I_t + \frac{2}{5}V_t = 0$$

$$\Rightarrow \frac{V_t}{I_t} = 2.5$$

$$\Rightarrow R_{Th} = 2.5\ \Omega$$

Thus, the time constant of the RC circuit is as follows.

$$\tau = 2.5 \times 2$$

$$\Rightarrow \tau = 5\ \text{sec}$$

Choice (4) is the answer.

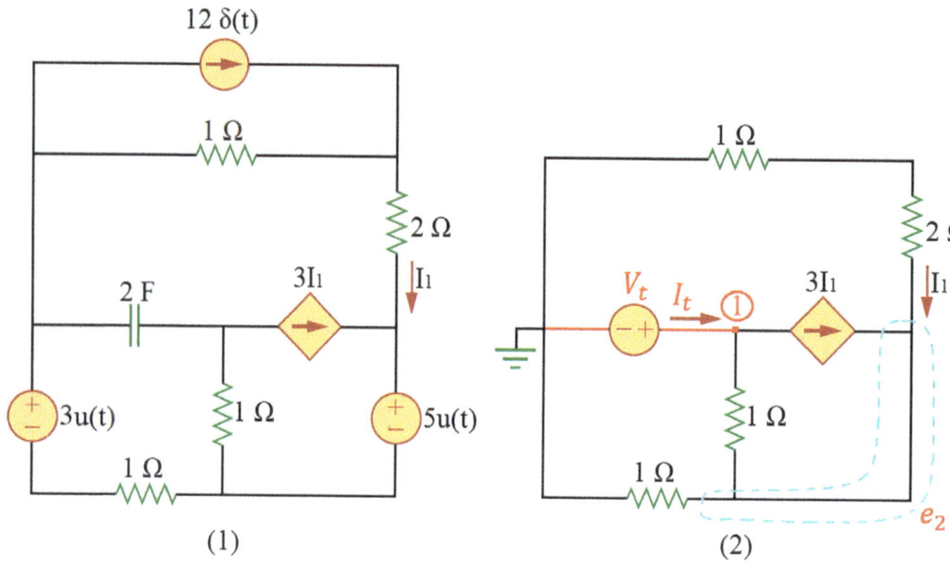

Fig. 10.3 The circuits of problem 10.4

10.2 Source-Free Response of an RC Circuit

10.6. As can be seen, there is no independent power source (independent voltage or current source) in the circuit. Therefore, the type of response of the circuit is source-free.

The general form of source-free response of an RC circuit is as follows.

$$V_C(t) = V_C(0^+)e^{-\frac{t}{\tau}}$$

Herein, $V_C(0^+)$ is the initial voltage of the capacitor and $\tau = R_{Th}C$ is the time constant of the circuit.

Calculating the initial voltage of the capacitor: Based on the information given in the problem, we have:

$$V_C(0^+) = 10\ \text{V}$$

Calculating the time constant of the circuit: The Thevenin resistance seen by the capacitor can be calculated as follows. See Fig. 10.4.2.

$$R_{Th} = 9 + (4 + 8)||(3 + 1)$$

$$\Rightarrow R_{Th} = 9 + \frac{12 \times 4}{12 + 4}$$

$$\Rightarrow R_{Th} = 12\ \Omega$$

Therefore:

$$\tau = 12 \times 3 = 36\,\text{sec}$$

Calculating the source-free response of the RC circuit:

$$V_C(t) = 10e^{-\frac{t}{36}}\ \text{V}$$

Choice (1) is the answer.

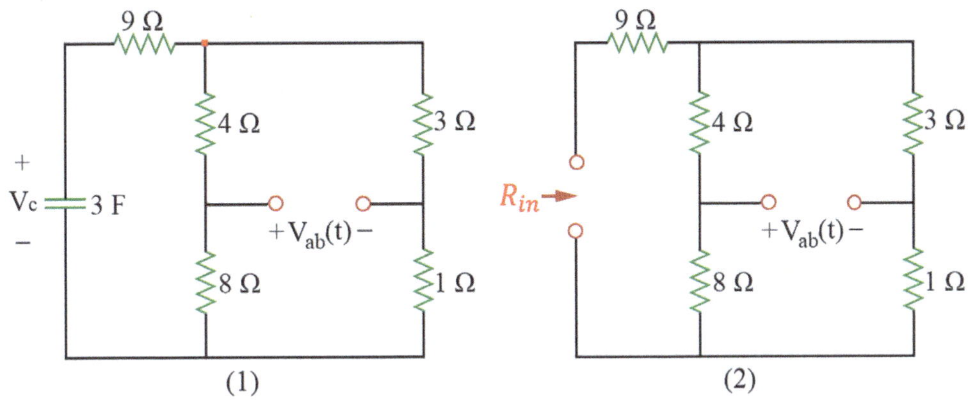

Fig. 10.4 The circuits of problem 10.6

10.7. From Problem 10.6, we have:

$$V_C(t) = 10e^{-\frac{t}{36}} V$$

Applying voltage division rule for points "a" and "b" in the circuit of Fig. 10.5:

$$V_{ab}(t) = V_a(t) - V_b(t) = \frac{8}{8+4} V_x(t) - \frac{1}{1+3} V_x(t)$$

$$\Rightarrow V_{ab}(t) = \frac{5}{12} V_x(t) \qquad (1)$$

Applying voltage division rule for node "x":

$$V_x(t) = \frac{(4+8)||(3+1)}{(4+8)||(3+1)+9} V_C(t)$$

$$\Rightarrow V_x(t) = \frac{\frac{12 \times 4}{12+4}}{\frac{12 \times 4}{12+4} + 9} \times 10e^{-\frac{t}{36}}$$

$$\Rightarrow V_x(t) = \frac{3}{3+9} \times 10e^{-\frac{t}{36}}$$

$$\Rightarrow V_x(t) = \frac{5}{2} e^{-\frac{t}{36}} \tag{2}$$

Solving (1) and (2):

$$V_{ab}(t) = \frac{5}{12} \times \frac{5}{2} e^{-\frac{t}{36}}$$

$$\Rightarrow V_{ab}(t) = \frac{25}{24} e^{-\frac{t}{36}}$$

$$\Rightarrow V_{ab}(t) = 1.04e^{-\frac{t}{36}} V$$

Choice (2) is the answer.

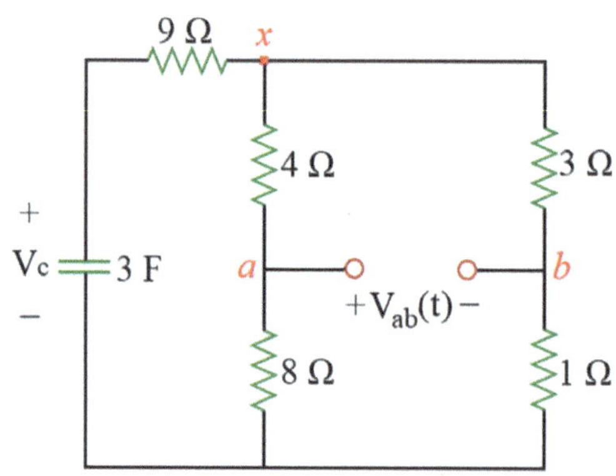

Fig. 10.5 The circuits of problem 10.7

10.9. Based on the information given in Problem 10.8, we have:

$$V_C(t) = 10e^{-t} V \tag{1}$$

Applying voltage-current relation of capacitors:

$$I_C(t) = C \frac{d}{dt} V_C(t)$$

$$\Rightarrow I_C(t) = 1 \times \frac{d}{dt}\left(10e^{-t}\right)$$

$$\Rightarrow I_C(t) = -10e^{-t} \tag{2}$$

KCL in the indicated node:

$$-5+I+I_C(t)+\frac{0-0}{2}=0$$

$$\Rightarrow -5+I+I_C(t)=0 \quad (3)$$

Solving (2) and (3):

$$-5+I-10e^{-t}=0$$

$$\Rightarrow I=5+10e^{-t} \quad (4)$$

Hence, the current of switch at $t=1\ sec$ is as follows.

$$I(t=1)=5+10e^{-1}$$

$$\Rightarrow I(t=1)=5\left(1+2e^{-1}\right)\ \text{A}$$

Choice (4) is the answer (Fig. 10.6)

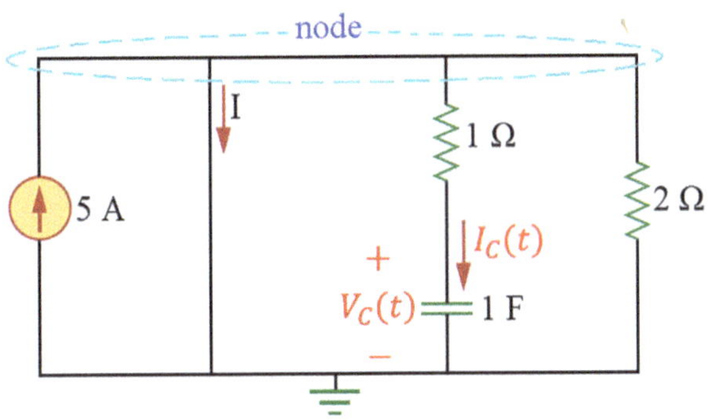

Fig. 10.6 The circuit of problem 10.9

10.10. Since there is no independent power source (independent voltage or current source) in the circuit, the type of response of the circuit is source-free.

The general form of source-free response of an RC circuit is as follows.

$$V_C(t)=V_C(0^+)e^{-\frac{t}{\tau}}$$

Herein, $V_C(0^+)$ is the initial voltage of the capacitor and $\tau=R_{Th}C$ is the time constant of the circuit.

Calculating the initial voltage of the capacitor: Based on the information given in the problem, we have:

$$V_C(0^-)=2\ \text{V} \quad (1)$$

Due to voltage continuity of capacitors in lack of a power source with an infinite size (e.g. impulse power source), it is concluded that:

$$V_C(0^+) = 2 \text{ V} \tag{2}$$

Calculating the time constant of the circuit: Defining i based on the node voltages:

$$i = \frac{V_t - e_1}{1}$$

$$\Rightarrow i = V_t - e_1 \tag{3}$$

KCL in the right-side node:

$$-i + \frac{e_1}{3} - 2i = 0$$

$$\Rightarrow e_1 = 9i \tag{4}$$

KCL in the left-side node:

$$-I_t + \frac{V_t}{2} + \frac{V_t - e_1}{1} = 0$$

$$\Rightarrow -I_t + \frac{3}{2}V_t - e_1 = 0 \tag{5}$$

Solving (3) and (4):

$$i = V_t - 9i$$

$$\Rightarrow i = \frac{1}{10}V_t \tag{6}$$

Solving (4)-(6):

$$-I_t + \frac{3}{2}V_t - \frac{9}{10}V_t = 0$$

$$\Rightarrow -I_t + \frac{6}{10}V_t = 0$$

$$\Rightarrow \frac{V_t}{I_t} = \frac{5}{3}$$

$$\Rightarrow R_{Th} = \frac{5}{3}\ \Omega$$

Therefore:

$$\tau = \frac{5}{3} \times 3 = 5 \text{ sec}$$

Calculating the source-free response of the RC circuit:

$$V_C(t) = 2e^{-\frac{t}{5}} \text{ V}$$

Choice (4) is the answer.

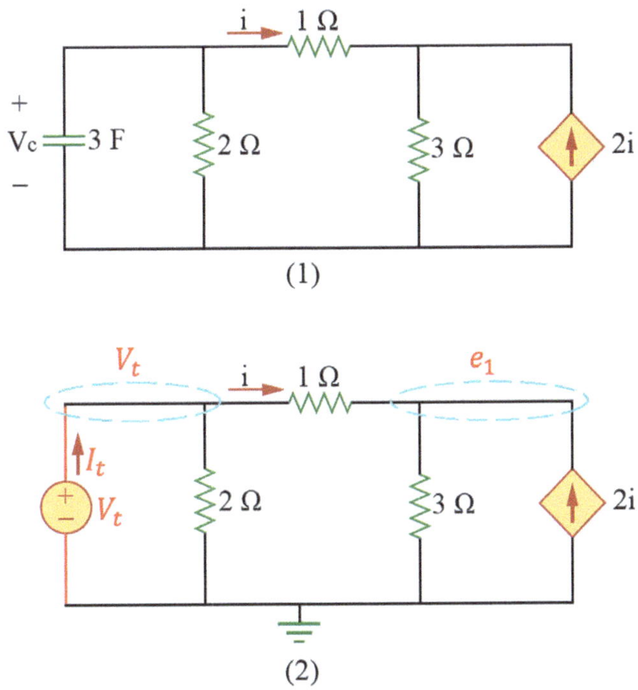

Fig. 10.7 The circuits of problem 10.10

10.11. Based on the information given in Problem 10.10, we have:

$$V_C(t) = 2e^{-\frac{t}{5}}\ \text{V} \tag{1}$$

$$V_C(0^+) = 2\ \text{V} \tag{2}$$

Hence:

$$V_C(t_0) = 0.5 \times 2 = 1\ \text{V} \tag{3}$$

Solving (1) and (3):

$$2e^{-\frac{t_0}{5}} = 1$$

$$\Rightarrow e^{-\frac{t_0}{5}} = 0.5$$

$$\Rightarrow -\frac{t_0}{5} = \ln(0.5)$$

$$\Rightarrow t_0 = 5 \ln 2\ \text{sec}$$

Choice (1) is the answer.

Notes

$$e^a = b \Rightarrow a = \ln b$$

10.3 Step Response of an RC Circuit

10.12. Since the initial voltage of the capacitor is zero and the power source (independent current source) is a step function, the response of the circuit is step response.

The general form of step response of an RC circuit is as follows.

$$V_C(t) = V_C(\infty)\left(1 - e^{-\frac{t}{\tau}}\right)$$

Herein, $V_C(\infty)$ is the voltage of the capacitor after a long time and $\tau = R_{Th}C$ is the time constant of the circuit.

Calculating the time constant of the circuit: As can be noticed from Fig. 10.8.2, the circuit is symmetric if the Thevenin resistance is calculated from terminal A-B. Thus, as is shown in Fig. 10.8.3, we can fold the circuit around the A-B axis. By doing this, the two ends of the 3-Ω resistors are short-circuited, therefore they are eliminated. However, the other resistors are paralleled two by two. Fig. 10.8.4 illustrates the simplified circuit of Fig. 10.8.3 Therefore:

$$R_{Th} = 0.5 + 1 + 0.5 = 2\ \Omega$$

And the time constant of the RC circuit can be calculated as follows.

$$\tau = 2 \times 2 = 4\ \text{sec}$$

Calculating the final voltage of the capacitor: The final voltage of the capacitor can be calculated by assuming the capacitor as the open circuit branch, as is shown in Fig. 10.8.5.

$$V_C(\infty) = 1 \times (0.5 + 1 + 0.5)$$

$$\Rightarrow V_C(\infty) = 2\ \text{V}$$

Calculating the step response of the RC circuit:

$$V_C(t) = 2\left(1 - e^{-\frac{t}{4}}\right) V, t \geq 0$$

Since the last equation is valid for $t \geq 0$, it can be written by using a unit step function as follows.

$$\Rightarrow V_C(t) = 2\left(1 - e^{-0.25t}\right)u(t)\ \text{V}$$

Choice (4) is the answer.

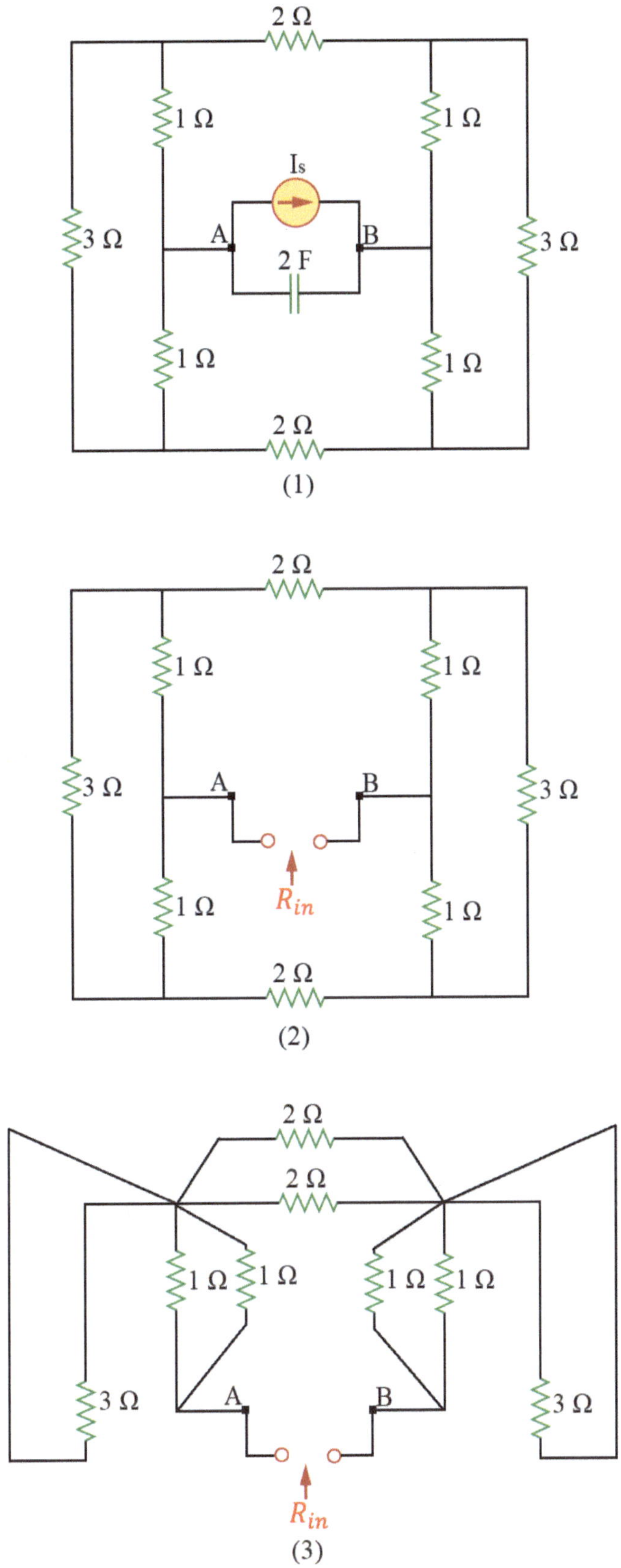

Fig. 10.8 The circuits of problem 10.12

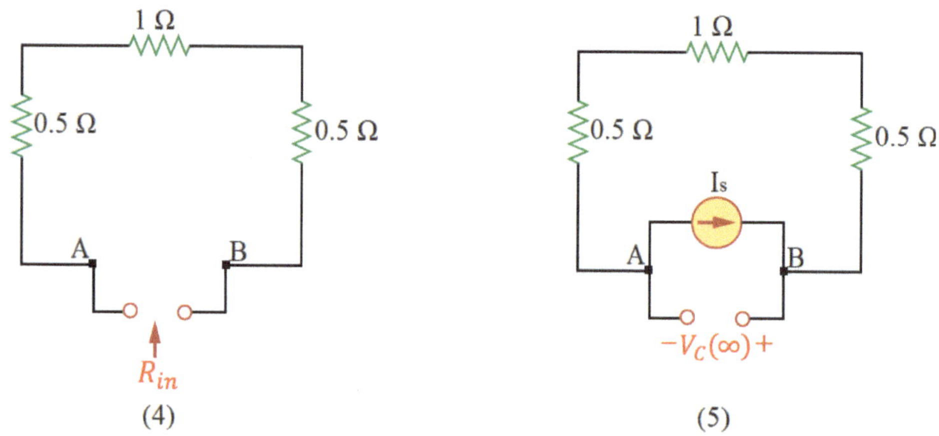

Fig. 10.8 (continued)

10.14. Since the initial voltage of the capacitor is zero and the power source (independent voltage source) is a step function, the response of the circuit is step response.

The general form of step response of an RC circuit is as follows.

$$V_C(t) = V_C(\infty)\left(1 - e^{-\frac{t}{\tau}}\right)$$

Herein, $V_C(\infty)$ is the voltage of the capacitor after a long time and $\tau = R_{Th}C$ is the time constant of the circuit.
To simplify the circuit, first, we should determine the Thevenin equivalent circuit seen by the capacitor. As is shown in Fig. 10.9.2, the Thevenin resistance of the circuit can be calculated by using series-parallel rule as follows.

$$R_{Th} = (1||1 + 0.5)||(1)$$

$$\Rightarrow R_{Th} = (0.5 + 0.5)||(1)$$

$$\Rightarrow R_{Th} = 0.5 \text{ k}\Omega$$

As can be seen in Fig. 10.9.3, we can apply voltage division rule twice, to calculate the open circuit voltage as follows.

$$e_1 = \frac{1||(0.5+1)}{1 + 1||(0.5+1)} \times 8$$

$$\Rightarrow e_1 = \frac{\frac{1 \times 1.5}{1 + 1.5}}{1 + \frac{1 \times 1.5}{1 + 1.5}} \times 8$$

$$\Rightarrow e_1 = \frac{\frac{1.5}{2.5}}{\frac{4}{2.5}} \times 8$$

$$\Rightarrow e_1 = 3 \text{ V}$$

$$V_{oc} = \frac{1}{1+0.5} e_1$$

$$\Rightarrow V_{oc} = \frac{1}{1.5} \times 3$$

$$\Rightarrow V_{oc} = 2 \text{ V}$$

$$\Rightarrow V_{Th} = 2 \text{ V}$$

The Thevenin equivalent circuit is illustrated in Fig. 10.9.4

Calculating the time constant of the circuit: By using the circuit of Fig. 10.9.4, we have:

$$\tau = 0.5 \times 10^3 \times 2 \times 10^{-6} = 1 \text{ ms}$$

Calculating the final voltage of the capacitor: Since the capacitor behaves like an open circuit in a steady state condition, the final voltage of the capacitor will be 2 V (See Fig. 10.9.4). In other words:

$$V_C(\infty) = 2 \text{ V}$$

Calculating the step response of the RC circuit:

$$V_C(t) = 2\left(1 - e^{-\frac{t}{10^{-3}}}\right)$$

$$\Rightarrow V_C(t) = 2\left(1 - e^{-10^3 t}\right)$$

For $t = 1 \text{ ms} = 10^{-3}$ s, we have:

$$V_C\left(t = 10^{-3}\right) = 2\left(1 - e^{-10^3 \times 10^{-3}}\right)$$

$$\Rightarrow V_C\left(t = 10^{-3}\right) = 2\left(1 - e^{-1}\right) \text{ V}$$

Choice (2) is the answer.

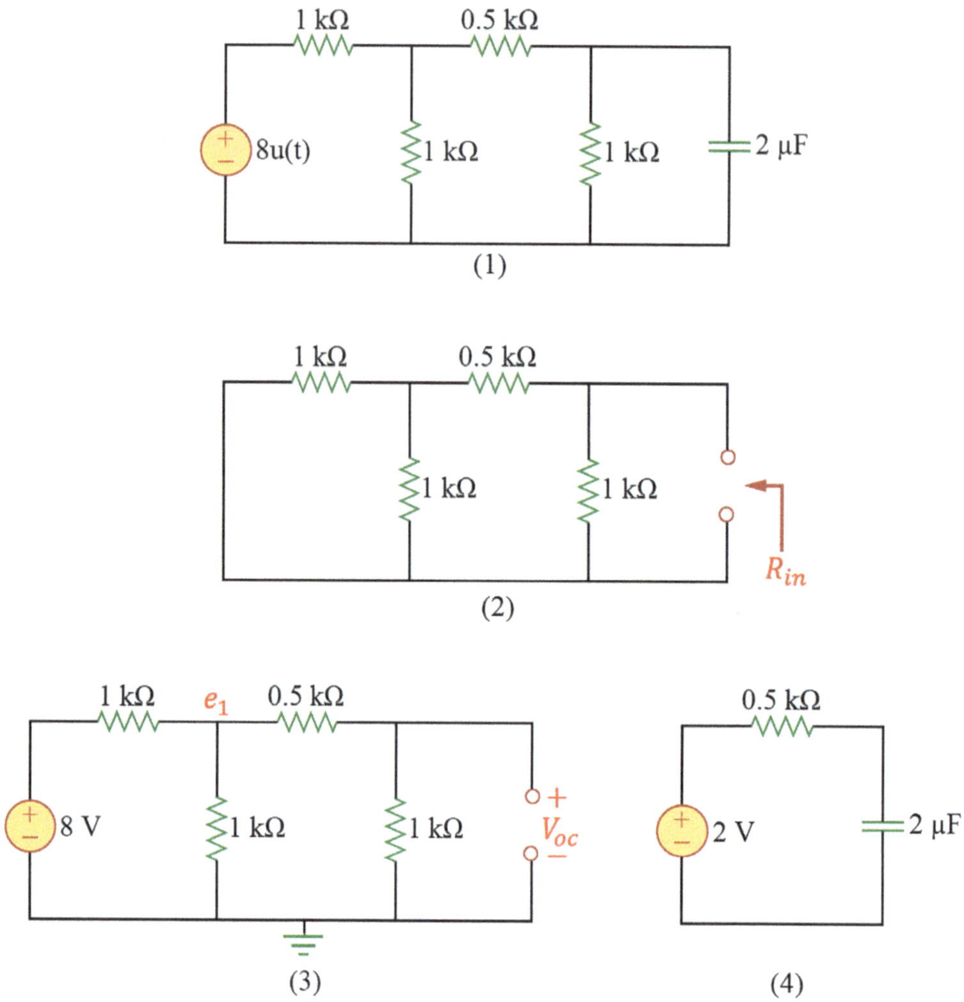

Fig. 10.9 The circuits of problem 10.14

10.15. The problem is concerned with the step response of the circuit because the initial voltage of the capacitor is zero and the power source (independent voltage source) is added to the circuit at a specific time.

The general form of step response of an RC circuit is as follows.

$$V_C(t) = V_C(\infty)\left(1 - e^{-\frac{t}{\tau}}\right)$$

Herein, $V_C(\infty)$ is the voltage of the capacitor after a long time and $\tau = R_{Th}C$ is the time constant of the circuit.

Calculating the time constant of the circuit: The time constant of the circuit can be calculated as follows by assuming that the independent voltage source is turned off in Fig. 10.10.2

$$\tau = (1 + 1) \times 1 = 2 \text{ sec}$$

Calculating the final voltage of the capacitor: As we know, the capacitor will behave like an open circuit after a long time. Thus, the current of the circuit will be zero. By applying KVL in the loop, we have:

$$V_C(\infty) = 300 - 1 \times 0 + 1 \times 0$$

$$\Rightarrow V_C(\infty) = 300 \text{ V}$$

Calculating the step response of the RC circuit:

$$V_C(t) = 300\left(1 - e^{-\frac{t}{2}}\right) \quad (1)$$

Based on the information given in the problem, we have:

$$V_C(t_0) = 0.5 \times V_C(\infty)$$

Therefore:

$$V_C(t_0) = 0.5 \times 300$$

$$\Rightarrow V_C(t_0) = 150 \quad (2)$$

To determine the value of t_0, we need to solve (1) and (2) as follows.

$$300\left(1 - e^{-\frac{t_0}{2}}\right) = 150$$

$$\Rightarrow 1 - e^{-\frac{t_0}{2}} = 0.5$$

$$\Rightarrow e^{-\frac{t_0}{2}} = 0.5$$

$$\Rightarrow -\frac{t_0}{2} = \ln(0.5)$$

$$\Rightarrow \frac{t_0}{2} = \ln 2$$

$$\Rightarrow t_0 = 2 \ln 2$$

Choice (4) is the answer.

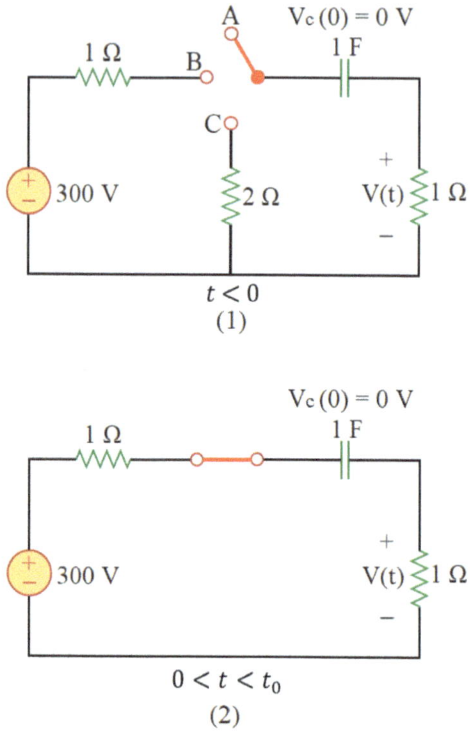

Fig. 10.10 The circuits of problem 10.15

Notes

$$e^a = b \Rightarrow a = \ln b$$

10.16. From Problem 10.15, we have:

$$t_0 = 2 \ln 2$$

Since there is no independent power source (independent voltage source) in the circuit, the type of response of the circuit is source-free.

The general form of source-free response of an RC circuit is as follows.

$$V_C(t) = V_C(0^+)e^{-\frac{t}{\tau}}$$

Herein, $V_C(0^+)$ is the initial voltage of the capacitor and $\tau = R_{Th}C$ is the time constant of the circuit.

Calculating the initial voltage of the capacitor: Based on the information given in the problem, we have:

$$V_C\left(t_0^-\right) = 150 \text{ V}$$

Due to voltage continuity of capacitors in lack of a power source with an infinite size (e.g. impulse power source), it is concluded that:

$$V_C(t_0^+) = 150 \text{ V}$$

Calculating the time constant of the circuit: As can be noticed from Fig. 10.11.2, we have:

$$R_{Th} = 2 + 1 = 3 \ \Omega$$

Therefore:

$$\tau = 3 \times 1 = 3 \text{ sec}$$

Calculating the source-free response of the RC circuit:

$$V_C(t) = 150e^{-\frac{t - t_0}{3}}$$

$$\Rightarrow V_C(t) = 150e^{-\frac{t - 2\ln 2}{3}}$$

Applying voltage division rule to calculate the voltage of 1-Ω resistor:

$$V(t) = -\frac{1}{1+2}V_C(t)$$

$$\Rightarrow V(t) = -50e^{-\frac{t - 2\ln 2}{3}}$$

Choice (4) is the answer.

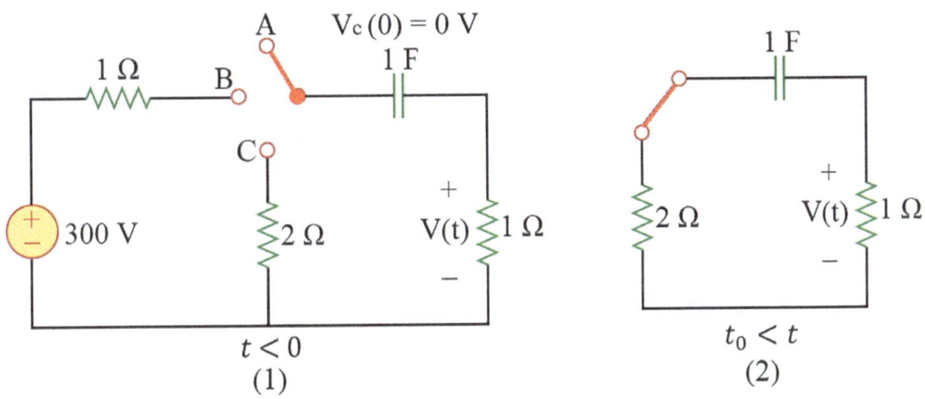

Fig. 10.11 The circuits of problem 10.16

10.18. The power source (independent voltage source) is a pulse function. Hence, the response of the circuit for $0 < t < 1$ is a step response (Part 1). However, the response of the circuit for $t > 1$ is a source-free response (Part 2) since the power source is separated from the circuit.

Part 1 ($0 < t < 1$): The general form of step response of an RC circuit is as follows.

$$V_C(t) = V_C(\infty)\left(1 - e^{-\frac{t}{\tau}}\right)$$

Herein, $V_C(\infty)$ is the voltage of the capacitor after a long time and $\tau = R_{Th}C$ is the time constant of the circuit.

Calculating the time constant of the circuit: The time constant of the circuit is as follows.

$$\tau = 0.5 \times 2 = 1 \text{ sec}$$

Calculating the final voltage of the capacitor: The final voltage of the capacitor will be 10 V since the capacitor will behave like an open circuit branch after a long time.

Calculating the step response of the RC circuit:

$$V_C(t) = 10(1 - e^{-t})V \text{ for } 0 < t < 1 \tag{1}$$

$$\Rightarrow V_C(t) = 10(1 - e^{-t})(u(t) - u(t-1)) \tag{2}$$

In (2), $u(t)$ has been applied to rewrite (1). The definition of unit step function is as follows.

$$u(t) = \begin{cases} 0 & t < 0 \\ 1 & t \geq 0 \end{cases}$$

Part 2 ($t > 1$): In this condition, no independent power source (independent voltage or current source) exists in the circuit. Thus, the type of response of the circuit is source-free.
The general form of source-free response of an RC circuit is as follows.

$$V_C(t) = V_C(0^+)e^{-\frac{t}{\tau}}$$

Herein, $V_C(0^+)$ is the initial voltage of the capacitor and $\tau = R_{Th}C$ is the time constant of the circuit.

Calculating the initial voltage of the capacitor: By using (2) for $t = 1^-$, we have:

$$V_C(1^-) = 10\left(1 - e^{-1}\right)\text{V}$$

Due to voltage continuity of capacitors in lack of a power source with an infinite size (e.g. impulse power source), it is concluded that:

$$V_C(1^+) = 10\left(1 - e^{-1}\right)\text{V}$$

Calculating the time constant of the circuit: There is no change in the time constant of the circuit.

Calculating the source-free response of the RC circuit:

$$V_C(t) = 10\left(1 - e^{-1}\right)e^{-(t-1)} \text{ for } t > 1 \tag{3}$$

$$\Rightarrow V_C(t) = 10\left(1 - e^{-1}\right)e^{-(t-1)}u(t-1) \tag{4}$$

$$\Rightarrow V_C(t) = \left(10e^{-(t-1)} - 10e^{-t}\right)u(t-1)$$

Choice (3) is the answer (Fig. 10.12).

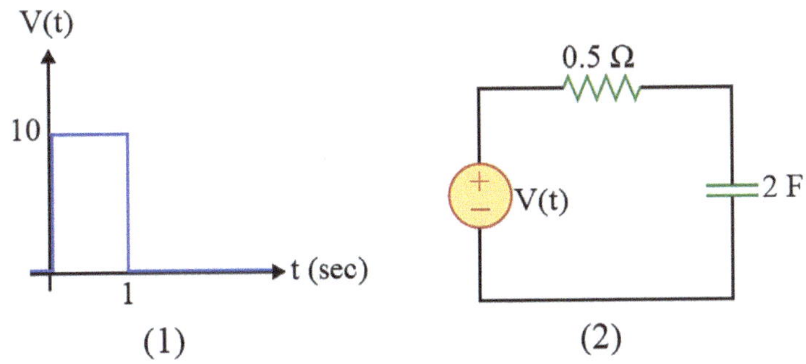

Fig. 10.12 The circuits of problem 10.18

10.20. As is shown in Fig. 10.13.2-3, we can use superposition theorem to solve the problem since the circuit includes more than one power source. The circuit shown in Fig. 10.13.2 includes a Wheatstone bridge; therefore, the capacitor will not be charged by the voltage source. Hence:

$$V_{C1}(t) = 0$$

Regarding the circuit of Fig. 10.13.3, since the initial voltage of the capacitor is zero (zero state before $t = 0$) and the power source (independent current source) is a step function, the response of the circuit is step response.
The general form of step response of an RC circuit is as follows.

$$V_C(t) = V_C(\infty)\left(1 - e^{-\frac{t}{\tau}}\right)$$

Herein, $V_C(\infty)$ is the voltage of the capacitor after a long time and $\tau = R_{Th}C$ is the time constant of the circuit.
Herein, it is better to simplify the circuit first, as is shown in Fig. 10.13.3-5 Moreover, it is convenient to use source transformation technique, as is shown in Fig. 10.13.6-9

Calculating the time constant of the circuit: The time constant of the circuit can be calculated by using Fig. 10.13.9 as follows.

$$\tau_2 = 2R \times C = 2\ \mathrm{RC}$$

Calculating the final voltage of the capacitor: The final voltage of the capacitor will be $-\frac{1}{2}V$ (see Fig. 10.13.9) since the capacitor will be an open circuit after a long time. In other words:

$$V_{C2}(\infty) = -\frac{1}{2}\ \mathrm{V}$$

Calculating the step response of the RC circuit:

$$V_{C2}(t) = -\frac{1}{2}\left(1 - e^{-\frac{t}{2RC}}\right)$$

Now, the total voltage of the capacitor can be calculated as follows.

$$V_C(t) = V_{C1}(t) + V_{C2}(t)$$

$$\Rightarrow V_C(t) = 0 - \frac{1}{2}\left(1 - e^{-\frac{t}{2RC}}\right)$$

$$\Rightarrow V_C(t) = -\frac{1}{2}\left(1 - e^{-\frac{t}{2RC}}\right)$$

Now, by using voltage-current relation of capacitors, we have:

$$I_C(t) = C\frac{d}{dt}V_C(t)$$

$$\Rightarrow I_C(t) = C\frac{d}{dt}\left(-\frac{1}{2}\left(1 - e^{-\frac{t}{2RC}}\right)\right)$$

$$\Rightarrow I_C(t) = \frac{C}{2} \times \left(-\frac{1}{2RC}\right)e^{-\frac{t}{2RC}}$$

$$\Rightarrow I_C(t) = -\frac{1}{4R}e^{-\frac{t}{2\,\mathrm{RC}}} \quad \text{for } t > 0$$

Since the result is valid for $t > 0$, , we can use unit step function ($u(t)$) in the last equation as follows.

$$I_C(t) = -\frac{1}{4R}e^{-\frac{t}{2RC}}u(t)$$

Choice (3) is the answer.

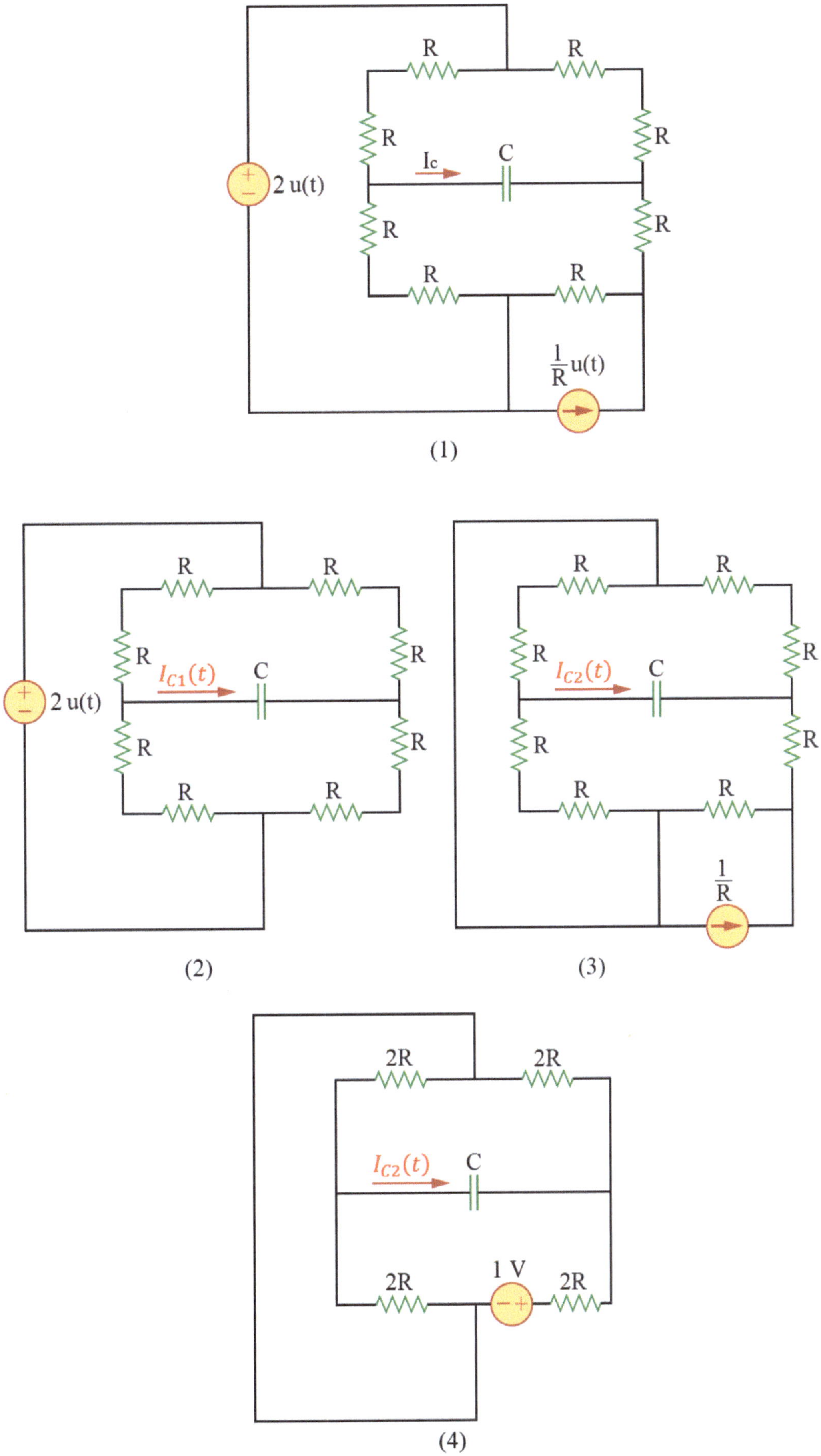

Fig. 10.13 The circuits of problem 10.20

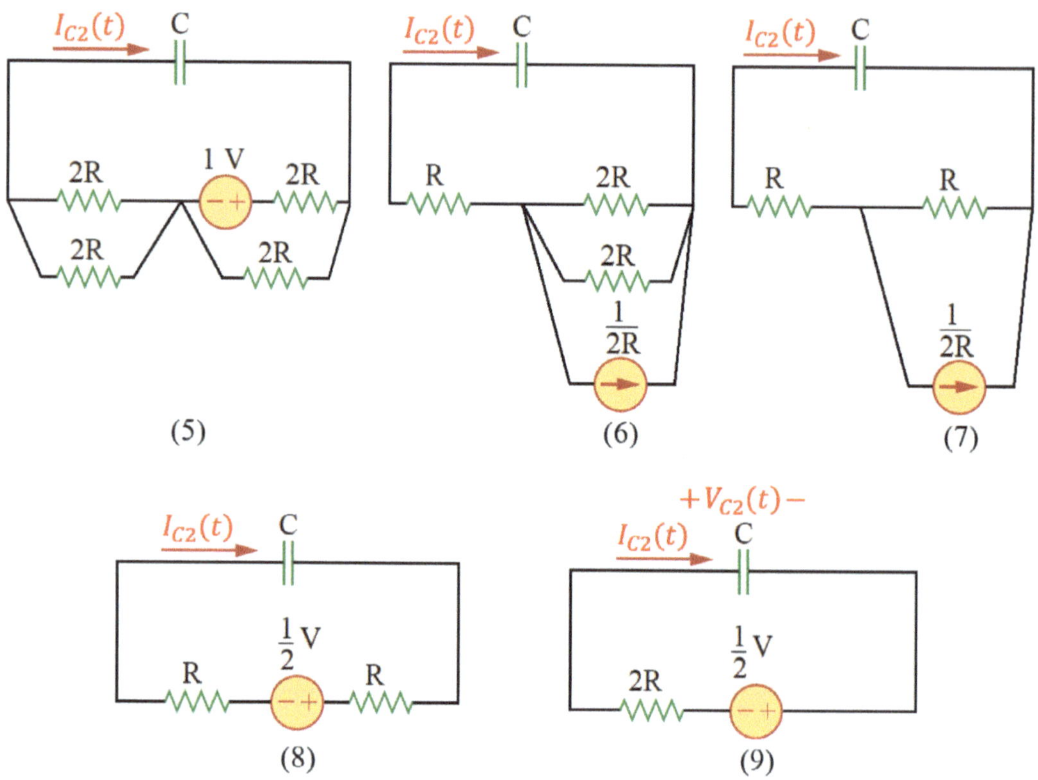

Fig. 10.13 (continued)

10.21. **Part 1**: Since the initial voltage of the capacitor is zero and the power source (independent current source) is a step function, the response of the circuit is step response.

The general form of step response of an RC circuit is as follows.

$$V_C(t) = V_C(\infty)\left(1 - e^{-\frac{t}{\tau}}\right) \quad (1)$$

Herein, $V_C(\infty)$ is the voltage of the capacitor after a long time and $\tau = R_{Th}C$ is the time constant of the circuit. Based on the information given in the problem, we have:

$$V(t) = \frac{4}{3}\left(1 - e^{-1.5t}\right)u(t) \quad (2)$$

By using the circuit of Fig. 10.14.1, we can calculate the final voltage of the capacitor as follows.

$$V(\infty) = 1 \times (2||R)$$

$$\Rightarrow V(\infty) = \frac{2R}{2+R} \quad (3)$$

Solving (1)-(3):

$$\frac{2R}{2+R}\left(1-e^{-\frac{\infty}{\tau}}\right)=\frac{4}{3}\left(1-e^{-1.5\times\infty}\right)u(\infty)$$

$$\Rightarrow\frac{2R}{2+R}=\frac{4}{3}$$

$$\Rightarrow 6R=8+4\ \mathrm{R}$$

$$\Rightarrow R=4\ \Omega \tag{4}$$

By comparing (1) and (2), we can calculate the time constant of the circuit as follows.

$$\frac{1}{\tau}=1.5$$

$$\Rightarrow\tau=\frac{2}{3}\ \sec \tag{5}$$

By using $\tau=R_{Th}C$ and (4) and (5), we have:

$$\frac{2}{3}=(2||4)C$$

$$\Rightarrow\frac{2}{3}=\frac{2\times4}{2+4}\mathrm{C}$$

$$\Rightarrow\frac{2}{3}=\frac{8}{6}\mathrm{C}$$

$$\Rightarrow C=\frac{1}{2}\ \mathrm{F}$$

Part 2: By replacing the 2-Ω resistor with a 0.5-F capacitor, the response of the circuit is still step response. The simplified circuit of Fig. 10.14.2 is illustrated in Fig. 10.14.3.

Calculating the time constant of the circuit:

$$\tau=1\times4=4\ \sec$$

Calculating the final voltage of the capacitor: The final voltage of the capacitor can be calculated by using Ohm's law as follows.

$$V(\infty)=1\times4=4\ \mathrm{V}$$

Calculating the step response of the RC circuit:

$$V(t)=4\left(1-e^{-0.25t}\right)$$

Choice (4) is the answer.

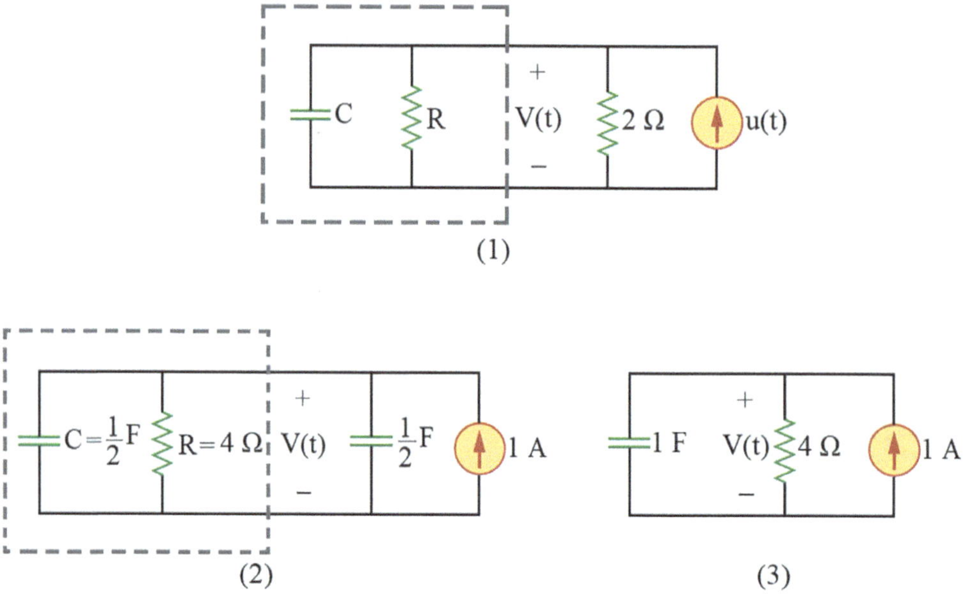

Fig. 10.14 The circuits of problem 10.21

10.23. The problem is concerned with the step response since the capacitor is free of charge and a power source (independent voltage source) is added to the circuit due to the switching.

The general form of step response of an RC circuit is as follows.

$$V_C(t) = V_C(\infty)\left(1 - e^{-\frac{t}{\tau}}\right)$$

Herein, $V_C(\infty)$ is the voltage of the capacitor after a long time and $\tau = R_{Th}C$ is the time constant of the circuit.

Calculating the final voltage of the capacitor: As is illustrated in Fig. 10.15.2, the final voltage of the capacitor can be calculated by using voltage division rule as follows.
Herein, the capacitor voltage will be the same as the 6-Ω resistor's voltage since there is no voltage drop across the top 2-Ω resistor.

$$V_C(\infty) = V_{6\Omega}(\infty) = \frac{6}{6+2} \times 12$$

$$\Rightarrow V_C(\infty) = 9 \text{ V}$$

Calculating the time constant of the circuit: Fig. 10.15.3-4 is used to calculate the Thevenin resistance seen by the capacitor as follows.

$$R_{Th} = 6||2 + 2$$

$$\Rightarrow R_{Th} = \frac{6 \times 2}{6+2} + 2$$

$$\Rightarrow R_{Th} = 3.5 \ \Omega$$

Thus, the time constant of the circuit is as follows.

$$\tau = 3.5 \times 1 = \frac{7}{2} \text{ sec}$$

Calculating the step response of the RC circuit:

$$V_C(t) = 9\left(1 - e^{-\frac{2}{7}t}\right)\ \mathrm{V}$$

Choice (1) is the answer.

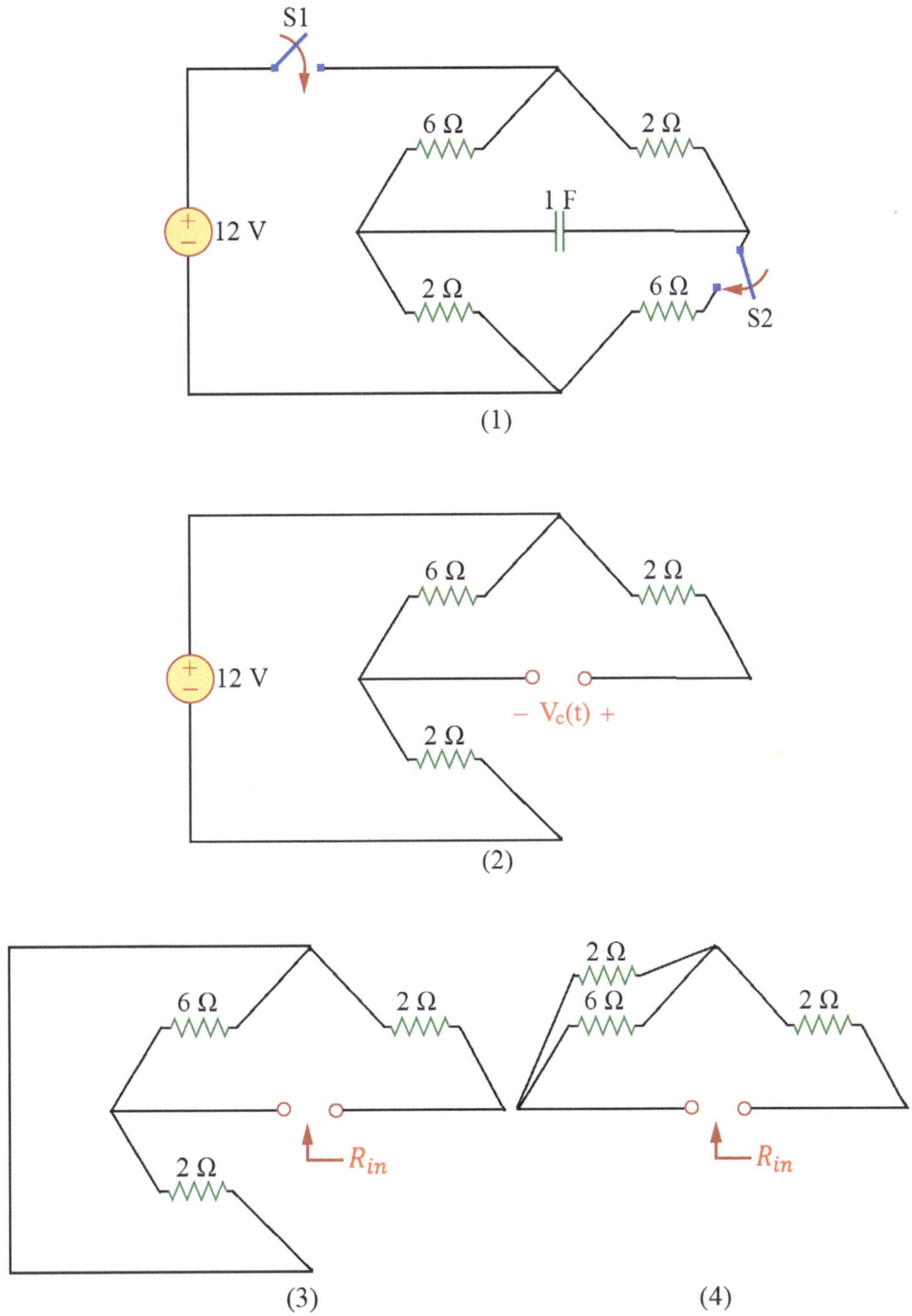

Fig. 10.15 The circuits of problem 10.23

10.4 Complete Response of an RC Circuit

10.24. From Partially Solved Exercise 9.17, we have:

$$V_C(t) = R_1\left(1 - e^{-\frac{t}{R_1 C}}\right) \tag{1}$$

Since the capacitor includes an initial voltage as well as the power source (independent current source) is available in the circuit and is a step function, the response of the circuit is complete.
The general form of complete response of an RC circuit is as follows.

$$V_C(t) = V_C(\infty) + (V_C(0^+) - V_C(\infty))e^{-\frac{t}{\tau}} \tag{2}$$

Herein, $V_C(0^+)$ is the initial voltage of the capacitor, $V_C(\infty)$ is the voltage of the capacitor after a long time, and $\tau = R_{Th}C$ is the time constant of the circuit.

Calculating the initial voltage of the capacitor: The voltage of the capacitor at $t = t_1$ can be calculated by using (1) as follows.

$$V_C(t_1) = R_1\left(1 - e^{-\frac{t_1}{R_1 C}}\right), t_1 = R_1 C$$

$$\Rightarrow V_C(t_1) = R_1\left(1 - e^{-\frac{R_1 C}{R_1 C}}\right)$$

$$\Rightarrow V_C(t_1) = R_1\left(1 - e^{-1}\right)$$

Due to voltage continuity of capacitors in lack of a power source with an infinite size (e.g. impulse power source), it is concluded that:

$$\Rightarrow V_C(t_1^+) = R_1\left(1 - e^{-1}\right) \tag{3}$$

Calculating the time constant of the circuit: No need to calculate it.

Calculating the final voltage of the capacitor: The final voltage of the capacitor can be determined by using Ohm's law and assuming the capacitor as an open circuit branch. See Fig. 10.16.2.

$$V_C(\infty) = 1 \times \frac{R_1 R_2}{R_1 + R_2}$$

$$\Rightarrow V_C(\infty) = \frac{R_1 R_2}{R_1 + R_2} \tag{4}$$

Calculating the complete response of the RC circuit:

$$V_C(t) = \frac{R_1 R_2}{R_1 + R_2} + \left(R_1\left(1 - e^{-1}\right) - \frac{R_1 R_2}{R_1 + R_2}\right)e^{-\frac{t - t_1}{\tau}} \tag{5}$$

From (5), it is concluded that the transient part of the voltage must be eliminated so that the voltage of the capacitor remains constant after the switching operation. In other words:

$$R_1\left(1-e^{-1}\right)-\frac{R_1R_2}{R_1+R_2}=0$$

$$\frac{R_1R_2}{R_1+R_2}=R_1\left(1-e^{-1}\right)$$

$$\Rightarrow\frac{R_2}{R_1+R_2}=\left(1-e^{-1}\right)$$

$$\Rightarrow R_2=R_1\left(1-e^{-1}\right)+R_2\left(1-e^{-1}\right)$$

$$\Rightarrow R_2e^{-1}=R_1\left(1-e^{-1}\right)$$

$$\Rightarrow R_2=R_1(e-1)$$

Choice (4) is the answer.

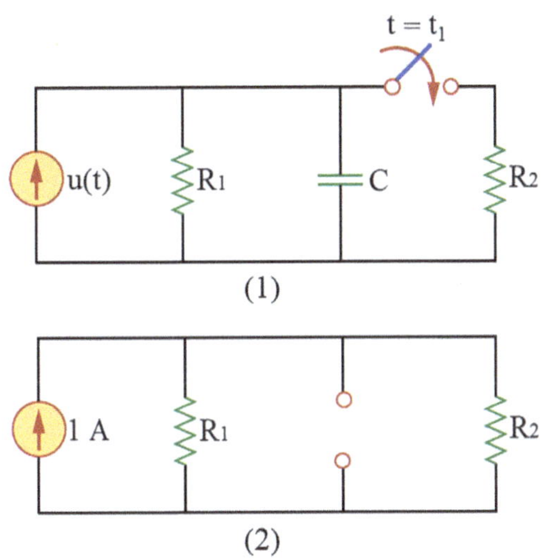

Fig. 10.16 The circuits of problem 10.24

10.26. The capacitor includes an initial voltage and, at least, an effective power source (independent voltage or current source) exists in the circuit; thus, the response of the circuit is complete.

The general form of complete response of an RC circuit is as follows.

$$V_C(t)=V_C(\infty)+\left(V_C(0^+)-V_C(\infty)\right)e^{-\frac{t}{\tau}} \quad (1)$$

Herein, $V_C(0^+)$ is the initial voltage of the capacitor, $V_C(\infty)$ is the voltage of the capacitor after a long time, and $\tau=R_{Th}C$ is the time constant of the circuit.

Calculating the initial voltage of the capacitor: Based on the information given in the problem, we have:

$$V_C(0^-) = 4\ \text{V}$$

Due to voltage continuity of capacitors in lack of a power source with an infinite size (e.g. impulse power source), it is concluded that:

$$V_C(0^+) = 4\ \text{V} \tag{2}$$

Calculating the time constant of the circuit: No need to calculate it.

Calculating the final voltage of the capacitor: As can be noticed form (1), to prevent the capacitor from charging and discharging, the transient term of the voltage equation must be removed. In other words:

$$V_C(0^+) = V_C(\infty) \tag{3}$$

Solving (2) and (3):

$$V_C(\infty) = 4\ \text{V} \tag{4}$$

On the other hand, superposition theorem and Fig. 10.17.2-3 can be used to calculate the final voltage of the capacitor as follows.
Using Ohm's law in Fig. 10.17.2:

$$V_{C1}(\infty) = (1||2) \times I_s$$

$$\Rightarrow V_{C1}(\infty) = \frac{2}{3} I_s \tag{5}$$

Using voltage division rule in Fig. 10.17.3:

$$V_{C2}(\infty) = \frac{1}{1+2} V_s$$

$$\Rightarrow V_{C2}(\infty) = \frac{1}{3} V_s \tag{6}$$

Thus, the total final voltage of the capacitor is as follows.

$$V_C(\infty) = V_{C1}(\infty) + V_{C2}(\infty)$$

$$\Rightarrow V_C(\infty) = \frac{2}{3} I_s + \frac{1}{3} V_s \tag{7}$$

Solving (4) and (7):

$$4 = \frac{2}{3} I_s + \frac{1}{3} V_s$$

$$\Rightarrow 2I_s + 3V_s = 12$$

Choice (4) is the answer.

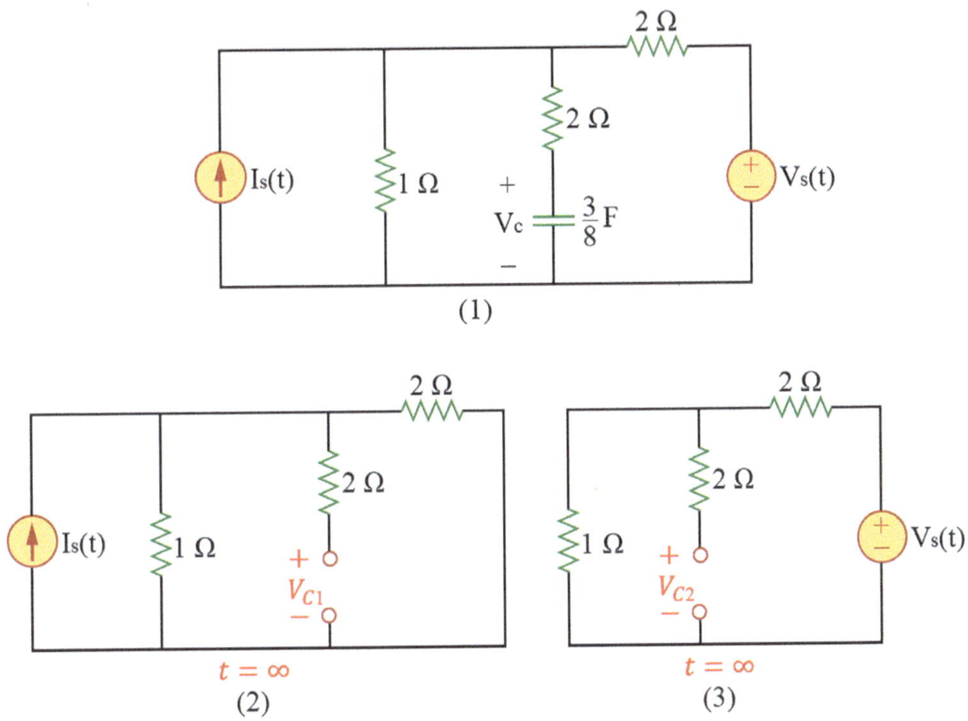

Fig. 10.17 The circuits of problem 10.26

10.27. The response of the circuit is complete since the capacitor includes an initial voltage as well as the power source (independent voltage source) is available in the circuit.

The general form of complete response of an RC circuit is as follows.

$$V_C(t) = V_C(\infty) + (V_C(0^+) - V_C(\infty))e^{-\frac{t}{\tau}}$$

Herein, $V_C(0^+)$ is the initial voltage of the capacitor, $V_C(\infty)$ is the voltage of the capacitor after a long time, and $\tau = R_{Th}C$ is the time constant of the circuit.
First, we need to determine the time that the voltage of the second switch reaches 9 V.
By writing a KVL in the loop of circuit of Fig. 10.18.2, we have:

$$V_S(t) = -2I_C(t) + 12 \quad (1)$$

From voltage-current relation of capacitors, we know that:

$$I_C(t) = C\frac{d}{dt}V_C(t) \quad (2)$$

From Problem 10.23, we know that

$$V_C(t) = 9\left(1 - e^{-\frac{2}{7}t}\right)V, t > 0 \quad (3)$$

Solving (1)-(3):

$$V_S(t) = -2\left(1 \times \frac{d}{dt}\left(9\left(1 - e^{-\frac{2}{7}t}\right)\right)\right) + 12$$

$$\Rightarrow V_S(t) = -2 \times \frac{18}{7} e^{-\frac{2}{7}t} + 12$$

$$\Rightarrow V_S(t) = -\frac{36}{7} e^{-\frac{2}{7}t} + 12 \quad (4)$$

Now, we can calculate the time that the voltage of the second switch is 9 V as follows.

$$-\frac{36}{7} e^{-\frac{2}{7}t} + 12 = 9$$

$$\Rightarrow \frac{36}{7} e^{-\frac{2}{7}t} = 3$$

$$\Rightarrow e^{-\frac{2}{7}t} = \frac{7}{12}$$

$$\Rightarrow -\frac{2}{7}t = -0.539$$

$$\Rightarrow t = 1.89 \text{ sec} \quad (5)$$

After $t = 1.89$ sec, the second switch is closed, and we have the circuit shown in Fig. 10.18.3.

Calculating the initial voltage of the capacitor: Solving (3) and (5):

$$V_C(1.89^-) = 9\left(1 - e^{-\frac{1.89}{3.5}}\right) = 3.75 \text{ V} \quad (6)$$

Due to voltage continuity of capacitors in lack of a power source with an infinite size (e.g. impulse power source), it is concluded that:

$$V_C(1.89^+) = 3.75 \text{ V}$$

Calculating the final voltage of the capacitor: Applying voltage division rule in the circuit of Fig. 10.18.4:

$$V_C(\infty) = V_+(\infty) - V_-(\infty)$$

$$\Rightarrow V_C(\infty) = \frac{6}{6+2} \times 12 - \frac{2}{2+6} \times 12$$

$$\Rightarrow V_C(\infty) = 6 \text{ V}$$

Calculating the time constant of the circuit: See Fig. 10.18.5-6.

$$R_{Th} = (6||2) + (6||2)$$

$$\Rightarrow R_{Th} = \frac{6 \times 2}{6 + 2} + \frac{6 \times 2}{6 + 2}$$

$$\Rightarrow R_{Th} = 3\ \Omega$$

$$\tau = 3 \times 1 = 3\ \text{sec}$$

Calculating the complete response of the RC circuit:

$$V_C(t) = 6 + (3.75 - 6)e^{-\frac{t - 1.89}{3}}$$

$$\Rightarrow V_C(t) = 6 - 2.25e^{-\frac{t - 1.89}{3}}$$

Choice (2) is the answer.

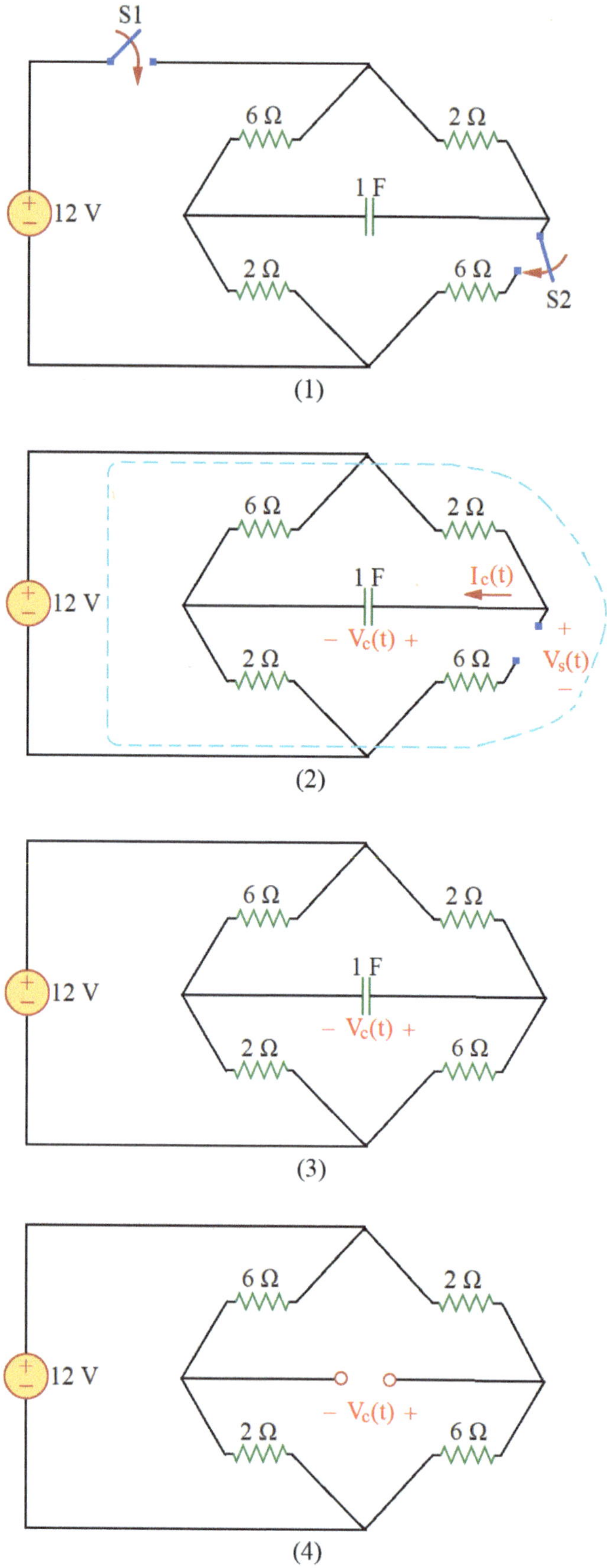

Fig. 10.18 The circuits of problem 10.27

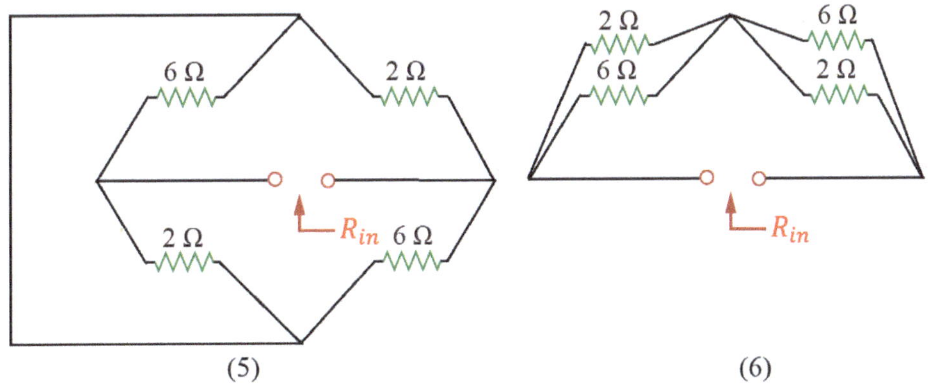

Fig. 10.18 (continued)

Notes

$$e^a = b \Rightarrow a = \ln b$$

10.5 Time Constant of First-Order RL Circuits

10.28. As we know, the time constant of a first-order RL circuit can be calculated as follows.

$$\tau = \frac{L_{eq}}{R_{Th}}$$

Herein, L_{eq} is the equivalent inductor and R_{Th} is the Thevenin resistance seen by the equivalent inductor. From previous chapters, we know that R_{Th} can be calculated as follows.
First, all the independent sources (independent voltage and current sources) are turned off (each independent voltage source is replaced by a short circuit branch and each independent current source is replaced by an open circuit branch). Then, if the circuit does not include any dependent source (dependent voltage or current source), the series-parallel rule is applied to calculate the input resistance of the circuit. However, if the circuit includes at least a dependent source, a test source (independent voltage or current source) with the voltage and current of V_t and I_t is connected to the input terminal of the circuit. Next, the circuit is analyzed by using a circuit analysis method to find the value of $\frac{V_t}{I_t}$. After that, we have:

$$R_{Th} = \frac{V_t}{I_t}$$

In this problem, nodal analysis is applied.
Applying Ohm's law to define i based on the node voltage:

$$i = \frac{V_t}{6} \quad (1)$$

KCL in the node:

$$-I_t - 2i + i + \frac{V_t}{3} = 0$$

$$\Rightarrow -I_t - i + \frac{V_t}{3} = 0 \quad (2)$$

Solving (1) and (2):

$$-I_t - \frac{V_t}{6} + \frac{V_t}{3} = 0$$

$$\Rightarrow -I_t + \frac{V_t}{6} = 0$$

$$\Rightarrow \frac{V_t}{I_t} = 6$$

$$\Rightarrow R_{Th} = 6\ \Omega$$

The time constant of the circuit is calculated as follows.

$$\tau = \frac{0.06}{6}$$

$$\Rightarrow \tau = 0.01 \text{ sec}$$

Choice (4) is the answer (Fig. 10.19)

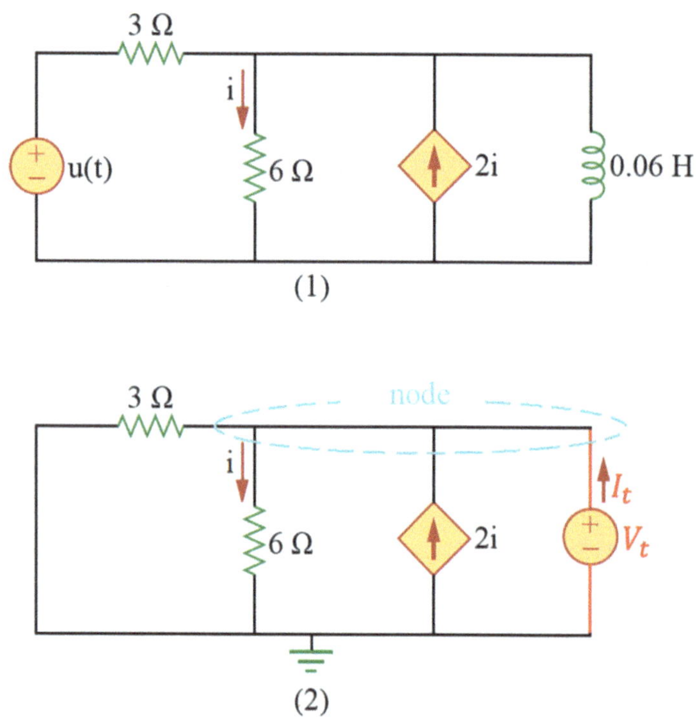

Fig. 10.19 The circuits of problem 10.28

10.6 Source-Free Response of an RL Circuit

10.32. Since there is no independent power source (independent voltage or current source) in the circuit, the type of response of the circuit is source-free.

The general form of source-free response of an RL circuit is as follows.

$$I_L(t) = I_L(0^+)e^{-\frac{t}{\tau}}$$

Herein, $I_L(0^+)$ is the initial current of the inductor and $\tau = \frac{L}{R_{Th}}$ is the time constant of the circuit.

Calculating the initial current of the inductor: Based on the information given in the problem, we have:

$$I_L(0) = 1 \text{ A}$$

Due to current continuity of inductor in lack of a power source with an infinite size (e.g. impulse power source):

$$I_L(0^+) = 1 \text{ A}$$

Calculating the time constant of the circuit: By applying the circuit of Fig. 10.20.2, we have:

$$R_{Th} = 1 + (1+3)||(2+4)$$

$$\Rightarrow R_{Th} = 1 + \frac{4 \times 6}{4+6} = 3.4 \ \Omega$$

The time constant of the RC circuit is as follows.

$$\tau = \frac{1}{3.4} \text{ sec}$$

Calculating the source-free response of the RL circuit:

$$I_L(t) = 1 \times e^{-3.4t}$$

$$\Rightarrow I_L(t) = e^{-3.4t} \text{ A}$$

Choice (1) is the answer.

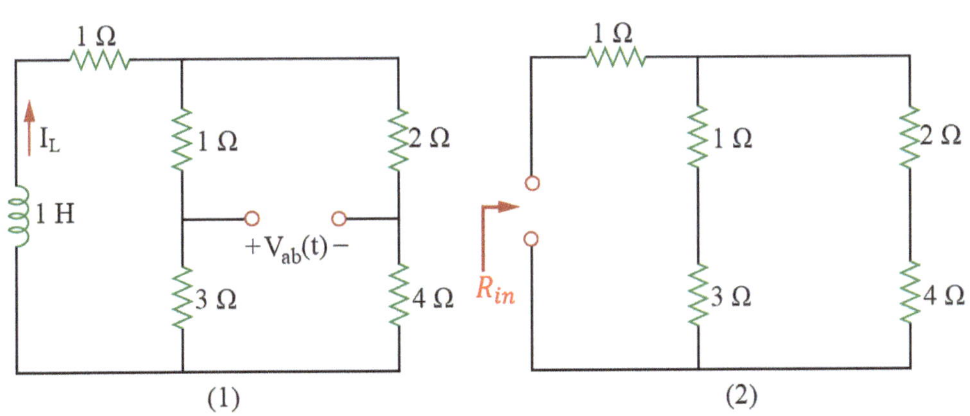

Fig. 10.20 The circuits of problem 10.32

10.33. By using current division rule and Ohm's law in the circuit of Fig. 10.21, we can calculate $V_{ab}(t)$ as follows.

$$I_{1,3} = \frac{2+4}{2+4+1+3} \times e^{-3.4t} = 0.6e^{-3.4t} \text{ A}$$

$$I_{2,4} = \frac{1+3}{2+4+1+3} \times e^{-3.4t} = 0.4e^{-3.4t} \text{ A}$$

$$V_{ab}(t) = V_a(t) - V_b(t)$$

$$\Rightarrow V_{ab}(t) = 3I_{1,3} - 4I_{2,4}$$

$$\Rightarrow V_{ab}(t) = 3 \times 0.6e^{-3.4t} - 4 \times 0.4e^{-3.4t}$$

$$\Rightarrow V_{ab}(t) = 0.2e^{-3.4t}\text{V}$$

Choice (2) is the answer.

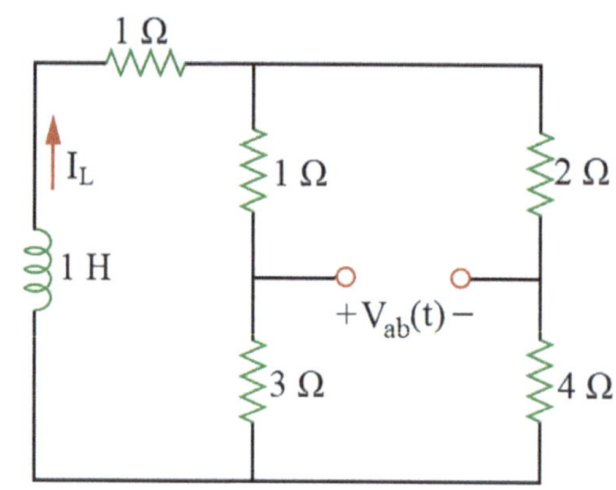

Fig. 10.21 The circuits of problem 10.33

10.35. Based on the information given in the problem, we have:

$$t = 4 \text{ sec}$$

$$R_2 = 0.5 \ \Omega$$

$$R_1 = 0.25 \ \Omega$$

$$I_s = 6 \text{ A}$$

$$L = 0.5 \text{ H}$$

Since the circuit with its power source (independent current source) is disconnected from the inductor, the type of response is source-free.

The general form of source-free response of an RL circuit is as follows.

$$I_L(t) = I_L(0^+)e^{-\frac{t}{\tau}}$$

Herein, $I_L(0^+)$ is the initial current of the inductor and $\tau = \frac{L}{R_{Th}}$ is the time constant of the circuit.

Calculating the initial current of the inductor: Since the switch has been open for a long time, the circuit has reached its steady state condition. Therefore, the inductor is like a short circuit branch before closing the switch, as is shown in Fig. 10.22.2. By using current division formula, we have:

$$I_L(0^-) = \frac{R_1}{R_1 + R_2} \times I_s$$

$$\Rightarrow I_L(0^-) = \frac{0.25}{0.25 + 0.5} \times 6$$

$$\Rightarrow I_L(0^-) = 2 \text{ A}$$

Due to current continuity of inductors in lack of a power source with an infinite size (e.g. impulse power source), it is concluded that:

$$I_L(0^+) = 2 \text{ A}$$

Calculating the time constant of the circuit: After the switching operation, the circuit is like the one shown in Fig. 10.22.3. As can be noticed, the Thevenin resistance seen by the inductor is zero because the inductor is short-circuited by the switch. Thus:

$$\tau = \frac{0.5}{0} = \infty$$

Calculating the source-free response of the RL circuit: The last relation implies that the current of inductor will remain constant forever. This fact can be noticed by using the general form of source-free response of the circuit as follows.

$$I_L(t) = 2e^{-\frac{t}{\infty}}, t > 0$$

$$\Rightarrow I_L(t) = 2A, t > 0$$

Now, by applying KCL in the node, we have:

$$-\frac{R_1}{R_1 + R_2} I_s + I_{sc} + I_L = 0$$

$$\Rightarrow I_{sc} = \frac{R_1}{R_1 + R_2} I_s - I_L$$

$$\Rightarrow I_{sc} = \frac{0.25}{0.25 + 0.5} \times 6 - 2$$

$$\Rightarrow I_{sc} = 0 \text{ A}$$

Choice (2) is the answer.

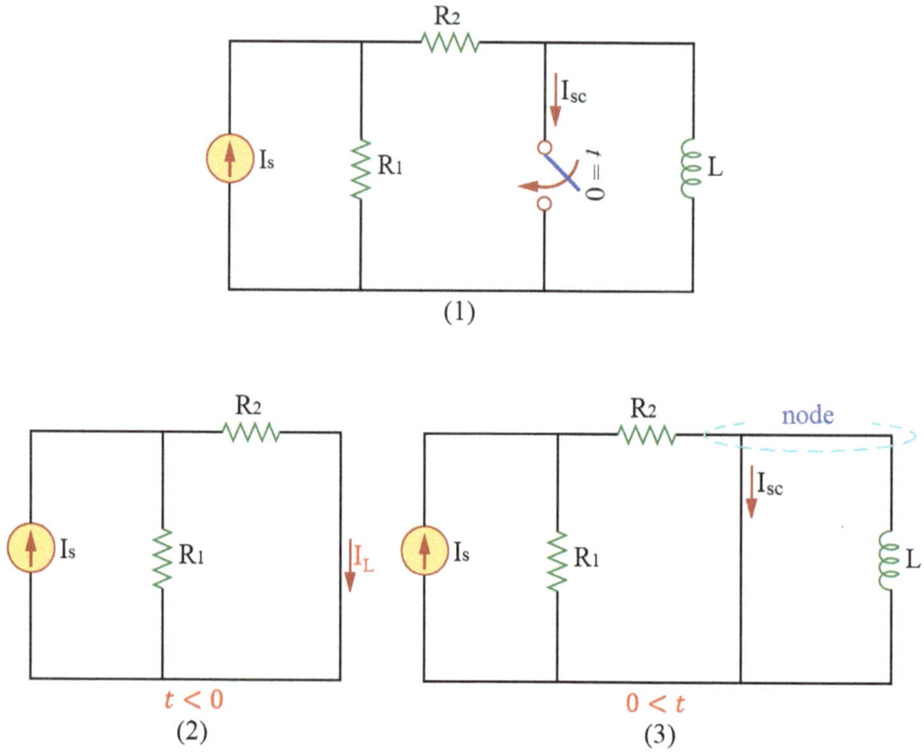

Fig. 10.22 The circuits of problem 10.35

10.36. The circuit seems to be a third-order circuit; however, it is a first-order RL circuit that needs to be simplified, as is shown in Fig. 10.23.2.

$$L_{eq} = 3||6 + 3$$

$$\Rightarrow L_{eq} = \frac{3 \times 6}{3 + 6} + 3$$

$$\Rightarrow L_{eq} = 5 \text{ H}$$

Since the power source (independent voltage source) is disconnected from the circuit, the type of response of the circuit is source-free.

The general form of source-free response of an RL circuit is as follows.

$$I_L(t) = I_L(0^+)e^{-\frac{t}{\tau}}$$

Herein, $I_L(0^+)$ is the initial current of the inductor and $\tau = \frac{L}{R_{Th}}$ is the time constant of the circuit.

Calculating the initial current of the inductor: The circuit for $t = 0^-$ is illustrated in Fig. 10.23.3, where the equivalent inductor has been replaced by a short circuit branch since the circuit is in a steady state condition. By using Ohm's law for the 3-Ω resistor, we can write as follows.

$$I_L(0^-) = \frac{15}{3} = 5 \text{ A}$$

Due to current continuity of inductors in lack of a power source with an infinite size (e.g. impulse power source), it is concluded that:

$$I_L(0^+) = 5\ \text{A}$$

Calculating the time constant of the circuit: The Thevenin resistance of the circuit can be calculated by using the circuit of Fig. 10.23.4 as follows.

$$R_{Th} = 3 + 7 = 10\ \Omega$$

Time constant of the circuit:

$$\tau = \frac{5}{10} = 0.5\ \text{sec}$$

Calculating the source-free response of the RL circuit:

$$I_L(t) = 5e^{-2t}$$

From the circuit of Fig. 10.23.4, it is noticed that:

$$I(t) = 5e^{-2t}\ \text{A}$$

Choice (2) is the answer.

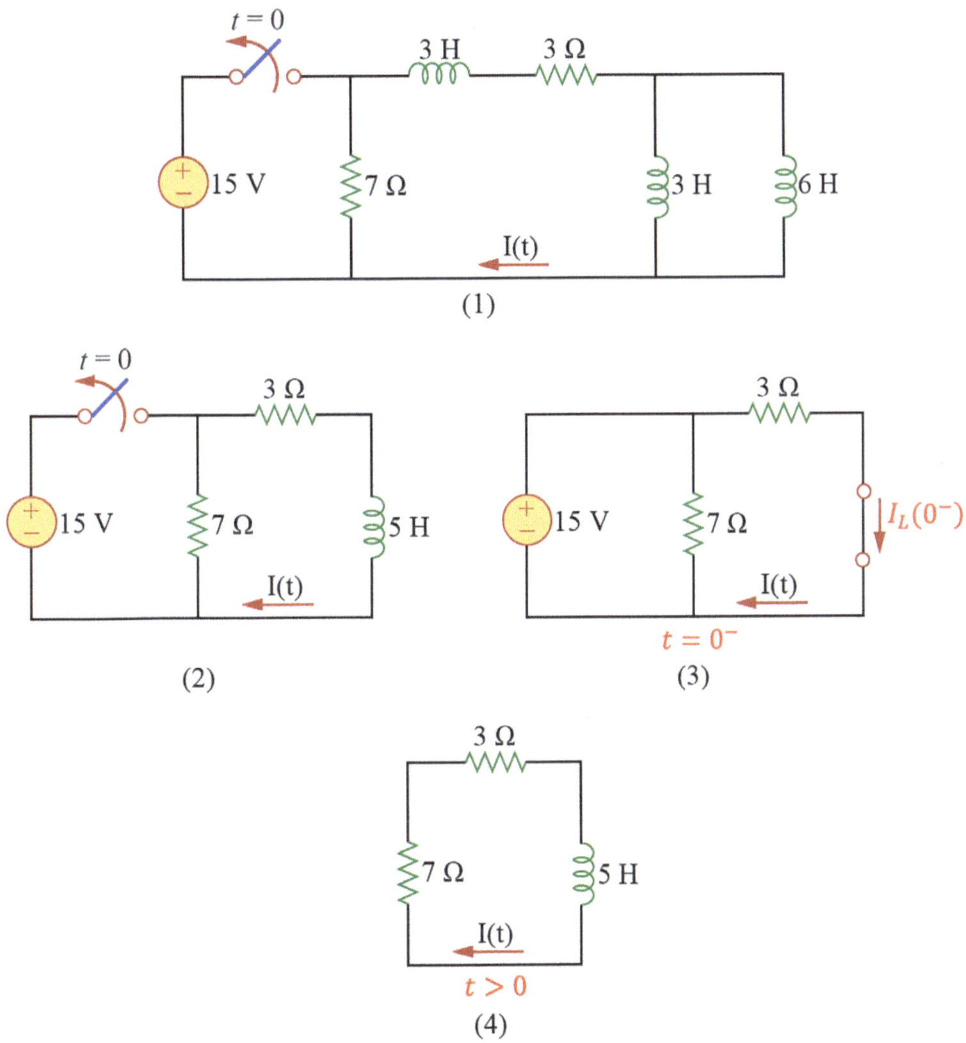

Fig. 10.23 The circuits of problem 10.36

10.37. After the switching operation, two inductors are connected in series. Therefore, the circuit is a first-order circuit although it seems to be a second-order circuit.

The power source (independent voltage source) is disconnected from the circuit; hence, the type of response of the circuit is source-free.
The general form of source-free response of an RL circuit is as follows.

$$I_L(t) = I_L(0^+)e^{-\frac{t}{\tau}}$$

Herein, $I_L(0^+)$ is the initial current of the inductor and $\tau = \frac{L}{R_{Th}}$ is the time constant of the circuit.

Calculating the initial current of the inductor: The switch has been in the closed position for a long time, therefore the circuit is in a steady state condition and the inductors are like short circuit branches, as can be seen in Fig. 10.24.2.
The current of the voltage source:

$$I(0^-) = \frac{9}{1+3||6} = \frac{9}{1+2} = 3 \text{ A}$$

Using current division rule:

$$I_{L1}(0^-) = \frac{6}{6+3} \times 3 = 2 \text{ A}$$

$$I_{L2}(0^-) = \frac{3}{6+3} \times 3 = 1 \text{ A}$$

After opening the switch, the circuit is converted to the one shown in Fig. 10.24.3. As can be seen, the inductors are connected in series, thus the rule of current continuity of inductors is not applicable. Instead, we must use the magnetic flux conservation law for the inductors as follows.

$$\varphi_{L,total}(0^+) = \varphi_{L,total}(0^-)$$

$$\Rightarrow L_{eq}I_L(0^+) = L_1 I_{L1}(0^-) + L_2 I_{L2}(0^-)$$

$$\Rightarrow (1+3) \times I_L(0^+) = 3 \times (-2) + 1 \times 1$$

$$\Rightarrow I_L(0^+) = -\frac{5}{4} \text{ A}$$

Calculating the time constant of the circuit: The circuit of Fig. 10.24.3 can be simplified like the one shown in Fig. 10.24.4. Now, we have a circuit with a single inductor (4 H) with the initial current of $-\frac{5}{4}$ A in series with two resistors. The time constant of the RL circuit can be calculated by using the circuit of Fig. 10.24.4.

$$\tau = \frac{4}{9} \text{ sec}$$

Calculating the source-free response of the RL circuit:

$$I_L(t) = -\frac{5}{4}e^{-\frac{9}{4}t}$$

To calculate the voltage of the 6-Ω resistor, Ohm's law can be used in the circuit of Fig. 10.24.4 as follows.

$$V(t) = 6I_L(t)$$

$$\Rightarrow V(t) = -7.5e^{-\frac{9}{4}t}$$

Choice (4) is the answer.

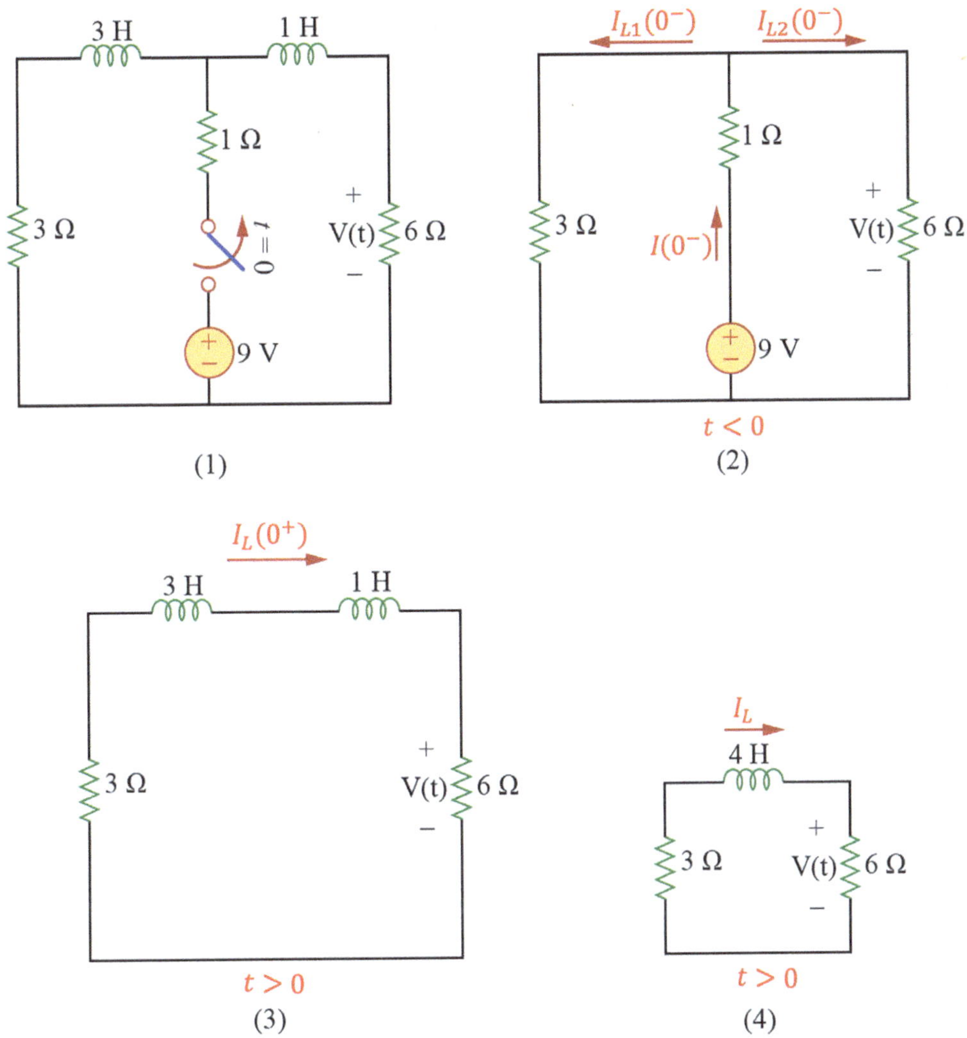

Fig. 10.24 The circuits of problem 10.37

Notes

The current of inductors may not be continuous if they are connected in series during a switching operation at $t = t_0$. Herein, the magnetic flux conservation law can be used to calculate their currents after the switching.

$$\varphi_{L,total}(t_0^+) = \varphi_{L,total}(t_0^-)$$

The voltage of capacitors may not be continuous if they are connected in parallel during a switching operation at $t = t_0$. Herein, the electric charge conservation law can be used to calculate their voltages after the switching.

$$Q_{C,total}\left(t_0^+\right) = Q_{C,total}\left(t_0^-\right)$$

Moreover, the current of inductors and voltage of capacitors may not be continuous if a power source with an infinite size (e.g. impulse power source) exists in the circuit.

10.7 Step Response of an RL Circuit

10.38. The circuit includes more than one power source, thus we can use superposition theorem to solve the problem, as can be seen in Fig. 10.25.2-3.

In the circuit of Fig. 10.25.2, the voltage source with any size is not able to excite the inductor since the inductor does not sense any voltage difference across itself. This type of circuit is called Wheatstone bridge if the condition $R_1 \times R_3 = R_2 \times R_4$ holds in the circuit of Fig. 10.25.4. Therefore:

$$I_{L1}(t) = 0, t > 0$$

Since the initial current of the inductor is zero and the power source (independent current source) is a step function, the response of the circuit is step response.
The general form of step response of an RL circuit is as follows.

$$I_L(t) = I_L(\infty)\left(1 - e^{-\frac{t}{\tau}}\right)$$

Herein, $I_L(\infty)$ is the current of the inductor after a long time and $\tau = \frac{L}{R_{Th}}$ is the time constant of the circuit.

Calculating the final current of the inductor: To calculate the final current of the inductor, we should simplify the circuit of Fig. 10.25.3, as is shown in Fig. 10.25.5-6. By using the circuit of Fig. 10.25.6, applying current division rule, and assuming that the inductor will be short circuit after a long time, we have:

$$I_{L2}(\infty) = \frac{1}{1+1} \times 5 = 2.5 \text{ A}$$

Calculating the time constant of the circuit: The time constant of circuit can be calculated by using the circuit of Fig. 10.25.6 as follows.

$$\tau_2 = \frac{2}{2} = 1 \text{ sec}$$

Calculating the step response of the RL circuit:

$$I_{L2}(t) = 2.5\left(1 - e^{-t}\right)A, t > 0$$

Based on superposition theorem, we have:

$$I_L(t) = I_{L1}(t) + I_{L2}(t)$$

$$\Rightarrow I_L(t) = 0 + 2.5\left(1 - e^{-t}\right) \text{ A}$$

$$\Rightarrow I_L(t) = 2.5\left(1 - e^{-t}\right) A, t > 0$$

Choice (2) is the answer.

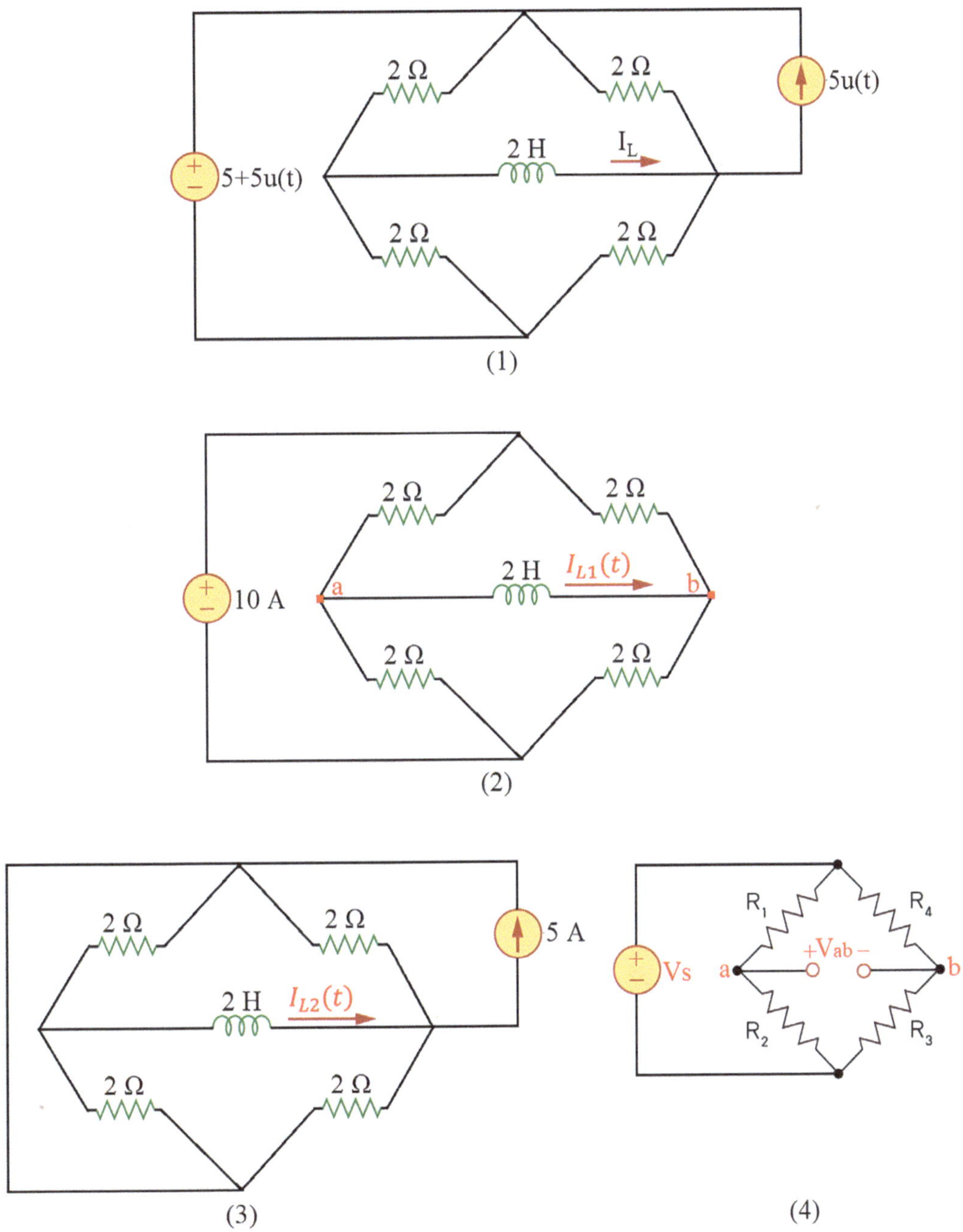

Fig. 10.25 The circuits of problem 10.38

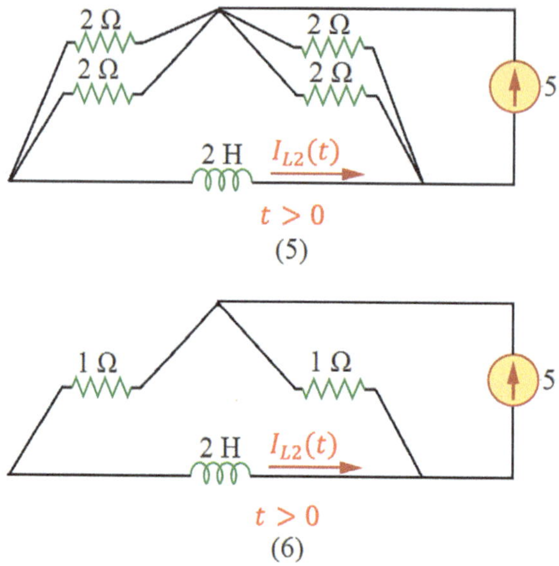

Fig. 10.25 (continued)

10.40. Based on the information given in the problem, we have:

$$V(t) = 2\left(1 - e^{-\frac{t}{2}}\right)u(t) \tag{1}$$

$$C = 1\ \text{F} \tag{2}$$

$$L = 3\ \text{H} \tag{3}$$

Since the network does not have any power source and the power source (independent voltage source) is a step function, the response of the circuit is step response.

Part 1: The general form of step response of an RC circuit is as follows.

$$V_C(t) = V_C(\infty)\left(1 - e^{-\frac{t}{\tau}}\right) \tag{4}$$

Herein, $V_C(\infty)$ is the voltage of the capacitor after a long time and the time constant of the circuit is as follows.

$$\tau = R_{Th}C \tag{5}$$

Calculating the time constant of the circuit: Comparing (1) and (4):

$$\tau = 2\ \text{sec} \tag{6}$$

Solving (2), (5), and (6):

$$R_{Th} \times 1 = 2$$

$$\Rightarrow R_{Th} = 2\ \Omega \tag{7}$$

Calculating the final voltage of the capacitor: Fig. 10.26.2 shows the Thevenin circuit of the network. Since the capacitor will behave like an open circuit after a long time, the final voltage of the capacitor will be equal to the Thevenin voltage of the circuit, as can be noticed from the circuit of Fig. 10.26.2. Therefore, by using (1), we have:

$$\Rightarrow V_{Th} = \lim_{t\to\infty} 2\left(1 - e^{-\frac{t}{2}}\right)u(t)$$

$$\Rightarrow V_{Th} = 2\ \text{V} \tag{8}$$

Part 2: The general form of step response of an RL circuit is as follows.

$$I_L(t) = I_L(\infty)\left(1 - e^{-\frac{t}{\tau}}\right) \tag{9}$$

Herein, $I_L(\infty)$ is the current of the inductor after a long time and the time constant of the circuit is calculated as follows.

$$\tau = \frac{L}{R_{Th}} \tag{10}$$

Calculating the time constant of the circuit: Solving (3), (7) and (10):

$$\tau = \frac{3}{2}\ \text{sec} \tag{11}$$

Calculating the final current of the inductor: The inductor will behave like a short circuit after a long time. Therefore, its final current can be calculated by using Ohm's law as follows (see Fig. 10.26.3).

$$I_L(\infty) = \frac{V_{Th}}{R_{Th}} = \frac{2}{2}$$

$$\Rightarrow I_L(\infty) = 1\ \text{A} \tag{12}$$

Calculating the step response of the RL circuit:

$$I_L(t) = 1 \times \left(1 - e^{-\frac{2}{3}t}\right)$$

$$\Rightarrow I_L(t) = 1 - e^{-\frac{2}{3}t} \tag{13}$$

Using the current-voltage relation of inductors:

$$V(t) = L\frac{d}{dt}I_L(t)$$

$$\Rightarrow V(t) = 3 \times \frac{d}{dt}\left(1 - e^{-\frac{2}{3}t}\right)$$

$$\Rightarrow V(t) = 3 \times \left(\frac{2}{3}\right) e^{-\frac{2}{3}t}$$

$$\Rightarrow V(t) = 2e^{-\frac{2}{3}t} V, t > 0 \quad (14)$$

Since the result is credible for $t > 0$, , unit step function ($u(t)$) can be applied in (14) as follows.

$$V(t) = 2e^{-\frac{2}{3}t} u(t) \text{ V}$$

Choice (2) is the answer.

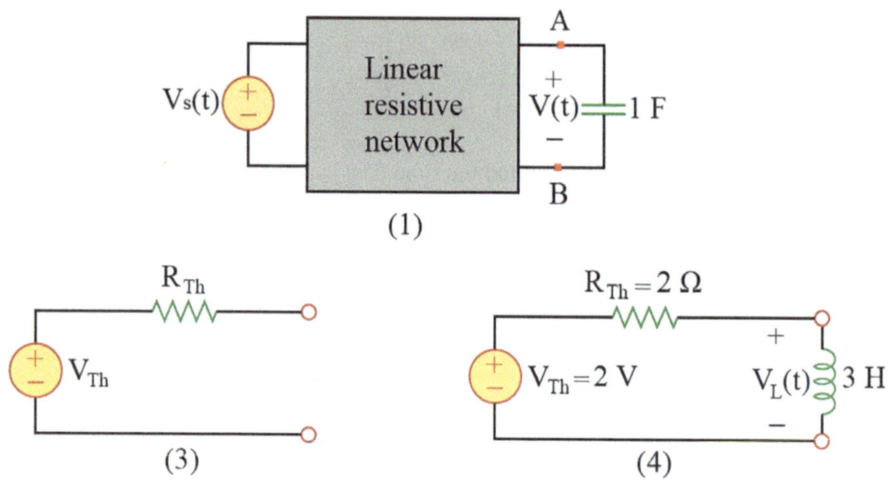

Fig. 10.26 The circuits of problem 10.40

10.8 Heuristic Methods for First-Order Circuits

10.43. Fig. 10.27.2 shows the status of the circuit before the switch is closed. As can be seen, the independent current source is replaced by an open circuit branch because $I = 0$ before closing the switch. Moreover, since the switch has been open for a long time, the capacitor has reached its steady state condition (open circuit). Therefore, the primary voltage of the capacitor is as follows

$$V_c(0^-) = 6 \text{ V}$$

After closing the switch, the configuration of the circuit is updated, as is illustrated in Fig. 10.27.3. Due to voltage continuity of capacitors in lack of a power source with an infinite size (e.g. impulse power source), it is concluded that:

$$V_c(0^+) = 6 \text{ V} \quad (1)$$

Now, by applying KCL in the indicated node of the circuit of Fig. 10.27.3, we have:

$$\frac{6-6}{6} + 5I(0^+) + I_c(0^+) + \frac{6-10}{4} = 0$$

$$\Rightarrow 5I(0^+) + I_c(0^+) - 1 = 0 \quad (2)$$

Defining $I(0^+)$ based on the node voltage:

$$I(0^+) = \frac{6 - 10}{4}$$

$$\Rightarrow I(0^+) = -1 \text{ A} \quad (3)$$

Solving (2) and (3):

$$5(-1) + I_c(0^+) - 1 = 0$$

$$\Rightarrow I_c(0^+) = 6 \text{ A}$$

Choice (2) is the answer.

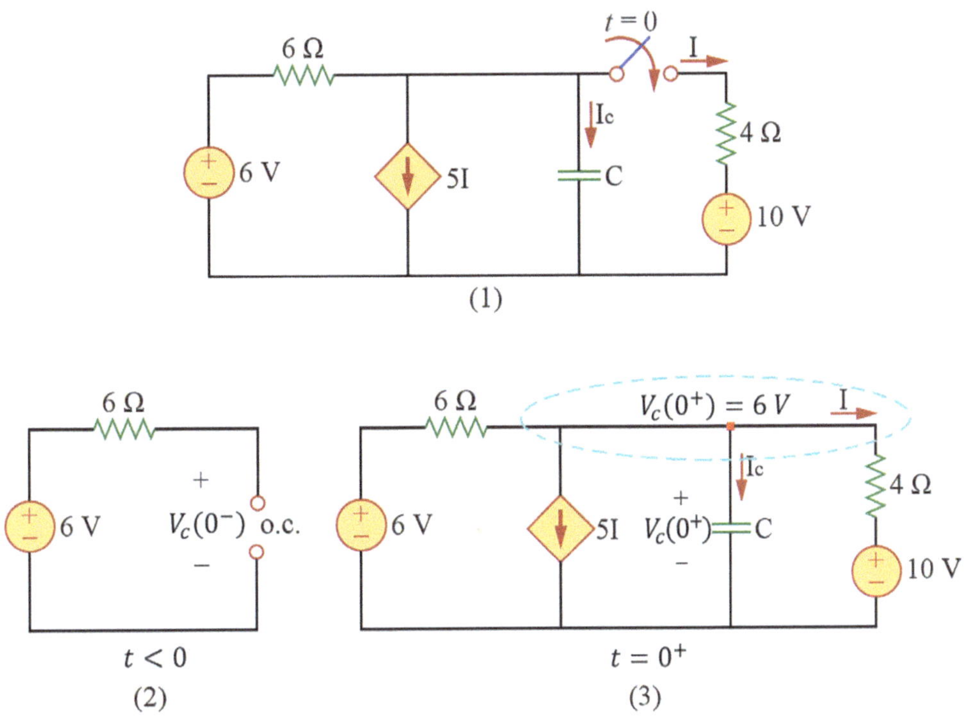

Fig. 10.27 The circuits of problem 10.43

10.44. Fig. 10.28.2 shows the circuit before opening the switch. Since the circuit has reached its steady state condition, the capacitor is like an open circuit branch.

Applying KVL in the right-side mesh:

$$-V_1 - \alpha V_1 + V_C(0^-) = 0$$

$$\Rightarrow V_C(0^-) = (1 + \alpha)V_1 \quad (1)$$

From the circuit, it is seen that:

$$V_1 = E \tag{2}$$

Solving (1) and (2):

$$V_C(0^-) = (1+\alpha)E \tag{3}$$

Due to voltage continuity of capacitors in lack of a power source with an infinite size (e.g. impulse power source), it is concluded that:

$$V_C(0^+) = (1+\alpha)E \tag{4}$$

For $t = 0^+$, we need to use the circuit of Fig. 10.28.3.
KVL in the mesh:

$$-V_1(0^+) - \alpha V_1(0^+) + V_C(0^+) = 0$$

$$\Rightarrow V_1(0^+) = \frac{V_C(0^+)}{1+\alpha} \tag{5}$$

Using Ohm's law for the resistor:

$$I_C(0^+) = -\frac{V_1(0^+)}{2} \tag{6}$$

Using voltage-current relation for the capacitor:

$$I_C(0^+) = \frac{1}{6}\frac{d}{dt}V_C(0^+)$$

$$\Rightarrow \frac{d}{dt}V_C(0^+) = 6I_C(0^+) \tag{7}$$

Solving (4) and (5):

$$V_1(0^+) = E \tag{8}$$

Solving (6) and (8):

$$I_C(0^+) = -\frac{E}{2} \tag{9}$$

Solving (7) and (9):

$$\frac{d}{dt}V_C(0^+) = 6\left(-\frac{E}{2}\right)$$

$$\Rightarrow \left.\frac{dV_C}{dt}\right|_{t=0^+} = -3E$$

Choice (2) is the answer.

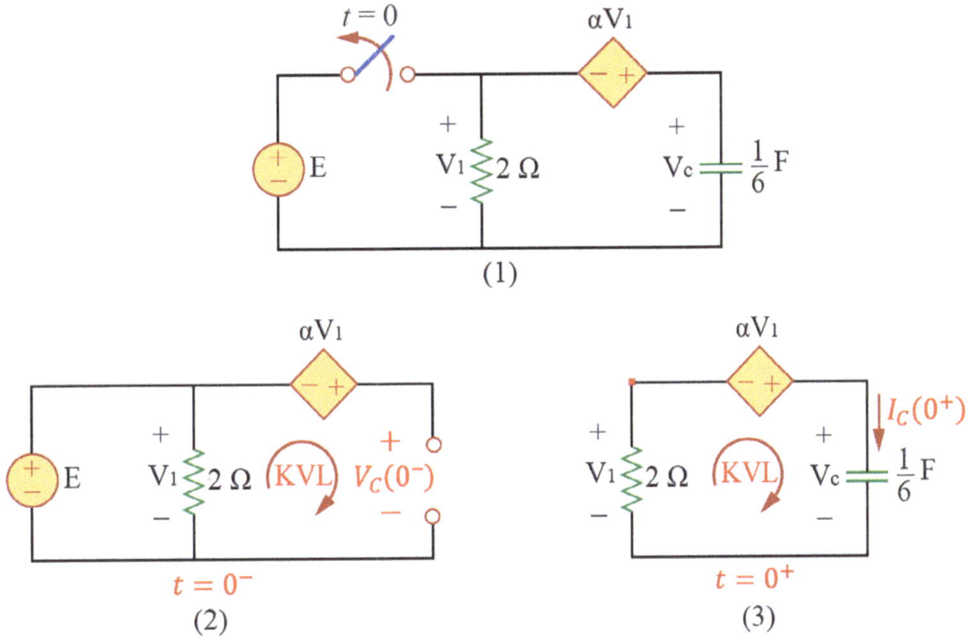

Fig. 10.28 The circuits of problem 10.44

10.9 First-Order Circuits Including Ramp and Singularity Functions

10.46. The circuit seems to be a second-order circuit, but it is the series connection of two separate first-order circuits. In other words, we can individually analyze the first-order circuits.

Using KVL in the indicated mesh in the circuit of Fig. 10.29:

$$-V_o(t) - V_C(t) + V_L(t) = 0$$

$$\Rightarrow V_o(t) = -V_C(t) + V_L(t) \quad (1)$$

Based on the information given in the problem, we have:

$$V_C(0^-) = 1 \text{ V} \quad (2)$$

Due to voltage continuity of capacitors in lack of a power source with an infinite size (e.g. impulse power source), it is concluded that:

$$V_C(0^+) = 1 \text{ V} \quad (3)$$

The final voltage of the capacitor will be zero since the RC circuit is not connected to any power source. In other words:

$$V_C(\infty) = 0 \quad (4)$$

The time constant of the RC circuit:

$$\tau = R_{Th}C = 1 \times 0.5 = 0.5 \text{ sec} \quad (5)$$

The general form of source-free response of an RC circuit is as follows.

$$V_C(t) = V_C(0^+)e^{-\frac{t}{\tau}}$$

Hence:

$$V_C(t) = e^{-2t} \tag{6}$$

Now, by using current-voltage relation of inductors, we have:

$$V_L(t) = L\frac{d}{dt}I_L(t)$$

$$\Rightarrow V_L(t) = 1\frac{d}{dt}r(t)$$

$$\Rightarrow V_L(t) = u(t) \tag{7}$$

In (7), $r(t)$ is ramp function with the following definition:

$$r(t) = \begin{cases} 0 & t < 0 \\ t & t \geq 0 \end{cases} \tag{8}$$

Solving (1), (6), and (7):

$$V_o(t) = -e^{-2t} + u(t)$$

$$\Rightarrow V_o(t) = 1 - e^{-2t}V, t \geq 0$$

Choice (2) is the answer.

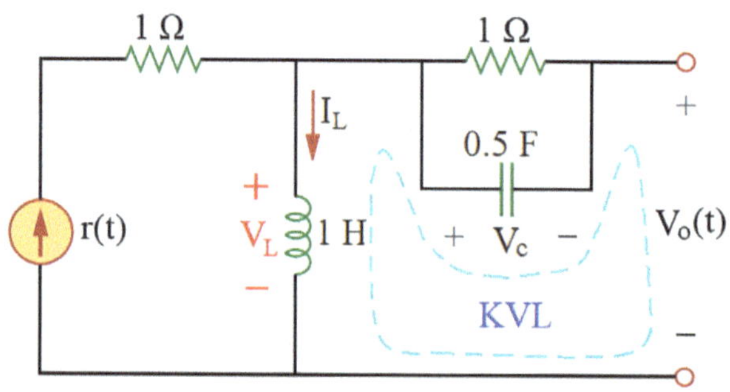

Fig. 10.29 The circuits of problem 10.45

10.47. Based on the information given in the problem, we have:

$$V_C(0^-) = 1 \text{ V} \tag{1}$$

In this problem, it should be noted that due to the presence of impulse voltage source ($3\delta(t)$) in the circuit, the voltage of the capacitor will not be continuous. Thus, to calculate the value of $V_C(0^+)$, we need to use the current-voltage relation of capacitors, as follows:

$$V_C(0^+) = V_C(0^-) + \frac{1}{C}\int_{0^-}^{0^+} I_C(t)dt$$

$$\overset{(1)}{\Rightarrow} V_C(0^+) = 1 + \frac{1}{0.5}\int_{0^-}^{0^+} I_C(t)dt$$

$$\Rightarrow V_C(0^+) = 1 + 2\int_{0^-}^{0^+} I_C(t)dt \quad (2)$$

Fig. 10.30.2 shows the circuit at $t = 0$. As we know, in the presence of an impulse source, a capacitor behaves like a short circuit branch. In addition, the step voltage source is turned off at $t = 0$ based on the definition below.

$$u(-t) = \begin{cases} 1 & t < 0 \\ 0 & t \geq 0 \end{cases} \quad (3)$$

Therefore, by using Ohm's law for the right-side mesh of the circuit of Fig. 10.30.2, we have:

$$I_C(t) = \frac{3\delta(t)}{2} \quad (4)$$

Solving (2) and (4):

$$V_C(0^+) = 1 + 2\int_{0^-}^{0^+} \frac{3\delta(t)}{2}dt$$

$$\Rightarrow V_C(0^+) = 1 + 3\int_{0^-}^{0^+} \delta(t)dt \quad (5)$$

The definition of an impulse function is as follows.

$$\delta(t) = \begin{cases} 0 & t \neq 0 \\ \infty & t = 0 \end{cases}, \text{ where } \int_{-\infty}^{\infty} \delta(t)dt = 1 \quad (6)$$

Solving (5) and (6):

$$V_C(0^+) = 1 + 3 \times 1$$

$$\Rightarrow V_C(0^+) = 4 \text{ V}$$

Choice (4) is the answer.

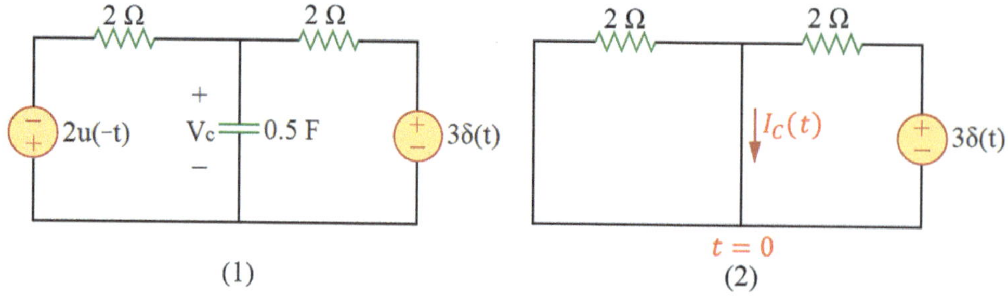

Fig. 10.30 The circuits of problem 10.47

Notes

$$u(t-t_0)=\begin{cases}0 & t<t_0\\ 1 & t\geq t_0\end{cases}$$

$$\delta(t-t_0)=\begin{cases}0 & t\neq t_0\\ \infty & t=t_0\end{cases},\ \text{where}\ \int_{-\infty}^{\infty}\delta(t-t_0)dt=1$$

The current of inductors and voltage of capacitors may not be continuous if a power source with an infinite size (e.g. impulse power source) exists in the circuit.

In the presence of an impulse source, a capacitor behaves like a short circuit branch and an inductor acts like an open circuit branch.

10.10 Differential Equation of First-Order Circuits

10.48. First, we should determine the Thevenin equivalent circuit seen by the capacitor.

Using current division rule and Ohm's law to calculate the open circuit voltage in the circuit of Fig. 10.31.2.

$$V_{Th}=V_{oc}^{+}-V_{oc}^{-}$$

$$\Rightarrow V_{Th}=\frac{I_s}{2}\times 1-\frac{I_s}{2}\times 1$$

$$\Rightarrow V_{Th}=0\ \text{V}$$

Calculating the Thevenin resistance of the circuit seen by the capacitor in the circuit of Fig. 10.31.3.

$$R_{Th}=R_{in}=(1+1)||(1+1)=2||2$$

$$\Rightarrow R_{Th}=1\ \Omega$$

Now, we have the Thevenin equivalent circuit shown in Fig. 10.31.4.

KCL in the indicated node:

$$I_C(t) + I_R(t) = 0$$

$$\Rightarrow 1 \times \frac{d}{dt}V_C(t) + \frac{V_C(t)}{1} = 0$$

$$\Rightarrow \frac{d}{dt}V_C(t) + V_C(t) = 0$$

Choice (3) is the answer.

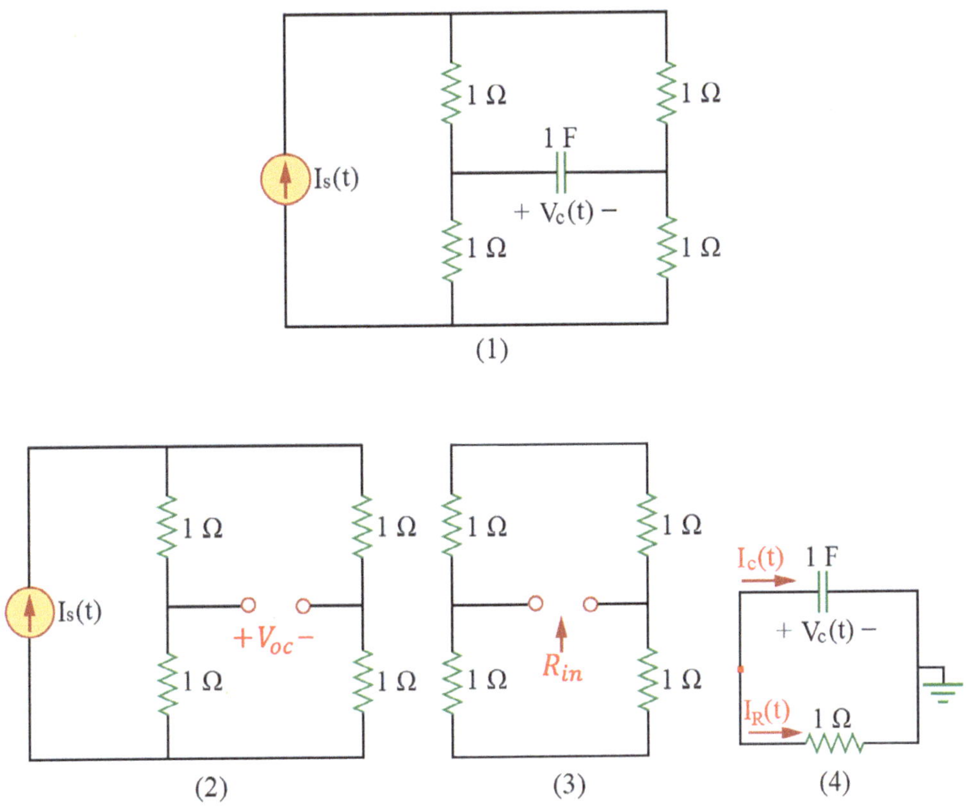

Fig. 10.31 The circuits of problem 10.48

10.49. **Part 1**: To solve the first part of the problem, we need to write the differential equation of the circuit.

From the graph of Fig. 10.32.2, the voltage source can be characterized as follows.

$$V_s(t) = 10\sin(2t), \; t = \left[0, \frac{\pi}{2}\right] \tag{1}$$

Writing a KVL in the mesh of the circuit of Fig. 10.32.3.

$$-V_s(t) + L\frac{d}{dt}I_L(t) + RI_R(t) = 0 \tag{2}$$

From the circuit, it is seen that:

$$I_R(t) = I_L(t) \tag{3}$$

Solving (1)-(3), and the given information in the problem:

$$-10\sin(2t) + 2.5\frac{d}{dt}I_L(t) + 5I_L(t) = 0$$

$$\Rightarrow \frac{d}{dt}I_L(t) + 2I_L(t) = 4\sin(2t) \tag{4}$$

Equation (4) is a first-order differential equation that has the solution below.

$$I_L(t) = \sin(2t) - \cos(2t) + e^{-2t} \tag{5}$$

Equation (5) is valid just for the period of $\left[0, \frac{\pi}{2}\right]$.

Part 2: After $t = \frac{\pi}{2}$ sec, the voltage source is turned off and the circuit is converted to the one shown in Fig. 10.32.4 Thus, the type of response of the circuit is source-free.

The general form of source-free response of an RL circuit is as follows.

$$I_L(t) = I_L(0^+)e^{-\frac{t}{\tau}}$$

Herein, $I_L(0^+)$ is the initial current of the inductor and $\tau = \frac{L}{R_{Th}}$ is the time constant of the circuit.

Calculating the initial current of the inductor: Calculating the primary current of the inductor:

$$I_L\left(t = \frac{\pi}{2}\right) = \sin(\pi) - \cos(\pi) + e^{-\pi}$$

$$\Rightarrow I_L\left(t = \frac{\pi}{2}\right) = 0 - (-1) + 0.0432$$

$$\Rightarrow I_L\left(t = \frac{\pi}{2}\right) = 1.0432 \text{ A}$$

Due to current continuity of inductors in lack of a power source with an infinite size (e.g. impulse power source), it is concluded that:

$$I_L\left(t = \frac{\pi^+}{2}\right) = 1.0432 \text{ A}$$

Calculating the time constant of the circuit: The time constant of the circuit is calculated as follows.

$$\tau = \frac{2.5}{5} = 0.5 \text{ sec}$$

Calculating the source-free response of the RL circuit: For $\frac{\pi}{2} < t < \infty$, we have:

$$I_L(t) = 1.0432e^{-\frac{t-\frac{\pi}{2}}{0.5}}, \quad \frac{\pi}{2} < t < \infty$$

To find $V_o(t)$, Ohm's law can be used for the resistor as follows.

$$V_o(t) = RI_L(t) = 5 \times 1.0432e^{-\frac{t-\frac{\pi}{2}}{0.5}}$$

$$\Rightarrow V_o(t) = 5.21e^{-\frac{t-\frac{\pi}{2}}{0.5}}, \frac{\pi}{2} < t < \infty$$

Calculating the output voltage at $t = 2.2\,sec$:

$$V_o(t = 2.2) = 5.21e^{-\frac{2.2-\frac{\pi}{2}}{0.5}}$$

$$\Rightarrow V_o(t = 2.2) = 1.48 \text{ V}$$

Choice (4) is the answer.

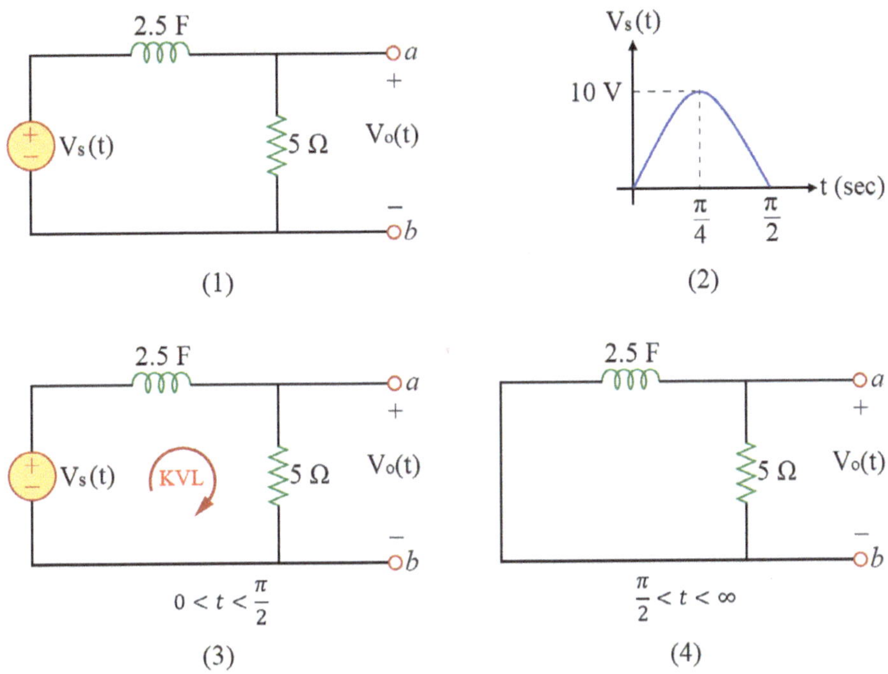

Fig. 10.32 The circuits of problem 10.49

Notes

$$\sin(\pi) = 0$$

$$\cos(\pi) = -1$$

10.11 Energy Stored in First-Order Circuits

10.50. Since the capacitor includes an initial voltage as well as the step power source (independent current source) is available in the circuit, the response of the circuit is complete.

The general form of complete response of an RC circuit is as follows.

$$V_C(t) = V_C(\infty) + (V_C(0^+) - V_C(\infty))e^{-\frac{t}{\tau}}$$

Herein, $V_C(0^+)$ is the initial voltage of the capacitor, $V_C(\infty)$ is the voltage of the capacitor after a long time, and $\tau = R_{Th}C$ is the time constant of the circuit.

Calculating the initial voltage of the capacitor: Based on the information given in the problem, we have:

$$V_C(0^-) = 4 \text{ V}$$

Due to voltage continuity of capacitors in lack of a power source with an infinite size (e.g. impulse power source), it is concluded that:

$$V_C(0^+) = 4 \text{ V} \tag{1}$$

Calculating the time constant of the circuit: The time constant of the circuit is as follows.

$$\tau = 0.5 \times 2 = 1 \text{ sec} \tag{2}$$

Calculating the final voltage of the capacitor: The final voltage of the capacitor can be calculated by using Ohm's law and assuming the fact that the capacitor will be open circuit after a long time. Thus:

$$V_C(\infty) = -0.5 \times 2 = -1\ V \tag{3}$$

Calculating the complete response of the RC circuit:

$$V_C(t) = -1 + (4 - (-1))e^{-\frac{t}{1}}$$

$$\Rightarrow V_C(t) = -1 + 5e^{-t} \tag{4}$$

Now, we need to determine the time that the voltage of the capacitor drops to 1 V. In fact, the equation below needs to be solved.

$$V_C(t_0) = 1 \text{ V} \tag{5}$$

Solving (4) and (5):

$$-1 + 5e^{-t_0} = 1$$

$$\Rightarrow 5e^{-t_0} = 2$$

$$\Rightarrow e^{-t_0} = \frac{2}{5}$$

$$\Rightarrow -t_0 = \ln\left(\frac{2}{5}\right)$$

$$\Rightarrow t_0 = 0.91 \text{ sec}$$

Choice (1) is the answer Fig. 10.33

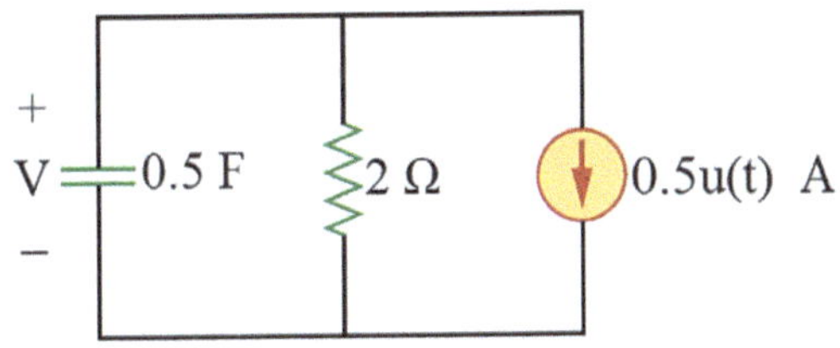

Fig. 10.33 The circuit of problem 10.50

Notes

$$e^a = b \Rightarrow a = \ln b$$

10.51. From Problem 10.50, we know have:

$$t_0 = 0.91 \text{ sec}$$

$$V_C(t) = V_{Source}(t) = -1 + 5e^{-t}$$

To calculate the energy of the independent current source, we need to integrate its power over the interval of $[0, t_0] = [0, 0.91]$, as follows:

$$W = \int_0^{t_0} VI$$

$$\Rightarrow W = \int_0^{0.91} \left(-1 + 5e^{-t}\right) \times \left(\frac{1}{2}\right) dt$$

$$\Rightarrow W = \int_0^{0.91} \left(-\frac{1}{2} + \frac{5}{2}e^{-t}\right) \times dt$$

$$\Rightarrow W = -\frac{t}{2} - \frac{5}{2}e^{-t}\Big|_0^{0.91}$$

$$\Rightarrow W = \left(-\frac{0.91}{2} - \frac{5}{2}e^{-0.91}\right) - \left(-\frac{0}{2} - \frac{5}{2}e^{0}\right)$$

$$\Rightarrow W = (-1.46) - \left(-\frac{5}{2}\right)$$

$$\Rightarrow W = 1.04 \text{ J}$$

Since the value of energy is positive, it is concluded that the independent current source has consumed energy during the period. Choice (4) is the answer.

Notes

$$\int t^n dt = \frac{1}{n+1} t^{n+1}$$

$$\int e^{at} dt = \frac{1}{a} e^{at}$$

When an associated reference direction for the voltage and current of a circuit component is considered, that is, when the current enters the positive terminal of the component:

(a) the component absorbs or consumes power if the value of multiplication of its voltage and current (VI) is positive, and
(b) the component delivers or generates power if the value of multiplication of its voltage and current (VI) is negative.

10.52. The energy of a capacitor will not change if its voltage remains constant, as can be noticed from (1).

$$W_C(t) = \frac{1}{2} C (V_C(t))^2 \tag{1}$$

Based on the information given in the problem, we have:

$$V_C(0^-) = 4 \text{ V} \tag{2}$$

Due to voltage continuity of capacitors in lack of a power source with an infinite size (e.g. impulse power source), it is concluded that:

$$V_C(0^+) = 4 \text{ V} \tag{3}$$

The general form of complete response of an RC circuit is as follows.

$$V_C(t) = V_C(\infty) + (V_C(0^+) - V_C(\infty)) e^{-\frac{t}{\tau}} \tag{4}$$

As can be noticed from (4), if the transient part is eliminated, the voltage of the capacitor will remain constant for $t \geq 0$. Therefore:

$$V_C(\infty) = V_C(0^+) \tag{5}$$

Solving (3) and (5):

$$V_C(\infty) = 4 \text{ V} \tag{6}$$

To calculate the final voltage of the capacitor, we need to assume that the capacitor is an open circuit branch, as can be seen in Fig. 10.34.2. This circuit can be analyzed by using superposition theorem as follows.
Using voltage division rule in Fig. 10.34.3:

$$V_{C1}(\infty) = \frac{4+2}{4+2+R} \times 2$$

$$\Rightarrow V_{C1}(\infty) = \frac{12}{6+R} \tag{5}$$

Using current division rule and Ohm's law in Fig. 10.34.4:

$$V_{C2}(\infty) = \left(\frac{4}{4+2+R} \times 3\right) \times R$$

$$\Rightarrow V_{C2}(\infty) = \frac{12R}{6+R} \tag{6}$$

Therefore:

$$V_C(\infty) = V_{C1}(\infty) + V_{C2}(\infty)$$

$$\Rightarrow V_C(\infty) = \frac{12}{6+R} + \frac{12R}{6+R}$$

$$\Rightarrow V_C(\infty) = \frac{12+12R}{6+R} \tag{7}$$

Solving (6) and (7):

$$\frac{12+12R}{6+R} = 4$$

$$\Rightarrow 12 + 12R = 24 + 4\text{ R}$$

$$\Rightarrow 8R = 12$$

$$\Rightarrow R = \frac{3}{2}\ \Omega$$

Choice (2) is the answer.

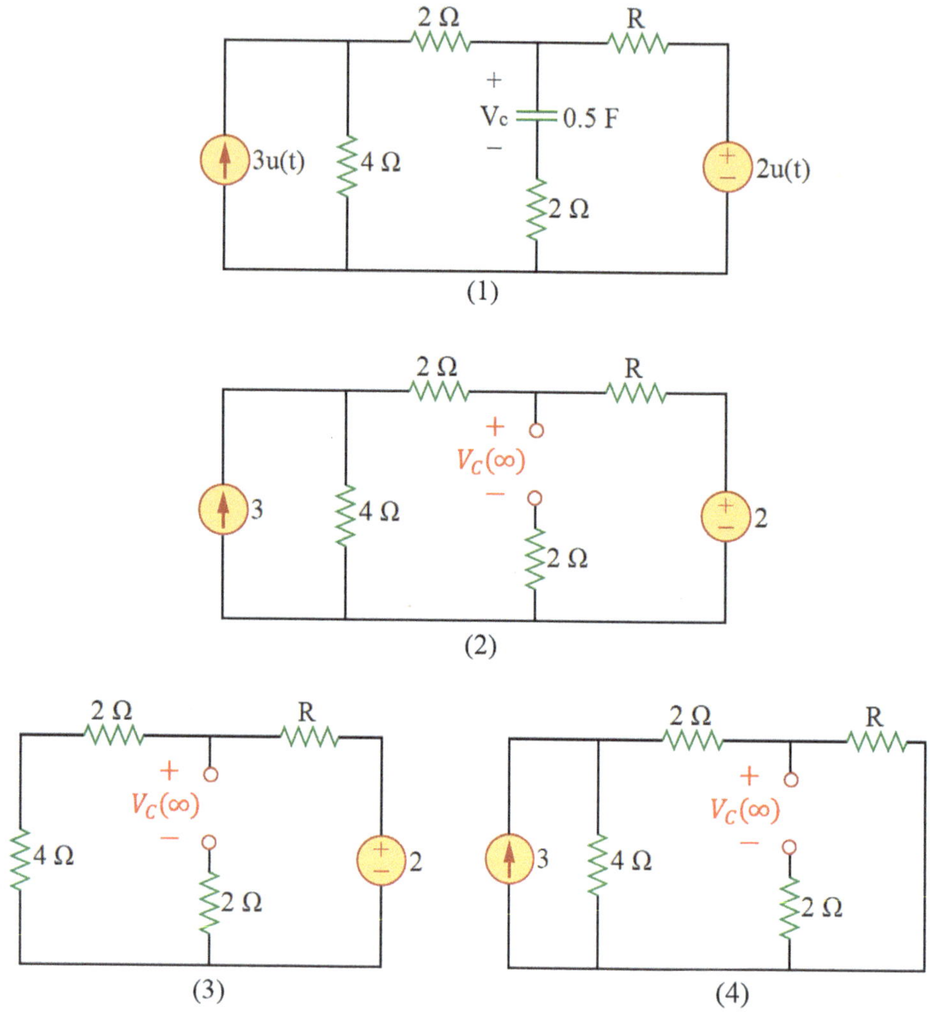

Fig. 10.34 The circuits of problem 10.52

References

1. Rahmani-Andebili, M., DC Electrical Circuit Analysis - Practice Problems, Methods, and Solutions (1st Ed.), Springer Nature, 2020.
2. Rahmani-Andebili, M., General Physics II - Practice Problems, Methods, and Solutions, Sringer Nature, 2025.
3. Rahmani-Andebili, M., General Physics I - Practice Problems, Methods, and Solutions, Springer Nature, 2025.
4. Rahmani-Andebili, M., Mathematics of Engineering and Science - Comprehensive Lessons, Springer Nature, 2024.
5. Rahmani-Andebili, M., Differential Equations – Practice Problems, Methods, and Solutions, Springer Nature, 2022.
6. Rahmani-Andebili, M., Calculus III – Practice Problems, Methods, and Solutions, Springer Nature, 2023.
7. Rahmani-Andebili, M., Calculus II – Practice Problems, Methods, and Solutions, Springer Nature, 2023.
8. Rahmani-Andebili, M., Calculus I (2nd Ed.) – Practice Problems, Methods, and Solutions, Springer Nature, 2023.
9. Rahmani-Andebili, M., Precalculus (2nd Ed.) – Practice Problems, Methods, and Solutions, Springer Nature, 2024.

11 Analysis of Second-Order Circuits: Part A

Abstract

In this chapter, the basic and advanced problems concerned with the analysis of second-order circuits are studied. The subjects include differential equation of second-order circuits, response types of second-order circuits, general response of second-order circuits, analysis of second-order circuits during switching operation, second-order circuits including singularity functions, and energy stored in second-order circuits. Herein, different types of problems and exercises are presented that are categorized as follows.

- ***Problems with detailed solution***: They have been designed to teach students the subjects in detail. Moreover, they have been categorized in different levels based on their difficulty levels (easy, normal, and hard) and calculation amounts (small, normal, and large).
- ***Partially solved exercises***: They have been designed to encourage students to practice problems while guiding them through the problem-solving procedure and hinting the required formulas.
- ***Exercises with final answer***: They have been designed to encourage students to practice more by themselves while hinting them by the final answer as well as to help instructors to give tests or quizzes.

11.1 Differential Equation of Second-Order Circuits

Problem

11.1 Determine the Time-Dependent Deferential Equation of I_R in the Circuit of Fig. 11.1 [1–9].

Difficulty level ○ Easy ● Normal ○ Hard

Calculation amount ○ Small ● Normal ○ Large

1) $L\frac{dI_R(t)}{dt} - RI_R(t) = L\frac{dI_S(t)}{dt} - V_S(t)$

2) $L\frac{dI_R(t)}{dt} + RI_R(t) = L\frac{dI_S(t)}{dt} + V_S(t)$

3) $LC\frac{d^2I_R(t)}{dt^2} + \frac{L}{R}\frac{dI_R(t)}{dt} + I_R(t) = L\frac{dI_S(t)}{dt} + V_S(t)$

4) $LC\frac{d^2I_R(t)}{dt^2} - \frac{L}{R}\frac{dI_R(t)}{dt} + I_R(t) = L\frac{dI_S(t)}{dt} - V_S(t)$

S. Manias, *DC Electrical Circuits Analysis*, https://doi.org/10.1007/978-3-032-00253-2_11

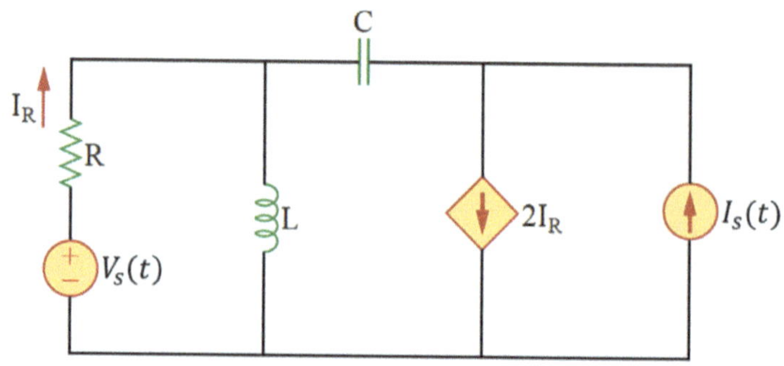

Fig. 11.1 The circuit of problem 11.1

Partially Solved Exercise

11.2. Find the Deferential Equation of $I_L(t)$ in the Circuit of Fig. 11.2.

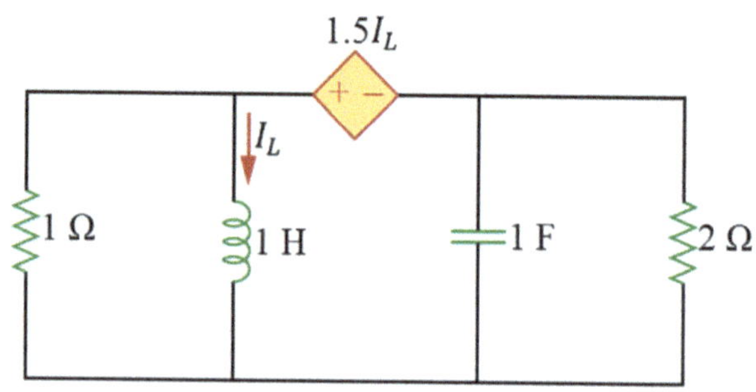

Fig. 11.2 The circuit of partially solved exercise 11.2

Solution

Applying KVL in the indicated mesh of the circuit of Fig. 12.3:

$$-V_L(t) + 1.5I_L(t) + V_C(t) = 0$$

$$\Rightarrow V_C(t) = (\quad) + (\quad)\ (1)$$

Applying KCL in the supernode:

$$I_L(t) + \frac{V_L(t)}{1} + I_C(t) + \frac{V_C(t)}{2} = 0\ (2)$$

Voltage-current relation of capacitors:

$$I_C(t) = C\frac{d}{dt}V_C(t) = (\quad) \times \frac{d}{dt}V_C(t)$$

$$\Rightarrow I_C(t) = \frac{d}{dt}V_C(t)\ (3)$$

Solving (2) and (3):

$$I_L(t) + V_L(t) + \frac{d}{dt}V_C(t) + \frac{V_C(t)}{2} = 0 \ (4)$$

Solving (1) and (4):

$$(\qquad) + (\qquad) + (\qquad) + (\qquad) = 0$$

$$\Rightarrow \frac{1}{4}I_L(t) + \frac{3}{2}V_L(t) + \frac{d}{dt}V_L(t) - \frac{3}{2}\frac{d}{dt}I_L(t) = 0 \ (5)$$

Current–voltage relation of inductors:

$$V_L(t) = L\frac{d}{dt}I_L(t) = (\qquad) \times \frac{d}{dt}I_L(t)$$

$$\Rightarrow V_L(t) = \frac{d}{dt}I_L(t) \ (6)$$

Solving (5) and (6):

$$(\qquad) + (\qquad) + (\qquad) + (\qquad) = 0$$

$$\Rightarrow \frac{d^2}{dt^2}I_L(t) + \frac{1}{4}I_L(t) = 0$$

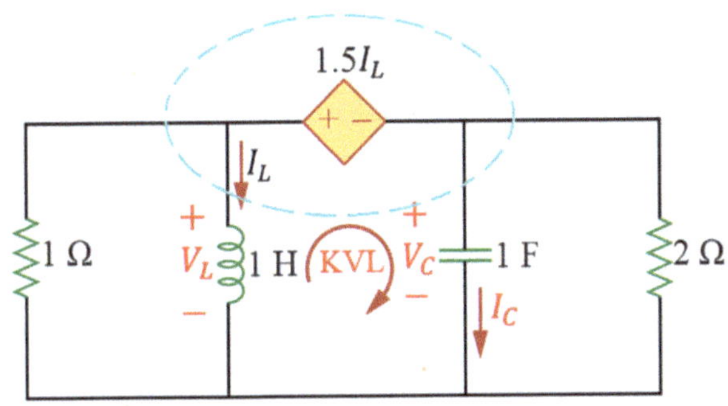

Fig. 11.3 The circuit of partially solved exercise 11.2

Exercise

11.3. Find the Deferential Equation of $I_L(t)$ in the Circuit of Fig. 11.4.

Final Answer

$$\frac{d^2}{dt^2}I_L(t) + 3\frac{d}{dt}I_L(t) + 2I_L(t) = 0$$

Note: *Read the "educational support" part of the preface of the textbook.*

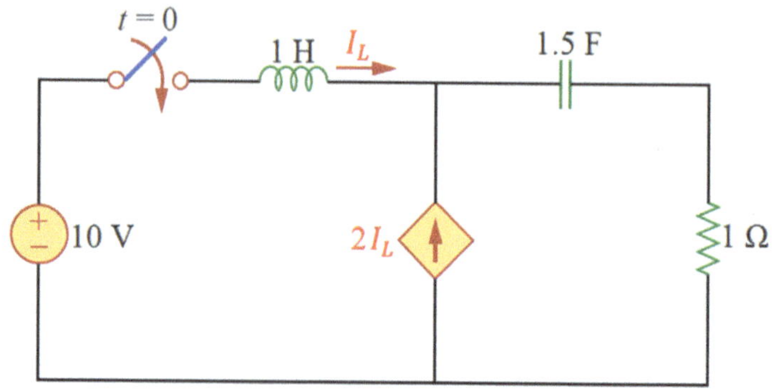

Fig. 11.4 The circuit of exercise 11.3

11.2 Response Types of Second-Order Circuits

Problem

11.4. The characteristic equation of a second-order circuit is as follows.

$$s^2 - 1 = 0$$

Identify the response type of the circuit.

Difficulty level ● Easy ○ Normal ○ Hard
Calculation amount ● Small ○ Normal ○ Large

1) Overdamped
2) Critically damped
3) Underdamped
4) Undamped (oscillating)
5) Unstable

Problem

11.5. The differential equation of a second-order circuit is as follows.

$$\frac{d^2}{dt^2}V_C(t) + 5\frac{d}{dt}V_C(t) + 6V_C(t) = 0$$

Identify the response type of the circuit.

Difficulty level ● Easy ○ Normal ○ Hard
Calculation amount ● Small ○ Normal ○ Large

1) Overdamped
2) Critically damped
3) Underdamped
4) Undamped (oscillating)
5) Unstable

Problem

11.6. The differential equation of a second-order circuit is as follows.

$$\frac{d^2}{dt^2}V_C(t) + 2\frac{d}{dt}V_C(t) + V_C(t) = 0$$

Identify the response type of the circuit.

Difficulty level ● Easy ○ Normal ○ Hard
Calculation amount ● Small ○ Normal ○ Large

1) Overdamped
2) Critically damped
3) Underdamped
4) Undamped (oscillating)
5) Unstable

Problem

11.7. The differential equation of a second-order circuit is as follows.

$$\frac{d^2}{dt^2}I_L(t) + 2\frac{d}{dt}I_L(t) + 2I_L(t) = 0$$

Identify the response type of the circuit.

Difficulty level ● Easy ○ Normal ○ Hard
Calculation amount ● Small ○ Normal ○ Large

1) Overdamped
2) Critically damped
3) Underdamped
4) Undamped (oscillating)
5) Unstable

Problem

11.8. In Problem 11.2, Identify the Response Type of the Second-Order Circuit for $I_L(t)$.

Difficulty level ● Easy ○ Normal ○ Hard
Calculation amount ● Small ○ Normal ○ Large

1) Overdamped
2) Critically damped
3) Underdamped
4) Undamped (oscillating)
5) Unstable

Exercise

11.9. The differential equation of a second-order circuit is as follows.

$$\frac{d^2}{dt^2}v(t) - 2\frac{d}{dt}v(t) + 2v(t) = 0$$

Identify the response type of the circuit.

Final Answer

Unstable

Note: Read the "educational support" part of the preface of the textbook.

Exercise

11.10. The characteristic equation of a second-order circuit is as follows.

$$s^2 - 2s + 1 = 0$$

Identify the response type of the circuit.

Final Answer

Unstable

Note: Read the "educational support" part of the preface of the textbook.

Exercise

11.11. The differential equation of a second-order circuit is as follows.

$$\frac{d^2}{dt^2}i(t) - 3\frac{d}{dt}i(t) + 2i(t) = 0$$

Identify the response type of the circuit.

Final Answer

Unstable

Note: Read the "educational support" part of the preface of the textbook.

Problem

11.12. In the circuit of Fig. 11.5, the switch is closed at $t = 0$. Determine the size of the inductor so that a critically damped response for $I_L(t)$ for $t > 0$ is achieved.

Difficulty level ○ Easy ○ Normal ● Hard
Calculation amount ○ Small ● Normal ○ Large

1) 0.08 H
2) 0.04 H
3) 0.16 H
4) 0.32 H

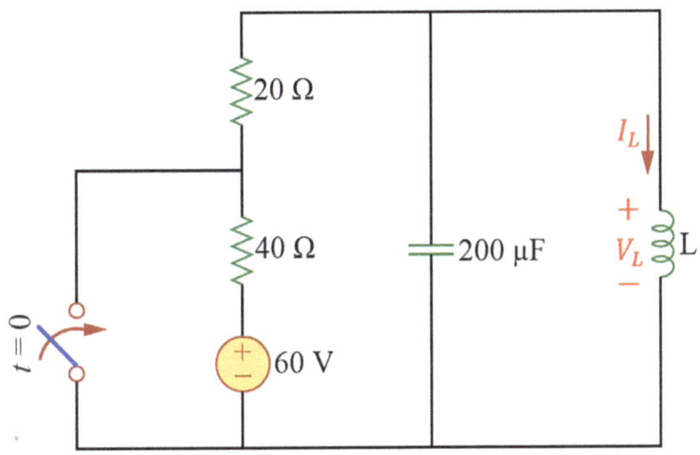

Fig. 11.5 The circuit of problem 11.12

Problem

11.13. In Fig. 11.6, calculate the value of k to set the response of the circuit ($I(t)$) in a critically damped condition.

Difficulty level ○ Easy ○ Normal ● Hard
Calculation amount ○ Small ○ Normal ● Large

1) 1
2) 2
3) 3
4) 4

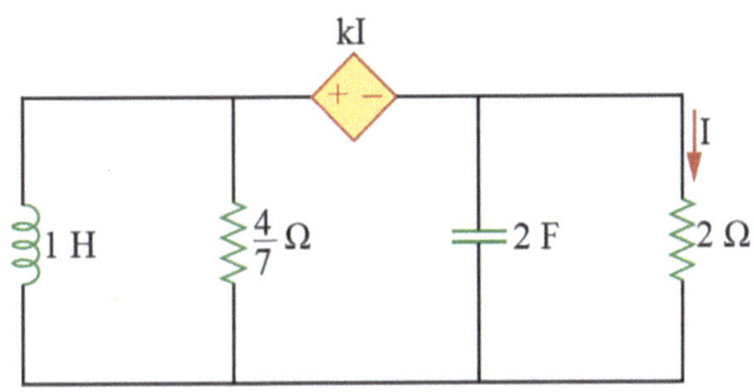

Fig. 11.6 The circuit of problem 11.13

Partially Solved Exercise

11.14. In the circuit of Fig. 11.7, calculate the value of R so that a critically damped response is achieved for the circuit after closing the switch at $t = 0$.

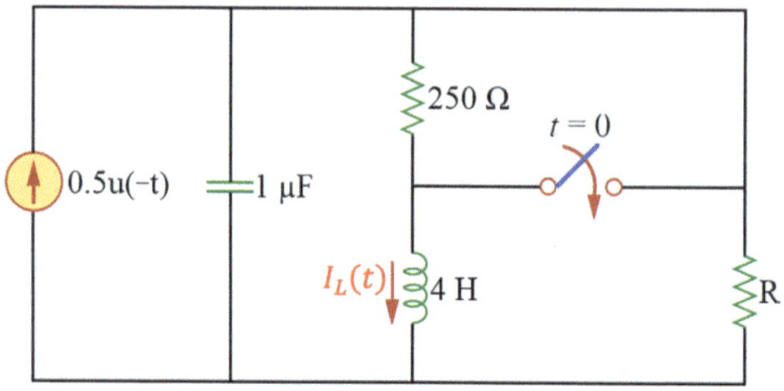

Fig. 11.7 The circuit of partially solved exercise 11.14

Solution

Step 1 (writing the differential equation of the circuit)

From the circuit of Fig. 17.8 which is for $t > 0$, it is seen that:

$$V_C(t) = V_L(t) \ (1)$$

Voltage-current relation of capacitors:

$$I_C(t) = C\frac{d}{dt}V_C(t) = (\quad) \times \frac{d}{dt}V_C(t)$$

$$\overset{(1)}{\Rightarrow} I_C(t) = 10^{-6}\frac{d}{dt}V_L(t) \ (2)$$

Current–voltage relation of inductors:

$$V_L(t) = L\frac{d}{dt}I_L(t)$$

$$\Rightarrow V_L(t) = (\quad) \times \frac{d}{dt}I_L(t) \ (3)$$

KCL in the node of the circuit:

$$I_C(t) + I_L(t) + \frac{1}{R}V_L(t) = 0 \ (4)$$

Applying (2) in (4):

$$(\quad) + (\quad) + (\quad) = 0$$

$$\overset{(3)}{\Rightarrow} 10^{-6}\frac{d}{dt}\left(4\frac{d}{dt}I_L(t)\right) + I_L(t) + \frac{1}{R}\left(4\frac{d}{dt}I_L(t)\right) = 0$$

$$\Rightarrow (\quad\quad)\frac{d^2}{dt^2}I_L(t) + (\quad\quad)I_L(t) + (\quad\quad)\frac{d}{dt}I_L(t) = 0$$

$$\Rightarrow \frac{d^2}{dt^2}I_L(t) + \frac{10^6}{R}\frac{d}{dt}I_L(t) + \frac{10^6}{4}I_L(t) = 0$$

Step 2 (writing the characteristic equation of the differential equation)

$$s^2 + (\quad\quad)s + (\quad\quad) = 0$$

Step 3

Based on the information given in the problem, the response type of the circuit is critically damped. Therefore, the discriminant of the characteristic equation of the circuit must be zero based on Table 11.1. In other words:

$$\Delta = 0$$

$$\Rightarrow (\quad\quad)^2 - 4(\quad\quad)(\quad\quad) = 0$$

$$\Rightarrow (\quad\quad)^2 = (\quad\quad)$$

$$\Rightarrow (\quad\quad) = \pm(\quad\quad)$$

$$\Rightarrow R = (\quad\quad)$$

Herein, just the positive value is acceptable. Hence:

$$R = 1000\ \Omega$$

Table 11.1 Determining the discriminant of the characteristic equation of the differential equation based on the response type of the circuit

Response type	Descrimination
Overdamped	$\Delta > 0$
Critically damped	$\Delta = 0$
Underdamped	$\Delta < 0$
Undamped (oscilating)	$\Delta < 0$
Unstable	Any

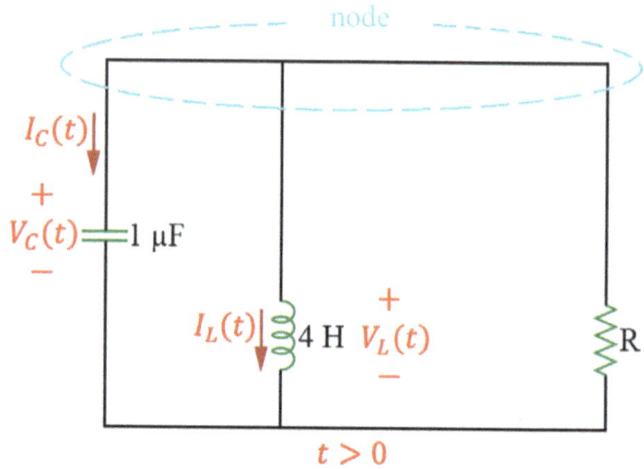

Fig. 11.8 The circuit of partially solved exercise 11.14

11.3 General Response of Second-Order Circuits

Problem

11.15. In Problem 11.14, calculate the value of $I_L(t)$ for $t > 0$.

Difficulty level ○ Easy ○ Normal ● Hard
Calculation amount ○ Small ○ Normal ● Large

1) $(0.4 + 2.25t)e^{-500t}u(t)$ A
2) $(0.4 + 0.225t)e^{-200t}u(t)$ A
3) $(0.4 + 225t)e^{-1000t}u(t)$ A
4) $(0.4 + 225t)e^{-500t}u(t)$ A

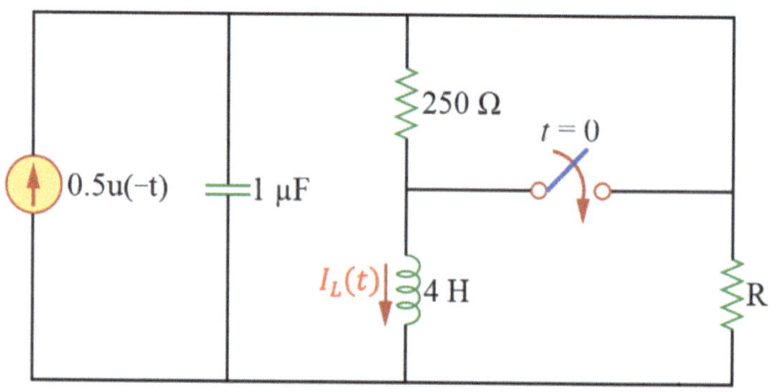

Fig. 11.9 The circuit of problem 11.15

Partially Solved Exercise

11.16. In the circuit of Fig. 11.10, the switch is closed at $t = 0$. Calculate the time-dependent equation of I_L considering $I_L(0^-) = 0$ and $V_C(0^-) = 0$.

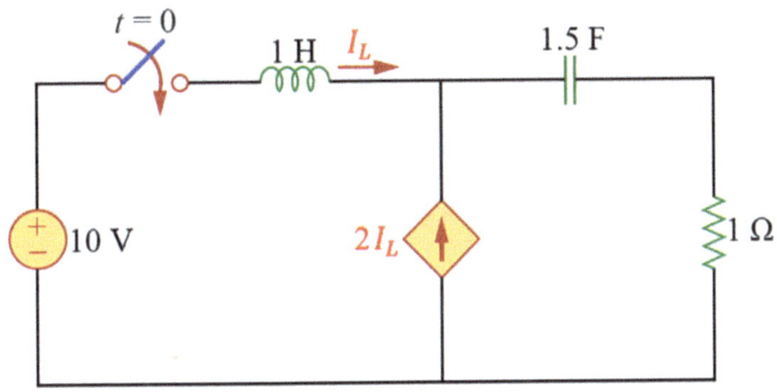

Fig. 11.10 The circuit of partially solved exercise 11.16

Solution

Based on the information given in the problem, we have:

$$I_L(0^-) = 0$$

$$V_C(0^-) = 0$$

Due to current continuity of inductors and voltage continuity of capacitors in lack of a power source with an infinite size (e.g. impulse power source), it is concluded that:

$$I_L(0^+) = (\quad) \text{ A } (1)$$

$$V_C(0^+) = (\quad) \text{ V } (2)$$

Step 1 (writing the differential equation of the circuit)

The equation of I_L has been requested in the problem. Thus, the differential equation must be written based on this variable. Applying KCL in the indicated node of the circuit of Fig. 11.11:

$$(\quad) + (\quad) + (\quad) = 0$$

$$\Rightarrow I_C(t) = 3I_L(t) \ (3)$$

Applying KVL in the indicated loop:

$$(\quad) + (\quad) + (\quad) + (\quad) = 0 \ (4)$$

$$\overset{\frac{d}{dt}}{\Rightarrow} \frac{d}{dt}V_L(t) + \frac{d}{dt}V_C(t) + \frac{d}{dt}I_C(t) = 0 \ (5)$$

Current–voltage relation of inductors:

$$V_L(t) = L\frac{d}{dt}I_L(t) = (\quad) \times \frac{d}{dt}I_L(t)$$

$$\Rightarrow V_L(t) = \frac{d}{dt}I_L(t) \ (6)$$

Voltage-current relation of capacitors:

$$I_C(t) = C\frac{d}{dt}V_C(t) = (\quad) \times \frac{d}{dt}V_C(t)$$

$$\Rightarrow \frac{d}{dt}V_C(t) = \frac{2}{3}I_C(t) \ (7)$$

Solving (3) and (5)-(7):

$$\frac{d^2}{dt^2}I_L(t) + 3\frac{d}{dt}I_L(t) + 2I_L(t) = 0 \ (8)$$

Step 2 (writing the characteristic equation of the differential equation and finding its roots)

$$(\quad)s^2 + (\quad)s + (\quad) = 0$$

$$\Rightarrow s_{1,2} = (\quad), (\quad) \ (9)$$

Step 3 (determining the general form of the solution of differential equation)

Based on the roots of the characteristic equation of the differential equation, the general form of the solution of differential equation is as follows (see Table 11.2).

$$I_L(t) = Ae^{-t} + Be^{-2t} \ (10)$$

Step 4 (calculating the value of parameters of the general form of the solution of differential equation)

Applying (1) in (10):

$$Ae^{-0} + Be^{-2\times 0} = 0$$

$$\Rightarrow A + B = 0 \ (11)$$

Solving (3) and (4) for $t = 0^+$:

$$(\quad) + (\quad) + (\quad) + (\quad) = 0$$

$$\overset{(1)(2)}{\Rightarrow} -10 + V_L(0^+) + 0 + 3 \times 0 = 0$$

$$\Rightarrow V_L(0^+) = 10\,\text{V} \ (12)$$

Solving (6) and (12) for $t = 0^+$:

$$\frac{d}{dt}I_L(0^+) = 10A/sec \ (13)$$

Calculating the first derivate of (10):

$$\frac{d}{dt}I_L(t) = (\quad) + (\quad) \ (14)$$

Solving (13) and (14):

$$10 = -Ae^{-0} - 2Be^{-2\times 0}$$

$$\Rightarrow 10 = -A - 2B \ (15)$$

Solving (11) and (15):

$$A = (\quad), B = (\quad) \ (16)$$

Step 5 (determining the solution of differential equation)

Solving (10) and (16):

$$I_L(t) = \left(10e^{-t} - 10e^{-2t}\right)u(t) \text{ A}$$

Table 11.2 Determining the form of general response of the circuit based on the features of the roots of the characteristic equation

Features of roots of characteristics equation			Roots	General response
The real part of the root is negative	The roots are real	The roots are distict	S_1, S_2	$y(t) = Ae^{s_1 t} + Be^{s_2 t}$
		The roots are the same	S, S	$y(t) = (A + Bt)e^{st}$
	The roots are imaginary	The roots are complex Conjugate of each other	$\alpha \pm j\beta$	$y(t) = e^{\alpha t}(A\cos(\beta t) + B\sin(\beta t))$
The real part of the root is Zero	The roots are Pure imaginary		$\pm j\beta$	$y(t) = A\cos(\beta t) + B\sin(\beta t)$

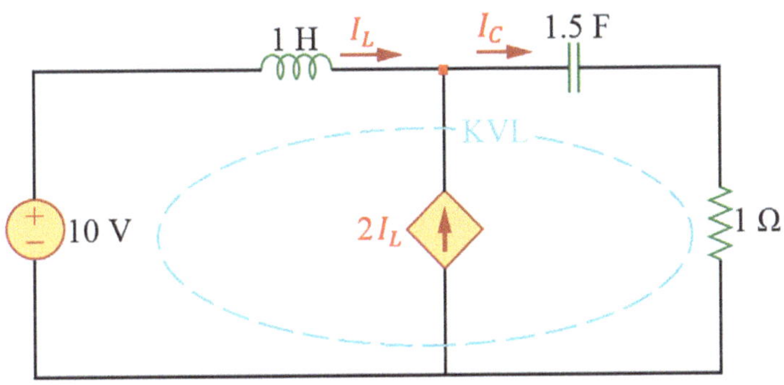

Fig. 11.11 The circuit of partially solved exercise 11.16

Problem

11.17. In the circuit of Fig. 11.12, if $I_L(0^-) = 2\ A$ and $V_C(0^-) = 4\ V$, calculate $I_L(t)$ for $t > 0$.

Difficulty level ○ Easy ○ Normal ● Hard
Calculation amount ○ Small ○ Normal ● Large

1) $-4e^{-t} + 6e^{-2t}$
2) $-4e^{-t} - 6e^{-2t}$
3) $4e^{-t} + 6e^{-2t}$
4) $4e^{-t} - 6e^{-2t}$

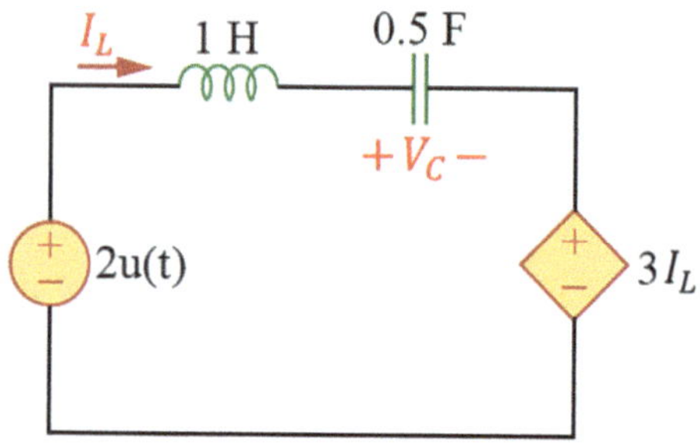

Fig. 11.12 The circuit of problem 11.17

Problem

11.18. In Problem 11.17, calculate $V_C(t)$ for $t > 0$.

Difficulty level ○ Easy ● Normal ○ Hard
Calculation amount ○ Small ● Normal ○ Large

1) $2 + 6e^{-t} + 4e^{-2t}$
2) $2 + 8e^{-t} - 6e^{-2t}$
3) $2 - 6e^{-t} + 6e^{-2t}$
4) $2 - 6e^{-t} + 8e^{-2t}$

Problem

11.19. In the circuit of Fig. 11.13, the switch has been open for a long time. Calculate the time-dependent equation of the current of the switch ($I(t)$) if it is closed at $t = 0$.

Difficulty level ○ Easy ○ Normal ● Hard
Calculation amount ○ Small ○ Normal ● Large

1) $I(t) = 10 - 2e^{-\frac{t}{4}} - 4\sin(2t)$ A
2) $I(t) = 8 + 2e^{-4t} + 4\sin(2t)$ A
3) $I(t) = 10 - 2e^{-4t} - 4\sin(2t)$ A
4) $I(t) = 10 - 2e^{-4t} + 4\sin(2t)$ A

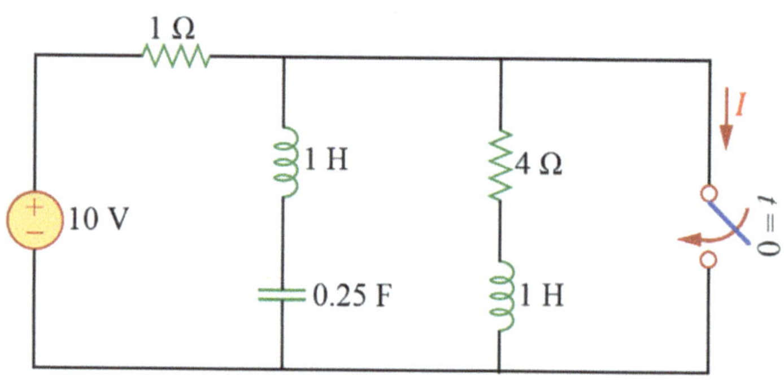

Fig. 11.13 The circuit of problem 11.19

Problem

11.20. In the circuit of Fig. 11.14, calculate the complete response of $V_C(t)$ for the unit step input while assuming $V_C(0^-) = 1V$ and $I_L(0^-) = 2A$.

Difficulty level ○ Easy ○ Normal ● Hard

Calculation amount ○ Small ○ Normal ● Large

1) $1+\frac{1}{3}\left(e^{-2t}-e^{-0.5t}\right)$ V
2) $1+3\left(e^{-2t}-e^{-0.5t}\right)$ V
3) $1+\frac{1}{3}\left(e^{-0.5t}-e^{-2t}\right)$ V
4) $1+3\left(e^{-0.5t}-e^{-2t}\right)$ V

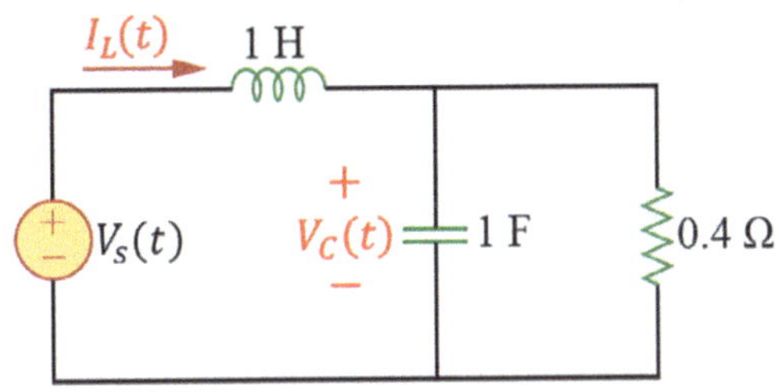

Fig. 11.14 The circuit of problem 11.20

11.4 Quality Factor of Second-Order Circuits

Problem

11.21. Calculate the quality factor (Q) of the circuit of Fig. 11.15.

Difficulty level ○ Easy ○ Normal ● Hard

Calculation amount ○ Small ○ Normal ● Large

1) $\frac{2\sqrt{3}}{19}$
2) $\frac{\sqrt{3}}{19}$
3) $\frac{24}{19}$
4) $\frac{12}{19}$

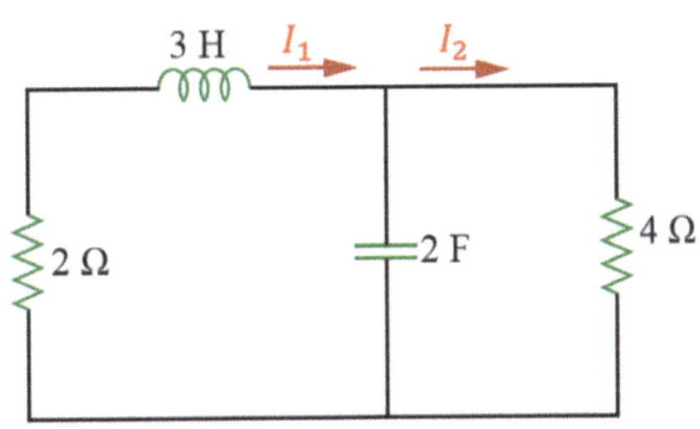

Fig. 11.15 The circuit of problem 11.21

Exercise

11.22. Calculate the quality factor (Q) of a circuit with the following characteristic equation.

$$s^2 + s + 4 = 0$$

Final Answer

$Q = 2$

Note: Read the "educational support" part of the preface of the textbook.

Exercise

11.23. Calculate the quality factor (Q) of a circuit with the following characteristic equation.

$$2s^2 + s + 18 = 0$$

Final Answer

$Q = 6$

Note: Read the "educational support" part of the preface of the textbook.

Problem

11.24. In the circuit of Fig. 11.16, all the resistors are 1 Ω and the circuit goes to infinity from each side. Calculate the size of the inductor so that the quality factor of circuit ($Q_.$) is 1.2.

Difficulty level ○ Easy ○ Normal

Calculation amount ○ Small ○ Normal

1) 0.5 H
2) 1 H
3) 1.2 H
4) 1.96 H

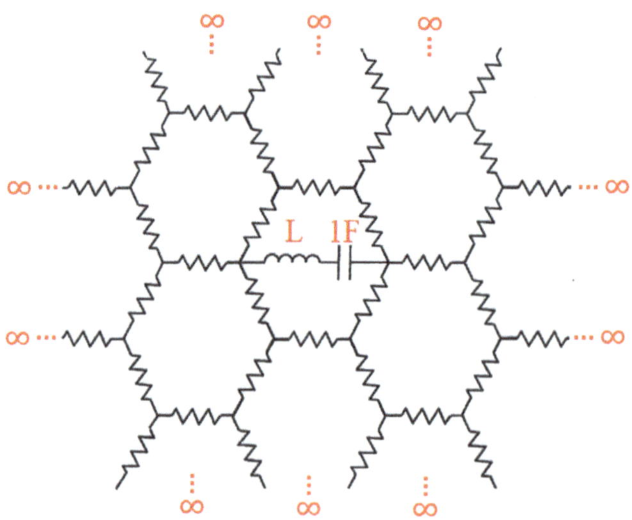

Fig. 11.16 The circuit of problem 11.24

Exercise

11.25. In the circuit of Fig. 11.17, the capacitance of the capacitor is 1 F, the inductance of the inductor is 1 H, the resistance of each resistor is 1 Ω, and the circuit goes to infinity from each side. Calculate the quality factor (Q) of the circuit.

Final Answer

$$Q = \frac{14}{12}$$

Note: *Read the "educational support" part of the preface of the textbook.*

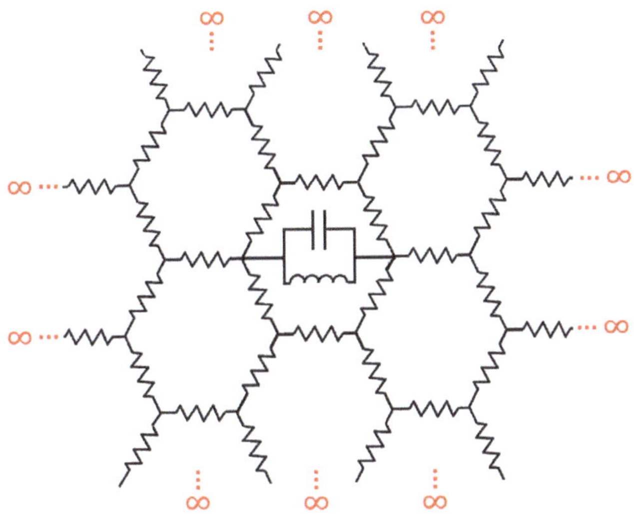

Fig. 11.17 The circuit of exercise 11.25

Exercise

11.26. Calculate the quality factor (Q) of a series RLC circuit and a parallel RLC circuit with the following parameters.

$$R = 0.5\Omega, L = 1H, C = 0.25\ F$$

Final Answer

$Q_1 = 4,\ Q_2 = 0.25$

Note: *Read the "educational support" part of the preface of the textbook.*

11.5 Analysis of Second-Order Circuits During Switching Operation

Problem

11.27. In the circuit of Fig. 11.18, we have:

$$I_L(0^+) = 0$$

$$V_C(0^+) = 0$$

$$\frac{dV_C(0^+)}{dt} = 10\ V/sec$$

Calculate the capacity of the capacitor.

Difficulty level ○ Easy ○ Normal ● Hard

Calculation amount ● Small ○ Normal ○ Large

1) 0.25 F
2) 0.5 F
3) 1 F
4) 2 F

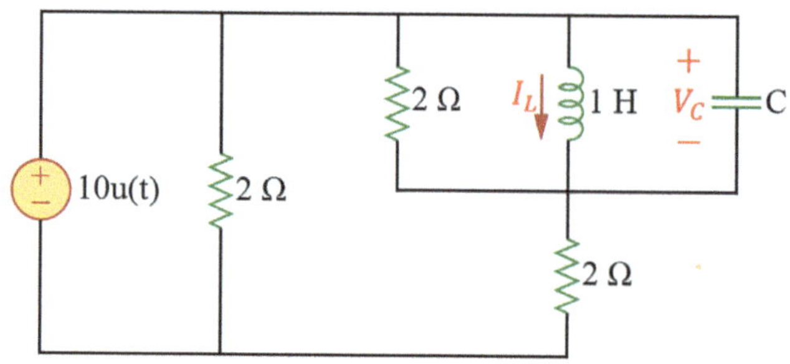

Fig. 11.18 The circuit of problem 11.27

Problem

11.28. In the circuit of Fig. 11.19, the switch has been in the “A” position for a long time, and then it is turned to the “B” position at $t = 0$. Calculate the value of $V_L(0^+)$.

Difficulty level ○ Easy ● Normal ○ Hard

Calculation amount ● Small ○ Normal ○ Large

1) 0 V
2) − 32 V
3) − 80 V
4) ∞

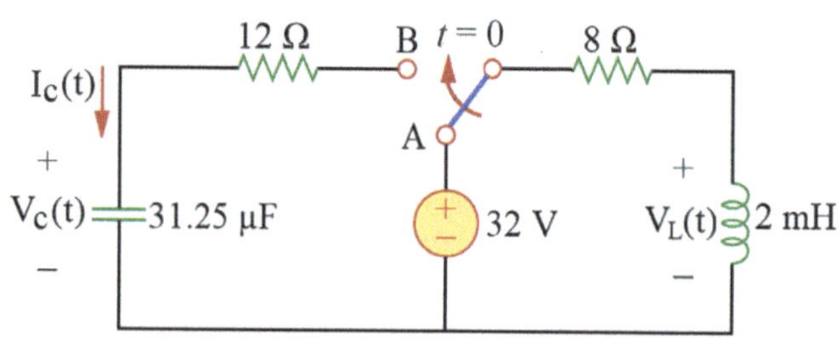

Fig. 11.19 The circuit of problem 11.28

Problem

11.29. Calculate the value of $I_C(0^+)$ and $V_L(0^+)$ in the circuit of Fig. 11.20 for the given waveform of $V_s(t)$.

Difficulty level ○ Easy ○ Normal ● Hard
Calculation amount ○ Small ● Normal ○ Large

1) $I_C(0^+) = 1$ A, $V_L(0^+) = 24$ V
2) $I_C(0^+) = 4$ A, $V_L(0^+) = 24$ V
3) $I_C(0^+) = 1$ A, $V_L(0^+) = 12$ V
4) $I_C(0^+) = 4$ A, $V_L(0^+) = 12$ V

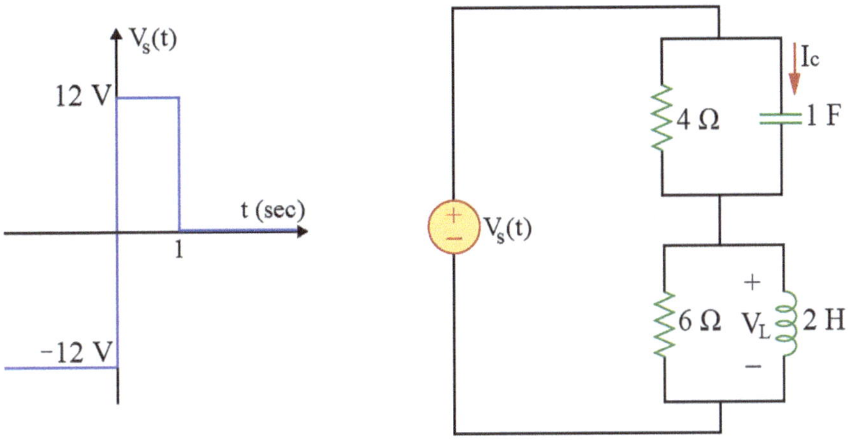

Fig. 11.20 The circuit of problem 11.29

Partially Solved Exercise

11.30. The circuit of Fig. 11.21 has reached its steady state condition before $t = 0$. Calculate the value of the following parameters.

$$V(0^+)$$

$$\frac{dI_L(0^+)}{dt}$$

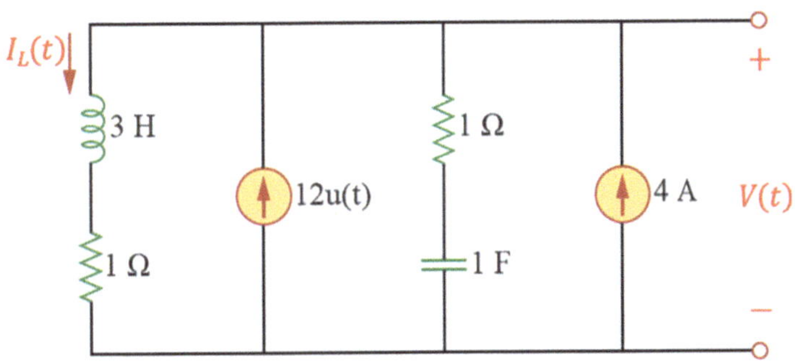

Fig. 11.21 The circuit of partially solved exercise 11.30

Solution

As can be seen in Fig. 11.22.2 that shows the circuit for $t = 0^-$, the whole current of the current source passes through the inductor. Therefore:

$$I_L(0^-) = 4 \text{ A } (1)$$

KVL in the left-side mesh:

$$-V_C(0^-) + 1 \times 4 = 0$$

$$\Rightarrow V_C(0^-) = 4 \text{ V } (2)$$

Due to voltage continuity of capacitors and current continuity of inductors in lack of power source with an infinite size (e.g. impulse power source), it is concluded that:

$$I_L(0^+) = (\quad) \text{ A } (3)$$

$$V_C(0^+) = (\quad) \text{ V } (4)$$

Fig. 11.22.3 shows the circuit for $t = 0^+$, where the 4-V voltage source and the 4-A current source are applied to model the primary voltage of the capacitor and the initial current of the inductor, respectively
KCL in the indicated node:

$$(\quad) + (\quad) + (\quad) + (\quad) = 0$$

$$\Rightarrow I_C(0^+) = 12 \text{ } (5)$$

KVL in loop 1:

$$(\quad) + (\quad) + (\quad) = 0$$

$$\stackrel{(5)}{\Rightarrow} -V(t) + 12 \times 1 + 4 = 0$$

$$\Rightarrow V(t) = 16 \text{ V } (6)$$

KVL in loop 2:

$$(\quad) + (\quad) + (\quad) = 0$$

$$\stackrel{(3)(6)}{\Rightarrow} -16 + V_L(0^+) + 4 \times 1 = 0$$

$$\Rightarrow V_L(0^+) = 12 \text{ V } (7)$$

Using current–voltage relation of inductors:

$$V_L(0^+) = L\frac{d}{dt}I_L(0^+) = (\quad)\frac{d}{dt}I_L(0^+)$$

$$\Rightarrow \frac{d}{dt}I_L(0^+) = \frac{V_L(0^+)}{(\quad)}$$

$$\stackrel{(7)}{\Rightarrow} \frac{d}{dt}I_L(0^+) = \frac{(\quad)}{(\quad)}$$

$$\Rightarrow \frac{d}{dt}I_L(0^+) = 4 \text{ A/sec}$$

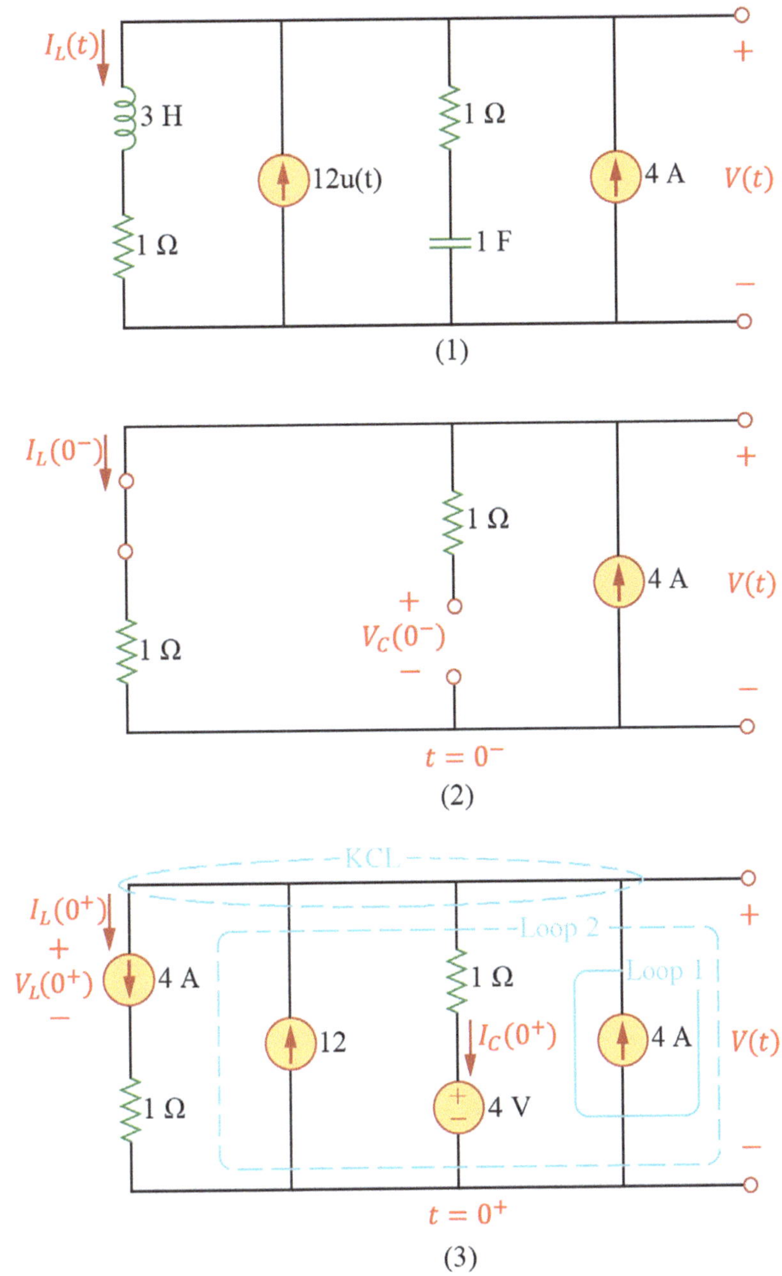

Fig. 11.22 The circuits of partially solved exercise 11.30

Problem

11.31. The circuit of Fig. 11.23 is in a steady state condition before $t = 0$. Calculate the value of the parameters below.

$$V_1(0^+)$$

$$\frac{dV_1(0^+)}{dt}$$

Difficulty level ○ Easy ○ Normal ● Hard
Calculation amount ○ Small ● Normal ○ Large

1) 0 V, 3 V/sec
2) 0 V, −3 V/sec
3) 3 V, 3 V/sec
4) 3 V, −3 V/sec

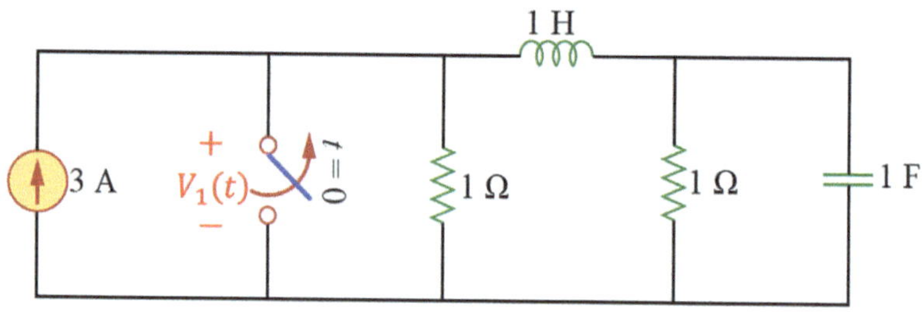

Fig. 11.23 The circuit of problem 11.31

Problem

11.32. In the circuit of Fig. 11.24, $I_L(0) = 5A$, $V_C(0) = -6V$. Calculate the value of $\frac{dI_L(0^+)}{dt}$.

Difficulty level ○ Easy ○ Normal ● Hard
Calculation amount ○ Small ● Normal ○ Large

1) $\frac{1}{3}$ A/sec
2) $\frac{1}{2}$ A/sec
3) $\frac{1}{4}$ A/sec
4) $\frac{1}{5}$ A/sec

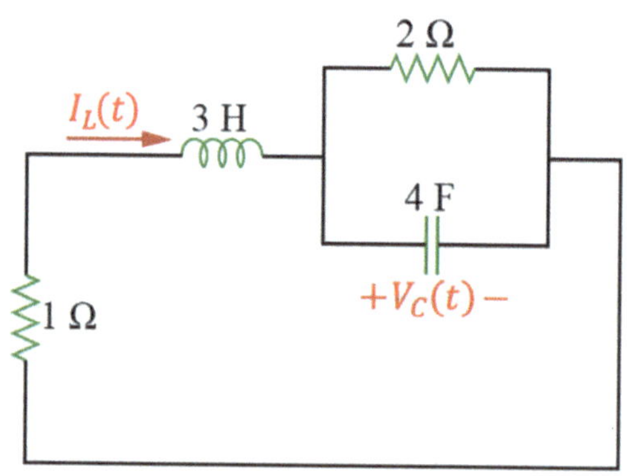

Fig. 11.24 The circuit of problem 11.32

Partially Solved Exercise

11.33. In the circuit of Fig. 11.25, calculate the value of $\frac{dV_C(0^+)}{dt}$.

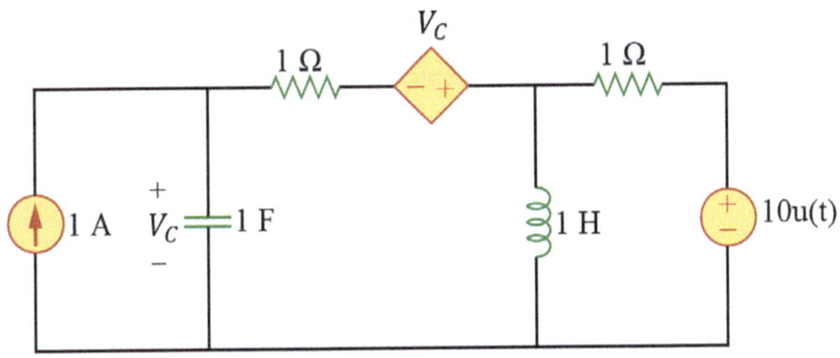

Fig. 11.25 The circuit of partially solved exercise 11.33

Solution

First, we need to analyze the circuit for $t = 0^-$, as is shown in Fig. 11.26.1. Since the circuit has reached its steady state condition, the capacitor is open circuit and the inductor is short circuit.
Using KVL in the mesh of the circuit of Fig. 11.26.1:

$$(\quad) + (\quad) + (\quad) = 0$$

$$\Rightarrow V_C(0^-) = \frac{1}{2}\text{ V } (1)$$

As can be seen in the circuit, the whole current of the current source flows through the inductor.

$$I_L(0^-) = 1\text{ A } (2)$$

Due to voltage continuity of capacitors and current continuity of inductors in lack of power source with an infinite size (e.g. impulse power source), it is concluded that:

$$V_C(0^+) = (\quad) V\ (3)$$

$$I_L(0^+) = (\quad) A\ (4)$$

Fig. 11.26.2 illustrates the circuit for $t = 0^+$, where the primary voltage of the capacitor and the primary current of the inductor are modeled by the voltage and current sources, respectively.
Now, by applying KVL in the indicated mesh, we have:

$$-V_C(0^+) + (1 - I_C(0^+)) \times 1 - V_C(0^+) + V_L(0^+) = 0$$

$$\Rightarrow -2V_C(0^+) + 1 - I_C(0^+) + V_L(0^+) = 0$$

$$\overset{(3)}{\Rightarrow} -2\left(\frac{1}{2}\right) + 1 - I_C(0^+) + V_L(0^+) = 0$$

$$\Rightarrow V_L(0^+) = I_C(0^+)\ (5)$$

KCL in the supernode:

$$(\quad)+(\quad)+(\quad)+(\quad)=0$$

$$\Rightarrow I_C(0^+)+V_L(0^+)-10=0\ (6)$$

Solving (5) and (6):

$$(\quad)+(\quad)+(\quad)=0$$

$$\Rightarrow I_C(0^+)=5\ \text{V}\ (7)$$

Voltage-current relation of capacitors:

$$I_C(t)=C\frac{d}{dt}V_C(t)=(\quad)\times\frac{d}{dt}V_C(t)$$

$$\Rightarrow\frac{d}{dt}V_C(t)=I_C(0^+)$$

$$\overset{(7)}{\Rightarrow}\frac{d}{dt}V_C(t)=5\ \text{V/sec}$$

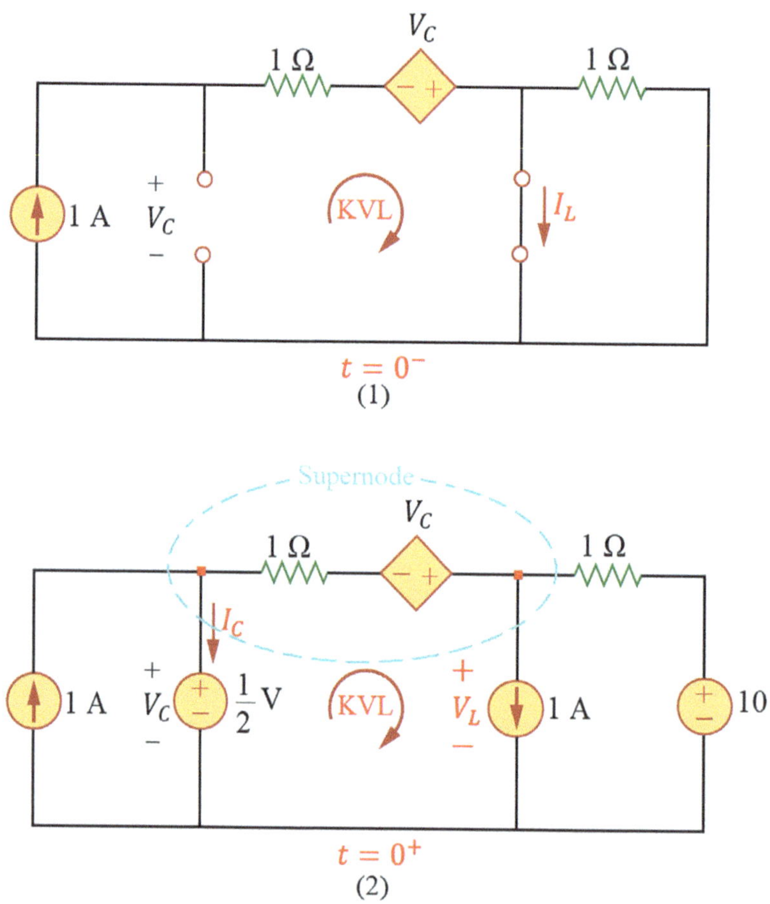

Fig. 11.26 The circuits of partially solved exercise 11.33

Problem

11.34. In the circuit of Fig. 11.27, the switch has been closed for a long time. Calculate the value of V_0 so that no impulse voltage is created across the switch when it is opened at $t = 0$.

Difficulty level ○ Easy ○ Normal ● Hard
Calculation amount ○ Small ● Normal ○ Large

1) 4 V
2) 12 V
3) 16 V
4) 24 V

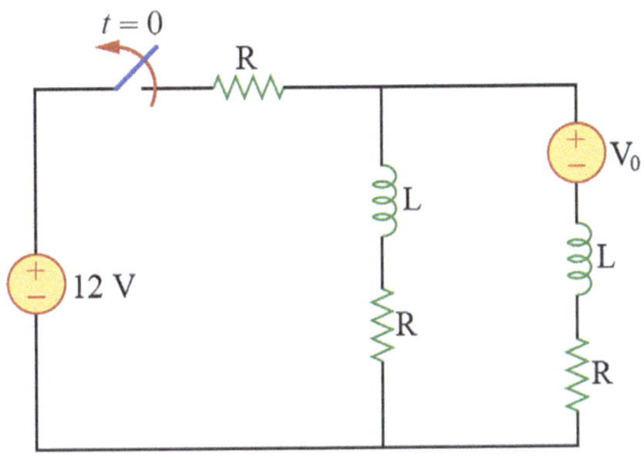

Fig. 11.27 The circuit of problem 11.34

Problem

11.35. In the circuit of Fig. 11.28, the circuit has reached its steady state condition. Calculate the voltage across the switch when it is opened.

Difficulty level ○ Easy ○ Normal ● Hard
Calculation amount ○ Small ● Normal ○ Large

1) 0 V
2) 2 V
3) 4 V
4) 8 V

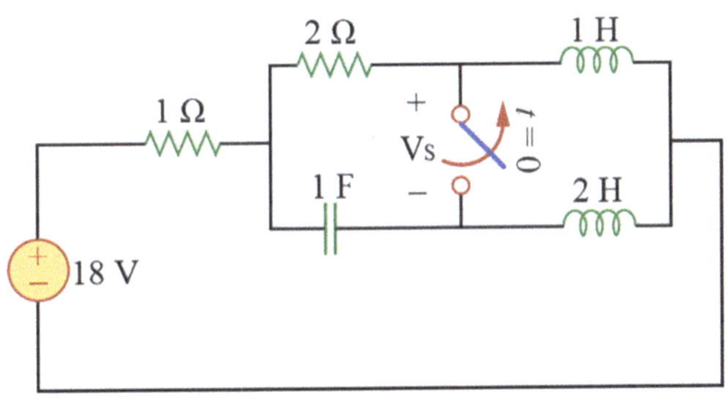

Fig. 11.28 The circuit of problem 11.35

Problem

11.36. In the circuit of Fig. 11.29, at $t = 0$, the switches S_1 and S_2 are closed and the switches S_3 and S_4 are opened simultaneously. Calculate the value of $\frac{dV_C}{dt}$ at $t = 0^+$.

Difficulty level ○ Easy ○ Normal ● Hard
Calculation amount ○ Small ● Normal ○ Large

1) $-\frac{7}{2}$ V/s
2) $-\frac{11}{4}$ V/s
3) $\frac{5}{2}$ V/s
4) $\frac{7}{4}$ V/s

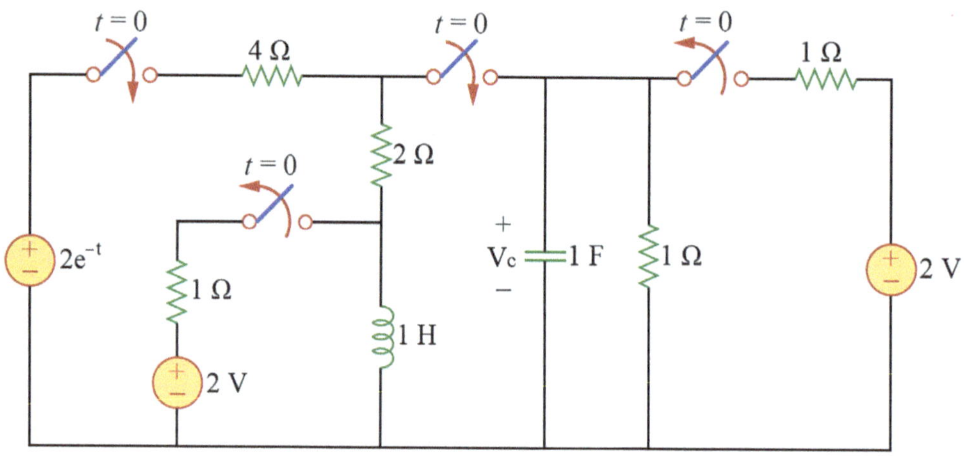

Fig. 11.29 The circuit of problem 11.36

Problem

11.37. In the circuit of Fig. 11.30, calculate the primary condition of the circuit $(V_{C1}(0), V_{C2}(0), I_L(0))$ to not excite any variable in the circuit.

Difficulty level ○ Easy ○ Normal ● Hard
Calculation amount ○ Small ● Normal ○ Large

1) $(1, 1, -1)$
2) $(1, 0, -1)$
3) $(1, -1, 1)$
4) $(1, 0, 0)$

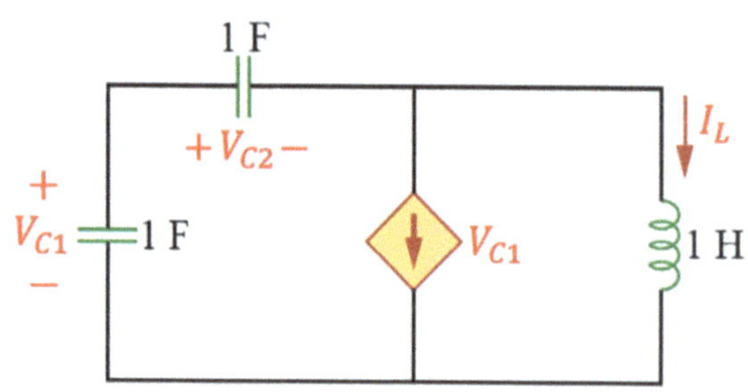

Fig. 11.30 The circuit of problem 11.37

Problem

38. The circuit of Fig. 11.31 is in a steady state condition. The switch is closed at $t = 0$. Calculate the value of parameters below.

$$e(0^+)$$

$$\frac{de(0^+)}{dt}$$

Difficulty level ○ Easy ○ Normal ● Hard
Calculation amount ○ Small ○ Normal ● Large

1) 8 V, 0 V/sec
2) 3 V, −12 V/sec
3) 8 V, −6 V/sec
4) 1 V, 1 V/sec

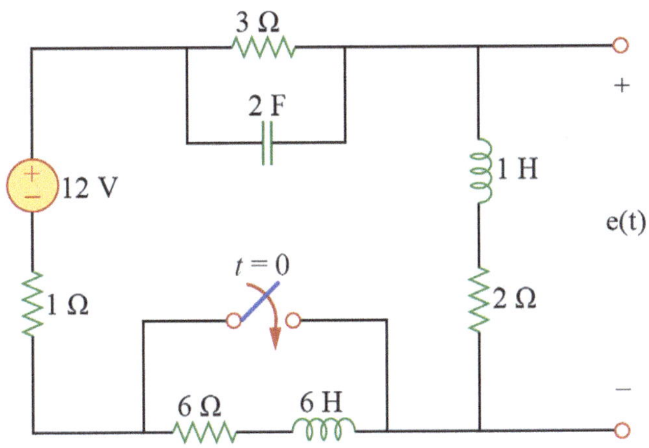

Fig. 11.31 The circuit of problem 11.38

Problem

11.39. In the circuit of Fig. 11.32, calculate the value of $\frac{d^2 I_L(0^+)}{dt^2}$ and $V_0(0^+)$ while assuming 0 V and 0 A for the initial voltage of capacitor and primary current of inductor, respectively.

Difficulty level ○ Easy ○ Normal ● Hard
Calculation amount ○ Small ○ Normal ● Large

1) $1\,\text{V},\ \frac{35}{2}\ (\text{A/sec})^2$
2) $1\,\text{V},\ \frac{35}{4}\ (\text{A/sec})^2$
3) $1\,\text{V},\ -\frac{35}{2}\ (\text{A/sec})^2$
4) $1\,\text{V},\ -\frac{35}{4}\ (\text{A/sec})^2$

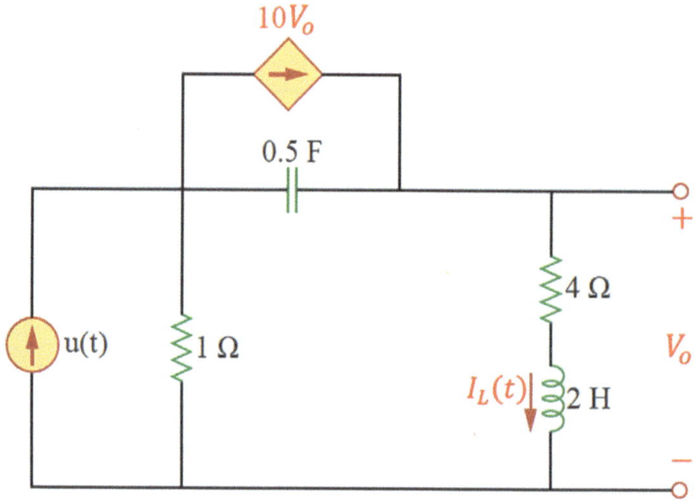

Fig. 11.32 The circuit of problem 11.39

Problem

11.40. In the circuit of Fig. 11.33, if $I(0^-) = 2$ A, $V_{C1}(0^-) = 3$ V, and $V_{C2}(0^-) = 1$ V, calculate the value of $\frac{dI(0^+)}{dt}$.

Difficulty level ○ Easy ○ Normal ● Hard

Calculation amount ○ Small ○ Normal ● Large

1) $-\frac{5}{4}$ A/sec
2) $-\frac{11}{12}$ A/sec
3) $\frac{11}{12}$ A/sec
4) $\frac{5}{4}$ A/sec

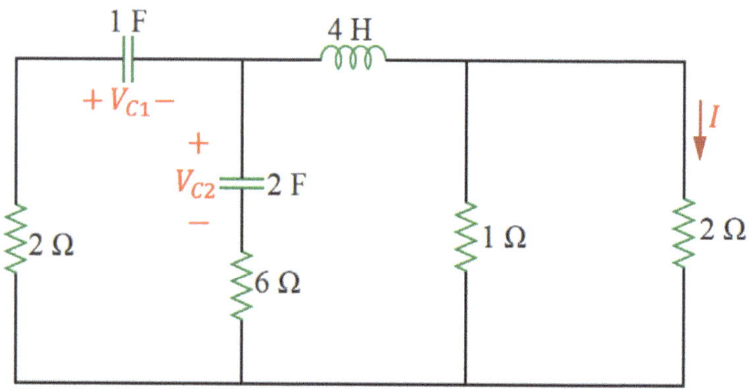

Fig. 11.33 The circuit of problem 11.40

11.6 Second-Order Circuits Including Singularity Functions

Problem

11.41. In the circuit of Fig. 11.34, calculate the value of R so that $I_R(t) = \frac{2}{3}\delta(t)$ for $V_s(t) = \delta(t)$.

Difficulty level ○ Easy ● Normal ○ Hard
Calculation amount ● Small ○ Normal ○ Large

1) 1 Ω
2) 2 Ω
3) $\frac{1}{2}$ Ω
4) $\frac{1}{3}$ Ω

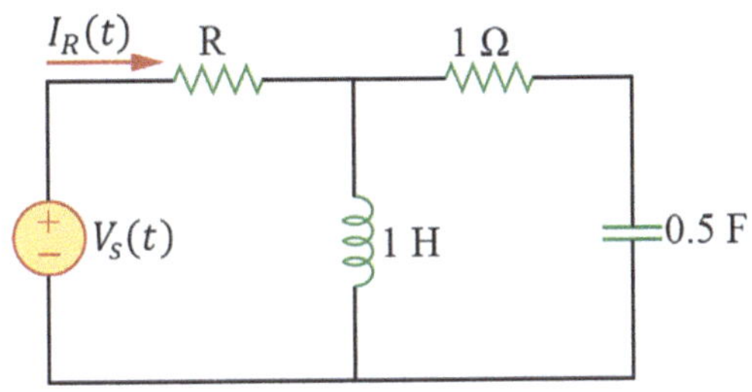

Fig. 11.34 The circuit of problem 11.41

Problem

11.42. The circuit shown in Fig. 11.35 is in zero state. Calculate the value of $I_o(0^+)$ if $V_s(t) = \delta(t)$.

Difficulty level ○ Easy ○ Normal ● Hard
Calculation amount ○ Small ● Normal ○ Large

1) $\frac{1}{3}$ A
2) $\frac{1}{4}$ A
3) $\frac{1}{6}$ A
4) $\frac{1}{8}$ A

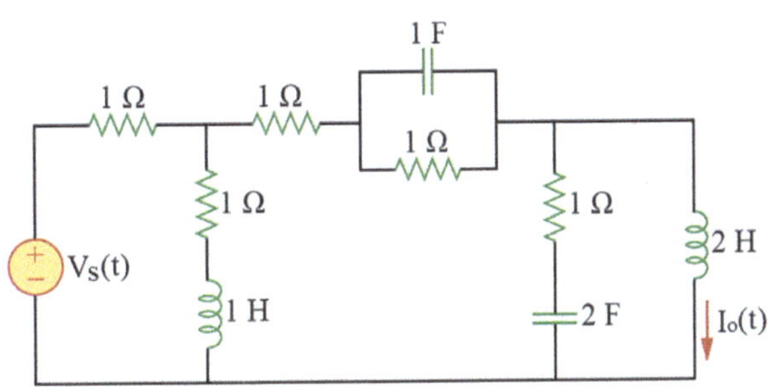

Fig. 11.35 The circuit of problem 11.42

11.7 Energy Stored in Second-Order Circuits

Problem

11.43. In the circuit of Fig. 11.36, calculate the amount of energy delivered to the RLC circuit by the current source in the interval of $[0, 1]$ sec. In this problem, assume that $C = 0.1\ F, L = 1H, R = 1.5\ \Omega, I_S(t) = 10$ t for t ≥ 0, and $V_C(0) = 20$ V.

Difficulty level ○ Easy ○ Normal ● Hard

Calculation amount ○ Small ● Normal ○ Large

1) 275 J
2) 325 J
3) 345 J
4) 375 J

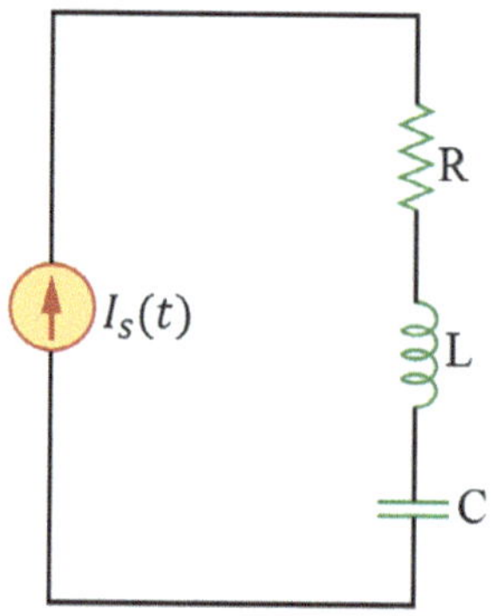

Fig. 11.36 The circuit of problem 11.43

Exercise

44. In Problem 11.43, calculate the amount of energy wasted by the resistor during the interval of $[0, 1]$ sec.

Final Answer

$E_R = 50$ J

Note: *Read the "educational support" part of the preface of the textbook.*

References

1. Rahmani-Andebili, M., DC Electrical Circuit Analysis - Practice Problems, Methods, and Solutions (1st Ed.), Springer Nature, 2020.
2. Rahmani-Andebili, M., General Physics II - Practice Problems, Methods, and Solutions, Sringer Nature, 2025.
3. Rahmani-Andebili, M., General Physics I - Practice Problems, Methods, and Solutions, Springer Nature, 2025.
4. Rahmani-Andebili, M., Mathematics of Engineering and Science - Comprehensive Lessons, Springer Nature, 2024.
5. Rahmani-Andebili, M., Differential Equations – Practice Problems, Methods, and Solutions, Springer Nature, 2022.
6. Rahmani-Andebili, M., Calculus III – Practice Problems, Methods, and Solutions, Springer Nature, 2023.
7. Rahmani-Andebili, M., Calculus II – Practice Problems, Methods, and Solutions, Springer Nature, 2023.
8. Rahmani-Andebili, M., Calculus I (2nd Ed.) – Practice Problems, Methods, and Solutions, Springer Nature, 2023.
9. Rahmani-Andebili, M., Precalculus (2nd Ed.) – Practice Problems, Methods, and Solutions, Springer Nature, 2024.

12 Analysis of Second-Order Circuits: Part B

Abstract
In this chapter, the problems of the eleventh chapter are fully solved, in detail, step-by-step, and with different methods. Herein, a "Notes Box" has been added after the solution of each problem that includes all the formulas and important notes that are needed to solve the problem. Moreover, a "Traps Box" has been added that discloses the traps that have been designed in the choices to challenge students' minds and encourage them to deeply think about the problem solutions as well as discourage them from careless selection of choices. The "Traps Box", in fact, predicts the possible confusions and tricks that might be encountered by students in real tests.

12.1 Differential Equation of Second-Order Circuits

12.1. We need to define all the variables of the differential equation of the circuit based on $I_R(t)$ [1–9].

KVL in the left-side mesh of the circuit of Fig. 12.1:

$$-V_S(t) + RI_R(t) + V_L(t) = 0 \quad (1)$$

Using current-voltage relation of inductors:

$$V_L(t) = L\frac{d}{dt}I_L(t) \quad (2)$$

KCL in the supernode:

$$-I_R(t) + I_L(t) + 2I_R(t) - I_S(t) = 0$$

$$\Rightarrow I_L(t) = -I_R(t) + I_S(t) \quad (3)$$

Solving (1) and (2):

$$-V_S(t) + RI_R(t) + L\frac{d}{dt}I_L(t) = 0$$

$$\overset{(3)}{\Rightarrow} -V_S(t) + RI_R(t) + L\frac{d}{dt}(-I_R(t) + I_S(t)) = 0$$

S. Manias, *DC Electrical Circuits Analysis*, https://doi.org/10.1007/978-3-032-00253-2_12

$$\Rightarrow -V_S(t) + RI_R(t) - L\frac{d}{dt}I_R(t) + L\frac{d}{dt}I_S(t) = 0$$

$$\Rightarrow L\frac{d}{dt}I_R(t) - RI_R(t) = L\frac{d}{dt}I_S(t) - V_S(t)$$

Choice (1) is correct.

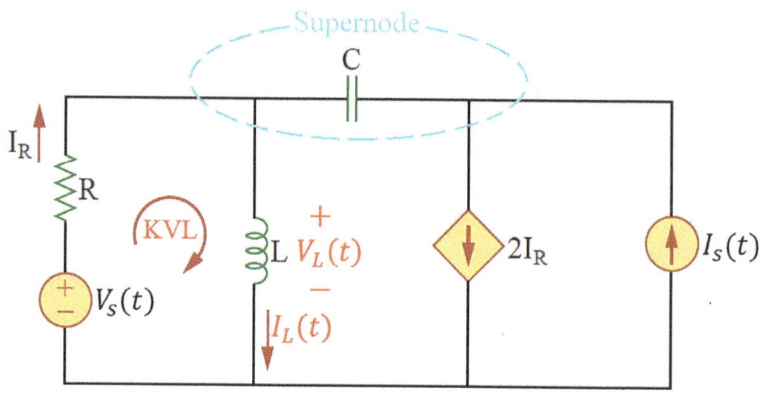

Fig. 12.1 The circuit of problem 12.1

12.2 Response Types of Second-Order Circuits

12.4. Based on the information given in the problem, we have:

$$s^2 - 1 = 0$$

$$\Rightarrow s^2 = 1$$

$$\Rightarrow s_{1,2} = \pm 1$$

See Table 12.1. Since the real part of at least one of the roots of the characteristic equation is positive, the response of the circuit is unstable. Choice (5) is correct.

Table 12.1 Determining the response type of a circuit based on the features of the roots of the characteristic equation of the differential equation of the desirable variable of the circuit

Features of roots of characteristic equation			Response type	Discriminant (A)
The real part of the roots is negative	The roots are real	The roots are distinct	Overdamped	Δ>0
		The roots are the same	Critically damped	Δ=0
	The roots are imaginary	The roots are complex conjugate of each other	Underdamped	Δ<0
The real part of the roots is zero	The roots are pure imaginary		Undamped (oscillating)	Δ<0
The real part of one or both roots is positive	Any (except pure imaginary roots)	Any	Unstable	Any

12.5. Based on the information given in the problem, we have:

$$\frac{d^2}{dt^2}V_C(t) + 5\frac{d}{dt}V_C(t) + 6V_C(t) = 0$$

Thus, the characteristic equation of the circuit is as follows.

$$s^2 + 5s + 6 = 0$$

$$\Rightarrow s_{1,2} = -2, -3$$

See Table 12.1. Since the roots are real, distinct, and their real part is negative, the response of the circuit is an overdamped response. Choice (1) is correct.

Notes

$$ax^2 + bx + c = 0 \Rightarrow x = \frac{-b \pm \sqrt{b^2 - 4ac}}{2a}$$

12.6. Based on the information given in the problem, we have:

$$\frac{d^2}{dt^2}V_C(t) + 2\frac{d}{dt}V_C(t) + V_C(t) = 0$$

Thus, the characteristic equation of the circuit is as follows.

$$s^2 + 2s + 1 = 0$$

$$(s + 1)^2 = 0$$

$$\Rightarrow s_{1,2} = -1, -1$$

See Table 12.1. Since the roots are real, the same, and their real part is negative, the response type of the circuit is critically damped. Choice (2) is correct.

12.7. Based on the information given in the problem, we have:

$$\frac{d^2}{dt^2}I_L(t) + 2\frac{d}{dt}I_L(t) + 2I_L(t) = 0$$

Thus, the characteristic equation of the circuit is as follows.

$$s^2 + 2s + 2 = 0$$

$$\Rightarrow s_{1,2} = -1 \pm j$$

See Table 12.1. Since the roots are imaginary, and the real part of the roots is negative, the response type of the circuit is underdamped. Choice (3) is correct.

Notes

$$ax^2 + bx + c = 0 \Rightarrow x = \frac{-b \pm \sqrt{b^2 - 4ac}}{2a}$$

12.8. *Step 1 (writing the differential equation of the circuit)*

From Problem 11.2, we have:

$$\frac{d^2}{dt^2}I_L(t) + \frac{1}{4}I_L(t) = 0$$

Step 2 (writing the characteristic equation of the differential equation and finding its roots)

$$s^2 + \frac{1}{4} = 0$$

$$\Rightarrow s^2 = -\frac{1}{4}$$

$$\Rightarrow s_{1,2} = \pm\frac{1}{2}j$$

Step 3 (determining the response type of the differential equation)

See Table 12.1. Since the characteristic equation includes pure imaginary roots, the response type of the circuit is undamped (oscillating). Choice (4) is correct.

12.12. The general response of a circuit depends only on the configuration of the circuit, and it is totally independent of the primary condition of the circuit. Therefore, we do not need to know the primary value of the variables. Figure. 12.2.2 illustrates the circuit for $t > 0$.

Step 1 (writing the differential equation of the circuit)

The differential equation must be written based on $I_L(t)$.

The capacitor and inductor are in parallel, thus:

$$V_C(t) = V_L(t) \tag{1}$$

Voltage-current relation of capacitors:

$$I_C(t) = C\frac{d}{dt}V_C(t) = 200 \times 10^{-6}\frac{d}{dt}V_C(t)$$

$$\overset{(1)}{\Rightarrow} I_C(t) = 2 \times 10^{-4} \frac{d}{dt} V_L(t) \tag{2}$$

Current-voltage relation of inductors:

$$V_L(t) = L \frac{d}{dt} I_L(t) \tag{3}$$

KCL in the supernode:

$$I_C(t) + I_L(t) + \frac{1}{20} V_L(t) = 0 \tag{4}$$

Solving (2) and (4):

$$2 \times 10^{-4} \frac{d}{dt} V_L(t) + I_L(t) + \frac{1}{20} V_L(t) = 0$$

$$\overset{(3)}{\Rightarrow} 2 \times 10^{-4} \frac{d}{dt} \left(L \frac{d}{dt} I_L(t) \right) + I_L(t) + \frac{1}{20} \left(L \frac{d}{dt} I_L(t) \right) = 0$$

$$\Rightarrow 2 \times 10^{-4} \times L \times \frac{d^2}{dt^2} I_L(t) + I_L(t) + \frac{L}{20} \frac{d}{dt} I_L(t) = 0$$

$$\Rightarrow \frac{d^2}{dt^2} I_L(t) + 250 \frac{d}{dt} I_L(t) + \frac{5000}{L} I_L(t) = 0$$

Step 2 (writing the characteristic equation of the differential equation and finding its roots)

Hence, the characteristic equation of the circuit is as follows.

$$s^2 + 250s + \frac{5000}{L} = 0 \tag{6}$$

Step 3

Based on the information given in the problem, we know that the response of the circuit is critically damped. Thus, the delta (discriminant) of the quadratic characteristic equation is zero based on Table 12.2. In other words:

$$\Delta = 0 \tag{7}$$

Solving (6) and (7):

$$(250)^2 - 4(1)\left(\frac{5000}{L}\right) = 0$$

$$\Rightarrow L = \left(\frac{4 \times 5000}{250^2}\right)$$

$$\Rightarrow L = 0.32 \text{ H}$$

Choice (4) is correct.

Table 12.2 Determining the discriminant of the characteristic equation of the differential equation based on the response type of the circuit

Response type	*Discriminant (Δ)*
Overdamped	$\Delta > 0$
Critically damped	$\Delta = 0$
Underdamped	$\Delta < 0$
Undamped (oscillating)	$\Delta < 0$
Unstable	Any

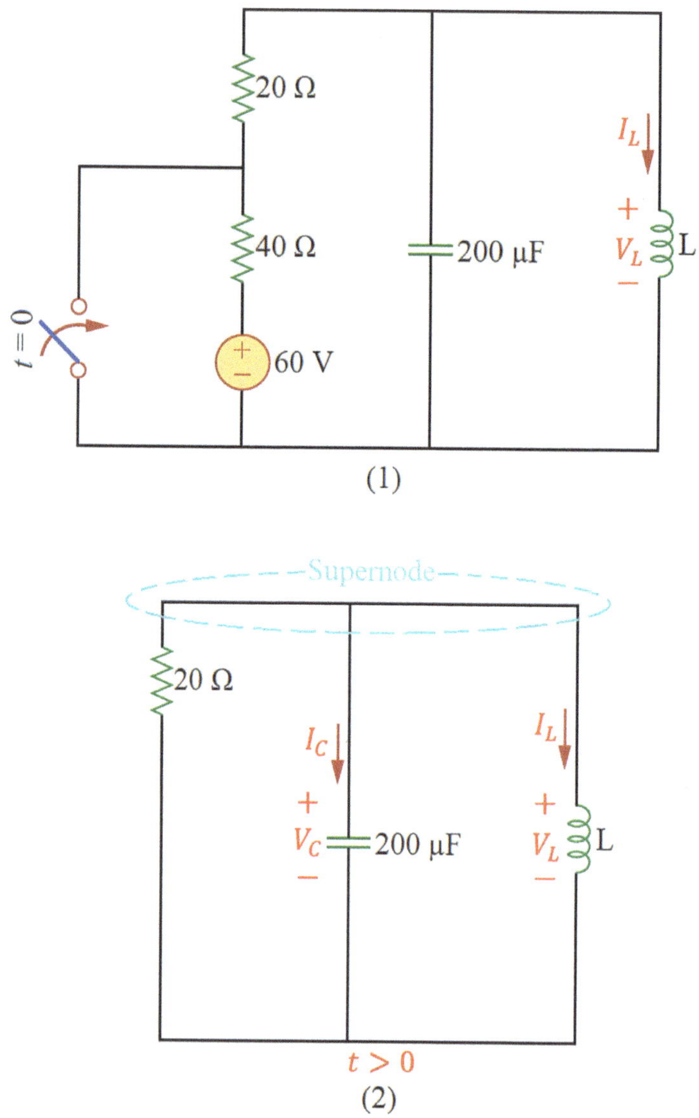

Fig. 12.2 The circuits of problem 12.4

Notes

$$ax^2 + bx + c = 0 \Rightarrow \Delta = b^2 - 4ac$$

12.13. *Step 1 (writing the differential equation of the circuit)*

The differential equation must be written based on $I(t)$.

Defining the voltage of the independent voltage source based on the node voltages in the circuit of Fig. 12.2:

$$kI(t) = V_L(t) - V_C(t)$$

$$\Rightarrow V_L(t) = kI(t) + V_C(t) \tag{1}$$

Applying Ohm's law for the 2-Ω resistor:

$$V_C(t) = 2I(t) \tag{2}$$

Applying KCL in the supernode:

$$I_L(t) + \frac{V_L(t)}{\frac{4}{7}} + I_C(t) + I(t) = 0$$

$$\Rightarrow I_L(t) + \frac{7}{4}V_L(t) + I_C(t) + I(t) = 0 \tag{3}$$

Voltage-current relation of capacitors:

$$I_C(t) = C\frac{d}{dt}V_C(t)$$

$$\Rightarrow I_C(t) = 2\frac{d}{dt}V_C(t) \tag{4}$$

Voltage-current relation of inductors:

$$I_L(t) = I_L(0^-) + \frac{1}{L}\int_{0^-}^{t} V_L(t)dt$$

$$\Rightarrow I_L(t) = 0 + \frac{1}{1}\int_{0^-}^{t} V_L(t)dt$$

$$\Rightarrow I_L(t) = \int_{0^-}^{t} V_L(t)dt \tag{5}$$

In (5), $I_L(0^-)$ is assumed to be zero since the response type of a circuit does not depend on its primary condition. Applying (4) and (5) in (3):

$$\int_{0^-}^{t} V_L(t).dt + \frac{7}{4}V_L(t) + 2\frac{d}{dt}V_C(t) + I(t) = 0$$

$$\overset{\frac{d}{dt}}{\Rightarrow} V_L(t) + \frac{7}{4}\frac{d}{dt}V_L(t) + 2\frac{d^2}{dt^2}V_C(t) + \frac{d}{dt}I(t) = 0$$

$$\overset{(1)}{\Rightarrow} kI(t) + V_C(t) + \frac{7}{4}\frac{d}{dt}(kI(t) + V_C(t)) + 2\frac{d^2}{dt^2}V_C(t) + \frac{d}{dt}I(t) = 0$$

$$\Rightarrow kI(t) + V_C(t) + \left(\frac{7}{4}k + 1\right)\frac{d}{dt}I(t) + \frac{7}{4}\frac{d}{dt}V_C(t) + 2\frac{d^2}{dt^2}V_C(t) = 0$$

$$\overset{(2)}{\Rightarrow} kI(t) + 2I(t) + \left(\frac{7}{4}k + 1\right)\frac{d}{dt}I(t) + \frac{7}{4}\frac{d}{dt}(2I(t)) + 2\frac{d^2}{dt^2}(2I(t)) = 0$$

$$\Rightarrow 4\frac{d^2}{dt^2}I(t) + \left(\frac{7}{4}k + \frac{9}{2}\right)\frac{d}{dt}I(t) + (k+2)I(t) = 0$$

$$\Rightarrow \frac{d^2}{dt^2}I(t) + \left(\frac{7}{16}k + \frac{9}{8}\right)\frac{d}{dt}I(t) + \frac{k+2}{4}I(t) = 0$$

Step 2 (writing the characteristic equation of the differential equation and finding its roots)

$$s^2 + \left(\frac{7}{16}k + \frac{9}{8}\right)s + \frac{k+2}{4} = 0$$

Step 3

To adjust the circuit response in the critically damped condition, the delta (discriminant) of the quadratic characteristic equation must be zero based on Table 12.2. Therefore:

$$\Delta = \left(\frac{7}{16}k + \frac{9}{8}\right)^2 - 4(1)\left(\frac{k+2}{4}\right) = 0$$

$$\Rightarrow \frac{49}{256}k^2 + \frac{63}{64}k + \frac{81}{64} - k - 2 = 0$$

$$\Rightarrow \frac{49}{256}k^2 - \frac{1}{64}k - \frac{47}{64} = 0$$

$$\Rightarrow k = \frac{\frac{1}{64} \pm \sqrt{\left(\frac{1}{64}\right)^2 + 4\left(\frac{49}{256}\right)\left(\frac{47}{64}\right)}}{\frac{98}{256}}$$

$$\Rightarrow k = \frac{\frac{1}{64} \pm \sqrt{\frac{9}{16}}}{\frac{98}{256}}$$

$$\Rightarrow k = \frac{\frac{1}{64} \pm \frac{3}{4}}{\frac{98}{256}}$$

$$\Rightarrow k = \frac{4 \pm 192}{98}$$

$$\Rightarrow k = -\frac{94}{49}, 2$$

As can be seen, $k = 2$ is one of the answers which is available among the choices. Choice (2) is correct Fig. 12.3.

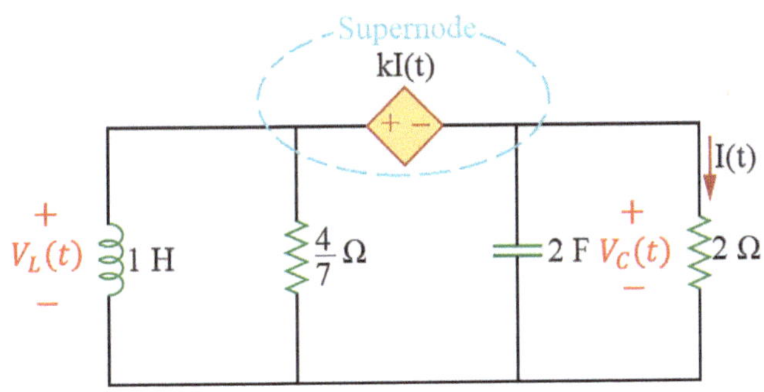

Fig. 12.3 The circuits of problem 12.13

Notes

$$ax^2 + bx + c = 0 \Rightarrow \Delta = b^2 - 4ac$$

$$ax^2 + bx + c = 0 \Rightarrow x = \frac{-b \pm \sqrt{b^2 - 4ac}}{2a}$$

12.3 General Response of Second-Order Circuits

12.15. From Partially Solved Exercise 11.14, we have:

$$R = 1000\ \Omega$$

$$s^2 + \frac{10^6}{R}s + \frac{10^6}{4} = 0$$

Step 1 (finding the roots of the characteristic equation)

$$s^2 + \frac{10^6}{1000}s + \frac{10^6}{4} = 0$$

$$\Rightarrow s^2 + 1000s + 250000 = 0$$

$$\Rightarrow (s + 500)^2 = 0$$

$$\Rightarrow s = -500,\ -500$$

Step 2 (determining the general form of the solution of differential equation)

Based on the roots of the characteristic equation of the differential equation, the general form of the solution of differential equation is as follows (see Table 12.3).

$$I_L(t) = (A + Bt)e^{-500t}u(t) \quad (1)$$

Step 3 (calculating the value of parameters of the general form of the solution of differential equation)

Using current division rule in the circuit of Fig. 12.4.2:

$$I_L(0^-) = \frac{1000}{1000 + 250} \times 0.5$$

$$\Rightarrow I_L(0^-) = 0.4\ \text{A} \quad (2)$$

Due to current continuity of inductor in lack of a power source with an infinite size (e.g. impulse power source), it is concluded that:

$$I_L(0^+) = 0.4\ \text{A} \quad (3)$$

Using Ohm's law for the 250-Ω resistor in the circuit of Fig. 12.4.2:

$$V_C(0^-) = I_L(0^-) \times 250 = 0.4 \times 250$$

$$\Rightarrow V_C(0^-) = 100\ \text{V} \quad (4)$$

Due to voltage continuity of capacitors in lack of a power source with an infinite size (e.g. impulse power source), it is concluded that:

$$V_C(0^+) = 100\ \text{V} \quad (5)$$

Applying (3) in (1):

$$0.4 = (A + B \times 0^+)e^{-500\times 0^+}$$

$$\Rightarrow A = 0.4 \quad (6)$$

From the circuit of Fig. 12.4.3, it is seen that:

$$V_L(0^+) = V_C(0^+)$$

Thus, from (4), it is concluded that:

$$V_L(0^+) = 100 \text{ V} \tag{7}$$

Applying (7) in the current-voltage relation of the inductor for $t = 0^+$:

$$100 = 4\frac{d}{dt}I_L(0^+)$$

$$\Rightarrow \frac{d}{dt}I_L(0^+) = 25 \text{ A/sec} \tag{8}$$

Calculating the first derivative of (1):

$$\frac{d}{dt}I_L(t) = Be^{-500t} - 500e^{-500t}(A + Bt)$$

$$\stackrel{t=0^+}{\Rightarrow} \frac{d}{dt}I_L(0^+) = Be^{-500\times0^+} - 500e^{-500\times0^+}(A + B\times0^+)$$

$$\Rightarrow \frac{d}{dt}I_L(0^+) = B - 500 \text{ A} \tag{9}$$

Solving (6), (8) and (9):

$$25 = B - 500 \times 0.4$$

$$\Rightarrow B = 225\,(10)$$

Step 4 (determining the solution of differential equation)

Applying (6) and (10) in (1):

$$I_L(t) = (0.4 + 225t)e^{-500t}u(t) \text{ A}$$

Choice (4) is correct.

Table 12.3 Determining the form of general response of the circuit based on the features of the roots of the characteristic equation

Features of roots of characteristic equation			Roots	General response
The real part of the roots is negative	The roots are real	The roots are distinct	s_1, s_2	$y(t) = Ae^{s_1t} + Be^{s_2t}$
		The roots are the same	S, S	$y(t) = (A + Bt)e^{st}$
	The roots are imaginary	The roots are complex conjugate of each other	$\alpha \pm j\beta$	$y(t) = e^{\alpha t}(A\cos(\beta t) + B\sin(\beta t))$
The real part of the roots is zero	The roots are pure imaginary		$\pm j\beta$	$y(t) = A\cos(\beta t) + B\sin(\beta t)$

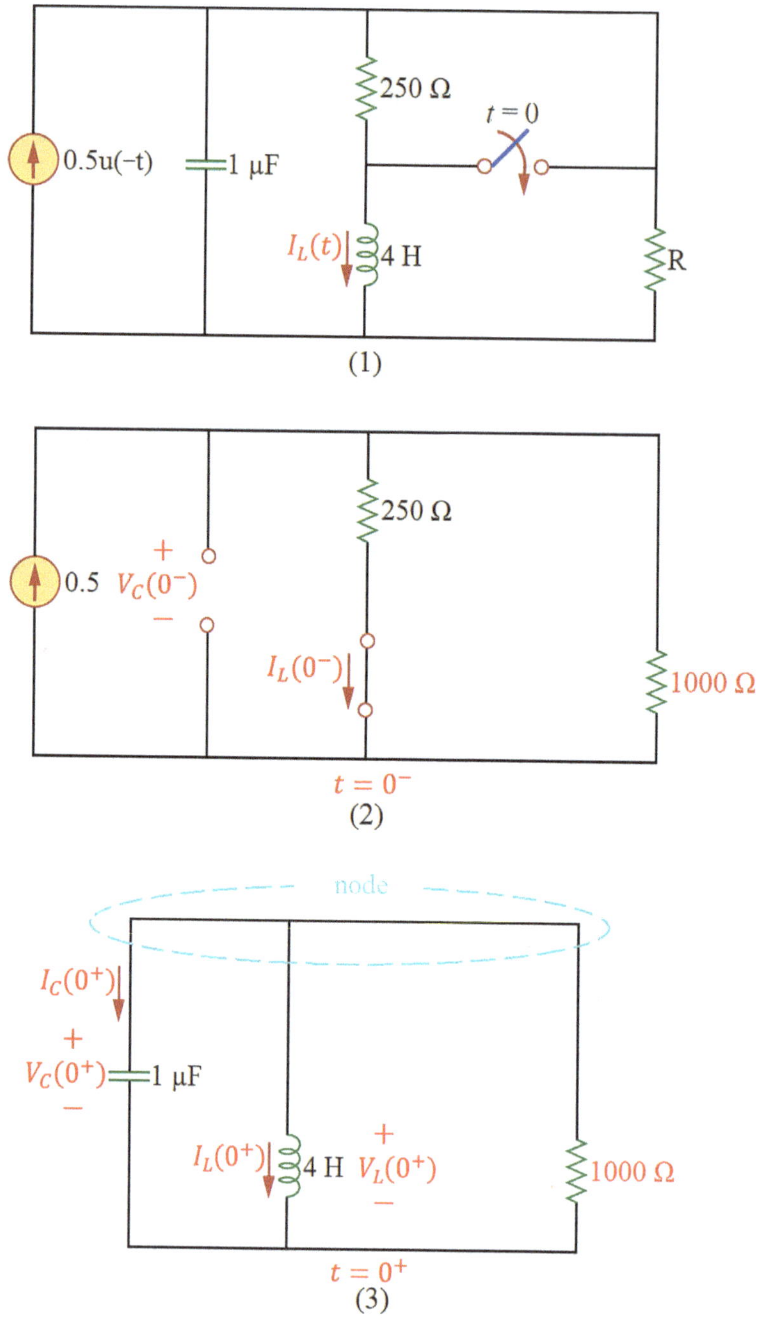

Fig. 12.4 The circuits of problem 12.15

12.17. ***Step 1 (writing the differential equation of the circuit)***

Based on the information given in the problem, we have:

$$V_C(0^-) = 4 \text{ V} \quad (1)$$

$$I_L(0^-) = 2 \text{ A} \quad (2)$$

Due to voltage continuity of capacitors in lack of a power source with an infinite size (e.g. impulse power source), it is concluded that:

$$V_C(0^+) = 4 \text{ V} \tag{3}$$

Due to current continuity of inductors in lack of a power source with an infinite size (e.g. impulse power source), it is concluded that:

$$I_L(0^+) = 2 \text{ A} \tag{4}$$

KVL in the loop:

$$-2 + V_L(t) + V_C(t) + 3I_L(t) = 0 \tag{5}$$

Current-voltage relation of inductors:

$$V_L(t) = L\frac{d}{dt}I_L(t) = 1 \times \frac{d}{dt}I_L(t)$$

$$\Rightarrow V_L(t) = \frac{d}{dt}I_L(t) \tag{6}$$

Current-voltage relation of capacitors:

$$V_C(t) = V_C(0^-) + \frac{1}{C}\int_{0^-}^{t} I_C(t)dt$$

$$\Rightarrow V_C(t) = 4 + \frac{1}{0.5}\int_{0^-}^{t} I_C(t)dt$$

$$\Rightarrow V_C(t) = 4 + 2\int_{0^-}^{t} I_C(t)dt \tag{7}$$

Solving (5)-(7):

$$-2 + \frac{d}{dt}I_L(t) + 4 + 2\int_{0^-}^{t} I_C(t).dt + 3I_L(t) = 0$$

$$\overset{\frac{d}{dt}}{\Rightarrow} \frac{d^2}{dt^2}I_L(t) + 2I_C(t) + 3\frac{d}{dt}I_L(t) = 0 \tag{8}$$

From the circuit, it is noticed that:

$$I_C(t) = I_L(t) \tag{9}$$

Solving (8) and (9):

$$\frac{d^2}{dt^2}I_L(t) + 3\frac{d}{dt}I_L(t) + 2I_L(t) = 0 \tag{10}$$

Step 2 (writing the characteristic equation of the differential equation and finding its roots)

$$s^2 + 3s + 2 = 0$$

$$\Rightarrow s = -1, -2 \tag{11}$$

Step 3 (determining the general form of the solution of differential equation)

Based on the roots of the characteristic equation of the differential equation, the general form of the solution of differential equation is as follows (see Table 12.3).

$$I_L(t) = Ae^{-t} + Be^{-2t} \tag{14}$$

Step 4 (calculating the value of parameters of the general form of the solution of differential equation)

Applying (4) in (14):

$$2 = Ae^{-0^+} + Be^{-2\times 0^+}$$

$$\Rightarrow A + B = 2 \tag{15}$$

Solving (5) for $t = 0^+$:

$$-2 + V_L(0^+) + V_C(0^+) + 3I_L(0^+) = 0$$

$$\Rightarrow V_L(0^+) = 2 - V_C(0^+) - 3I_L(0^+)$$

$$\overset{(3)(4)}{\Rightarrow} V_L(0^+) = 2 - 4 - 3 \times 2$$

$$\Rightarrow V_L(0^+) = -8 \tag{16}$$

Solving (6) and (14):

$$V_L(t) = \frac{d}{dt}I_L(t) = \frac{d}{dt}\left(Ae^{-t} + Be^{-2t}\right)$$

$$\Rightarrow V_L(t) = -Ae^{-t} - 2Be^{-2t} \tag{17}$$

Solving (16) and (17):

$$-8 = -Ae^{-0^+} - 2Be^{-2\times 0^+}$$

$$\Rightarrow A + 2B = 8 \tag{18}$$

Solving (15) and (18):

$$A + 2(2 - A) = 8$$

$$\Rightarrow A = -4, B = 6 \qquad (19)$$

Step 5 (determining the solution of differential equation)

Applying (19) in (14):

$$I_L(t) = -4e^{-t} + 6e^{-2t}, \ t > 0$$

Choice (1) is correct.

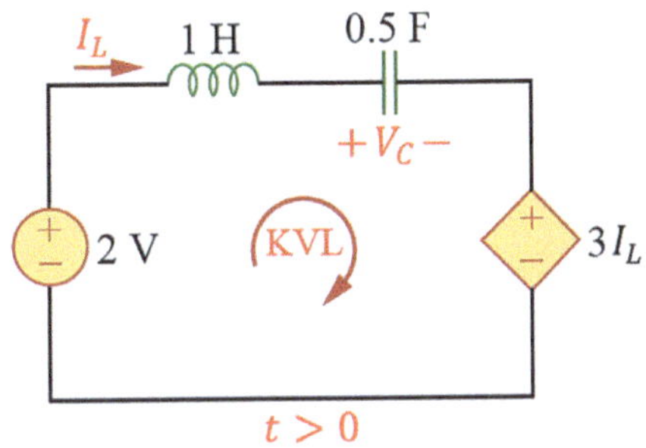

Fig. 12.5 The circuits of problem 12.17

12.18. From Problem 12.17, we have:

$$I_L(t) = -4e^{-t} + 6e^{-2t}, \ t > 0 \qquad (1)$$

From the circuit of Fig. 12.5, it is seen that:

$$I_C(t) = I_L(t) \qquad (2)$$

Current-voltage relation of capacitors:

$$V_C(t) = V_C(0^-) + \frac{1}{C}\int_{0^-}^{t} I_C(t)dt$$

$$\Rightarrow V_C(t) = 4 + \frac{1}{0.5}\int_{0^-}^{t} I_C(t)dt$$

$$\Rightarrow V_C(t) = 4 + 2\int_{0^-}^{t} I_C(t)dt \qquad (3)$$

Solving (2) and (3):

$$V_C(t) = 4 + 2\int_{0^-}^{t} I_L(t)dt \tag{4}$$

Solving (1) and (4):

$$V_C(t) = 4 + 2\int_{0^-}^{t} (-4e^{-t} + 6e^{-2t})dt$$

$$V_C(t) = 4 + 2\left[-4(-e^{-t}) + 6\left(-\frac{1}{2}e^{-2t}\right)\right]_{0^-}^{t}$$

$$\Rightarrow V_C(t) = 4 + 2\left[4e^{-t} - 3e^{-2t} - (4-3)\right]$$

$$\Rightarrow V_C(t) = 2 + 8e^{-t} - 6e^{-2t},\ t > 0$$

Choice (2) is correct.

Notes

$$\int e^{at}dt = \frac{1}{a}e^{at}$$

12.19. By applying KCL in the indicated node of Fig. 12.6.2, we have:

$$-I_0(t) + I_{L1}(t) + I_{L2}(t) + I(t) = 0$$

$$\Rightarrow I(t) = I_0(t) - I_{L1}(t) - I_{L2}(t) \tag{1}$$

Therefore, we need to calculate the value of $I_0(t)$, $I_{L1}(t)$, and $I_{L2}(t)$.

As can be noticed from Fig. 12.6.2, due to the existence of the short circuit branch, the circuit can be considered as three separate circuits including zero-order circuit, first-order circuit, and second-order circuit.

The status of the circuit for $t = 0^-$ is illustrated in Fig. 12.6.3. As can be seen, the capacitor and the inductors are open circuit and short circuits, respectively since the circuit has reached its steady state condition.

KVL in the indicated loop:

$$-10 + I_{L2}(0^-) \times 1 + 4 \times I_{L2}(0^-) = 0$$

$$\Rightarrow I_{L2}(0^-) = \frac{10}{5} = 2\ \text{A} \tag{2}$$

Using Ohm's law for the 4-Ω resistor:

$$V_C(0^-) = 4I_{L2}(0^-) = 4 \times 2$$

$$\Rightarrow V_C(0^-) = 8 \text{ V} \tag{3}$$

As can be noticed from the circuit of Fig. 12.6.3, the current of the left-side inductor is zero.

$$I_{L1}(0^-) = 0 \text{ A} \tag{4}$$

Due to current continuity of inductors and voltage continuity of capacitors in lack of a power source with an infinite size (e.g. impulse power source), it is concluded that:

$$I_{L1}(0^+) = 0 \text{ A} \tag{5}$$

$$I_{L2}(0^+) = 2 \text{ A} \tag{6}$$

$$V_C(0^+) = 8 \text{ V} \tag{7}$$

Zero-order circuit: The zero-order circuit includes the series combination of the voltage source and the 1-Ω resistor, as can be seen in Fig. 12.6.4.

Applying Ohm's law for the 1-Ω resistor:

$$I_0(t) = \frac{10 - 0}{1}$$

$$\Rightarrow I_0(t) = 10 \text{ A} \tag{8}$$

First-order circuit: The first-order circuit includes the series combination of the 4-Ω resistor and the right-side inductor, as can be seen in Fig. 12.6.5.

The primary current of the inductor:

$$I_{L2}(0^+) = 2 \text{ A} \tag{9}$$

The final current of the inductor is zero since its circuit does not include any power source.

$$I_{L2}(\infty) = 0 \text{ A} \tag{10}$$

The time constant of the first-order circuit:

$$\tau = \frac{L}{R_{Th}} = \frac{1}{4} \Omega \tag{11}$$

Hence, by using the general form of complete response of an RL circuit, we have:

$$I_{L2}(t) = I_{L2}(\infty) + (I_{L2}(0^+) - I_{L2}(\infty))e^{-\frac{t}{\tau}}$$

$$\Rightarrow I_{L2}(t) = 0 + (2 - 0)e^{-4t}$$

$$\Rightarrow I_{L2}(t) = 2e^{-4t} \tag{12}$$

Second-order circuit: The second-order circuit includes the series combination of the left-side inductor and the capacitor, as can be seen in Fig. 12.6.6.

Step 1 (writing the differential equation of the circuit)

Applying KVL in the mesh:

$$V_{L1}(t) + V_C(t) = 0 \tag{13}$$

$$\Rightarrow L_1 \frac{d}{dt} I_{L1}(t) + V_C(t) = 0$$

$$\Rightarrow 1 \times \frac{d}{dt} I_{L1}(t) + V_C(t) = 0$$

$$\overset{\frac{d}{dt}}{\Rightarrow} \frac{d^2}{dt^2} I_{L1}(t) + \frac{d}{dt} V_C(t) = 0 \tag{14}$$

As can be noticed, we have:

$$I_C(t) = I_{L1}(t) \tag{15}$$

Using voltage-current relation of capacitors:

$$I_C(t) = C \frac{d}{dt} V_C(t) = 0.25 \frac{d}{dt} V_C(t)$$

$$\Rightarrow \frac{d}{dt} V_C(t) = 4 I_C(t)$$

$$\overset{(15)}{\Rightarrow} \frac{d}{dt} V_C(t) = 4 I_{L1}(t) \tag{16}$$

Solving (14) and (16):

$$\frac{d^2}{dt^2} I_{L1}(t) + 4 I_{L1}(t) = 0 \tag{17}$$

Step 2 (writing the characteristic equation of the differential equation and finding its roots)

$$s^2 + 4 = 0$$

$$\Rightarrow s = \pm 2j \tag{18}$$

Step 3 (determining the general form of the solution of differential equation)

Based on the roots of the characteristic equation of the differential equation, the general form of the solution of differential equation is as follows (see Table 12.3).

$$I_{L1}(t) = A\cos(2t) + B\sin(2t) \tag{19}$$

Step 4 (calculating the value of parameters of the general form of the solution of differential equation)

Solving (4) and (19):

$$0 = A\cos(2 \times 0^+) + B\sin(2 \times 0^+)$$

$$\Rightarrow 0 = A \times 1 + B \times 0$$

$$\Rightarrow A = 0 \tag{20}$$

Solving (7) and (13):

$$V_{L1}(0^+) = -8 \text{ V} \tag{21}$$

Using current-relation of inductors:

$$V_{L1}(t) = L_1 \frac{d}{dt} I_{L1}(t) = 1 \times \frac{d}{dt} I_{L1}(t)$$

$$\Rightarrow V_{L1}(t) = \frac{d}{dt} I_{L1}(t) \tag{22}$$

Solving (19) and (22):

$$V_{L1}(t) = \frac{d}{dt}(A\cos(2t) + B\sin(2t))$$

$$\Rightarrow V_{L1}(t) = -2A\sin(2t) + 2B\cos(2t) \tag{23}$$

Solving (21) and (23):

$$-8 = -2A\sin(2 \times 0^+) + 2B\cos(2 \times 0^+)$$

$$\Rightarrow -8 = -2A \times 0 + 2B \times 1$$

$$\Rightarrow B = -4 \tag{24}$$

Step 5 (determining the solution of differential equation)

Solving (19), (20), and (24):

$$I_{L1}(t) = -4\sin(2t) \tag{25}$$

Final answer: Now, by solving (1), (8), (12), and (25), we have:

$$I(t) = 10 - (-4\sin(2t)) - \left(2e^{-4t}\right)$$

$$\Rightarrow I(t) = 10 - 2e^{-4t} + 4\sin(2t)$$

Choice (4) is the answer. *It was a real engineering problem. Wasn't it?*

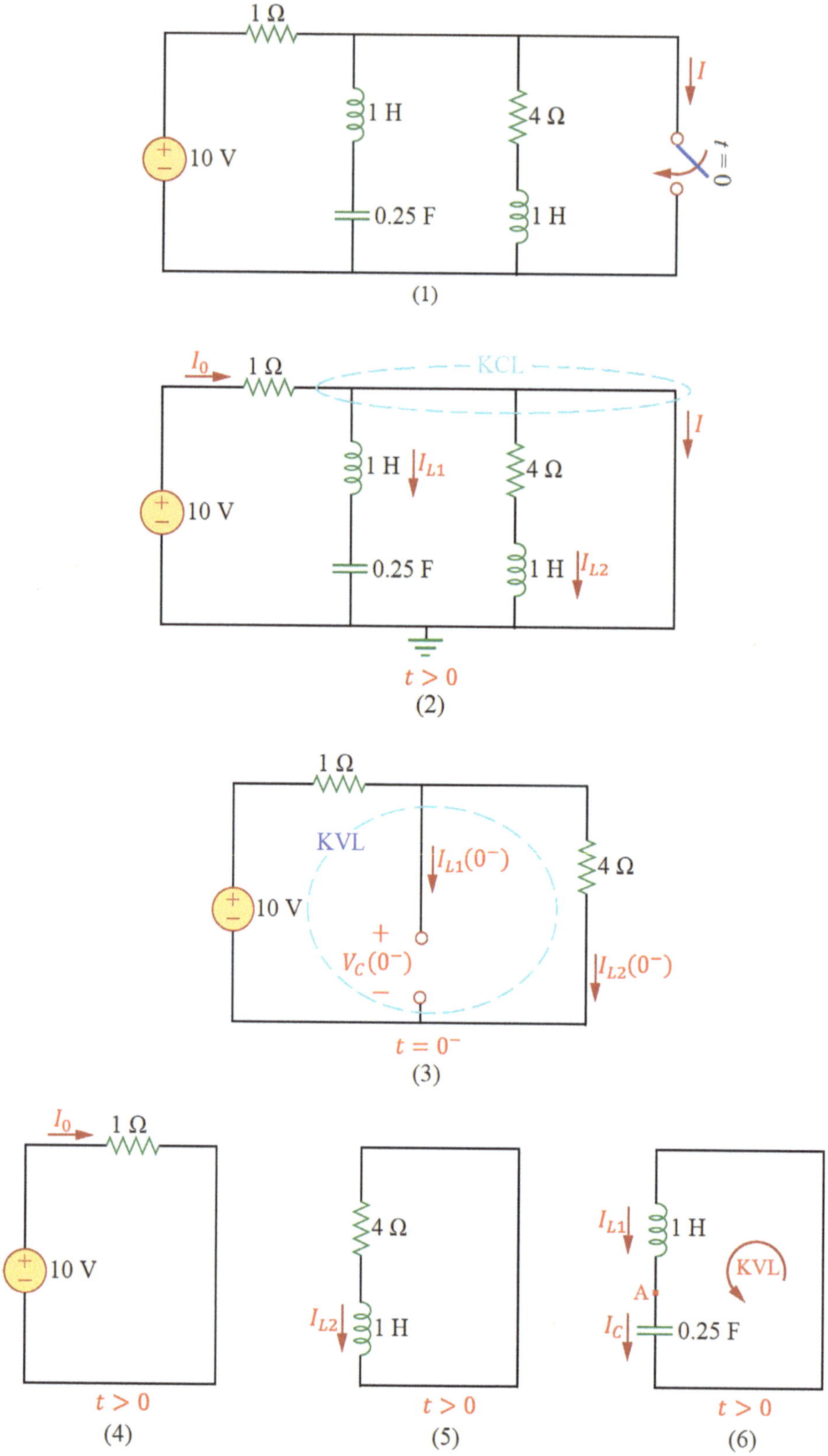

Fig. 12.6 The circuits of problem 12.19

12.20. *Step 1 (writing the differential equation of the circuit)*

We need to write the differential equation of the circuit based on the voltage of capacitor. Figure. 12.7 illustrates the circuit for $t > 0$.

Applying KCL in node "A":

$$-I_L(t) + I_C(t) + \frac{V_C(t)}{0.4} = 0$$

$$\Rightarrow -I_L(t) + I_C(t) + 2.5V_C(t) = 0 \tag{1}$$

$$\overset{\frac{d}{dt}}{\Rightarrow} -\frac{d}{dt}I_L(t) + \frac{d}{dt}I_C(t) + 2.5\frac{d}{dt}V_C(t) = 0 \tag{2}$$

Voltage-current relation of capacitors:

$$I_C(t) = C\frac{d}{dt}V_C(t) = 1 \times \frac{d}{dt}V_C(t)$$

$$\Rightarrow I_C(t) = \frac{d}{dt}V_C(t) \tag{3}$$

Based on the information given in the problem, we have:

$$V_s(t) = 1,\ t > 0 \tag{4}$$

KVL in the indicated mesh:

$$-1 + V_L(t) + V_C(t) = 0 \tag{5}$$

Current-voltage relation of inductors:

$$V_L(t) = L\frac{d}{dt}I_L(t) = 1 \times \frac{d}{dt}I_L(t)$$

$$\Rightarrow V_L(t) = \frac{d}{dt}I_L(t) \tag{6}$$

Solving (5) and (6):

$$\frac{d}{dt}I_L(t) = 1 - V_C(t) \tag{7}$$

Solving (2), (3), and (7):

$$-(1 - V_C(t)) + \frac{d}{dt}\left(\frac{d}{dt}V_C(t)\right) + 2.5\frac{d}{dt}V_C(t) = 0$$

$$\Rightarrow \frac{d^2}{dt^2}V_C(t) + 2.5\frac{d}{dt}V_C(t) + V_C(t) = 1 \tag{8}$$

Step 2 (writing the characteristic equation of the differential equation and finding its roots)

The characteristic equation of the differential equation is as follows.

$$s^2 + 2.5s + 1 = 0$$

$$\Rightarrow s_{1,2} = -0.5, -2 \tag{9}$$

Step 3 (determining the general form of the solution of differential equation)

Equation (8) is a second-order nonhomogeneous differential equation; thus, it has both common solution ($V_{C,c}(t)$) and particular solution ($V_{C,p}(t)$).

Based on the roots of the characteristic equation of the differential equation, the general form of the solution of differential equation is as follows (see Table 12.3).

$$V_{C,c}(t) = Ae^{-0.5t} + Be^{-2t} \tag{10}$$

Additionally, since "1" is a constant value in (8), a constant quantity is chosen as the particular solution of the differential equation (8) as follows:

$$V_{C,p}(t) = k \tag{11}$$

Now, by solving (8) and (11), we have:

$$\frac{d^2}{dt^2}k + 2.5\frac{d}{dt}k + k = 1$$

$$\Rightarrow 0 + 0 + k = 1$$

$$\Rightarrow k = 1 \tag{12}$$

Solving (11) and (12):

$$V_{C,p}(t) = 1 \tag{13}$$

The complete solution is sum of the common solution and particular solution. Hence:

$$\Rightarrow V_C(t) = V_{C,c}(t) + V_{C,p}(t)$$

$$\Rightarrow V_C(t) = Ae^{-0.5t} + Be^{-2t} + 1 \tag{14}$$

Step 4 (calculating the value of parameters of the general form of the solution of differential equation)

Based on the information given in the problem, we have:

$$V_C(0^-) = 1 \text{ V} \tag{15}$$

$$I_L(0^-) = 2 \text{ A} \tag{16}$$

Due to voltage continuity of capacitors and current continuity of inductor currents in lack of a power source with an infinite size (e.g. impulse power source), it is concluded that:

$$V_C(0^+) = 1 \text{ V} \tag{17}$$

$$I_L(0^+) = 2 \text{ A} \tag{18}$$

Applying (17) in (14):

$$1 = Ae^{-0.5 \times 0^+} + Be^{-2 \times 0^+} + 1$$

$$\Rightarrow A + B = 0 \tag{19}$$

Solving (1) for $t = 0^+$:

$$-I_L(0^+) + I_C(0^+) + 2.5V_C(0^+) = 0$$

$$\Rightarrow I_C(0^+) = I_L(0^+) - 2.5V_C(0^+) \tag{20}$$

Solving (17), (18), and (20):

$$I_C(0^+) = 2 - 2.5 \times 1$$

$$\Rightarrow I_C(0^+) = -0.5 \text{ A} \tag{21}$$

Solving (3) and (14):

$$I_C(t) = \frac{d}{dt}V_C(t) = \frac{d}{dt}\left(Ae^{-0.5t} + Be^{-2t} + 1\right)$$

$$\Rightarrow I_C(t) = -0.5Ae^{-0.5t} - 2Be^{-2t}$$

$$\overset{\text{for } t=0^+}{\Rightarrow} I_C(0^+) = -0.5Ae^{-0.5 \times 0^+} - 2Be^{-2 \times 0^+}$$

$$\Rightarrow I_C(0^+) = -0.5A - 2B \tag{22}$$

Solving (21) and (22):

$$-0.5 = -0.5A - 2B \tag{23}$$

Solving (19) and (23):

$$-0.5 = -0.5A - 2(-A)$$

$$\Rightarrow -0.5 = 1.5A$$

$$\Rightarrow A = -\frac{1}{3},\ B = \frac{1}{3} \tag{24}$$

Step 5 (determining the solution of differential equation)

Solving (14) and (24):

$$V_C(t) = 1 + \frac{1}{3}\left(-e^{-0.5t} + e^{-2t}\right) \text{ V}$$

Choice (1) is the answer.

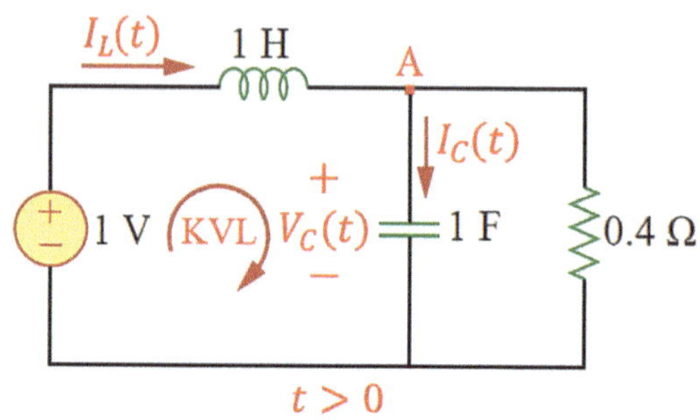

Fig. 12.7 The circuits of problem 12.20

12.4 Quality Factor of Second-Order Circuits

12.21. The quality factor of a circuit (Q) can be calculated by using two methods as follows.

Method 1: First, the characteristic equation with the form $s^2 + 2\alpha s + \omega_0^2 = 0$ is extracted from the differential equation of the circuit, and then the quality factor of the circuit is calculated as follows.

$$Q = \frac{\omega_0}{2\alpha} \quad (1)$$

Method 2: First, the equivalent resistance of the circuit seen by the LC circuit is calculated. Then, if the circuit is a series RLC circuit, the quality factor of the circuit is calculated by using (2). Otherwise, if the circuit is a parallel RLC circuit, the quality factor of the circuit is calculated by using (3).

$$Q = \frac{1}{R_{eq}}\sqrt{\frac{L}{C}} \quad (2)$$

$$Q = R_{eq}\sqrt{\frac{C}{L}} \quad (3)$$

In this problem, the first method is more appropriate.

Applying KVL in the left-side mesh of the circuit of Fig. 12.8:

$$2I_1 + 3\frac{d}{dt}I_1 + V_C = 0$$

$$\overset{\frac{d}{dt}}{\Rightarrow} 2\frac{d}{dt}I_1 + 3\frac{d^2}{dt^2}I_1 + \frac{d}{dt}V_C = 0 \tag{4}$$

KCL in node "A":

$$I_C = I_1 - I_2 \tag{5}$$

Voltage-current relation of capacitors:

$$I_C = C\frac{d}{dt}V_C = 2\frac{d}{dt}V_C$$

$$\Rightarrow \frac{d}{dt}V_C = \frac{1}{2}I_C$$

$$\overset{(5)}{\Rightarrow} \frac{d}{dt}V_C = \frac{1}{2}(I_1 - I_2) \tag{6}$$

Solving (4) and (7):

$$2\frac{d}{dt}I_1 + 3\frac{d^2}{dt^2}I_1 + \frac{1}{2}(I_1 - I_2) = 0$$

$$\Rightarrow 4\frac{d}{dt}I_1 + 6\frac{d^2}{dt^2}I_1 + I_1 - I_2 = 0 \tag{7}$$

KVL in the right-side mesh of the circuit:

$$4I_2 - V_C = 0$$

$$\overset{\frac{d}{dt}}{\Rightarrow} 4\frac{d}{dt}I_2 - \frac{d}{dt}V_C = 0$$

$$\overset{(7)}{\Rightarrow} 4\frac{d}{dt}I_2 - \frac{1}{2}(I_1 - I_2) = 0$$

$$\Rightarrow I_1 = 8\frac{d}{dt}I_2 + I_2 \tag{8}$$

Solving (7) and (8):

$$4\frac{d}{dt}\left(8\frac{d}{dt}I_2 + I_2\right) + 6\frac{d^2}{dt^2}\left(8\frac{d}{dt}I_2 + I_2\right) + \left(8\frac{d}{dt}I_2 + I_2\right) - I_2 = 0$$

$$\Rightarrow\ 32\frac{d^2}{dt^2}I_2 + 4\frac{d}{dt}I_2 + 48\frac{d^3}{dt^3}I_2 + 6\frac{d^2}{dt^2}I_2 + 8\frac{d}{dt}I_2 = 0$$

$$\Rightarrow 48\frac{d^3}{dt^3}I_2 + 38\frac{d^2}{dt^2}I_2 + 12\frac{d}{dt}I_2 = 0$$

$$\overset{\int dt}{\Rightarrow} 48\frac{d^2}{dt^2}I_2 + 38\frac{d}{dt}I_2 + 12I_2 = 0$$

$$\Rightarrow \frac{d^2}{dt^2}I_2 + \frac{19}{24}\frac{d}{dt}I_2 + \frac{1}{4}I_2 = 0 \tag{9}$$

Therefore, the characteristic equation of the circuit is as follows.

$$s^2 + \frac{19}{24}s + \frac{1}{4} = 0 \tag{10}$$

Now, by comparing (10) with the standard form of the characteristic equation, the following results are achieved.

$$2\alpha = \frac{19}{24},\ \omega_0 = \frac{1}{2}$$

Therefore:

$$Q = \frac{\frac{1}{2}}{\frac{19}{24}}$$

$$\Rightarrow Q = \frac{12}{19}$$

Choice (4) is the answer.

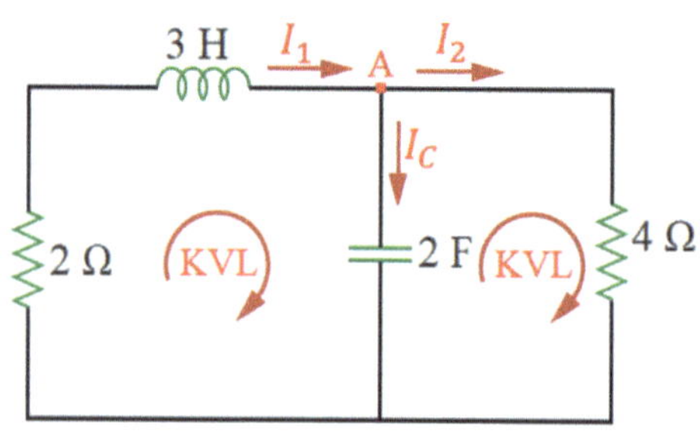

Fig. 12.8 The circuits of problem 12.21

12.24. The quality factor of a circuit (Q) can be calculated by using two methods as follows.

Method 1: First, the characteristic equation with the form $s^2 + 2\alpha s + \omega_0^2 = 0$ is extracted from the differential equation of the circuit, and then the quality factor of the circuit is calculated as follows.

$$Q = \frac{\omega_0}{2\alpha} \tag{1}$$

Method 2: First, the equivalent resistance of the circuit seen by the LC circuit is calculated. Then, if the circuit is a series RLC circuit, the quality factor of the circuit is calculated by using (2). Otherwise, if the circuit is a parallel RLC circuit, the quality factor of the circuit is calculated by using (3).

$$Q = \frac{1}{R_{eq}}\sqrt{\frac{L}{C}} \tag{2}$$

$$Q = R_{eq}\sqrt{\frac{C}{L}} \tag{3}$$

In this problem, the second method is more appropriate.

To calculate the equivalent resistance of the circuit, we need to apply a test source (e.g. a test current source with the voltage and current of V_t and I_t), then $R_{eq} = \frac{V_t}{I_t}$, as is shown in Fig. 12.9.2.

Since the circuit has infinite dimension, instead of connecting the current source between nodes “A” and “D” (see Fig. 12.9.2), we can use superposition theorem and connect the current source to each node separately, as is shown in Fig. 12.9.3-4.

In Fig. 12.9.3, due to the infinite size of the circuit, the current I_t entering node “A” will be equally divided. Thus:

$$I_{AB} = \frac{1}{3}I_t \tag{4}$$

Moreover, due to the same reason, I_{AB} will be equally divided. Therefore:

$$I_{BC} = \frac{1}{2}I_{AB} = \frac{1}{2}\times\frac{1}{3}I_t = \frac{1}{6}I_t \tag{5}$$

Similarly:

$$I_{CD} = \frac{1}{2}I_{BC} = \frac{1}{2}\times\frac{1}{6}I_t = \frac{1}{12}I_t \tag{6}$$

Now, by applying KVL in the loop, we have:

$$-V_{t1} + \frac{1}{3}I_t \times 1 + \frac{1}{6}I_t \times 1 + \frac{1}{12}I_t \times 1 = 0$$

$$\Rightarrow V_{t1} = \frac{7}{12}I_t \tag{7}$$

Additionally, in Fig. 12.9.4, the current of I_t exiting from node “D” will be equally divided because the circuit has an infinite distance from each side. Therefore:

$$I_{CD} = \frac{1}{3}I_t \tag{8}$$

Likewise:

$$I_{BC} = \frac{1}{2}I_{CD} = \frac{1}{2}\times\frac{1}{3}I_t = \frac{1}{6}I_t \tag{9}$$

$$I_{AB} = \frac{1}{2} I_{BC} = \frac{1}{2} \times \frac{1}{6} I_t = \frac{1}{12} I_t \tag{10}$$

Applying KVL in the loop:

$$-V_{t2} + \frac{1}{12} I_t \times 1 + \frac{1}{6} I_t \times 1 + \frac{1}{3} I_t \times 1 = 0$$

$$\Rightarrow V_{t2} = \frac{7}{12} I_t \tag{11}$$

Based on superposition theorem, we can write as follows.

$$V_t = V_{t1} + V_{t2}$$

$$\Rightarrow V_t = \frac{7}{12} I_t + \frac{7}{12} I_t = \frac{14}{12} I_t$$

$$\Rightarrow \frac{V_t}{I_t} = \frac{14}{12}$$

$$\Rightarrow R_{eq} = \frac{14}{12}\ \Omega \tag{12}$$

Based on the information given in the problem, we have:

$$Q = 1.2 \tag{13}$$

Solving (2) and (13):

$$1.2 = \frac{1}{\frac{14}{12}} \sqrt{\frac{L}{1}}$$

$$\Rightarrow 1.4 = \sqrt{L}$$

$$\Rightarrow L = 1.96\ \text{H}$$

Choice (4) is the answer.

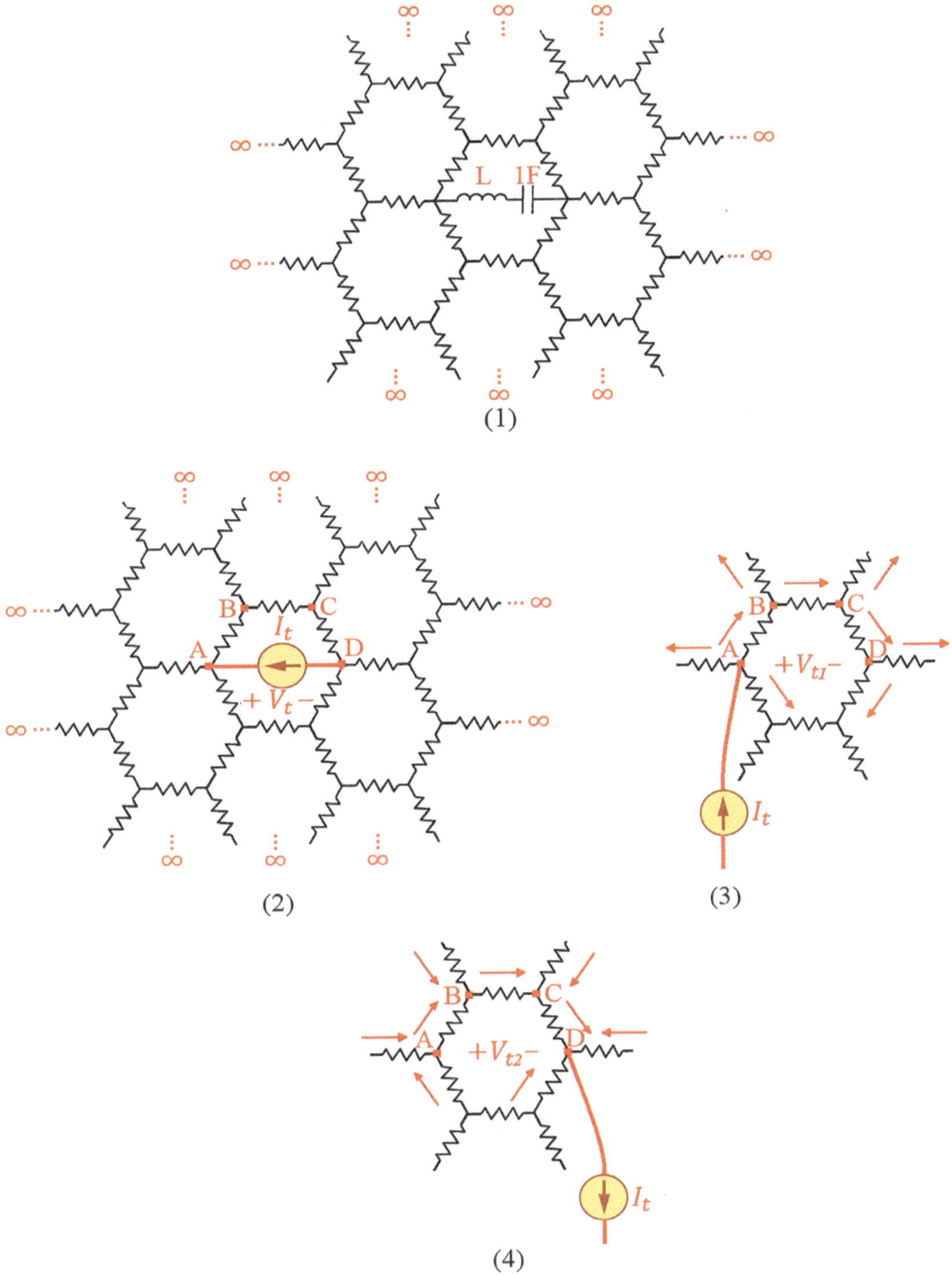

Fig. 12.9 The circuits of problem 12.24

12.5 Analysis of Second-Order Circuits During Switching Operation

12.27. Figure 12.10.2 shows the circuit for $t = 0^+$. As can be seen, the capacitor has been replaced by a short circuit branch since it has zero primary voltage. Moreover, the inductor has been replaced by an open circuit branch because it has zero primary current.

Based on the information given in the problem, we have:

$$\frac{d}{dt}V_C(0^+) = 10\ \text{V/sec} \tag{1}$$

Using voltage-current relation of capacitors:

$$I_C(0^+) = C\frac{d}{dt}V_C(0^+)$$

$$\overset{(1)}{\Rightarrow} I_C(0^+) = 10\,C \tag{2}$$

Applying KVL in the indicated loop:

$$-10 + 2I_C(0^+) = 0$$

$$\Rightarrow I_C(0^+) = 5\ \text{A} \tag{3}$$

Solving (2) and (3):

$$10C = 5$$

$$\Rightarrow C = 0.5\ \text{F}$$

Choice (2) is the answer.

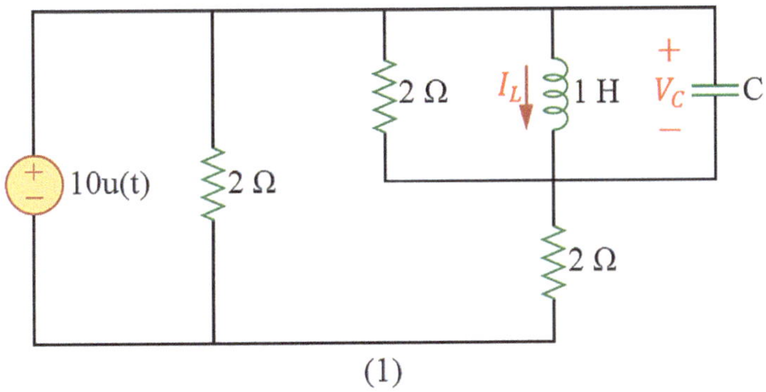

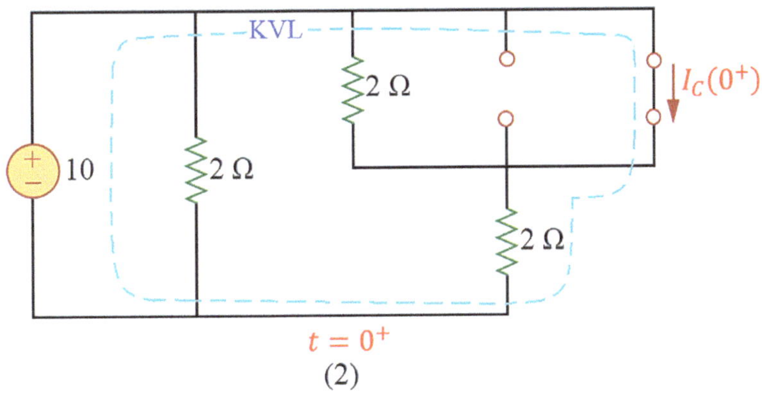

Fig. 12.10 The circuits of problem 12.27

12.28. First, we need to analyze the circuit for $t < 0$, shown in the circuit of Fig. 12.11.2, where the inductor has been replaced by a short circuit branch.

Using Ohm's law:

$$I_L(0^-) = \frac{32}{8} = 4 \text{ A} \tag{1}$$

Moreover, since the capacitor is not connected to a power source, its primary voltage is zero.

$$V_C(0^-) = 0 \tag{2}$$

Due to voltage continuity of capacitors and current continuity of inductors in lack of power source with an infinite size (e.g. impulse power source), we can conclude that:

$$I_L(0^+) = 4 \text{ A} \tag{3}$$

$$V_C(0^+) = 0 \text{ V} \tag{4}$$

To analyze the circuit at $t = 0^+$, we can model the capacitor and inductor by using a voltage source (with the size of 0 V) and a current source (with the size of 4 A), respectively, as is shown in Fig. 12.11.3.

Applying KVL in the loop:

$$-V_L(0^+) - 8 \times 4 - 12 \times 4 = 0$$

$$\Rightarrow V_L(0^+) = -80 \text{ V}$$

Choice (3) is the answer.

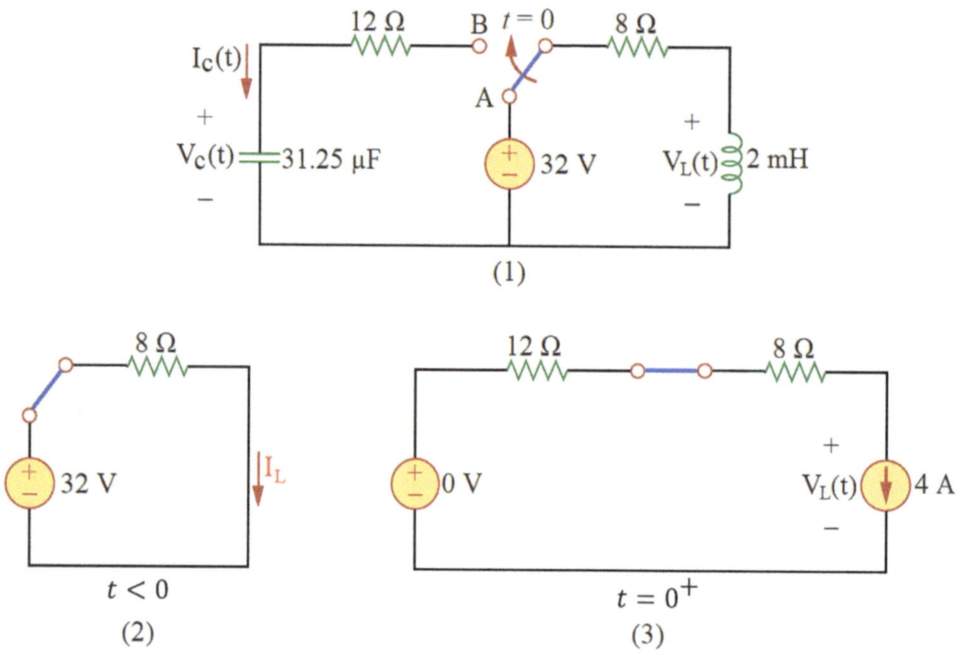

Fig. 12.11 The circuits of problem 12.28

12.29. First, we need to analyze the circuit for $t = 0^-$. Since the circuit has reached its steady state, the capacitor and inductor are open circuit and short circuit, respectively, as is shown in Fig. 12.12.3. Herein, the 6-Ω resistor is eliminated because its two ends are short-circuited.

Using Ohm's law for the whole circuit:

$$I_L(0^-) = \frac{-12}{4+0} = -3 \text{ A} \tag{1}$$

Using Ohm's law for the 4-Ω resistor:

$$V_C(0^-) = 4 \times (-3) = -12 \text{ V} \tag{2}$$

Due to voltage continuity of capacitors and current continuity of inductors in lack of power source with an infinite size (e.g. impulse power source), it is concluded that:

$$I_L(0^+) = -3 \text{ A} \tag{3}$$

$$V_C(0^+) = -12 \text{ V} \tag{4}$$

The circuit for $t = 0^+$ is shown in Fig. 12.12.4, where the primary voltage of the capacitor and the initial current of the inductor have been modelled by using the $-12V$ voltage source and $-3A$ current source, respectively.

Applying KVL in the indicated mesh:

$$-12 + (-12) + V_L(0^+) = 0$$

$$\Rightarrow V_L(0^+) = 24 \text{ V} \tag{5}$$

KCL in the indicated node:

$$-I_C(0^+) - \frac{-12}{4} + (-3) + \frac{V_L(0^+)}{6} = 0$$

$$\overset{(5)}{\Rightarrow} -I_C(0^+) + 3 - 3 + \frac{24}{6} = 0$$

$$\Rightarrow -I_C(0^+) + 4 = 0$$

$$\Rightarrow I_C(0^+) = 4 \text{ A} \tag{6}$$

Choice (2) is the answer.

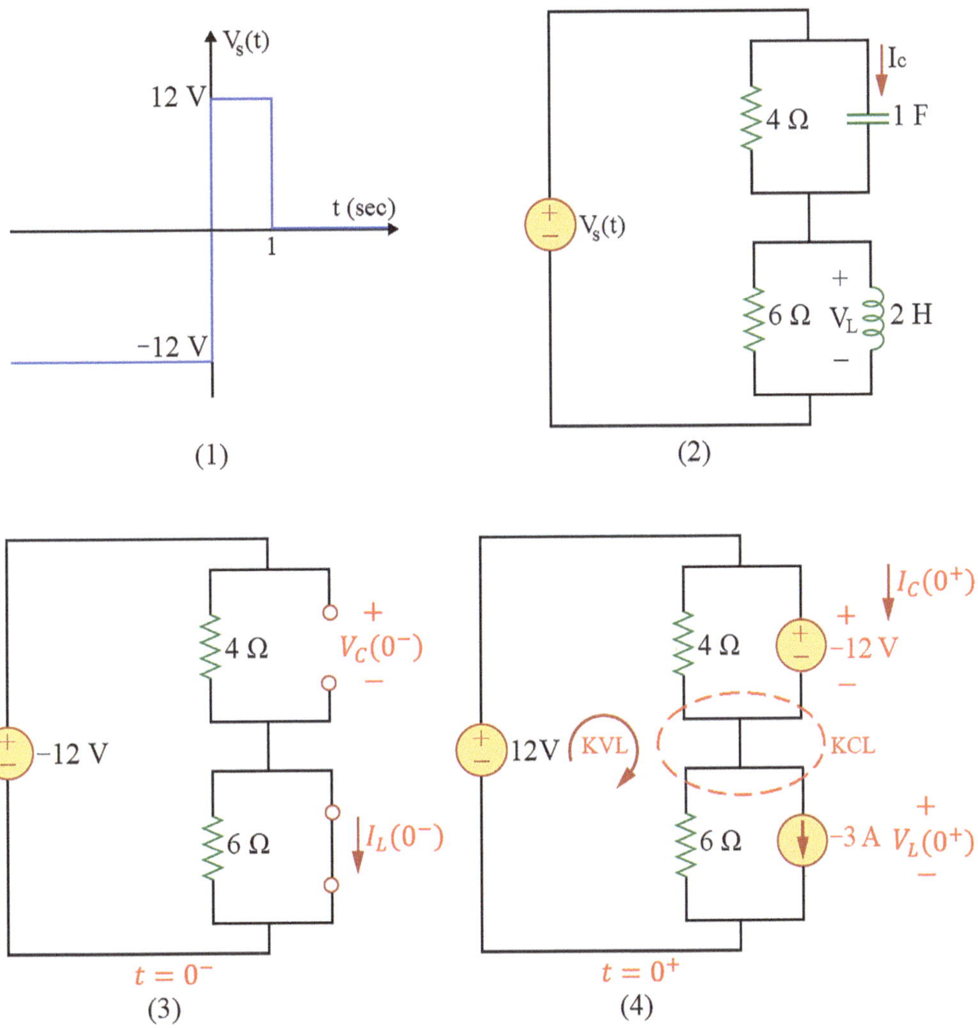

Fig. 12.12 The circuits of problem 12.29

12.31. Figure 12.13.2 illustrates the circuit for $t = 0^-$. As can be seen, since the power source of the circuit is short-circuited, the initial voltage of the capacitor and the primary current of the inductor are zero. In other words:

$$I_L(0^-) = 0 \text{ A} \tag{1}$$

$$V_C(0^-) = 0 \text{ V} \tag{2}$$

Due to voltage continuity of capacitors and current continuity of inductors in lack of power source with an infinite size (e.g. impulse power source), it is concluded that:

$$I_L(0^+) = 0 \text{ A} \tag{3}$$

$$V_C(0^+) = 0 \text{ V} \tag{4}$$

Therefore, the capacitor and the inductor need to be modeled by a short circuit branch and an open circuit branch for $t = 0^+$, respectively, as is shown in Fig. 12.13.3.

In Fig. 12.13.3, the whole current of the current source flows through the left-side 1-Ω resistor. Therefore:

$$V_1(0^+) = 3 \times 1$$

$$\Rightarrow V_1(0^+) = 3 \text{ V} \tag{5}$$

KVL in the indicated loop of Fig. 12.13:

$$-V_1(0^+) + V_L(0^+) + 0 = 0$$

$$\Rightarrow V_L(0^+) = V_1(0^+)$$

$$\overset{(5)}{\Rightarrow} V_L(0^+) = 3 \text{ V} \tag{6}$$

Current-voltage relation of inductors:

$$V_L(0^+) = L\frac{d}{dt}I_L(0^+) = 1 \times \frac{d}{dt}I_L(0^+)$$

$$\overset{(6)}{\Rightarrow} \frac{d}{dt}I_L(0^+) = 3 \text{ A/sec} \tag{7}$$

The circuit for $t > 0$ is shown in Fig. 12.13.4. By applying KCL in node "A", we have:

$$-3 + \frac{V_1(t)}{1} + I_L(t) = 0$$

$$\overset{\frac{d}{dt}}{\Rightarrow} \frac{d}{dt}V_1(t) + \frac{d}{dt}I_L(t) = 0$$

$$\Rightarrow \frac{d}{dt}V_1(t) = -\frac{d}{dt}I_L(t)$$

$$\overset{(7)}{\Rightarrow} \frac{d}{dt}V_1(t) = -3 \text{ V/sec}$$

Choice (4) is the answer.

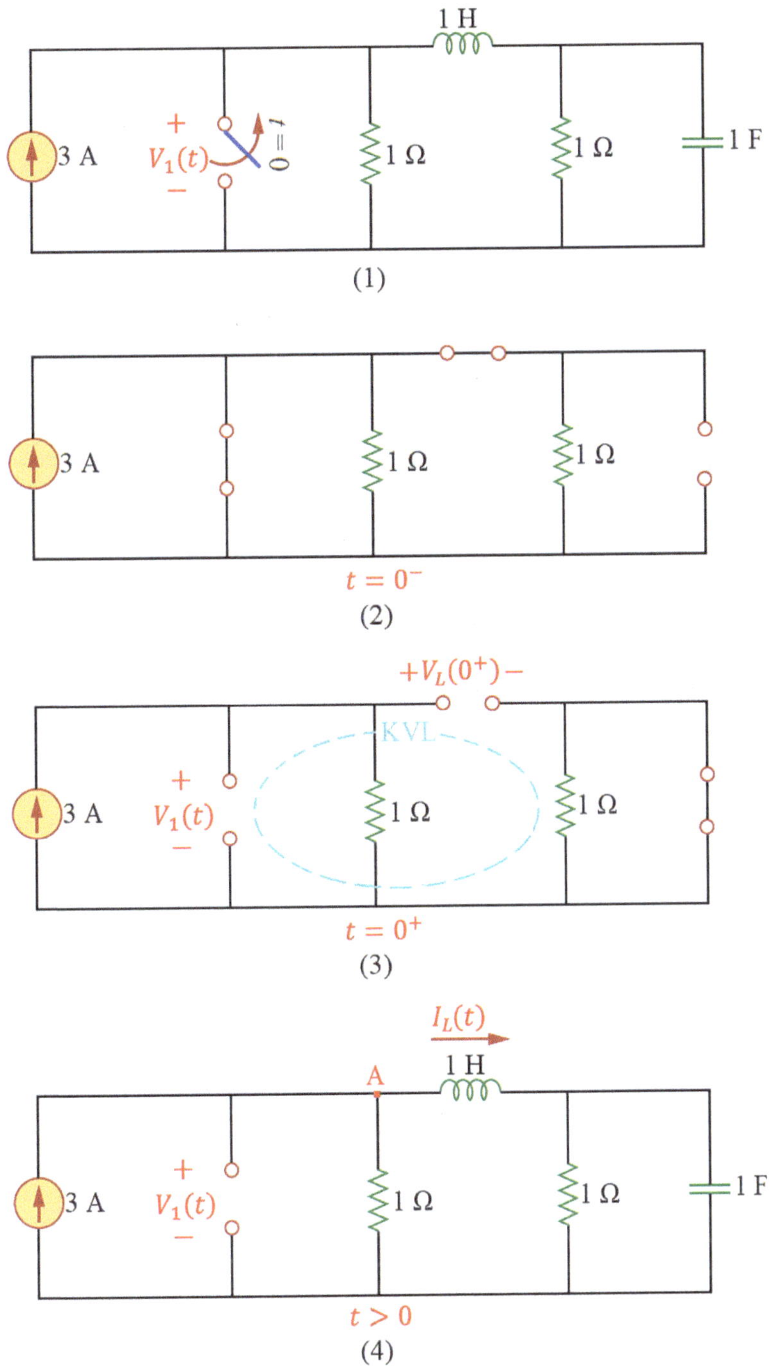

Fig. 12.13 The circuits of problem 12.31

12.32. Based on the information given in the problem, we have:

$$V_C(0) = -6 \text{ V} \tag{1}$$

$$I_L(0) = 5 \text{ A} \tag{2}$$

Due to voltage continuity of capacitors and current continuity of inductors in lack of power source with an infinite size (e.g. impulse power source), it is concluded that:

$$V_C(0^+) = -6 \text{ V} \tag{3}$$

$$I_L(0^+) = 5 \text{ A} \tag{4}$$

KVL in the indicated mesh of the circuit of Fig. 12.14:

$$I_L(t) \times 1 + V_L(t) + V_C(t) = 0 \tag{5}$$

Current-voltage relation of inductors:

$$V_L(t) = L\frac{d}{dt}I_L(t) = 3\frac{d}{dt}I_L(t) \tag{6}$$

Solving (5) and (6) for $t = 0^+$:

$$I_L(0^+) + 3\frac{d}{dt}I_L(0^+) + V_C(0^+) = 0$$

$$\overset{(3),(4)}{\Rightarrow} 5 + 3\frac{d}{dt}I_L(0^+) + (-6) = 0$$

$$\Rightarrow \frac{d}{dt}I_L(0^+) = \frac{1}{3} \text{ A/sec}$$

Choice (1) is the answer.

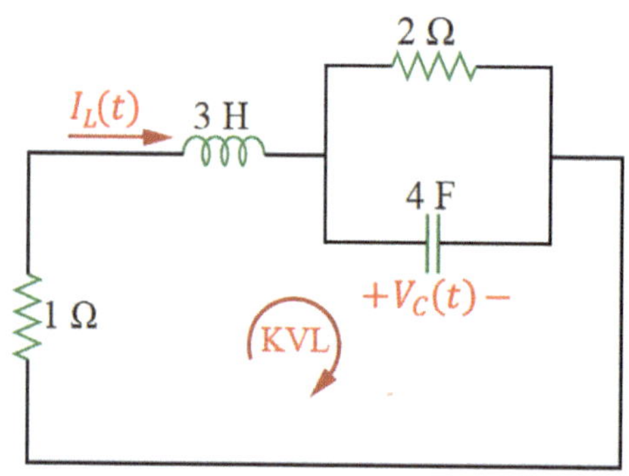

Fig. 12.14 The circuits of problem 12.32

12.34. Before the switching operation, the circuit is in the steady state condition. Therefore, the inductors can be replaced by the short circuit branches, as can be seen in Fig. 12.15.2. To calculate the inductors' currents before opening the switch, we can use both superposition theorem and current division rule in Fig. 12.15.2 as follows.

$$I_{L1}(t=0^-) = \frac{R}{R+R} \times \frac{12}{R+\frac{R}{2}} + \frac{R}{R+R} \times \frac{V_0}{R+\frac{R}{2}}$$

$$\Rightarrow I_{L1}(t=0^-) = \frac{12}{3R} + \frac{V_0}{3R}$$

$$\Rightarrow I_{L1}(t=0^-) = \frac{12+V_0}{3R} \quad (1)$$

$$I_{L2}(t=0^-) = \frac{R}{R+R} \times \frac{12}{R+\frac{R}{2}} - \frac{V_0}{R+\frac{R}{2}}$$

$$\Rightarrow I_{L2}(t=0^-) = \frac{12}{3R} - \frac{V_0}{\frac{3}{2}R}$$

$$\Rightarrow I_{L2}(t=0^-) = \frac{12-2V_0}{3R} \quad (2)$$

From the circuit of Fig. 12.15.3, it is noticed that, right after the switching ($t=0^+$), the inductors' currents are equal in magnitude but opposite in direction. In other words:

$$I_{L1}(t=0^+) = -I_{L2}(t=0^+) \quad (3)$$

To avoid having any impulse voltage across the switch, the above-mentioned relation between them must be held before the switching operation ($t=0^-$) as well. In other words:

$$I_{L1}(t=0^-) = -I_{L2}(t=0^-) \quad (4)$$

Solving (1), (2), and (4):

$$\frac{12+V_0}{3R} = -\frac{12-2V_0}{3R}$$

$$\Rightarrow 12+V_0 = -12+2V_0$$

$$\Rightarrow V_0 = 24 \text{ V}$$

Choice (4) is the answer.

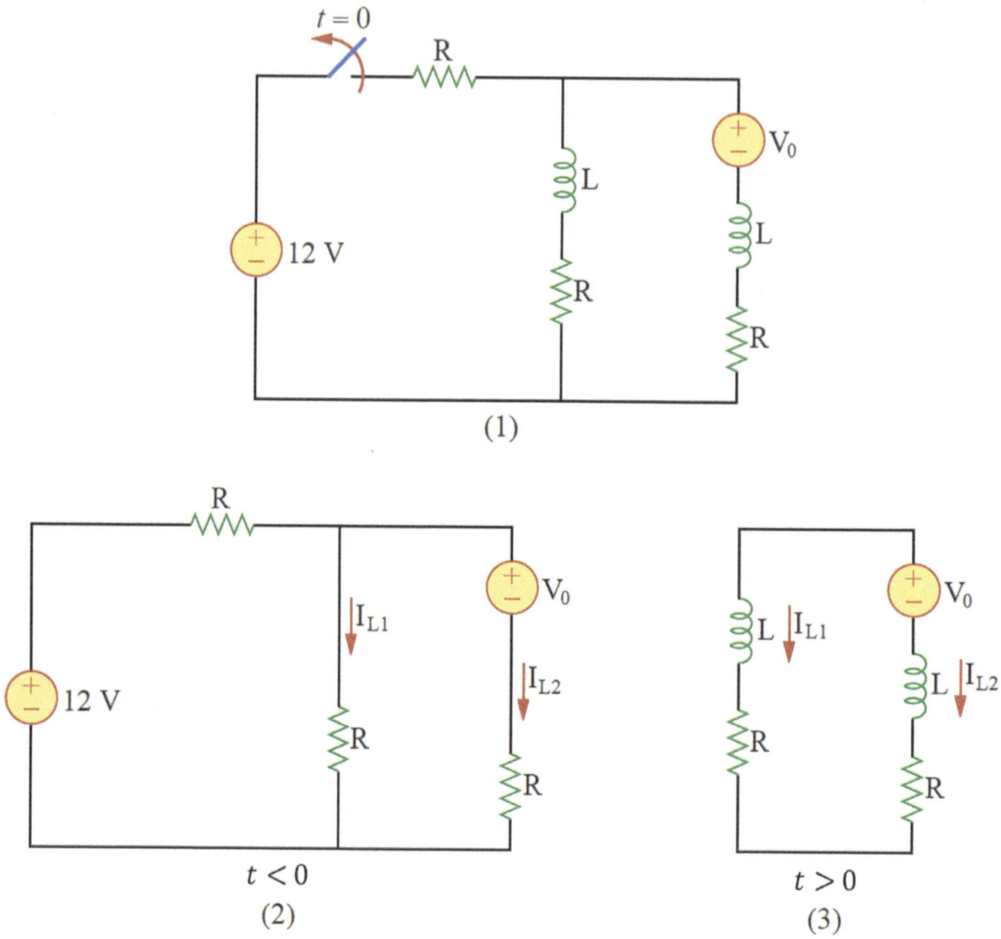

Fig. 12.15 The circuits of problem 12.34

12.35. Before the switching operation, since the circuit is in the steady state condition, the capacitor and inductors behave like an open circuit branch and short circuit branch, respectively, as is shown in Fig. 12.16.2.

By simplifying the circuit of Fig. 12.16.2 and showing in Fig. 12.16.3, the total current of the circuit can be calculated as follows.

$$I(0^-) = \frac{18}{1+2} = 6 \text{ A} \tag{1}$$

Current division for the inductors (see Fig. 12.16.2):

$$I_{L1}(0^-) = \frac{2}{1+2} \times 6 = 4 \text{ A} \tag{2}$$

$$I_{L2}(0^-) = \frac{1}{1+2} \times 6 = 2 \text{ A} \tag{3}$$

Using (1) and Ohm's law for the 2-Ω resistor in the circuit of Fig. 12.16.4:

$$V_C(0^-) = 2I(0^-) = 2 \times 6$$

$$\Rightarrow V_C(0^-) = 12 \text{ V} \tag{4}$$

Due to voltage continuity of capacitors and current continuity of inductors in lack of power source with an infinite size (e.g. impulse power source), it is concluded that:

$$I_{L1}(0^+) = 4 \text{ A} \tag{5}$$

$$I_{L2}(0^+) = 2 \text{ A} \tag{6}$$

$$V_C(0^+) = 12 \text{ V} \tag{7}$$

For a very short interval, that is, $[0, 0^+]$, we can model the capacitor and the inductors by using a voltage source and current sources, as is illustrated in Fig. 12.16.4.

Now, by applying KVL in the indicated mesh, we can calculate the voltage of the switch right after the switching operation as follows.

$$V_S(0^+) = -2 \times 4 + 12$$

$$\Rightarrow V_S(0^+) = 4 \text{ V}$$

Choice (3) is the answer.

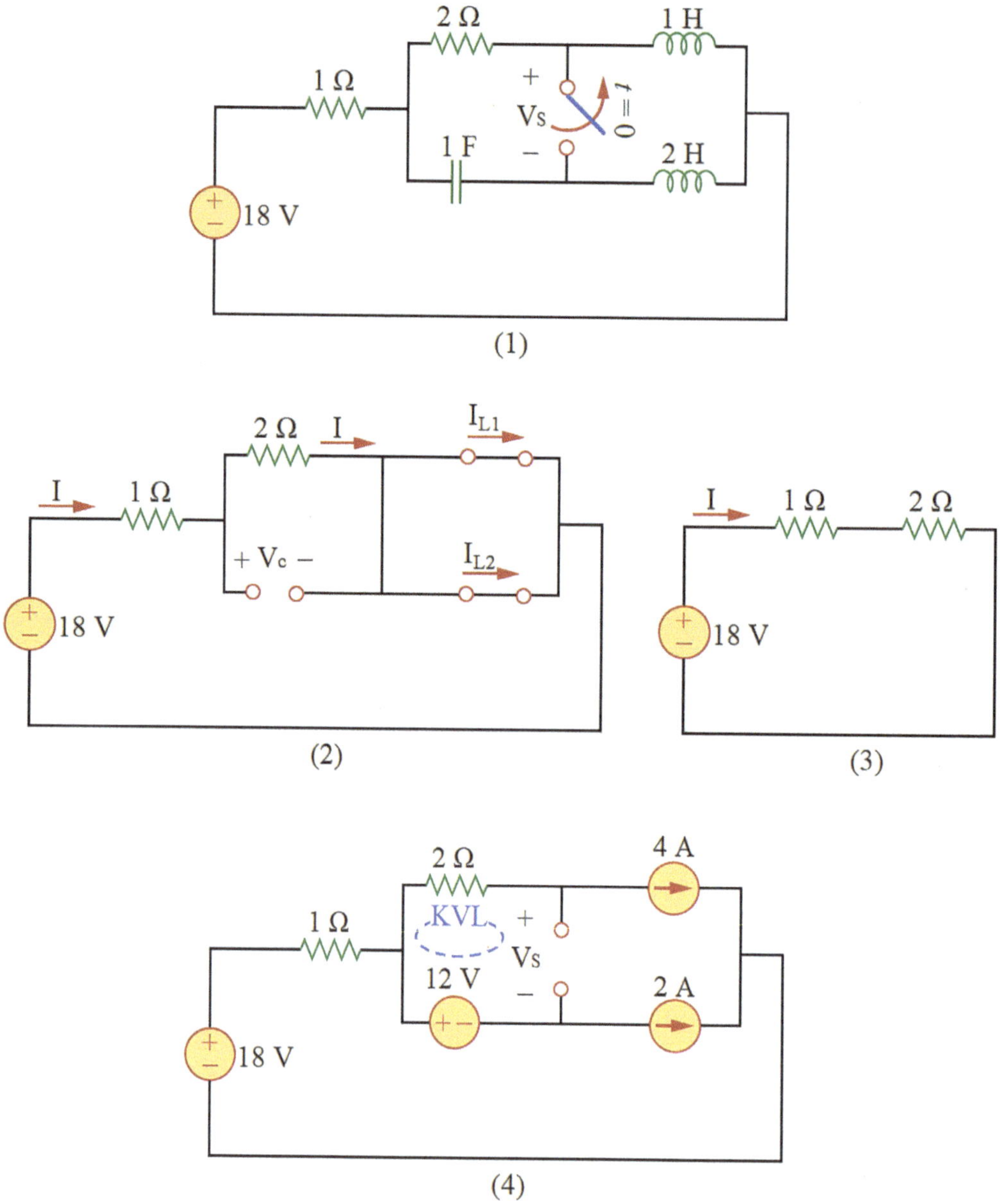

Fig. 12.16 The circuits of problem 12.35

12.36. The circuit for $t = 0^-$ is illustrated in Fig. 12.17.2. Since the circuit is in the steady state condition, the capacitor and inductor are open circuit and short circuit, respectively.

Using Ohm's law for the left-side circuit:

$$I_L(0^-) = \frac{2}{1}$$

$$\Rightarrow I_L(0^-) = 2 \text{ A} \quad (1)$$

Using voltage division formula for the right-side circuit:

$$V_C(0^-) = \frac{1}{1+1} \times 2$$

$$\Rightarrow V_C(0^-) = 1 \text{ V} \qquad (2)$$

Due to voltage continuity of capacitors and current continuity of inductors in lack of power source with an infinite size (e.g. impulse power source), it is concluded that:

$$I_L(0^+) = 2 \text{ A} \qquad (3)$$

$$V_C(0^+) = 1 \text{ V} \qquad (4)$$

Fig 12.17.3 shows the circuit for $t = 0^+$, while modeling the primary current of the inductor and initial voltage of the capacitor by the $2 - A$ independent current source and the $1 - V$ independent voltage source, respectively.

Now, by applying KCL in the indicated node, we have:

$$\frac{1-2}{4} + 2 + I_C(0^+) + \frac{1}{1} = 0$$

$$\Rightarrow I_C(0^+) = -\frac{11}{4} \text{ A} \qquad (5)$$

Using voltage-current relation of capacitors:

$$I_C(0^+) = C\frac{d}{dt}V_C(0^+)$$

$$\overset{(5)}{\Rightarrow} -\frac{11}{4}A = 1 \times \frac{d}{dt}V_C(0^+)$$

$$\Rightarrow \frac{d}{dt}V_C(0^+) = -\frac{11}{4} \text{ V/s}$$

Choice (2) is the answer.

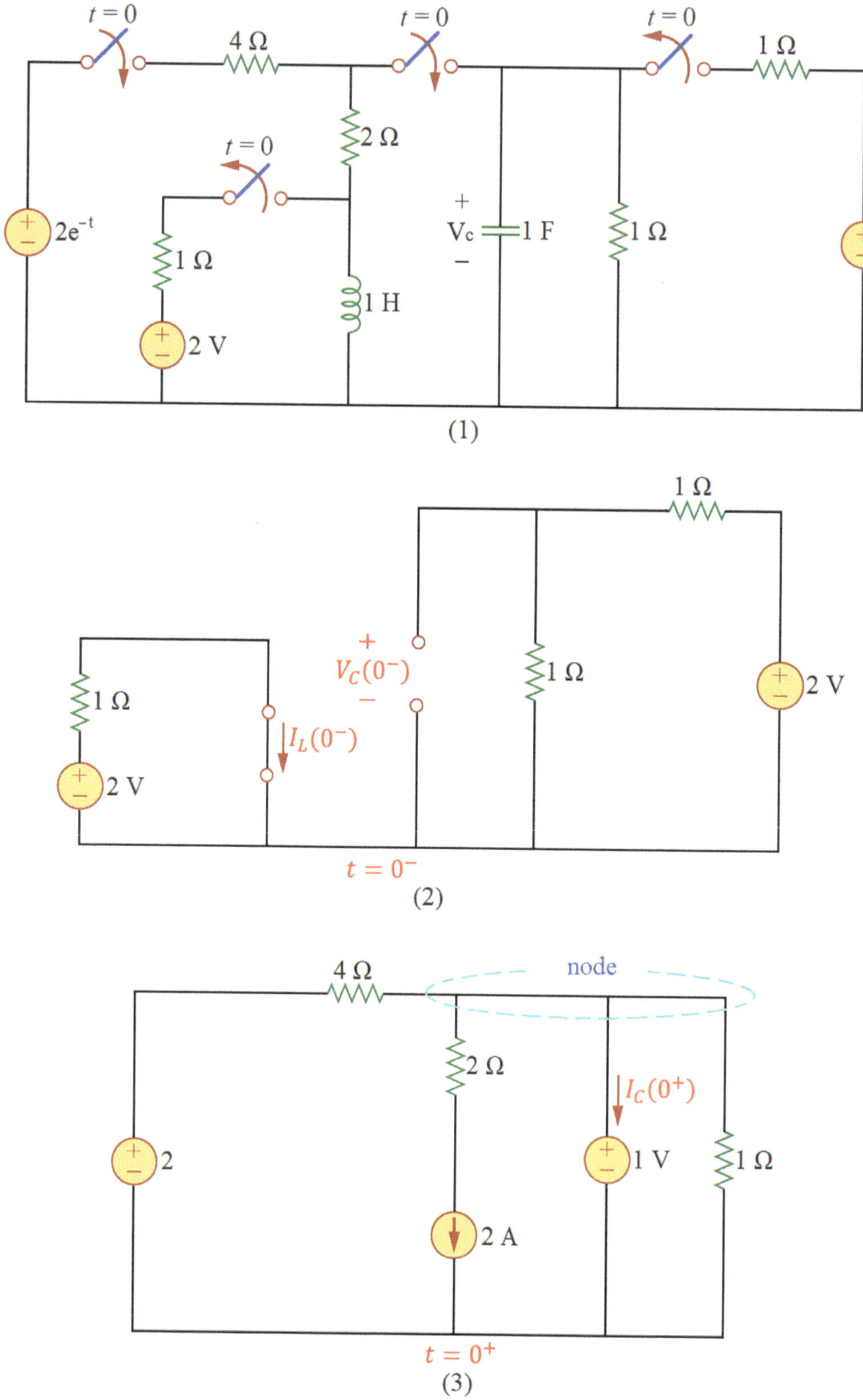

Fig. 12.17 The circuits of problem 12.36

12.37. To prevent the problem variables form any excitation, their first derivative must be adjusted to zero. In other words:

$$\frac{d}{dt}V_{C1}(t) = 0 \tag{1}$$

$$\frac{d}{dt}V_{C2}(t) = 0 \tag{2}$$

$$\frac{d}{dt}I_L(t) = 0 \tag{3}$$

KVL in the indicated loop of the circuit of Fig. 12.18:

$$-V_{C1}(0) + V_{C2}(0) + V_L(0) = 0 \tag{4}$$

Current-voltage relation of inductors:

$$V_L(t) = L\frac{d}{dt}I_L(t)$$

$$\overset{(3)}{\Rightarrow} V_L(t) = 0 \tag{5}$$

Solving (4) and (5):

$$-V_{C1}(0) + V_{C2}(0) + 0 = 0$$

$$\Rightarrow V_{C1}(0) = V_{C2}(0) \tag{6}$$

KCL in the indicated node of the circuit:

$$-I_{C2}(0) + V_{C1}(0) + I_L(0) = 0 \tag{7}$$

Applying voltage-current relation of capacitors for the second capacitor:

$$I_{C2}(0) = C_2\frac{d}{dt}V_{C2}(0)$$

$$\overset{(2)}{\Rightarrow} I_{C2}(0) = 0 \tag{8}$$

Solving (7) and (8):

$$0 + V_{C1}(0) + I_L(0) = 0$$

$$\Rightarrow V_{C1}(0) = -I_L(0) \tag{9}$$

Solving (6) and (9):

$$V_{C1}(0) = V_{C2}(0) = -I_L(0) \tag{10}$$

The last relation has an infinite number of solutions that one of them is as follows.

$$(V_{C1}(0), V_{C2}(0), I_L(0)) = (1, 1, -1)$$

Choice (1) is the answer.

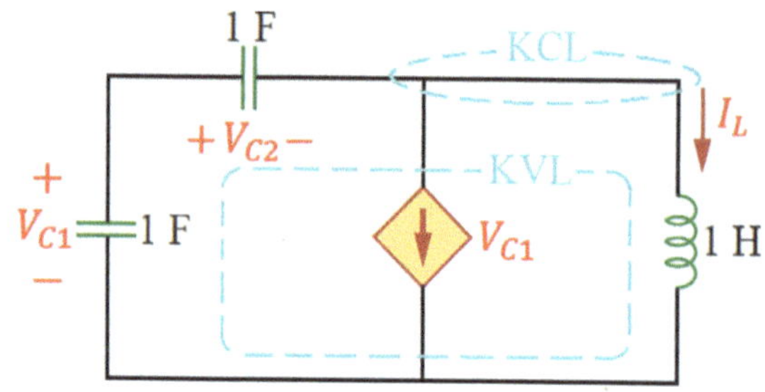

Fig. 12.18 The circuits of problem 12.37

12.38. First, we need to analyze the circuit for $t = 0^-$. Since the circuit is in a steady state condition, the capacitor and the inductors are open circuit branch and short circuit branch, respectively, as is shown in Fig. 12.19.2.

Using Ohm's law for the whole circuit:

$$I_{L1}(0^-) = I_{L2}(0^-) = \frac{12}{3+2+6+1}$$

$$\Rightarrow I_{L1}(0^-) = I_{L2}(0^-) = 1 \text{ A} \quad (1)$$

Using Ohm's law for the 3-Ω resistor:

$$V_C(0^-) = 3I_{L1}(0^-) = 3 \times 1$$

$$\Rightarrow V_C(0^-) = 3 \text{ V} \quad (2)$$

Due to voltage continuity of capacitors and current continuity of inductors in lack of power source with an infinite size (e.g. impulse power source), it is concluded that:

$$I_{L1}(0^+) = 1 \text{ A} \quad (3)$$

$$V_C(0^+) = 3 \text{ V} \quad (4)$$

For $t = 0^+$, the switch is closed, and the circuit is updated, as is illustrated in Fig. 12.19.3.
Applying KVL in the indicated mesh:

$$e(0^+) + 1 \times 1 - 12 + 3 = 0$$

$$\Rightarrow e(0^+) = 8 \text{ V} \quad (5)$$

To determine the value of $\frac{d}{dt}e(0^+)$, we need to analyze the circuit for $t > 0$, as can be seen in Fig. 12.19.4.

Applying KVL in the left-side mesh:

$$e(t) + 1 \times I_{L1}(t) - 12 + V_C(t) = 0 \tag{6}$$

$$\overset{\frac{d}{dt}}{\Rightarrow} \frac{d}{dt}e(t) + \frac{d}{dt}I_{L1}(t) + \frac{d}{dt}V_C(t) = 0$$

$$\overset{t=0^+}{\Rightarrow} \frac{d}{dt}e(0^+) + \frac{d}{dt}I_{L1}(0^+) + \frac{d}{dt}V_C(0^+) = 0 \tag{7}$$

KCL in the node for $t = 0^+$:

$$I_{L1}(0^+) - I_C(0^+) - \frac{V_C(0^+)}{3} = 0$$

$$\overset{(3)(4)}{\Rightarrow} 1 - I_C(0^+) - \frac{3}{3} = 0$$

$$\Rightarrow I_C(0^+) = 0 \text{ A} \tag{8}$$

Using voltage-current relation of capacitors:

$$I_C(t) = C\frac{d}{dt}V_C(t)$$

$$\overset{t=0^+}{\Rightarrow} I_C(0^+) = 2\frac{d}{dt}V_C(0^+)$$

$$\Rightarrow \frac{d}{dt}V_C(0^+) = \frac{1}{2}I_C(0^+)$$

$$\Rightarrow \frac{d}{dt}V_C(0^+) = \frac{1}{2} \times 0 = 0 \tag{9}$$

Applying KVL in the right-side mesh:

$$e(t) - 2 \times I_{L1}(t) - V_{L1}(t) = 0$$

$$\overset{t=0^+}{\Rightarrow} e(0^+) - 2I_{L1}(0^+) - V_{L1}(0^+) = 0$$

$$\overset{(3)(5)}{\Rightarrow} 8 - 2(1) - V_L(0^+) = 0$$

$$\Rightarrow V_{L1}(0^+) = 6 \text{ V} \tag{10}$$

Using voltage-current relation of capacitors:

$$V_{L1}(t) = L_1\frac{d}{dt}I_{L1}(t)$$

$$\Rightarrow V_{L1}(t) = 1 \times \frac{d}{dt}I_{L1}(t)$$

$$\overset{t=0^+}{\Rightarrow} V_{L1}(0^+) = \frac{d}{dt}I_{L1}(0^+) \tag{11}$$

Solving (10) and (11):

$$\frac{d}{dt}I_{L1}(0^+) = 6 \text{ A/sec} \tag{12}$$

Solving (7), (9), and (12):

$$\frac{d}{dt}e(0^+) + 6 + 0 = 0$$

$$\Rightarrow \frac{d}{dt}e(0^+) = -6 \text{ V/sec}$$

Choice (3) is the answer.

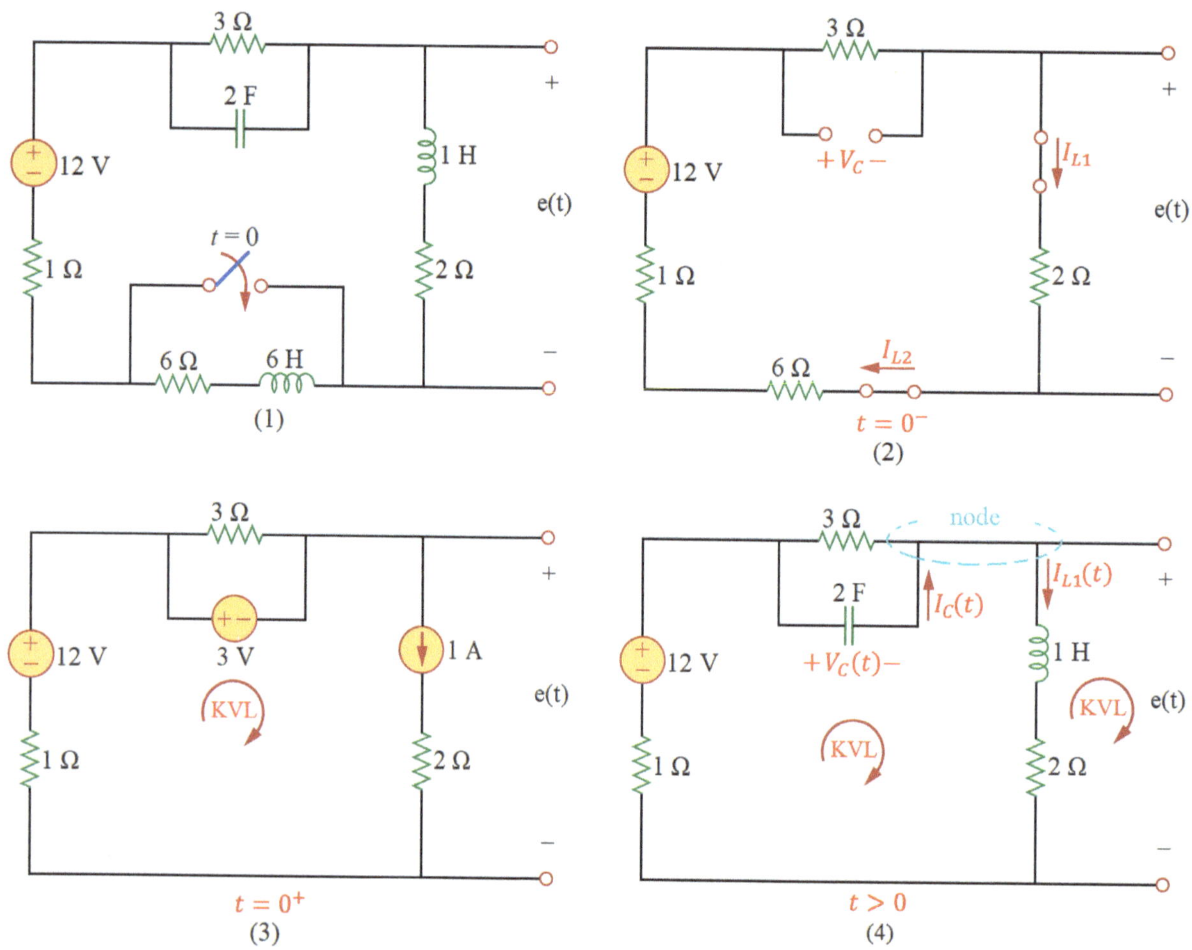

Fig. 12.19 The circuits of problem 12.38

12.39. Based on the information given in the problem, we have:

$$V_C(0^-) = 0 \text{ V} \quad (1)$$

$$I_L(0^-) = 0 \text{ A} \quad (2)$$

Due to voltage continuity of capacitors and current continuity of inductors in lack of power source with an infinite size (e.g. impulse power source), it is concluded that:

$$V_C(0^+) = 0 \text{ V} \quad (3)$$

$$I_L(0^+) = 0 \text{ A} \quad (4)$$

Figure 12.20.2. shows the circuit for $t = 0^+$. As can be seen, the capacitor and the inductor have been modeled by a short circuit branch and an open circuit branch, respectively, due to the relations (3) and (4).
Applying KCL in the supernode:

$$-1 + I_{R=1\,\Omega}(0^+) + 0 = 0$$

$$\Rightarrow I_{R=1\,\Omega}(0^+) = 1 \quad (5)$$

Applying KVL in the loop:

$$-V_o(0^+) + 0 + 1 \times I_{R=1\,\Omega}(0^+)$$

$$\Rightarrow -V_o(0^+) + 1 \times 1 = 0$$

$$\Rightarrow V_o(0^+) = 1 \text{ V} \quad (6)$$

Figure 12.20.3. shows the circuit for $t > 0$. Using current-voltage relation of inductors:

$$V_L(t) = L\frac{d}{dt}I_L(t) = 2\frac{d}{dt}I_L(t) \quad (7)$$

Applying KVL in the right-side mesh:

$$-V_o(t) + 4I_L(t) + V_L(t) = 0$$

$$\overset{(7)}{\Rightarrow} -V_o(t) + 4I_L(t) + 2\frac{d}{dt}I_L(t) = 0$$

$$\overset{t=0^+}{\Rightarrow} \frac{d}{dt}I_L(0^+) = \frac{1}{2}V_o(0^+) - 2I_L(0^+)$$

$$\overset{(4)(6)}{\Rightarrow} \frac{d}{dt}I_L(0^+) = \frac{1}{2} \times 1 - 2 \times 0 = \frac{1}{2} \text{ A/sec} \quad (8)$$

Applying KCL in supernode 1:

$$-1 + I_{R=1\,\Omega}(t) + I_L(t) = 0$$

$$\Rightarrow I_{R=1\,\Omega}(t) = 1 - I_L(t) \tag{9}$$

Applying KCL in node 2

$$-10V_o(t) - I_C(t) + I_L(t) = 0$$

$$\Rightarrow I_C(t) = I_L(t) - 10V_o(t) \tag{10}$$

Applying KVL in the middle mesh:

$$-I_{R=1\,\Omega}(t) \times 1 + V_C(t) + 4I_L(t) + V_L(t)$$

$$\overset{(9)}{\Rightarrow} -(1 - I_L(t)) + V_C(t) + 4I_L(t) + V_L(t) = 0$$

$$\Rightarrow -1 + V_C(t) + 5I_L(t) + V_L(t) = 0 \tag{11}$$

Using current-voltage relation of capacitors and (1):

$$V_C(t) = V_C(0^-) + \frac{1}{C}\int_{0^-}^{t} I_C(t).dt$$

$$\Rightarrow V_C(t) = 0 + \frac{1}{0.5}\int_{0^-}^{t} I_C(t).dt$$

$$\Rightarrow V_C(t) = 2\int_{0^-}^{t} I_C(t).dt$$

$$\overset{(10)}{\Rightarrow} V_C(t) = 2\int_{0^-}^{t} (-10V_o(t) + I_L(t)).dt \tag{12}$$

Applying (7) and (11) in (12):

$$-1+2\int_{0^-}^{t}(-10V_o(t)+I_L(t)).dt+5I_L(t)+2\frac{d}{dt}I_L(t)=0$$

$$\overset{\frac{d}{dt}}{\Rightarrow}-20V_o(t)+2I_L(t)+5\frac{d}{dt}I_L(t)+2\frac{d^2}{dt^2}I_L(t)=0$$

$$\overset{t=0^+}{\Rightarrow}-20V_o(0^+)+2I_L(0^+)+5\frac{d}{dt}I_L(0^+)+2\frac{d^2}{dt^2}I_L(0^+)=0$$

$$\overset{(4),(6),(8)}{\Rightarrow}-20\times 1+2\times 0+5\times\frac{1}{2}+2\frac{d^2}{dt^2}I_L(0^+)=0$$

$$\Rightarrow -\frac{35}{2}+2\frac{d^2}{dt^2}I_L(0^+)=0$$

$$\Rightarrow \frac{d^2}{dt^2}I_L(0^+)=\frac{35}{4}(\mathrm{A/sec})^2$$

Choice (2) is correct.

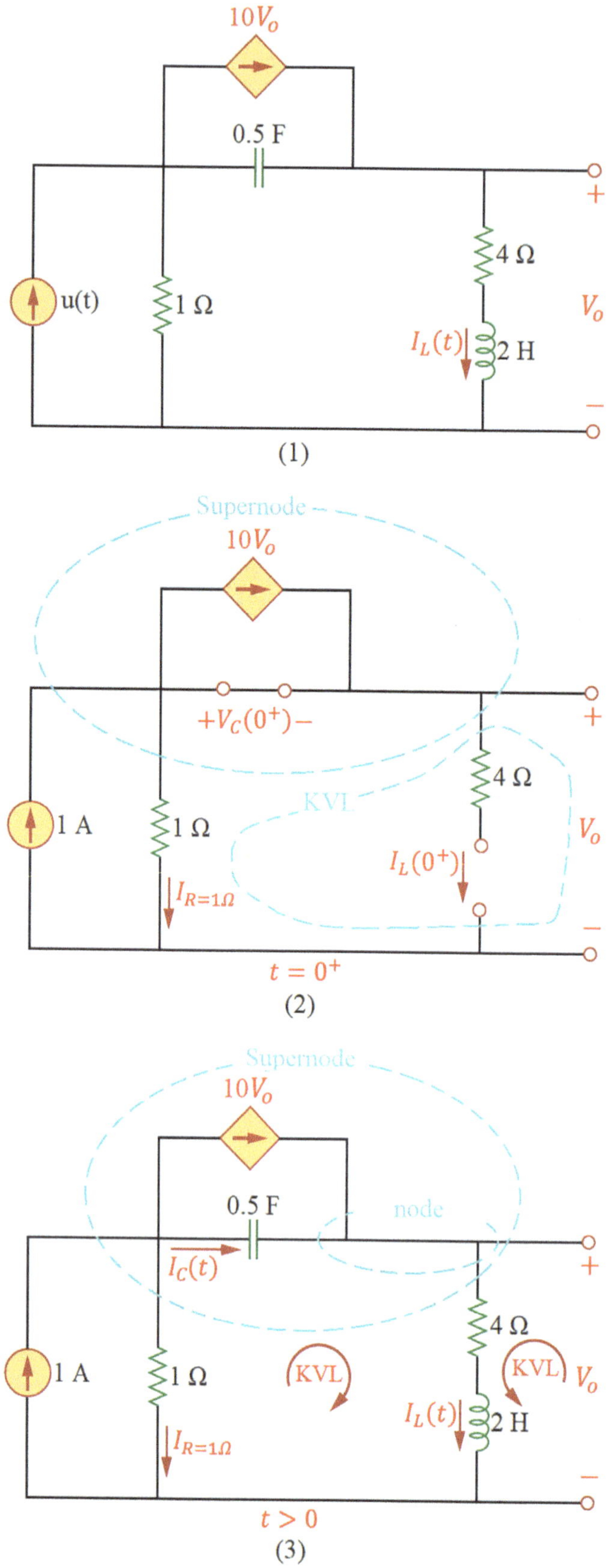

Fig. 12.20 The circuits of problem 12.39

12.40. Based on the information given in the problem, we have:

$$I(0^-) = 2\ \text{A} \tag{1}$$

$$V_{C1}(0^-) = 3\ \text{V} \tag{2}$$

$$V_{C2}(0^-) = 1\ \text{V} \tag{3}$$

First, it should be noted that the circuit is not in the steady state condition for $t = 0^-$. Now, by using current division rule in node "B" in the circuit of Fig. 12.21.2, we can write as follows.

$$I(0^-) = \frac{1}{1+2} \times I_L(0^-) \tag{4}$$

Solving (1) and (4):

$$2 = \frac{1}{1+2} \times I_L(0^-)$$

$$\Rightarrow I_L(0^-) = 6\ \text{A} \tag{5}$$

Due to voltage continuity of capacitors and current continuity of inductors in lack of power source with an infinite size (e.g. impulse power source), it is concluded that:

$$V_{C1}(0^+) = 3\ \text{V} \tag{6}$$

$$V_{C2}(0^+) = 1\ \text{V} \tag{7}$$

$$I_L(0^+) = 6\ \text{A} \tag{8}$$

Figure 12.21.3 shows the circuit for $t = 0^+$. As can be seen, the capacitors have been modeled by the voltage sources with the sizes of 3 V and 1 V. Additionally, the inductor has been modeled by a 6 A current source.

Applying KCL in node "A":

$$\frac{V_A(0^+) - (-3)}{2} + \frac{V_A(0^+) - 1}{6} + 6 = 0$$

$$\Rightarrow \frac{3V_A(0^+) + 9 + V_A(0^+) - 1 + 36}{6} = 0$$

$$\Rightarrow 4V_A(0^+) + 44 = 0$$

$$\Rightarrow V_A(0^+) = -11\ \text{V} \tag{9}$$

Using current division rule in node "B":

$$I(0^+) = \frac{1}{1+2} \times I_L(0^+)$$

$$\overset{(8)}{\Rightarrow} I(0^+) = \frac{1}{3} \times 6$$

$$\Rightarrow I(0^+) = 2 \text{ A} \quad (10)$$

Using Ohm's law for the right-side 2-Ω resistor:

$$V_B(0^+) = 2I(0^+)$$

$$\Rightarrow V_B(0^+) = 4 \text{ V} \quad (11)$$

Defining the voltage of the inductor based on the node voltages:

$$V_L(0^+)$$

$$\overset{(9)(11)}{\Rightarrow} V_L(0^+) = -11 - 4$$

$$\Rightarrow V_L(0^+) = -15 \text{ V} \quad (12)$$

Using current-voltage relation of inductors for $t = 0^+$:

$$V_L(0^+) = L\frac{d}{dt}I_L(0^+) = 4\frac{d}{dt}I_L(0^+)$$

$$\Rightarrow \frac{d}{dt}I_L(0^+) = \frac{V_L(0^+)}{4}$$

$$\overset{(12)}{\Rightarrow} \frac{d}{dt}I_L(0^+) = \frac{-15}{4} \quad (13)$$

Using current division rule in node "B":

$$I(t) = \frac{1}{1+2} \times I_L(t)$$

$$\overset{\frac{d}{dt}}{\Rightarrow} \frac{d}{dt}I(t) = \frac{1}{3} \times \frac{d}{dt}I_L(t)$$

$$\overset{t=0^+}{\Rightarrow} \frac{d}{dt}I(0^+) = \frac{1}{3} \times \frac{d}{dt}I_L(0^+)$$

$$\overset{(13)}{\Rightarrow} \frac{d}{dt}I(0^+) = \frac{1}{3} \times \left(\frac{-15}{4}\right)$$

$$\Rightarrow \frac{d}{dt}I(0^+) = \frac{-5}{4} \text{ A/sec}$$

Choice (1) is correct.

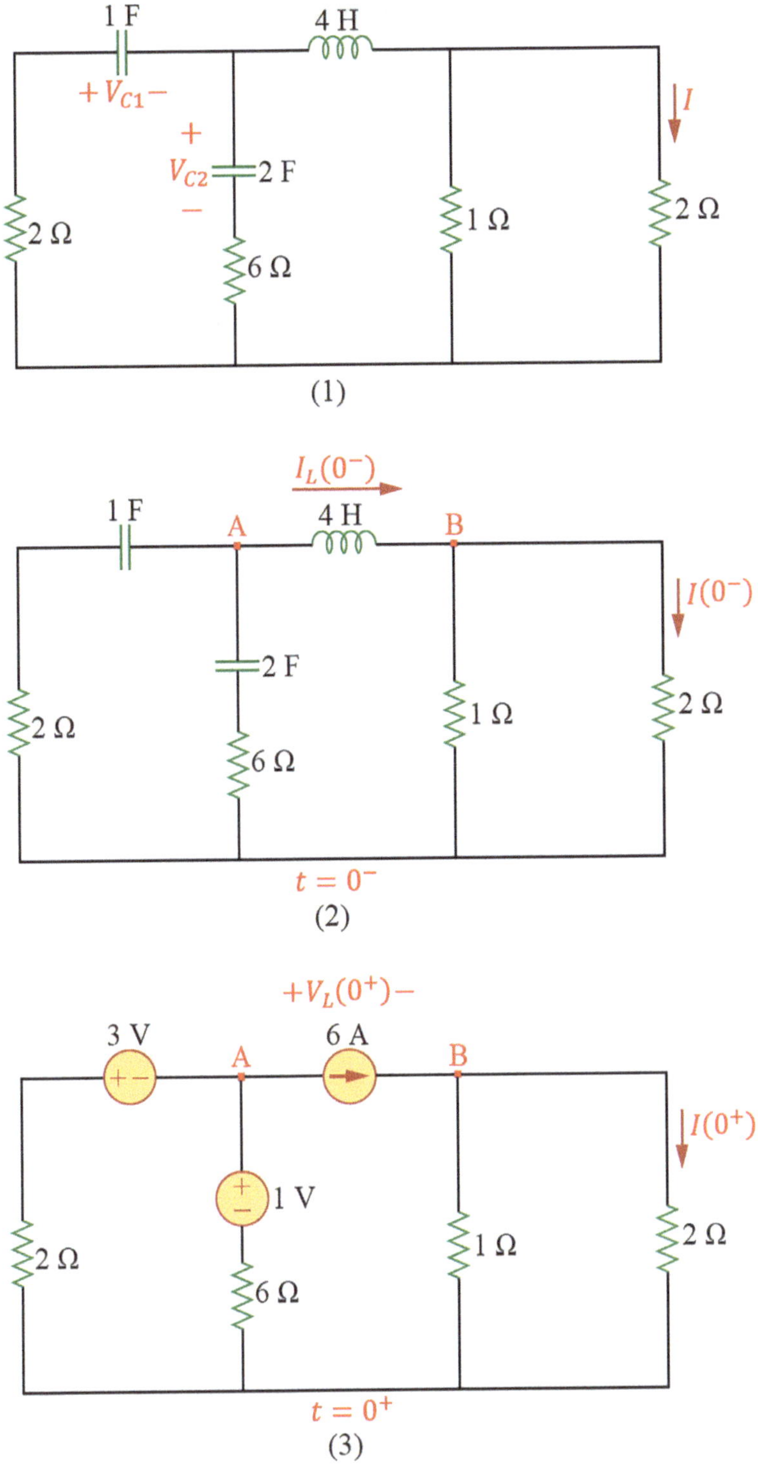

Fig. 12.21 The circuits of problem 12.40

12.6 Second-Order Circuits Including Singularity Functions

12.41. Based on the information given in the problem, we have:

$$V_s(t) = \delta(t) \tag{1}$$

$$I_R(t) = \frac{2}{3}\delta(t) \tag{2}$$

Since the voltage source is an impulse function, the inductor and capacitor will behave like an open circuit and short circuit, respectively, as is shown in Fig. 12.22.2.
Applying KVL in the circuit:

$$-V_s(t) + RI_R(t) + 1 \times I_R(t) = 0$$

$$\Rightarrow I_R(t) = \frac{V_s(t)}{R+1} \tag{3}$$

Solving (1)-(3):

$$\frac{2}{3}\delta(t) = \frac{\delta(t)}{R+1}$$

$$\Rightarrow \frac{2}{3} = \frac{1}{R+1}$$

$$\Rightarrow 2R + 2 = 3$$

$$\Rightarrow R = \frac{1}{2}\ \Omega$$

Choice (3) is correct.

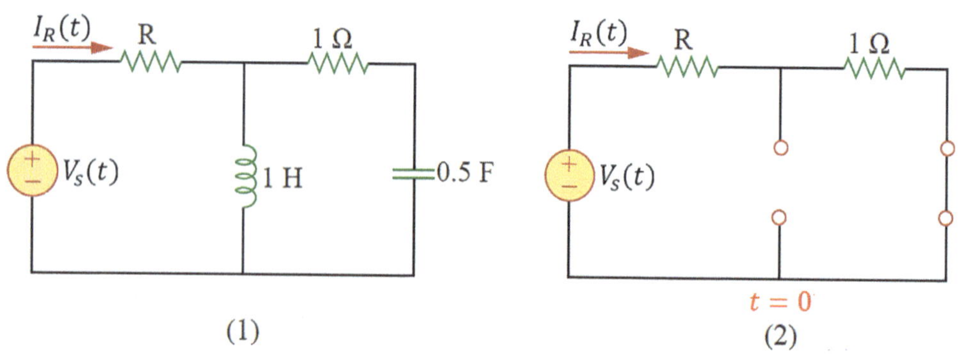

Fig. 12.22 The circuits of problem 12.41

12.42. Based on the information given in the problem, we have:

$$V_s(t) = \delta(t) \tag{1}$$

Due to the presence of an impulse voltage source in the circuit, the capacitors and inductors will behave like the short circuit and open circuit branches, respectively, as can be seen in Fig. 12.23.2.

The requested output is the current of 2-H inductor. To calculate its current, we need to use voltage-current relation of inductors for that inductor as follows:

$$I_L(t) = \frac{1}{L} \int_{-\infty}^{t} V_L(t).dt$$

$$\Rightarrow I_{2H}(0^+) = \frac{1}{2} \left\{ \int_{-\infty}^{0^-} 0.dt + {}_{0^-}^{0^+} V_{2H}(t)dt \right\}$$

$$\Rightarrow I_{2H}(0^+) = \frac{1}{2} \int_{0^-}^{0^+} V_{2H}(t)dt \quad (2)$$

Using voltage division rule in the circuit of Fig. 12.23.3, which is the simplified circuit of Fig. 12.23.2:

$$V_{2H}(t) = \frac{1}{1+1+1} \delta(t)$$

$$\Rightarrow V_{2H}(t) = \frac{1}{3} \delta(t) \quad (3)$$

Solving (2) and (3):

$$\Rightarrow I_{2H}(0^+) = \frac{1}{2} \int_{0^-}^{0^+} \frac{1}{3} \delta(t)dt$$

$$\Rightarrow I_{2H}(0^+) = \frac{1}{6} \int_{0^-}^{0^+} \delta(t)dt$$

$$\Rightarrow I_{2H}(0^+) = \frac{1}{6} \times 1$$

$$\Rightarrow I_o(0^+) = \frac{1}{6} \text{ A}$$

Choice (3) is the answer.

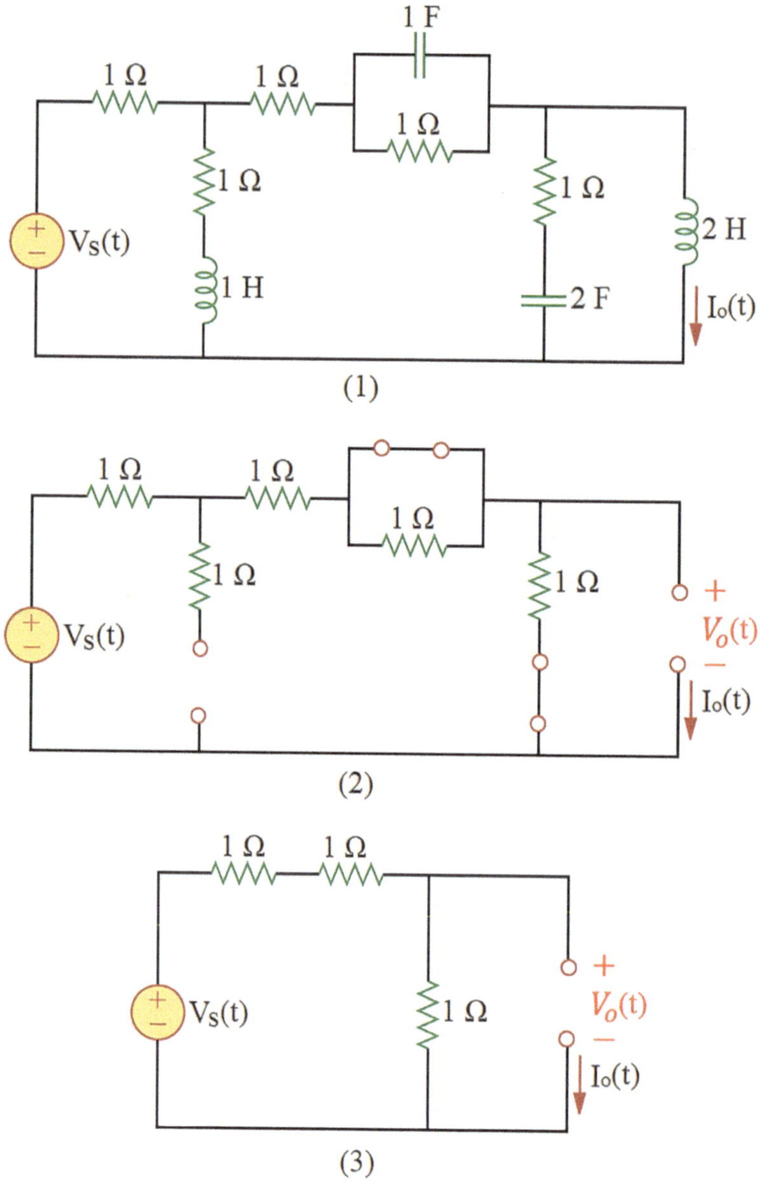

Fig. 12.23 The circuits of problem 12.42

Notes

$$\int_{0^-}^{0^+} \delta(t)dt = 1$$

12.7 Energy Stored in Second-Order Circuits

12.43. Based on the information given in the problem, we have:

$$C = 0.1 \text{ F}$$

$$L = 1 \text{ H}$$

$$R = 1.5 \ \Omega$$

$$I_S(t) = 10t, \ t \geq 0$$

$$V_C(0) = 20 \text{ V}$$

$$t = [0, 1]\text{sec}$$

To calculate the energy of a current source, its time-dependent voltage equation must be known.

KVL in the loop of the circuit:

$$-V_S(t) + V_R(t) + V_L(t) + V_C(t) = 0 \tag{1}$$

Current-voltage relation of resistor:

$$V_R(t) = RI_R(t) = RI_S(t) = 1.5 \times 10t$$

$$\Rightarrow V_R(t) = 15t \text{ V} \tag{2}$$

Current-voltage relation of inductors:

$$V_L(t) = L\frac{d}{dt}I_L(t) = L\frac{d}{dt}I_S(t) = 1 \times \frac{d}{dt}(10t)$$

$$\Rightarrow V_L(t) = 10 \text{ V} \tag{3}$$

Current-voltage relation of capacitors:

$$V_C(t) = V_C(0) + \frac{1}{C}\int_0^t I_C(t)dt$$

$$\Rightarrow V_C(t) = 20 + \frac{1}{0.1}\int_0^t 10t.dt$$

$$\Rightarrow V_C(t) = 20 + 10\left(10\frac{t^2}{2}\right)$$

$$V_C(t) = 20 + 50t^2 \text{ V} \tag{4}$$

Applying (2)-(4) in (1):

$$-V_S(t) + 15t + 10 + 20 + 50t^2 = 0$$

$$\Rightarrow V_S(t) = 15t + 30 + 50t^2 \tag{5}$$

Now, we can calculate the energy of the current source for the given interval, that is, [0, 1]:

$$E = \int_0^t V_S(t) I_S(t) dt$$

$$\Rightarrow E = \int_0^1 \left(15t + 30 + 50t^2\right) \times 10t dt$$

$$\Rightarrow E = \int_0^1 \left(150t^2 + 300t + 500t^3\right) dt$$

$$\Rightarrow E = \left[50t^3 + 150t^2 + 125t^4\right]_0^1$$

$$\Rightarrow E = (50 + 150 + 125) - (0 + 0 + 0)$$

$$\Rightarrow E = 325 \text{ J}$$

Choice (2) is correct Fig. 12.24.

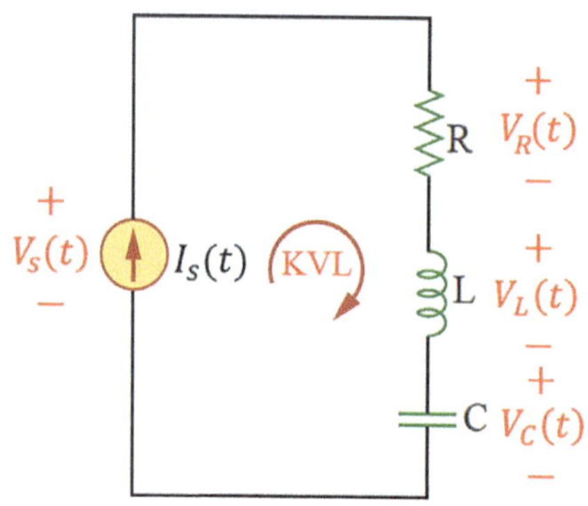

Fig. 12.24 The circuits of problem 12.43

Notes

$$\int t^n dt = \frac{1}{n+1} t^{n+1}$$

References

1. Rahmani-Andebili, M., DC Electrical Circuit Analysis - Practice Problems, Methods, and Solutions (1st Ed.), Springer Nature, 2020.
2. Rahmani-Andebili, M., General Physics II - Practice Problems, Methods, and Solutions, Sringer Nature, 2025.
3. Rahmani-Andebili, M., General Physics I - Practice Problems, Methods, and Solutions, Springer Nature, 2025.
4. Rahmani-Andebili, M., Mathematics of Engineering and Science - Comprehensive Lessons, Springer Nature, 2024.
5. Rahmani-Andebili, M., Differential Equations – Practice Problems, Methods, and Solutions, Springer Nature, 2022.
6. Rahmani-Andebili, M., Calculus III – Practice Problems, Methods, and Solutions, Springer Nature, 2023.
7. Rahmani-Andebili, M., Calculus II – Practice Problems, Methods, and Solutions, Springer Nature, 2023.
8. Rahmani-Andebili, M., Calculus I (2nd Ed.) – Practice Problems, Methods, and Solutions, Springer Nature, 2023.
9. Rahmani-Andebili, M., Precalculus (2nd Ed.) – Practice Problems, Methods, and Solutions, Springer Nature, 2024.

Index

M. Rahmani-Andebili, *DC Electrical Circuits Analysis*, https://doi.org/10.1007/978-3-032-00253-2

FSC
www.fsc.org
MIX
Papier | Fördert gute Waldnutzung
FSC® C083411

Kolibri 360 GmbH
Ferdinand-Jühlke-Straße 7
99095 Erfurt, Deutschland
produktsicherheit@kolibri360.de